# Molecular Symmetry in Chemistry Via Group Theory

# Molecular Symmetry in Chemistry Via Group Theory


**Prof. U.C. Agarwala**
*Retd. Professor*
*Department of Chemistry*
*I.I.T., Kanpur*

**Prof. H.L. Nigam**
*Retd. Vice Chancellor*
*A.P.S. University*
*Reewa*

**Dr. Sudha Agrawal**
*Lecturer*
*Department of Chemistry*
*P.P.N.(P.G.) College, Kanpur*

**Dr. S.S. Kalra**
*Lecturer*
*Department of Chemistry*
*D.A.V.(P.G.) College, Kanpur*



**Ane Books Pvt. Ltd.**

New Delhi ♦ Chennai

**Molecular Symmetry in Chemistry Via Group Theory**
U.C. Agarwala, H.L. Nigam, Sudha Agrawal and S.S. Kalra



*Published by*

**Ane Books Pvt. Ltd.**

4821, Parwana Bhawan, 1st Floor, 24 Ansari Road, Darya Ganj,
**New Delhi** - 110 002, Tel.: +91(011) 23276843-44, Fax: +91(011) 23276863
e-mail: kapoor@anebooks.com, Website: www.anebooks.com

*Branch Office:*

- Avantika Niwas, 1st Floor, 19 Doraiswamy Road, T. Nagar,
  **Chennai** - 600 017, Tel.: +91(044) 28141554, 28141209
  e-mail: anebookschennai@gmail.com, rathinam@anebooks.com



ISBN : 978-93-8116-292-7



Printed at : Sagar Colorscan, New Delhi.

*Dedicated to*

*Aayusha Agarwal*

*(Grandson of U.C. Agarwala)*

*and*

*(Mrs.) Sneh Lata Nigam (Late)*

*(Wife of H.L. Nigam)*

# Preface

This book is written to primarily motivate and help students who are devoid of the beautiful essence of science behind excess in mathematical abstractions. For them, theoreticians appeared as awe because of their tendency to present a formalism which soars to ethereal heights of mathematical punctuations without providing any physical interpretations to their formalism. This head-raising effect remained prevalent until the middle of the twentieth century. Science was compartmentalized into artificial segments. Unity of scientific disciplines was little known and science became an exclusive preserve of specialists. It is only towards the mid-twentieth century that mathematisation of science began taking its shape. A deeper understanding of any descriptive phenomena is incomplete without some insight into the concepts of modern mathematics. One of the very powerful universal mathematical structures that has been exploited profitably by all scientists in the past fifty years or so is symmetry which has been and is being exploited in the unification of science as one discipline. Although generations of chemists have been fascinated by crystal symmetry, the molecular symmetry etc; but its role has been recently recognized and is now sought to prevail in every area of art, science and life. Symmetry is a general unifying principle of the universe. Some of the areas of importance are atomic and molecular orbital symmetry, electronic, vibrational spectroscopy and stereochemistry etc. It is interesting to realize that a discipline like descriptive chemistry is bombarded with the applications of symmetry properties to analyze concerted reaction besides ligand field and molecular orbital theories. Since nature is one, so science has to be one. Possibly, symmetry of nature should be a unifying factor among all phenomena.

It is our own experience that in our own student days, if we met any reference of symmetry, during our studies, we had to refer a number of books to obtain a rudimentary knowledge. Because of the latter feeling among various groups of scientists, a number of excellent books on group theory and its applications have appeared; but no standard text covering all the areas of chemistry has come from Indian authors. Some books from foreign authors are available but the Indian students can hardly comprehend their presentation. Those are perhaps, just too difficult for undergraduate to assimilate the concept in the time allowed to graduate. This calls for a book with a presentation, keeping in mind the following points:

1. Symmetry and group theoretical concepts are a little too abstract and it requires different presentations from different authors. To make the abstract subject more intelligible, the presentation should be followed by proper examples related to the concept. The presentations should give a chance to students to solve the questions (short answer, multiple choice and others).We have followed this concept by providing a good number of questions at the end of the book.
2. It is our experience that a majority of students identify symmetry with the group theory. It is not the group theory but the symmetry properties that are applied to analyze the phenomenon. Group theory acts only as a tool to analyze symmetry, so that it can be exploited to study symmetry. This differentiation must be made very explicit.
3. Without numbers, symmetry remains only a means of appreciation of beauty and harmony. The relationship has been brought out by the abstract group theory, through its role of representation theory. We have restricted ourselves to discrete groups only which gives enough experience to readers to deal with the basic tools of group theory including character tables, reducible and irreducible representations, great orthogonality theorem etc. which are extremely important for symmetry applications.
4. In most of the books of the group theory, concepts are described rather inadequately perhaps, because the authors take it for granted that the students know them fully.We have dealt with the basic concepts of group theory, projection operators, vector spaces etc. in separate small chapters in the book adequately to prepare the student to be able to apply them for different applications.

Symmetry not only includes molecular symmetry but it covers large number of fields like particle physics, super symmetry, quasi symmetry and gauge symmetry; we have described these very briefly in chapter two. These are the hot topics for future development of science, and these involve continuous groups. These are introduced extremely superficially (only up to definition level) in the present book.

Chapters one to seven titled as (1) Introducing groups (2) Introducing Symmetries (3) Vector spaces and Representations (4) Properties of Irreducible Representations (5) Bases, Projection Operators and SALCs (6) Direct Products and Vanishing Integrals (7) Symmetry in Quantum Chemistry and Hybridisation describe the basic concepts of the group theory and symmetry. The rest of the chapters eight to twelve titled as (8) Symmetry Properties of Molecular Orbitals of

$AX_n$ type Molecules (9) Symmetry Aspects of Molecular Orbital Theory of Multicentre $A_MB_N$ Type Molecules (10) Symmetry in Reactivity (11) Ligand Field Theory and Angular Overlap Model (12) Application of Symmetry Properties, Molecular Vibration Analysis illustrate the very important role that symmetry provide in exploring molecular properties.

We have provided questions and problems at the end of each chapter; some are multiple choice, some are short answer type, some numerical, some involve interpretation of data or some logical analysis and some are those that may induce discussion among the readers. We have attempted to give hints to some of the problems. We have not given the full answer of all the questions as an effective educational technique to help the students in developing a confidence in judging the right and the wrong side of the question. We firmly believe that wide, logical conclusion of any data requires capability of critical examination of a problem. When all is said and done, regardless of who may read this text, we hope that the reader finds it interesting, educative and informative.

Though the readers may not find it to be very exhaustive but we must take the blame ourselves for what little remains and for what the readers expect us to write. We most welcome ideas, comments, criticisms and errors (conceptual and typographical) from users of this book. Such criticisms and comments will certainly stimulate and encourage us in the preparation of the second error-free text and to seek better ways to present the subject.

Grateful thanks are due to Prof. R.B. Mishra (Retd. Prof. of Mathematics of A.P.S. University, Reewa) for his critical comments. One of the authors (U.C.A) would like to express his gratitude to the authorities and librarian of I.I.T. Kanpur for providing the facility of consultation

In last our thanks are due to Archana Shukla, Sumedha Agrawal, Kshitij Gupta and Pankaj Chauhan for their assistance in preparing the manuscript.

**Authors**

# Contents

# 1

# Introducing Groups

Group is one among the many abstract mathematical concepts. Its fundamental importance lies not only in modern mathematics but also in the related scientific fields like physics, chemistry, engineering *etc.* One can gauge its importance by its applications that lie behind it. It provides a formal framework for the different types of symmetries that pervade in science. In a nutshell, the two concepts (symmetry and group theory) are inseparable. In this context, it becomes natural and indispensable to learn group and its theory prior to discussing about its applications in detail. The group theory, in short, is a mathematical tool for shaping and simplifying the scientific problems through symmetry. In this chapter, we introduce the definition of a group and present several examples of both finite and continuous groups. In future, we plan to discuss more about its theory, as and when necessary, by using symmetry of various systems as examples.

## 1.1 WHAT IS A GROUP?

Group is an abstract mathematical structure. It is formed by a non-empty set containing finite or infinite number of elements[1] on which one binary operation[2] (represented as $\circ$) is defined. Satisfying the following four conditions:

1. The nature of elements is undefined. It may be anything (objects, operators, transformations or even numbers: real or complex).
2. The property ascertains the combination $a \circ a$ (even for repeated elements $a$, $a$) too. If, in addition, for any two elements $a$, $b$ (usually distinct from each other) there also holds $a \circ b = b \circ a$, the binary operation becomes commutative.

***(i)* Closure [2]($C$ ∘):** For any two elements of the set say $a$ and $b$, both the products $a \circ b$ and $b \circ a$ are also elements of the set, *i.e.*,

$$\{a, b \in G; a \circ b \text{ and } b \circ a \in G\} \qquad ...(1.1)$$

Accordingly, the operation ∘ is called binary.

The combination $a \circ b$ is single-valued. There can never be any ambiguity about the result of a binary operation. In general, $a \circ b$ and $b \circ a$ are not the same. They are equal when the operation ∘ is commutative.

***(ii)* Associative law ($A$ ∘):** If for any three elements of the set, $\{a, b, c, d, \ldots, n\}$, there hold the condition.

$$(a \circ b) \circ c \equiv a \circ (b \circ c) \qquad ...(1.2)$$

The operation becomes associative and the result is called associative law.

This condition implies that the order in which we combine the pair of elements is immaterial : either (1) by having first the binary operation (∘) between $b$ and $c$ (*viz.* $b \circ c$) to get an element $p$ of the set, followed by the operation ($a \circ p$) corresponding to $a \circ (b \circ c)$ to get the final element '$m$' (say) or (2) Leading the same final element must also be obtained by having the first combining $a$ and $b$ (*viz.* $a \circ b$) resulting in the element '$d$' of the set and then combining $d$ and $c$ (*i.e.*, $d \circ c$) corresponding the net operation $(a \circ b) \circ c$. The final result must be identical to that of $a \circ (b \circ c)$, [*viz.* ($m$)] of the set. We thus have the following equation,

$$a \circ (b \circ c) \equiv (a \circ b) \circ c \equiv a \circ b \circ c \qquad ...(1.3)$$

The above relationship is not limited to three elements only, but it holds good for any number of binary operations. Thus, we have

$$\begin{aligned} a \circ b \circ (c \circ d) \circ f \circ \ldots &\equiv a \circ b \circ c \circ d \circ (f \circ g) \circ \ldots \\ &\equiv (a \circ b) \circ c \circ d \circ f \circ g \circ \ldots \\ &\equiv (a \circ b \circ c) \circ (d \circ f \circ g) \circ \ldots \textit{ etc.} \end{aligned} \qquad ...(1.4)$$

One might question here that once we have defined a binary operation among two elements, there should be no problem about having the binary operations among any number of elements together. This is not the case. Let us consider the triple product $(a \circ b \circ c)$. Because we have defined a binary operation, *i.e.*, operation between a pair of elements only, we have to select first a pair of elements among, $a$, $b$, and $c$ and find the result of the product of these two elements. Thus, we have a choice between $(a \circ b) \circ c$ and $a \circ (b \circ c)$. But suppose $(a \circ b) = c$ and $(b \circ c) = d$, then the results of our triple product is $(c \circ c)$ and $(a \circ d)$. From these results, it is by no means evident that these two binary products (*viz.* $c \circ c$ and $a \circ d$) are equal unless this equality is introduced as a requirement of the group. This is just what the statement, "binary operation (or multiplication) is associative", is supposed to fulfill. It means that it must be true that

$$(a \circ b) \circ c \equiv a \circ (b \circ c) \equiv a \circ b \circ c \qquad ...(1.5)$$

It implies that the binary operation is operative among the string of elements $(a, b, c, d \dots g, h)$. That is,

$$(a \circ b) \circ c \circ d \circ \dots = a \circ b \circ (c \circ d) \circ g \circ h \circ \dots = a \circ b \circ c \circ (d \circ g) \circ h \dots \textit{ etc.}$$

**(*iii*) Identity ($E \circ$ or $I \circ$):** The set of elements $\{a, b, c \dots\}$ of the group must contain one and only one element designated as identity element. It is represented as $E$ (or $I$)[1]. If there exists some elements, say $e$ and $e'$ in a non-empty set where in a binary operation is defined so that for any elements a of the set there hold the conditions:

$$a \circ e = a \text{ (right identity law)}, \; e' \circ a = a \text{ (left identity law)}$$

then $e$ (respectively $e'$) is called the right (respectively left) identity element in the set with regard to binary operation $\circ$. When $e = e'$ we drop the prefixes left or right. For example, in the set of integers 0 (zero) is the identity element with respect to operation (+) while 1 is the identity element with respect to multiplication (.) in the same set. It will be shown later that the identity element is imagining with respect to a given operation in a group formed by a non-empty set with respect to that operation.

**Note:** [1]We will use $E$ for identity operation or element throughout this book.

This element is such that for every element $\{a, b, c \dots E\}$, of the set, $E \circ$ (element of the set) $\equiv$ (element of the set) $\circ E \equiv$ *element of the set.*

*For example:* $E \circ a = a \circ E = a$; $E \circ b = b \circ E = b$; $E \circ E = E \circ E = E \dots$ *etc.*

The presence of the identity element in the set implies that the binary operation between $E$ and any element of the set gives the same element itself with which the identity element combine. One may think that such a member of the set ($E$) should not be of any importance to the theory. But it is not true because this concept is mathematically very useful in the development and applications of the group theory. We will realise it a little later.

**(*iv*) Inverse ($i \circ$):** If for every element of the set, there exists a unique element in the set which is termed as the inverse of the element the two elements are called inverses to each other. For example, for the elements $\{a, b, c \dots\}$ of the set, there will be an inverse element for each element, [*viz.* $\{a^{-1}, b^{-1}, c^{-1} \dots\}$] their combinations yield $(a \circ a^{-1}) = (a^{-1} \circ a) = E$ *etc.* It implies that for every element of the set (say $m$), there must be some other element in the set (say $x$) such that their combinations between $m$ and $x$, *e.g.* $[(m \circ x$ and $x \circ m)]$ are equal: $(m \circ x) = (x \circ m) = E$. In this statement, it is not asserted that $m$ and $x$ are necessarily distinct. These two elements may be the same which means that in some cases an element may be the inverse of itself (that is $X.X^{-1} = E$). For example, inverse of an identity element is the same element. It must also be realised that each element of the set has a unique inverse. There cannot be more that one inverse of one element in the set.

Now, we are able to define a group as a set of elements (called group elements) with the binary operation ($\circ$) [*i.e.*, $\{G, \circ\}$] satisfying all the four properties forms an algebraic group. In this manner, we consider a set of elements and a rule of combination written as ($\circ$) by which two elements, $g_i$ and $g_j$ or any ordered couple in the set can be combined so as to yield a unique element which is again present in the set. This binary composition and the resulting element must follow the following conditions. The set of elements along with binary operations form the group. The postulates are,

Postulate 1: (Closure under $\circ$) ($C$)

Postulate 2: (Associativity of elements under $\circ$) ($A$)

Postulate 3: (Unique identity $E$ under $\circ$) ($N$)

Postulate 4: (Existence of unique inverse of all elements under $\circ$) ($I$)

And, in mathematical language we can define a group as "A group is an ordered pair ($G$, $\circ$) with the following properties ($C$, $A$, $N$, $I$) where,

($i$) $C = \forall\ a, b \in G, (a \circ b) \in G$,

($ii$) $A = \forall\ a, b, c \in G, (a \circ b) \circ c = a \circ (b \circ c)$,

($iii$) $N = \forall\ a \in G\ \exists\ E \in G$ such that $a \circ E = E \circ a = a$ and

($iv$) $I = \forall\ a \in G\ \exists\ y \in G$ such that $a \circ y = y \circ a = E$.

Group is denoted as $\{G, \circ\}$. Where the underlying set $G$ has elements $\{g_1, g_2, g_3, g_4, \ldots, g_n\}$.

## 1.2 ABELIAN GROUPS

Amongst the various groups, there are special groups which are termed as 'Abelian Groups'. In such groups, besides the four conditions (or postulates) imposed for defining interrelationships among the elements by the binary operation defining the group, there is a fifth condition or postulate that has to be fulfilled. It is the condition of "Commutativity". In postulate 1, we have not differentiated between combination of two elements $a$ and $b$ if $(a \circ b) \equiv (b \circ a)$. We have only indicated the combinations as $a \circ b$ or $b \circ a$. Now, the question arises if the combination $a \circ b$ is identical to $b \circ a$. In algebra or in arithmetic, the answer is yes for addition. That is $a \circ b = b \circ a$. But in group theory, this law of combination cannot be generalized. Thus, the combination $(a \circ b)$ may give the element ($c$), while that of $(b \circ a)$ may give a different element ($d$) of the group where $c$ and $d$ are two different elements of the group. The groups where this law of combination (*viz.* $a \circ b = b \circ a$) holds are known as **"Abelian Groups".** Therefore for the 'Abelian group', we have a fifth postulate which says that in "Abelian groups" the binary law of combination is "Commutative" for all the members of the group. In *non-Abelian groups*, the fifth postulate does not hold good. However, it should be emphasized here that in non-Abelian groups, there are some elements for which the commutative law holds good, *e.g.*, Identity commutes with every element of the group.

## 1.3 LEFT AND RIGHT-HANDED COMBINATION AND OTHER PROPERTIES

In general, in many binary combinations it is sometimes necessary to state whether an element '$b$' is to be binary combined by the element '$a$' in the sense $(a \circ b)$ or $(b \circ a)$. In the first case, we say that the element '$a$' is left combined or left multiplied by the element '$b$'. In the second case $(b \circ a)$, the element '$a$' is right combined or right multiplied by the element '$b$'.

At this point, we would also like to state a theorem (without giving any proof), which concerns "reciprocal" or the inverse of the elements. The theorem is

"The reciprocal of any combination (or product) of two or more than two elements of the group is equal to the combination of their reciprocals in the reverse order." Thus, we have (eq. 1.6) (inverse and reciprocal will be used synonymously in this book)

$$(a \circ b \circ c \circ \ldots m \circ n)^{-1} = [n^{-1} \circ m^{-1} \ldots c^{-1} \circ b^{-1} \circ a^{-1}) \qquad \ldots(1.6)$$

There are two kinds of groups: (1) finite (2) infinite. Finite groups are those which contain a limited number of elements. Infinite groups contain an infinite number of elements in the set forming a group.

The number of elements in a finite group is called the "*order*" of the group. The usual symbol of the order for a group is '$h$'.

To illustrate the above abstract concepts defining a group, we will give below some examples.

## 1.4 EXAMPLES OF GROUPS, BINARY COMBINATION AND ABSTRACT MULTIPLICATION TABLES AND SOME MORE PROPERTIES OF BINARY COMBINATIONS

Prior to giving examples of groups, we will henceforth use the term "Multiplication" of two or more elements in place of binary combination. This is the usual convention which is followed in most books on group theory. Thus, the product of two elements as a result of multiplication is the result of binary combination of two elements of the group.

While defining a group, we have used some words like binary combination, inverse, identity element, group elements associativity and commutativity *etc.*

In order to clarify their meaning of operations; we give below some examples of binary operations before exemplifying the group concept.

**(*a*) Examples of Binary Combination (Multiplication):** Some of the examples of binary combinations are given in Table 1.1. As stated earlier a binary combination is combination (∘) on two objects from a set to produce a unique third object by applying the combination (∘).

**Table 1.1:** Examples of binary combinations

| *S. No.* | *Example* | *Binary combination* |
|---|---|---|
| 1. | $(-5) \times (+3) = -15$ | Multiplication (×) |
| 2. | (6, 9) (=) 54 | Multiplication (×) |
| 3. | (6, 9) (=) –3 | Subtraction |
| 4. | (6, 9) (=) 6/9 | Division |
| 5. | (12,18) (HCF)6 | Taking HCF of 12 and 18 |
| 6. | (12,18) = 12 or $\left.\begin{matrix}(x \circ y) = x \\ x < y\end{matrix}\right\}$ | Taking the lower number out of two. |
| 7. | Matrices addition | Addition of elements of two matrices. |
| 8. | For any two points $A$ and $B$ on a line we define $A \circ B = C$ where $C$ is the mid point between $A$, and $B$ | Geometric operation |
| 9. | $\overline{A} \circ \overline{B} = \overline{C}$, where $\overline{A}, \overline{B}, \overline{C}$ are vectors | Vector addition |
| 10. | $\begin{vmatrix} 1 & 2 & 3 \\ 2 & 3 & 1 \end{vmatrix}$ | Permutation operation |
| 11. | Rotation of figure | (123) → (321) Rotation |
| 12. | $R_1 \circ R_2 = R_3$ | Successive rotation of a figure ($R_2$) then ($R_1$) (First $R_2$ then $R_1$) |

**(*b*) Closure:** We have used the word closure. By closure we mean that a binary operation (∘) on a pair of elements (Table 1.1) of any set ($S$) is closed, if for every possible choice of $x$ and $y$ elements in $S$, the product $(x \circ y)$ also lies in the set. For example,

(*i*) We have a set of positive integers {1, 2, 3....∞) and our binary combination (∘) in the set is addition of any two integers (*e.g.*, $x \circ y = (x + y) = z$). $z$ is also one of the positive integers and which is also a member of the set. Hence, the set is closed for the addition (binary operation). However, for the same set if we take subtraction as binary operation, then the set is not closed because, for $y > x$, with subtraction as binary operation. $(x \circ y) = (x - y) = -z$.

The product $(-z)$ is not a member of the set of positive integers and hence the set is not closed.

Let us take another example of rotations. We have a figure and we rotate it about an axis (*e.g.*, $z$) passing through a point which lies in the figure and the rotation $\alpha$ is about this axis which is perpendicular to the plane of the figure. We have a set of 24 rotation operations, through angle {15°, 30°, 45°, 60°, 75°, 90°, 105°, 120°, 135°, 150°, 165°, 180°, 195°, 210°, 225°, 240°, 255°, 270°, 285°, 300°, 315°, 330°, 345°, 360°}. This is a set of rotation of a figure. Here, the elements of the set are rotations $(R_1 \circ R_2)$ on the same figure. That is, we apply a set of operations on a figure. Is this set closed? Now, we examine $(x \circ y)$ which means that we first rotate through an angle '$y$' followed by another rotation through an angle '$x$'. Let us take $x$ = rotation of the figure through an angle 330° and $y$ = rotation of the same figure by 15°. Then $x \circ y$ = {Rotation by (330°)} $\circ$ {Rotation by 15°) = (Rotation of figure by 345°), thus, the product of two operations on the figure is also a member of set. Hence, the set is closed.

***(c)* Associativeness:** A binary operation $(\circ)$ is said to be associative if for all the elements of the set the property $(a \circ b) \circ c = a \circ (b \circ c)$ (Here $a$, $b$, $c$ are any three elements of the set).

It can easily be shown that a product of $(n)$ number of elements given in a definite order has also a unique meaning, so that the brackets can be inserted at will so long as the order of the elements remains unaltered. Thus,

$$\{a_1 \circ a_2 \circ a_3 \circ \ldots\ldots a_n\} = (a_1 \circ a_2 \circ a_3 \circ \ldots\ldots a_m)(a_{m+1} \circ a_{m+2} \circ \ldots\ldots a_n) \quad \ldots(1.7)$$

where $a_1, a_2, a_3 \ldots\ldots a_n \in$ set and $1 < m < n$.

If $p > n > m > 1$ then,

$$(a_1 \circ a_2 \circ \ldots\ldots a_n)(a_{n+1} \circ \ldots \circ a_p) = (a_1 \circ a_2 \circ \ldots\ldots \circ a_m)(a_{m+1} \circ a_{m+2} \circ \ldots \circ a_p) \quad \ldots(1.8)$$

This implies that any two modes of bracket lead to the same result. If $a_1$ $a_2$ $a_3 \ldots a_p$ are the same, elements (*viz.* $a$) and or the same binary operation $(\circ)$ is carried out on the same elements $p$ times then, we can write

$$\{a \circ a \circ a \ldots\ldots a\} = a^p \text{ ($p$-times)} \quad \ldots(1.9)$$

or, $$(a \circ a \circ a \circ \ldots\ldots m)(a \circ a \circ a \circ \ldots\ldots n) = a^m \circ a^n = a^{m+n} \ldots\ldots \quad \ldots(1.10)$$

Hence, the familiar law of indices (equations 1.9 and 1.10) is the outcome of the associative law of multiplication (binary operation).

In case the two elements $a$ and $b$ of the set do not commute [*i.e.*, $(a \circ b) \neq (b \circ a)$], then we have,

$$(a \circ b)^n \neq (b \circ a)^n \neq b^n \circ a^n\} \quad \ldots(1.11)$$

***(d)* Identity:** In the preceding paragraphs, we have defined identity $E$ as an element in the set with binary operation ($\circ$) such that for all the elements of the set (*e.g.*, $x_i$), we have,

$$E \circ x_i = x_i \circ E = x_i \qquad ...(1.12)$$

(Here $x_i$ is one of the elements of the set).

Let us examine this property. We consider the set of integral numbers along with zero as one of the elements with subtraction as binary operation. Let us see it (eq. 1.12) is true. For this set, we will have for all the elements $x_i$ of this set.

$$x_i - 0 = x_i \qquad ...(1.13)$$

so that zero is a right identity. However, we will also have a left identity *viz.*

$$0 - x_i = -x_i \qquad ...(1.14)$$

In that case, zero is not the identity because $(-x_i)$ is not a member of the set. Hence, zero is not a left identity. [*i.e.*, $x_i - 0 \neq 0 - x_i$].

Similar is case for the set of real numbers with division ($\div$) as binary operation and 1 as the identity because

$$1 \div x_i = 1/x_i \neq x_i, \qquad ...(1.15)$$

while $\qquad x \div 1 = x_i$

So 1 is the right identity and not the left identity in this set.

The discussions in the preceding paragraph indicate systems, which have either left- or right-identity. A set whose members obey associative law must contain an element $E$, which is both left- and right-identity. In view of this, the associative law plays a vital role in defining a group. Thus, the statement that if a system possesses either *a left- or a right-identity and not both, asserts that either the binary operation is not associative or identity is not properly defined.*

**(*e*) Inverse:** This is the fourth requirement of the group. Thus, if a set has a binary combination ($\circ$) and identity element $E$ then for every element $a$, there can be found an element $b$, such that $a \circ b = b \circ a = E$. '$b$' then is known as inverse of '$a$'. Thus, the inverse element '$b$' depends upon '$a$'. In some cases, inverse of $y$ may be $y$ itself

$$y \circ y = E$$

Similar to identity element, and associativeness there are sets where only "Right inverses" are found (*i.e.*, $a \circ b = E \neq b \circ a$). There are also sets where only "Left inverses" are found in this set (*i.e.*, $b \circ a = E \neq a \circ b$). *But analogous to identity case, the sets where associative law holds, both "Right and Left" inverse must exist (i.e.,* $a \circ b = b \circ a = E$*).*

## EXERCISE

**1.** Prove that for a set of positive integers {1, 2, 3...$N$} provided with the binary operation $(\circ) = \left[\dfrac{x+y}{(x-y)}\right] (x > y)$ is not a closed set.

**2.** Prove that on a set of real numbers associated with a binary operation $x \circ y = (x^3 + y^3)$, the associative law does not hold good.

**3.** In the multiplication Table 1.2 (given below) we have a set of five objects ($x_1$, $x_2$, $x_3$, $x_4$, $x_5$) (do not bother what they are). These elements combine according to a binary operation ($\circ$). (Do not bother what it is). Is this binary operation "Associative"? (Try by taking $x_3 \circ x_4 \circ x_5$) and ($x_2 \circ x_3 \circ x_4$) *etc.*

**Table 1.2**

| $\circ$ | $x_1$ | $x_2$ | $x_3$ | $x_4$ | $x_5$ |
|---|---|---|---|---|---|
| $x_1$ | $x_1$ | $x_2$ | $x_3$ | $x_4$ | $x_5$ |
| $x_2$ | $x_2$ | $x_3$ | $x_5$ | $x_1$ | $x_4$ |
| $x_3$ | $x_3$ | $x_1$ | $x_4$ | $x_5$ | $x_2$ |
| $x_4$ | $x_4$ | $x_5$ | $x_1$ | $x_2$ | $x_3$ |
| $x_5$ | $x_5$ | $x_2$ | $x_3$ | $x_4$ | $x_1$ |

**Answer:** *Not associative*. There will be 120 combinations (5!) and some of combinations are not associative (Note even if a single combination $(A \circ B) \circ C = A \circ (B \circ C)$ is not equal, then the set is not associative).

**4.** Suppose we have three things, we number them as 1, 2, 3. We permute them. There are following six possible permutation operations.

1. $$\pi_1 = \begin{vmatrix} 1 & 2 & 3 \\ 2 & 1 & 3 \end{vmatrix}, \pi_2 = \begin{vmatrix} 1 & 2 & 3 \\ 3 & 2 & 1 \end{vmatrix}, \pi_3 = \begin{vmatrix} 1 & 2 & 3 \\ 1 & 3 & 2 \end{vmatrix},$$

$$\pi_4 = \begin{vmatrix} 1 & 2 & 3 \\ 3 & 1 & 2 \end{vmatrix}, \pi_5 = \begin{vmatrix} 1 & 2 & 3 \\ 2 & 3 & 1 \end{vmatrix} \text{ and } \pi_6 = \begin{vmatrix} 1 & 2 & 3 \\ 1 & 2 & 3 \end{vmatrix}$$

Product of these permutations is defined as for example

$$\pi_3 \circ \pi_4 = \begin{vmatrix} 1 & 2 & 3 \\ 1 & 3 & 2 \end{vmatrix} \circ \begin{vmatrix} 1 & 2 & 3 \\ 3 & 1 & 2 \end{vmatrix} = \begin{vmatrix} 1 & 2 & 3 \\ 3 & 2 & 1 \end{vmatrix} = \pi_2.$$

Prove that the binary operation of this set obey (1) associativity law (2) right-inverse = left-inverse. Which one of the elements is identity? Prove that Left identity = Right identity.

5. $\overline{AB}, \overline{CD}$ are two vectors in a two-dimensional space. Suppose that the addition of two vectors, *viz.* $\overline{AB} + \overline{CD} = \overline{AD}$, is the binary combination of the set of vectors (infinite in number). Prove that this set of vectors obey (*i*) Associative law (*ii*) Contain identity element and the inverse element and (*iii*) Right identity ≡ Left identity (*iv*) Right inverse = Left inverse. (*v*) What is the inverse of the vector $\overline{AB}$?
6. See Table 1.3 below showing the binary operations between elements of the set (*a*, *b*, *c*, *d*, *e*). Find out (*i*) the identity element (*ii*) is right-and left-identity and right-and left inverses are equal. (*iii*) Does it follow the law of Associativeness?

**Table 1.3:** Binary operations between the set (*a*, *b*, *c*, *d*, *e*)

| ∘ | ***a*** | ***b*** | ***c*** | ***d*** | ***e*** |
|---|---|---|---|---|---|
| *a* | *a* | *b* | *c* | *d* | *e* |
| *b* | *b* | *c* | *e* | *a* | *d* |
| *c* | *c* | *a* | *d* | *e* | *b* |
| *d* | *d* | *e* | *a* | *b* | *c* |
| *e* | *e* | *d* | *b* | *c* | *a* |

**Answers:** 1. (*a*) 2. No 3. No.

## Examples of Groups

Prior to giving a few examples of groups, let us first explain the role of "Multiplication Table" in the abstract theory of groups. It will show without any difficulty that if a particular set of elements under the binary combination for which the multiplication table is given forms a group.

**(*a*) Multiplication Table:** While defining a group, it was asserted that the elements of the set have no physical meaning. No sooner we associate a meaning to the elements (*e.g.*, transformation, substitution, permutation *etc.*), the group of such elements becomes a "*Realisation of the particular abstract group with a certain binary combination.*" In the abstract group, one only knows that all the possible products (binary combinations) of the elements like ($a \circ b$, $a, b \in$ set) according to certain well defined rule. If the order of group is $g$ (total number of elements in the group), then one finds $g^2$ such products (or combinations). These products ($g^2$) are written in the tabular form as suggested by Caley in 1854. (For the groups of order 2, 3, 4 and 5), these products are written as follows. (We will examine from these that all the four requirements of a group are satisfied).

**1. Groups of order 2:** The elements are $\{E, a\}$ and the only possible table is

First operation (columns); →Second operation (rows)

| $\circ$ | ***E*** | ***a*** |
|---|---|---|
| $E$ | $E$ | $a$ |
| $a$ | $a$ | $E$ |

Multiplication $g_i\, g_j$ are,

1. $E \circ E = E$ 2. $a \circ E = a$
3. $E \circ a = a$ 4. $a \circ a = E$.

**2. Groups of order 3 $\{E, a, b\}$**

| $\circ$ | ***E*** | ***a*** | ***b*** |
|---|---|---|---|
| $E$ | $E$ | $a$ | $b$ |
| $a$ | $a$ | $b$ | $E$ |
| $b$ | $b$ | $E$ | $a$ |

In this table multiplications of two elements are given as,

(*i*) $E \circ E = E$; $E \circ a = a$; $E \circ b = b$; $a \circ E = a$; $a \circ a = b$; $a \circ b = E$;

$b \circ E = b$; $b \circ a = E$; $b \circ b = a$; *etc.*

There is another alternate possibility of binary combination which is given below.

| | ***E*** | ***a*** | ***b*** |
|---|---|---|---|
| $E$ | $E$ | $a$ | $b$ |
| $a$ | $a$ | $E$ | $a$ |
| $b$ | $a$ | $b$ | $E$ |

(show that it is not a group)

**3. Groups of order 4 $\{E, a, b, c\}$:** There are three possible modes of combination. These are given below.

| $\circ$ | ***E*** | ***a*** | ***b*** | ***c*** |
|---|---|---|---|---|
| $E$ | $E$ | $a$ | $b$ | $c$ |
| $a$ | $a$ | $E$ | $c$ | $b$ |
| $b$ | $b$ | $c$ | $E$ | $a$ |
| $c$ | $c$ | $b$ | $a$ | $E$ |

(I) inverse of $a, b, c$ are elements themselves ($c_{2h}$)

| $\circ$ | ***E*** | ***a*** | ***b*** | ***c*** |
|---|---|---|---|---|
| $E$ | $E$ | $a$ | $b$ | $c$ |
| $a$ | $a$ | $b$ | $c$ | $E$ |
| $b$ | $b$ | $c$ | $E$ | $a$ |
| $c$ | $c$ | $E$ | $a$ | $b$ |

(II) cyclic group ($c_4$)

| $\circ$ | ***E*** | ***a*** | ***b*** | ***c*** |
|---|---|---|---|---|
| $E$ | $E$ | $a$ | $b$ | $c$ |
| $a$ | $a$ | $b$ | $c$ | $b$ |
| $b$ | $b$ | $c$ | $E$ | $a$ |
| $c$ | $c$ | $E$ | $a$ | $E$ |

(III) is not a group because $a \circ (b \circ c) \neq (a \circ b) \circ c$

Only two groups of order 4 are possible

Show that the set $\{E, a, b, c\}$ whose multiplication tables are given above form the group in the tables (I) and (II) while (III) does not form the group, so there will be two groups of order 4.

**4. Groups of order 5 {*E, a, b, c, d*}:** It can form only one group (5 is a prime number).

| ∘ | ***E*** | ***a*** | ***b*** | ***c*** | ***d*** |
|---|---|---|---|---|---|
| *E* | *E* | *a* | *b* | *c* | *d* |
| *a* | *a* | *b* | *c* | *d* | *E* |
| *b* | *b* | *c* | *d* | *E* | *a* |
| *c* | *c* | *d* | *E* | *a* | *b* |
| *d* | *d* | *E* | *a* | *b* | *c* |

Cyclic group ($C_5$)
(All the above groups of order one to five are 'Abelian groups' *i.e.*, $x \circ y = y \circ x$)

| ∘ | ***E*** | ***a*** | ***b*** | ***c*** | ***d*** |
|---|---|---|---|---|---|
| *E* | *E* | *a* | *b* | *c* | *d* |
| *a* | *a* | *E* | *d* | *b* | *c* |
| *b* | *b* | *c* | *E* | *d* | *a* |
| *c* | *c* | *d* | *a* | *E* | *b* |
| *d* | *d* | *b* | *c* | *a* | *E* |

This is not a group because

$$\left.\begin{aligned}(a \circ b) \circ c = a\\ a \circ (b \circ c) = c\end{aligned}\right\}$$ Associative law does not hold good.

In general, any set whose number of the elements is a prime number always forms only one group.

**5. Groups of order 6 {*E, a, b, c, d, f*}:**

| ∘ | ***E*** | ***a*** | ***b*** | ***c*** | ***d*** | ***f*** |
|---|---|---|---|---|---|---|
| *E* | *E* | *a* | *b* | *c* | *d* | *f* |
| *a* | *a* | *b* | *E* | *f* | *c* | *d* |
| *b* | *b* | *E* | *a* | *d* | *f* | *c* |
| *c* | *c* | *d* | *f* | *E* | *a* | *b* |
| *d* | *d* | *f* | *c* | *b* | *E* | *a* |
| *f* | *f* | *c* | *d* | *a* | *b* | *E* |

I
Non-Abelian
because $a \circ c = f$ and $c \circ a = d$

| ∘ | ***E*** | ***a*** | ***b*** | ***c*** | ***d*** | ***f*** |
|---|---|---|---|---|---|---|
| *E* | *E* | *a* | *b* | *c* | *d* | *f* |
| *a* | *a* | *b* | *c* | *d* | *f* | *E* |
| *b* | *b* | *c* | *d* | *f* | *E* | *a* |
| *c* | *c* | *d* | *f* | *E* | *a* | *b* |
| *d* | *d* | *f* | *E* | *a* | *b* | *c* |
| *f* | *f* | *E* | *a* | *b* | *c* | *d* |

II
Abelian group
(cyclic group)

(Find out other groups of order 6).

By the above-given procedure the multiplication tables for all groups of any order (not infinite) can be written. It is to be noted that these are square tables in which each row and each column consists of the same set of elements but in a different arrangement and further in any row or column, no element of the group appears more than once (Rearrangement theorem). One can therefore write a number of tables for a particular set of elements. These tables are known as "*Latin Squares*". Thus, a multiplication table of a group is a "Latin Square" But one must bear in mind that every "Latin Square" is not a group, however the multiplication table of a group is a Latin Square. (See example of order 5 or 6). Only those Latin squares form groups for which group-axioms are satisfied (*viz.* (*a*) Closure (*b*) Left and right-handed identity (*c*) Associativeness of the elements and (*d*) Right-and Left-handed inverse).

In writing a multiplication table of a group, the following conventions are followed,

1. The identity element $E$, always appears at the cross point of the first row and first column.
2. The orders of the elements in the first row and in the first column are the same.
3. To find the product $(m \circ n)$, we first look the column $n$ and then the row $m$, at the cross point of column '$n$' and row '$m$', the result of product $[m \circ n]$ is found.

(*b*) **Examples of Groups:** We now give some examples of groups.

1. Group of order 2 consisting of two numbers $\{1, -1\}$ with multiplication as binary operation.
2. Group of order 4 consisting of the number $\{1, i, -1, -i\}$ [here $i^2 = -1$]
3. Discrete infinite groups of all real integers (+, –, and zero) with addition as binary operation). *(discrete infinite groups are those groups which have infinite number of elements and each element is discrete like 1, 2, 3 etc. Compare them with infinite continuous groups where elements are not discrete, but fractional).* (Discrete vs. Continuous). All finite groups are discrete.
4. The set of all positive real numbers (zero is not a member) with multiplication (×) as binary composition (∘).
5. The non-empty set containing {1} as the only member with multiplication as binary composition.
6. The set of two matrices $\left\{\begin{vmatrix} 1 & 0 \\ 0 & 1 \end{vmatrix}, \begin{vmatrix} -1 & 0 \\ 0 & -1 \end{vmatrix}\right\}$ with matrix multiplication as binary composition (∘).
7. The set of all non-singular (determinant ≠ 0) square $|n \times n|$ matrices with matrix multiplication as binary combination (∘). ($n$ is +ve integer ).
8. The set of all complex number $(a + ib)$ under addition as binary composition (∘).
9. The set of all non-zero complex numbers under scalar multiplication as binary composition (∘).

10. A set of six functions $\left(f_1(x)=(x), f_2(x)=(1-x), f_3(x)=\dfrac{x}{(x-1)},\right.$

$\left. f_4(x)=\dfrac{1}{x}, f_5(x)=\dfrac{1}{(1-x)}, f_6(x)=\dfrac{(x-1)}{x}\right)$ with binary composition of two functions as the substitutions of a function into another (*i.e.*, function of a function). For example, $(f_5 \circ f_3)(x) = f_5 \circ \{f_3(x)\}$

$= f_5\left\{\dfrac{x}{(x-1)}\right\} = \dfrac{1}{\left(1-\dfrac{x}{x-1}\right)} = (1-x) = f_2(x)$ here $f_5(x) = \dfrac{1}{(1-x)}$. [Now

substitute $\dfrac{x}{(x-1)}$ for $x$ as $f_5(x) = f\left(\dfrac{1}{1-x}\right)$. Let us take another binary

composition *viz.* $(f_3 \circ f_4)\ (x) = f_3 \circ \left[\dfrac{1}{x}\right] = \dfrac{1/x}{(1/x)-1} = \dfrac{1}{(1-x)} = f_5(x)$.

[Here, we substitute $1/x$ in place of $x$ in $f_3(x)$.] Also $f_3 \circ f_3$

$= f_3 \circ [x/(x-1)] = \dfrac{x/(x-1)}{x/(x-1)-1} = x = f_1(x)$.

11. Let $x$ and $y$ be the elements of set $\{x, y\}$ such that $x^2 = E$; $y^3 = E$ and $y \circ x \circ y = x$. Now, we consider the set of elements $\{x, y, y^2\ (= y \circ y), xy, xy^2, E\}$ with $E$ as identity element, we can construct the multiplication table of these elements [We write $x \circ x = x^2$; $y \circ y = y^2$; $y \circ y \circ y = y^3 = E$. Thus, $y^2 \circ y = E$, *i.e.*, $y^2$ is inverse of $y$. Hence $(x \circ y^2) \circ x = (xy^2)(y \circ x \circ y) = xy^3\, xy = x$, $Exy = x^2y = Ey = y$]. Using these relationships construct the multiplication table (Table 1.4).

**Table 1.4:** Multiplication table of group

| $(\circ)$ | $E$ | $x$ | $y$ | $y^2$ | $xy$ | $xy^2$ |
|---|---|---|---|---|---|---|
| $E$ | $E$ | $x$ | $y$ | $y^2$ | $xy$ | $xy^2$ |
| $x$ | $x$ | $E$ | $xy$ | $xy^2$ | $y$ | $y^2$ |
| $y$ | $y$ | $xy^2$ | $y^2$ | $E$ | $x$ | $xy$ |
| $y^2$ | $y^2$ | $xy$ | $E$ | $y$ | $xy^2$ | $x$ |
| $xy$ | $xy$ | $y^2$ | $xy^2$ | $x$ | $E$ | $y$ |
| $xy^2$ | $xy^2$ | $y$ | $x$ | $xy$ | $y^2$ | $E$ |

**Table 1.5:** Multiplication table of permutation of three identical particle

| $S_3$ | **E** | **A** | **B** | **C** | **D** | **F** |
|---|---|---|---|---|---|---|
| *E* | *E* | *A* | *B* | *C* | *D* | *F* |
| *A* | *A* | *B* | *E* | *D* | *F* | *C* |
| *B* | *B* | *E* | *A* | *F* | *C* | *D* |
| *C* | *C* | *F* | *D* | *E* | *B* | *A* |
| *D* | *D* | *C* | *F* | *A* | *E* | *B* |
| *F* | *F* | *D* | *C* | *B* | *A* | *E* |

## EXERCISE

**1.** Prove that the following "Latin Square" does not represent a group.

| | **E** | **a** | **b** | **c** | **d** |
|---|---|---|---|---|---|
| *E* | *E* | *a* | *b* | *c* | *d* |
| *a* | *a* | *c* | *d* | *b* | *E* |
| *b* | *b* | *E* | *c* | *d* | *a* |
| *c* | *c* | *d* | *a* | *E* | *b* |
| *d* | *d* | *b* | *E* | *a* | *c* |

**Answer:** Inverse law is not followed.

## Permutation Groups

We have another class of interesting groups known as permutation groups. These are very important groups and used in quantum mechanics of identical particles like electrons, protons, *etc.* To have a better concept of these kinds of groups, let us take a concrete example. Consider a system of three identical particles, (*e.g.*, three electrons). If we interchange any two or more of these particles, the resulting configuration of the system remains invariant. The total numbers of possible exchanges of three particles are 6 (*i.e.*, 3!). These are,

| (1) $\begin{pmatrix}123\\123\end{pmatrix} = E$<br>(No change in the position of electrons *i.e.*,)<br>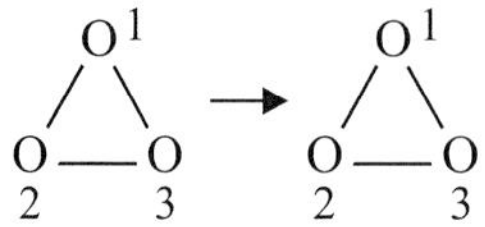 | (2) $\begin{pmatrix}123\\231\end{pmatrix} = A$<br>(Electron no. 1 $\rightarrow$ 2, electron 2 $\rightarrow$ 3, electron 3 $\rightarrow$ 1) *i.e.*,<br>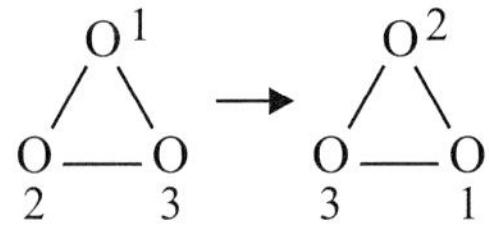 | (3) $\begin{pmatrix}123\\312\end{pmatrix} = B$<br>(Electron no. 1 $\rightarrow$ 3, electron 2 $\rightarrow$ 1, electron 3 $\rightarrow$ 2) *i.e.*,<br>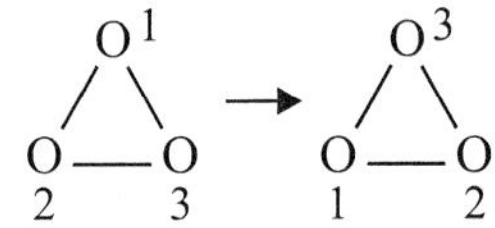 |
|---|---|---|
| (4) $\begin{pmatrix}123\\213\end{pmatrix} = C$<br>(electron 1 $\rightarrow$ 2, electron 2 $\rightarrow$ 1, electron 3 $\rightarrow$ 3) *i.e.*,<br>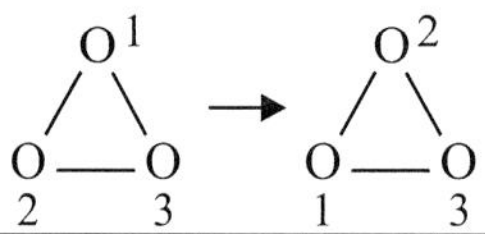 | (5) $\begin{pmatrix}123\\321\end{pmatrix} = D$<br>(Electron no. 1 $\rightarrow$ 3, electron 3 $\rightarrow$ 1, electron 2 $\rightarrow$ 2) *i.e.*,<br>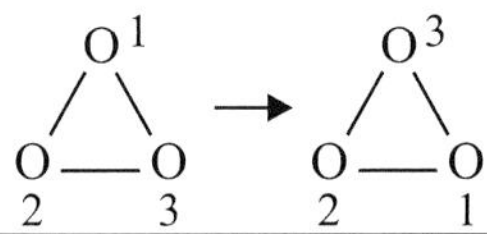 | (6) $\begin{pmatrix}123\\132\end{pmatrix} = F$<br>(Electron no. 1 $\rightarrow$ 1, electron 2 $\rightarrow$ 3, electron 3 $\rightarrow$ 2) *i.e.*,<br>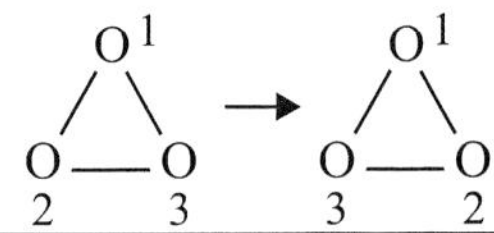 |

**Note:** Here 1, 2, 3 refer to the positions of electrons, rather than the electrons themselves. It means the electron 1, means one of the electrons in position 1, and another electron is in position 2 and electron 3 in position 3. Here electrons are not numbered, but their positions are numbered. We exchange the positions of electrons.

Thus, the three electrons have six possible exchanges (or permutations). We name them as *E*, *A*, *B*, *C*, *D* and *F*. Let us say that three electrons have six possible states which are obtained by the permutation operators (*E*, *A*, *B*, *C*, *D*, *F*) $\phi_1 = (1, 2, 3)$, $\phi_2 = (2, 3, 1)$, $\phi_3 = (3, 1, 2)$, $\phi_4 = (2, 1, 3)$, $\phi_5 = (3, 2, 1)$ and $\phi_6 = (1, 3, 2)$. The operators operate on these states and their operations are interpreted as follows.

$$\hat{C}\phi_2 = \begin{vmatrix} 1 & 2 & 3 \\ 2 & 1 & 3 \end{vmatrix} (2, 3, 1) = (3, 2, 1) = \phi_5 \qquad ...(1.16)$$

This operation $\hat{C} = (2, 1, 3)$ means that the electron in position 1 is to be put in position 2 and electron in position 2 is to be put in position 1, while the electron in position 3 is left as it is (eq. 1.16).

It can be shown by carrying out these permutation operations on the three electrons, that the set of six operations form a group. (The multiplication table of these permutation operations (given in Table 1.5) is worked out as follows.

Let us find the combinations,

$$(i)\ B \circ (C \circ \phi_2) = \begin{pmatrix} 1 & 2 & 3 \\ 3 & 1 & 2 \end{pmatrix}\begin{pmatrix} 1 & 2 & 3 \\ 2 & 1 & 3 \end{pmatrix}(231)$$

$$= \begin{pmatrix} 1 & 2 & 3 \\ 3 & 1 & 2 \end{pmatrix}(321) = (132) = \phi_6. \qquad ...(1.17)$$

$$(ii)\ D \circ \phi_2 = \begin{pmatrix} 1 & 2 & 3 \\ 3 & 2 & 1 \end{pmatrix}(231) = (132) = \phi_6 \qquad ...(1.18)$$

From eqs. (1.17) and (1.18), we get,

$$(B \circ C) \circ \phi_2 = D \circ \phi_2$$

or,

$$B \circ C = D \qquad ...(1.19)$$

(*See* Multiplication Table (1.5) for permutation group of three objects.)

One can now generalise the above concept for $n$-identical objects. If we interchange any two identical objects in this set of $n$-objects, the resulting configuration remains unaltered from the original one. We can consider all such interchanges of all $n$-objects as permutation transformations of the system. All such transformations form a set which form a group under permutation operations as binary compositions and under which the system is invariant. Since, the total number of permutations for $n$-objects is ($n$!), the group formed by $n$! such transformations will be of the order $n$! ($n$ factorial). This group is known as "Permutation Group" of '$n$' number of objects (not symmetry group of $n$-objects) and is termed as *Symmetric group* of degree $n$. It is denoted as $S_n$. Thus, group of three objects is $S_3$.

One can also show that each permutation of these objects can be expressed as successive transpositions of two objects taken at a time. We define a transposition ($pq$) on a $n$-identical objects as the operation in which the object in the positions $p$ and $q$ are interchanged while leaving all the remaining objects as they are. Such $S_n$ groups of $n$-objects can be generated by only ($n$ – 1) transpositions like (1 2), (1 3), (1 4) = 1...(1 $n$) transpositions. As an example, a set of generators of $S_3$ group are the two transposition (1 2) and (1 3). All the elements of $S_3$ can be the written as suitable products of these two transposition. Thus $A$ = (1 3) (1 2), $B$ = (1 3) (1 2) (1 3), $C$ = (1 2), *etc.* (check these).

Now, if a particular permutation consists of an even number of transpositions, it is called an even permutation and if it contains an odd number of transpositions, it is odd permutation. Thus, in our case of $S_3$, $E\ A\ B$ are even and $C\ D\ F$ are odd permutations. Further the product of even and even (*e.g.*, $A.B$) and odd and odd ($C.D$) are even permutation, while the product of odd and even or *vice versa* is an odd permutation (*e.g.*, $A.C$ or $B.D$, *etc.*).

The second point about the even and odd permutations is that the set containing of all even permutations form a group (or a subgroup). Thus, the set ($E$, $A$, $B$) forms a subgroup. Consider two non-empty sets $G_1$ and $G_2$ both for using an algebraic group with respect to a binary operation $O$ (say). If $G_1$ is a subset of $G_2$ then ($G$, 0) is called a subgroup of group ($G_2$, 0). Example, set of integers forms a subgroup of a group formed by the set of real numbers with respect to addition as binary operation of $S_3$ and is termed as an Alternating group of order $n!/2$ (*i.e.*, 6/2).

## Groups of Transformations

These types of groups are of special importance to chemists and physicists. By transformation we mean an operator which when operates on a physical system like triangle, square, *etc.*, leaves them invariant. We will learn in the next chapter that any transformation that leaves the system invariant is called a "*Symmetry transformation*". The following is an example which illustrates a group of transformations."

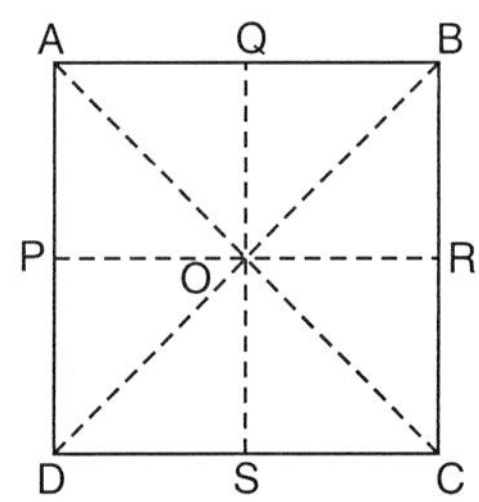


**Fig. 1.1** Square with its Corners, Edges and Centre Points.

### $d_1$ Group of Symmetry Transformations of a Square

Let us take a square as shown in Fig. 1.1 with its centre, $O$ and corners $A$, $B$, $C$, $D$ and mid points of edges as $P$, $Q$, $R$ and $S$.

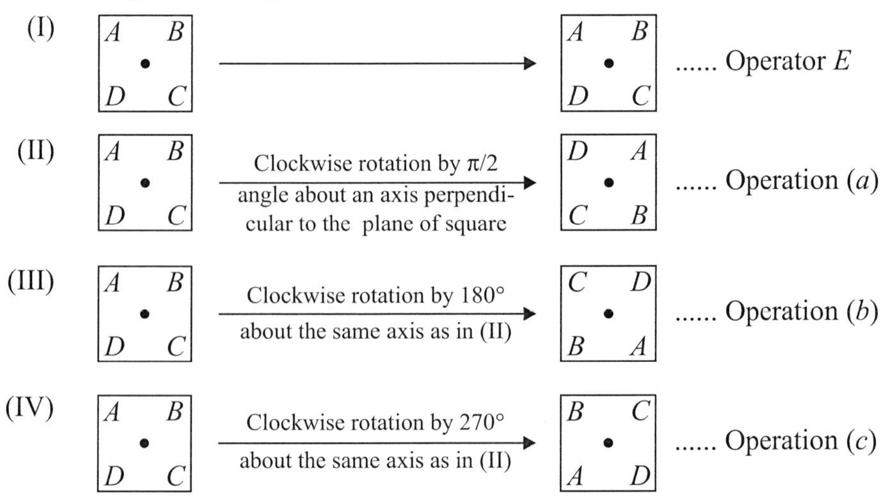

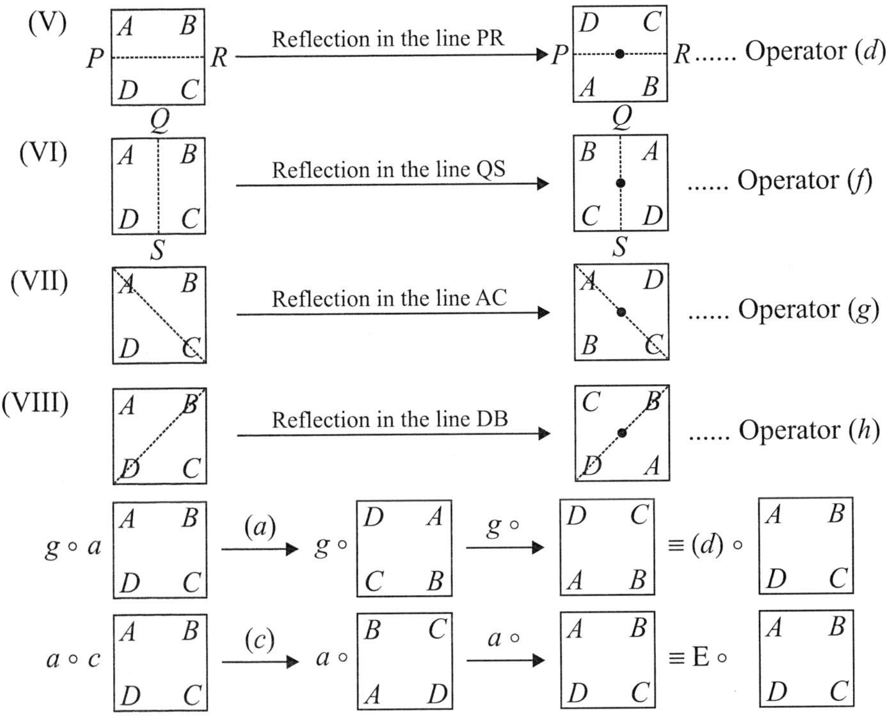


**Fig. 1.2** Operations on a Square which leave the Square Invariant.

We carry out the operations listed in Fig. 1.2 on a square (Fig. 1.1). You will notice that all these operations leave the square invariant. One cannot distinguish the square after the operation from the original one. (no distinguishing mark present on the square). Further you will notice that these are the only operations which will leave the square invariant (For the notation of these operations, *see* chapter two) one of the operations in the list is "Reflection in line (*e.g.*, *PR*). It simply means reflection in a plane perpendicular to the plane of square and which is passing through the line *PR*). One may think of another operation, *viz.* inversion through the centre of square O; but one should realise that this operation is identical to operation (*b*).

It can be shown that the set of these eight transformations $\{E, a, b, c, d, f, g, h\}$ form a group (group of symmetries of a square). Thus, let us find the product of two operations, [*e.g.*, $(g \circ a)$]. This can be carried out in the following way $(g \circ a)$ on a square:

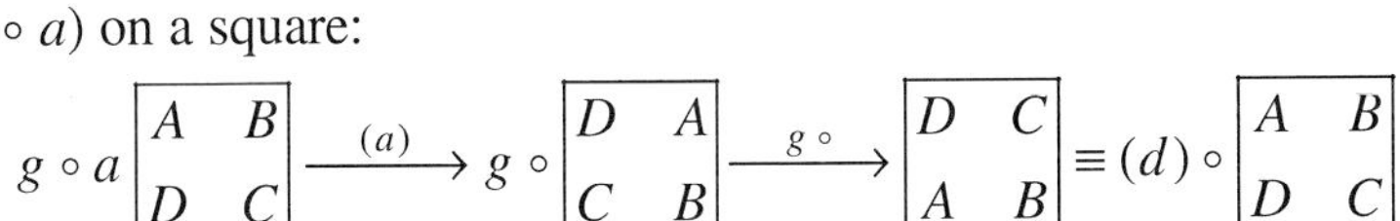


Thus, $(g \circ a) = (d)$ One can easily find the products of all possible binary combinations and the result will be one of the possible operations given in Fig. 1.2.

Inverse of each operation can also be picked up from the multiplication table (given below). Thus, consider the successive operations of $a \circ c$ on the square (given below).

$$g \circ c \boxed{\begin{matrix} A & B \\ D & C \end{matrix}} \xrightarrow{c} a \circ \boxed{\begin{matrix} B & C \\ A & D \end{matrix}} \xrightarrow{a} \boxed{\begin{matrix} A & B \\ D & C \end{matrix}} = E \boxed{\begin{matrix} A & B \\ D & C \end{matrix}}$$

The same result will be obtained if we consider the successive operations in the reverse order (*viz.* $c \circ a$). Thus,

or, $$\left.\begin{matrix} a \circ c = c \circ a = E \\ (a)^{-1} = c \quad \text{and} \quad (c)^{-1} = (a) \end{matrix}\right\} \tag{1.20}$$

From the given multiplication Table 1.6, one will notice that each element of the group has an inverse. We give below the multiplication table for this group. (Verify all the entries in the Table 1.6.)

**Table 1.6:** Multiplication table for the symmetry operations of square (Two ways of writing)

**Multiplication Table**

| (∘) | **E** | **a** | **b** | **c** | **D** | **f** | **g** | **h** |
|---|---|---|---|---|---|---|---|---|
| $E$ | $E$ | $a$ | $b$ | $c$ | $D$ | $f$ | $g$ | $h$ |
| $a$ | $a$ | $b$ | $c$ | $E$ | $G$ | $h$ | $f$ | $d$ |
| $b$ | $b$ | $c$ | $E$ | $a$ | $F$ | $d$ | $h$ | $g$ |
| $c$ | $c$ | $E$ | $a$ | $b$ | $H$ | $g$ | $d$ | $f$ |
| $d$ | $d$ | $h$ | $f$ | $g$ | $E$ | $b$ | $c$ | $a$ |
| $f$ | $f$ | $g$ | $d$ | $h$ | $B$ | $E$ | $a$ | $c$ |
| $g$ | $g$ | $d$ | $h$ | $f$ | $A$ | $c$ | $E$ | $b$ |
| $h$ | $h$ | $f$ | $g$ | $d$ | $C$ | $a$ | $b$ | $E$ |

**First Operation**

or

| ∘ (Second operation) | **E** | **a** | **b** | **c** | **d** | **f** | **g** | **h** |
|---|---|---|---|---|---|---|---|---|
| $E$ | $E$ | $a$ | $b$ | $c$ | $d$ | $f$ | $g$ | $h$ |
| $c$ | $c$ | $E$ | $a$ | $b$ | $h$ | $g$ | $d$ | $f$ |
| $b$ | $b$ | $c$ | $E$ | $a$ | $f$ | $d$ | $h$ | $g$ |
| $a$ | $a$ | $b$ | $c$ | $E$ | $g$ | $h$ | $f$ | $d$ |
| $d$ | $d$ | $h$ | $f$ | $g$ | $E$ | $b$ | $c$ | $a$ |
| $f$ | $f$ | $g$ | $d$ | $h$ | $b$ | $E$ | $a$ | $c$ |
| $g$ | $g$ | $d$ | $h$ | $f$ | $a$ | $c$ | $E$ | $b$ |
| $h$ | $h$ | $f$ | $g$ | $d$ | $c$ | $a$ | $b$ | $E$ |

(Second way)

We have shown the multiplication table of the symmetry operations of the square in two ways. (The ordering of rows and columns in writing down the table is immaterial.) In the second way of writing the table, the ordering is such that elements in the first column are the inverses of the corresponding elements in the first row. By writing the table in this fashion, all the elements that appear on the principal diagonal of the table are $E$. (*Normal group*). The advantage of writing the table in this form will be clear later in the book.

We have shown in the example of square that all the symmetry operations form a group. This is, in general, true and can be shown that the set of all the symmetry operations (or transformations) of a system constitutes a group (Chapter 2). We will be forming groups of different systems and studying their symmetry properties.

## 1.5 GENERATORS OF A FINITE GROUP

It is possible to generate all the elements of a group, starting from a certain few elements of set by subjecting them to certain relations. First, we consider the set of those elements whose powers ($x^n$) and product ($x^m y^n$) generate all the elements of the group. Let us suppose that we start with a single element $A$ and assume that all its integral powers are elements in the group. Thus, we generate new elements $A^2, A^3, A^4, \ldots$ of the group and we stop at a point when $A^n = E$ because the higher than $n$ power will not generate any new elements. This group will be formed by the set $\{E, A, A^2, A^3, \ldots, A^{n-1}\}$. The element $A$ will be of order $n$. It is a cyclic group. In the same manner, every element of the group will be of same order. This is called the order of the element. In some cases, the binary operation of two elements generates all the elements of the group. Let us assume that the two elements $A$ and $B$ generate all the elements of a group. Let these be subjected to relations $A^2 = E$ and $B^3 = E$ (*i.e.*, the orders of elements of $A$ and $B$ are 2 and 3 respectively). Accordingly the group will contain elements $E, A, B, B^2$. It must also contain $AB$ and $AB^2$. If it is non commutative group, then it must also have $BA$ and $B^2A$ as elements. Thus, in the non-Abelian group $A \circ B$ and $B \circ A$ are distinct elements. Regarding $(AB) \circ B = AB^2$, we have a relation $(A \circ B^2) = E$ and from this relationship $(AB)^{-1} = B^{-1}A^{-1} = AB$, or $B^{-1}A$ since $A^2 = E$. But from $B^3 = E$. We have $B^{-1} = B^2$. Hence, $(AB)^{-1} = B^2A$. Using this, we get $(AB)(B) = B^2AB = B^2B^2A = BA$, which belong to the set. Hence, the desired group is (E, $A$, $B$, $B^2$, $AB$, $BA$) and is of order 6.

In general, the generators of the group are not unique and can be chosen in a number of ways.

After giving a number of examples of groups, we now turn our attention towards various classifications of the elements of the group. We will now develop the concepts of classes, subgroups, cosets and factor groups.

## 1.6 SUBGROUPS, COSETS, FACTOR GROUPS AND CLASSES

There are several ways of sorting out the elements of a group into different categories. The important ones are subgroups, cosets, factor groups and classes.

### [a] Subgroups

To understand the concept of subgroups, let us start with a group ($G$) of symmetries of a square, reproduced in Table 1.7. Consider the four elements $\{E, a, b, c\}$ of the group with the same binary operation as that of the whole group $G$. Let this subset be termed as '$H$', we will notice that the elements of this set with the binary operation of the group $G$ satisfy all the postulates of the group ($E$, closure, associativity and inverse). (Multiplication Table is given within the thick square of the multiplication Table of $G$). This subset is termed as a subgroup $H$ of the group $G$. Thus, if a smaller set ($H$) within a group $G$ exists which follow the group axioms, under the same law of binary operation as that of group $G$, then the small set is said to be a subgroup of the group $G$. Thus, $H$ is a subgroup of $G$.

**Table 1.7:** Multiplication table

| ∘ | **E** | **a** | **b** | **c** | **d** | **f** | **g** | **h** |
|---|---|---|---|---|---|---|---|---|
| $E$ | $E$ | $a$ | $b$ | $c$ | $d$ | $f$ | $g$ | $h$ |
| $a$ | $a$ | $b$ | $c$ | $e$ | $g$ | $h$ | $f$ | $d$ |
| $b$ | $b$ | $c$ | $E$ | $a$ | $f$ | $d$ | $h$ | $g$ |
| $c$ | $c$ | $E$ | $a$ | $b$ | $h$ | $g$ | $d$ | $f$ |
| $d$ | $d$ | $h$ | $f$ | $g$ | $E$ | $b$ | $c$ | $a$ |
| $f$ | $f$ | $g$ | $d$ | $h$ | $b$ | $E$ | $a$ | $c$ |
| $g$ | $g$ | $d$ | $h$ | $f$ | $a$ | $c$ | $E$ | $b$ |
| $h$ | $h$ | $f$ | $g$ | $d$ | $c$ | $a$ | $b$ | $E$ |

| ∘ | **E** | **b** | **d** | **f** |
|---|---|---|---|---|
| $E$ | $E$ | $b$ | $d$ | $f$ |
| $b$ | $b$ | $E$ | $f$ | $d$ |
| $d$ | $d$ | $f$ | $E$ | $b$ |
| $f$ | $f$ | $d$ | $b$ | $E$ |

,

| ∘ | **E** | **h** |
|---|---|---|
| $E$ | $E$ | $h$ |
| $h$ | $h$ | $E$ |

..... *etc.*

There are more subgroups of $G$ besides $H$. Thus, we have (1) $\{E, b, d, f\}$, (2) $\{E, b\}$ (3) $\{E, g\}$ (4) $\{E, d\}$ and (5) $\{E, f\}$. Their multiplication tables are also given. Besides these subgroups, there are two more subgroups, $\{E\}$ and the total group $\{G\}$. These groups are present in all the groups and termed as Trivial subgroups.

**1. Order of Subgroups:** In all subgroups, the order of the subgroup is always lower than that of group $G$. An important theorem of the order of the subgroups is that if a subgroup is of order $H$ (*e.g.*, order of $H$ is $h$) and the order of group $G$ (main group) is $g$; then $g$ is an integral multiple of $h$. In our case $G$ is of order 8, then its subgroups will be either of order 4 or of order 2 because 8 is an integral multiple of 4 (*i.e.*, $2 \times 4$) or of 2 (*i.e.*, $4 \times 2$). We will not give the proof of the theorem (see any book on group theory). The integer $n\left(= \dfrac{\text{order of group } G}{\text{order of subgroup}}\right)$ (*e.g.*, $G/H$) is called the "INDEX" of the subgroup.

Thus, the number 2 is the INDEX of $H$ (= 8/4).

**2. Order of an element of a group:** We have indicated earlier that the order $p$ of an element $A$ of a group $G$ is given by the number $p$, such that $A^p \equiv E$. For example, in our group $G$ the order of the element $a$ is 4 because $a = a$, $a^2 = b$, $a^3 = c$, $a^4 = E$ (*i.e.*, $a^p = a^4 = E$) so that order of $a$ is 4. One can prove that if an element $A$ of a finite group $G$ is of order $n$ then the set $(A, A^2, A^3, \ldots, A^n = E)$ is a subgroup of the group $G$ (order $g$). It follows that the order of every element of finite group must be an integral divisor of order $g$ of the group $G$. This is a general theorem.

Thus, in the case of our group (order 8) of the symmetries of a square, the order of the elements are

(*i*) The order of $E$ is 1 ($E = 1$)

(*ii*) The order of $a = 4$ ($a$, $a^2 = b$, $a^3 = c$, $a^4 = E$)

(*iii*) The order of $b = 2$ ($b = b$, $b^2 = E$)

(*iv*) The order of $c = 4$ ($c = c$, $c^2 = b$, $c^3 = a$, $c^4 = E$)

(*v*) The order of $d = 2$ ($d^1 = d$, $d^2 = E$)

(*vi*) The order of $f = 2$ ($f^1 = f$, $f^2 = E$)

(*vii*) The order of $g = 2$ ($g^1 = g$, $g^2 = E$)

(*viii*) The order of $h = 2$ ($h^1 = h$, $h^2 = E$)

**Table 1.8:** Group $G$ (of order 6) of permutation of three objects

| (∘) | ***E*** | ***A*** | ***B*** | ***C*** | ***D*** | ***F*** |
|---|---|---|---|---|---|---|
| $\begin{vmatrix}123\\123\end{vmatrix}$ $E$ | $E$ | $A$ | $B$ | $C$ | $D$ | $F$ |
| \|231\| $A$ | $A$ | $B$ | $E$ | $D$ | $F$ | $C$ |
| \|312\| $B$ | $B$ | $E$ | $A$ | $F$ | $C$ | $D$ |
| \|213\| $C$ | $C$ | $F$ | $D$ | $E$ | $B$ | $A$ |
| \|321\| $D$ | $D$ | $C$ | $F$ | $A$ | $E$ | $B$ |
| \|132\| $F$ | $F$ | $D$ | $C$ | $B$ | $A$ | $E$ |

### [b] Cosets

To exemplify the concept of cosets, we consider the group $G$ (order 6) of permutations of three objects. We reproduce the group $G$ (Table 1.8). This group has four subgroups ($H$, $P$, $Q$, $R$) as given below,

(*i*)$\{E, A, B\} = H$ (order 3); (*ii*) $\{E, C\} = P$ (order 2); (*iii*) $\{E, D\} = Q$ (order 2); (*iv*) $\{E, F\} = R$ (order 2). Let us take the subgroup $H$. *i.e.*, $\{E, A, B\}$. The remaining elements of $G$ are $\{C, D$ and $F\}$.

Now, we take one element of the group $G$ and multiply the elements belonging to $H$ on its right-hand side. Construct all the products with all the elements of $G$ on the right with all the elements of $H$. In this manner, we form the following Table 1.9. These are called Right Cosets of the subgroup $H$. Similarly construct all the products with all the elements of $G$, on the left-hand side with all the elements of $H$. These are known as Left Cosets of $H$. Both these cosets are given in Table 1.9. In the same table, we have also given the LHS and RHS products of the elements of $H$. with the elements of $H$ itself. These are not right and left cosets associated with the subgroup (*viz. H*). Cosets are respectively *the RHS and LHS products of the elements of H with the elements of group G which are NOT the members of the subgroup, (viz. H)*. Thus, we have right cosets associated with $H(E, A, B)$ as $\{(CDF), (DFC), (FCD)\}$ and the left cosets of $H$ as $\{(CDF)\ (DCF)(FDC)\}$. Cosets are the subset of the elements of group $G$[1].

1. Only *RHS* and *LHS* products of the elements of $H$ with the elements of group $G$ which are NOT the members of the subgroup $H$ are taken as cosets in this book. We have not included the elements in cosets which are also the member of the subgroup. For example, [*E.E*, *A.A*, *B.B*, *A.B*, *B.A etc.*]

**Table 1.9:** Right and left cosets of the subgroup $H$ ($E$, $A$, $B$)

| | | | | |
|---|---|---|---|---|
| **H.E** | $E.E = E$ | $A.E = A$ | $B.E = B$ | |
| **H.A** | $E.A = A$ | $A.A = B$ | $B.A = E$ | |
| **H.B** | $E.B = B$ | $A.B = E$ | $B.B = A$ | |
| **H.C** | $E.C = C$ | $A.C = D$ | $B.C = F$ | Right cosets |
| **H.D** | $E.D = D$ | $A.D = F$ | $B.D = C$ | |
| **H.F** | $E.F = F$ | $A.F = C$ | $B.F = D$ | |
| **E.H** | $E.E = E$ | $E.A = A$ | $E.B = B$ | |
| **A.H** | $A.E = A$ | $A.A = B$ | $A.B = E$ | |
| **B.H** | $B.E = B$ | $B.A = E$ | $B.B = A$ | |
| **C.H** | $C.E = C$ | $C.A = F$ | $C.B = D$ | Left cosets |
| **D.H** | $D.E = D$ | $D.A = C$ | $D.B = F$ | |
| **F.H** | $F.E = F$ | $F.A = D$ | $F.B = C$ | |

Finally the following results are obtained:

1. The number of both right and left cosets of the group are the same *viz*.($C$, $D$, $F$), though in different order. This observation may not be true in every case. In case the left and right cosets of a subgroup ($H$) are the same, then the subgroup is called the *Normal Subgroup or Invariant Subgroup of group G*. Thus, $H$ is normal or invariant subgroup of our group $G$ ($E$, $A$, $B$, $C$, $D$, $F$).

$$X \circ H = H \circ X \qquad ...(1.21)$$

or

$$X^{-1} XH = X^{-1} HX = H \qquad ...(1.22)$$

for all $X$ belonging to $G$.

The condition (eq. 1.22) can also be expressed by requiring that every element of set $XH$ must be equal to some elements of $H$X.

or

$$XH_i = H_jX \qquad ...(1.23)$$

which will give,

$$X^{-1}H_jX = H_i \qquad ...(1.24)$$

This is a conjugated relation between $H_i$ and $H_j$ (sec.1.6) or $H_i$ and $H_j$ are conjugated elements of the set of cosets and $H_i$ is also member of the normal subgroup of $H_j$. Thus, a normal subgroup consists of complete classes of the bigger group (sec.1.6).

The second important concept of the subgroups and cosets is that of the Factor group.

### [c] Factor Groups

Consider again the group of order eight, [*viz.* $\{E, a, b, c, d, f, g, h\}$ given in the preceding section (symmetry group of square)]. In this group, we choose one of its subgroup *e.g.*, $\{E, b\}$. The following are its left and right cosets (Table 1.10).

**Table 1.10:** Left and right cosets of the subgroup $H_1 = (E, b)$ of the group $G = \{E, a, b, c, d, f, g, h\}$

| ***Right Cosets*** | | | | | | | |
|---|---|---|---|---|---|---|---|
| $HE$ | $H_1a$ | $H_1b$ | $H_1c$ | $H_1d$ | $H_1f$ | $H_1g$ | $H_1h$ |
| $EE = E$ | $Ea = a$ | $Eb = b$ | $Ec = c$ | $Ed = d$ | $Ef = f$ | $Eg = g$ | $Eh = h$ |
| $bE = b$ | $ba = c$ | $bb = E$ | $bc = a$ | $bd = f$ | $bf = d$ | $bg = h$ | $bh = g$ |
| ***Left Cosets*** | | | | | | | |
| $E \circ H_1$ | $a \circ H_1$ | $b \circ H_1$ | $c \circ H_1$ | $d \circ H_1$ | $f \circ H_1$ | $g \circ H_1$ | $h \circ H_1$ |
| $E \circ E = E$ | $a \circ E = a$ | $b \circ E = b$ | $c \circ E = c$ | $d \circ E = d$ | $f \circ E = f$ | $g \circ E = g$ | $h \circ E = h$ |
| $Eb = b$ | $ab = c$ | $bb = E$ | $cb = a$ | $db = f$ | $fb = d$ | $gb = h$ | $h \circ b = g$ |

All the cosets, [Right and left cosets of the subgroup $\{E, b\}$] are given with the remaining elements of the group $G$. Four different cosets of $H_1$ including that of $H_1$ are cosets (1) $\{E, b\}$, (2) $\{a, c\}$, (3) $\{d, f\}$ and (4) $\{g, h\}$. Other cosets are $\{c, a\}$, $\{d, f\}$, $\{f, d\}$, $\{g, h\}$. Similarly left cosets are $\{E, b\}$, $\{a, c\}$, $\{b, E\}$, $\{c, a\}$, $\{d, f\}$, $\{f, d\}$, $\{g, h\}$, $\{h, g\}$.

Let us call coset 1, coset 2, coset 3, and coset 4 as $p_1$, $p_2$, $p_3$ and $p_4$ respectively. Now, we define the product of two cosets as [cf. Table 1.6]

$$[\text{Coset } (2) \circ \text{coset } (3)] = p_2 \circ p_3 = \{a, c\} \circ \{d, f\} = \{a \circ d, a \circ f, c \circ d, c \circ f\} \quad ...(1.25)$$

$$= \{g, h, h, g\} = \{2g, 2h\} = \{g, h\} \quad ...(1.26)$$

Here, we have multiplied two sets and obtained a set by multiplying each element of one set with every element of the second set (Eq. 1.25). In case of repeating, we have taken the element only once, and obtained a set given in eq. (1.26). This type of multiplication is a little different from the usual multiplications of two elements of a group.

If we now take the set of cosets $\{p_1, p_2, p_3, p_4\}$ and construct its multiplication table, it comes out that this set of cosets forms a group (Table 1.11). Thus, it can be seen that the set of cosets $\{p_1, p_2, p_3, p_4\}$, where each coset $p_i$ is taken as an element on a higher plane of abstraction form a group under the given law of multiplication as discussed above. This group is known as *Factor* Group. It is also called the Quotient Group of $G$, with respect to subgroup $H$, and is denoted as $(G/H_1)$ with index $g/h_1$ where $g$ is the order of $G$ (in our case 8) and $h_1$ is the order of $H_1$( = 2 in our case) [*i.e.*, 8/2 = 4]. This index $(g/h_1)$ is equal to the order of the group of cosets [in our case (*i.e.*, $p_1$, $p_2$, $p_3$, $p_4$)].

**Table 1.11:** Multiplication table $p_1\,p_2\,p_3\,p_4$

| | $p_1$ | $p_2$ | $p_3$ | $p_4$ |
|---|---|---|---|---|
| $p_1$ | $p_1$ | $p_2$ | $p_3$ | $p_4$ |
| $p_2$ | $p_2$ | $p_1$ | $p_4$ | $p_3$ |
| $p_3$ | $p_3$ | $p_4$ | $p_1$ | $p_2$ |
| $p_4$ | $p_4$ | $p_3$ | $p_2$ | $p_1$ |

The third point about the group of cosets is that generally the identity element of the group of cosets (*e.g.*, $p_i$) is the normal subgroup (*i.e.*, $p_1 = \{E, b\}$) of the bigger group $G$.

## [d] Classes

**1. General Approach:** This is another method by which we subclassify group elements. We divide the elements into different classes.

While defining a group it was necessary to recognise that a binary combination of two elements of a group could not be presumed to be commutative. It implies that it is not necessarily true that

$$A \circ B = B \circ A \qquad ...(1.27)$$

This condition may hold for some pairs of elements or operations within a group but not for others [*e.g.*, for $E$, we always have $E \circ A = A \circ E$ or in the group $\{E, a, b, c, d, f, g, h\}$ (given for the symmetry of square) $g \circ a = d$ and $a \circ g = f$, while in the same group $a \circ\ b = c$ and $b \circ a = c$]. Thus, $a$ and $b$ commute while the elements $g$ and $f$ do not commute. Or in the group of permutations $S_3$ $\{E, A, B, C, D, F\}$ it is truc for $E, A$ and $B$ but it is not true when these are combined with $C, D$ and $F$. We have also shown that groups for which it is true for all pairs of elements are known as Abelian groups. In giving a more precise meaning to the word class we start with the following equation where $A$, $B$ are two elements of the group and where we assume that this equation is good for at least some of the operations of the non-Abelian groups.

$$AB \neq BA \qquad ...(1.28)$$

If we multiply on the RHS of $AB$ and $BA$ by the operator $A^{-1}$, we get,

$$ABA^{-1} \neq BAA^{-1} \qquad ...(1.29)$$

or

$$ABA^{-1} \neq B \qquad ...(1.30)$$

In eq. (1.30) $ABA^{-1}$ is some other element of the group, but not equal to $B$. Let us suppose that it is equal to the element $D$. There is a *hidden symmetry* in eq. (1.30). We write equation (1.30) as

$$ABA^{-1} = D \text{ (but not } B) \qquad ...(1.31)$$

The transformation of eq. (1.31) is known as *similarity transformation*. Here $A$ is said to transform $B$ into $D$ by a similarity transformation. We can also write eq. (1.31) as,

$$(A^{-1}A)\,B(A^{-1}A) = A^{-1}DA$$

or

$$B = A^{-1}DA \qquad ...(1.32)$$

That is, $A$ is also said to transform $D$ into $B$. Because of this relationship between $B$ and $D$, these elements are said to be *Conjugate* elements of the group. For the formation of conjugacy between $B$ and $D$, we have picked up the element $A$, at random, As a matter of fact, we could have picked up the element $C$ in place of $A$, and could have got the following relationship,

$$CBC^{-1} = F. \qquad ...(1.33)$$

(There is no theorem which would require that $CBC^{-1} = D$)

From eq. (1.33), we can get, [similar to eq. (1.32)],

$$B = C^{-1}\,FC \qquad ...(1.34)$$

That is, $B$ and $F$ are also related through conjugacy. Or $B$ and $F$ are also conjugated elements. It implies that $B$, $D$ and $F$ are also conjugate to each other as shown below. Consider,

$$A^{-1}\,DA = B = C^{-1}FC$$

or

$$(AA^{-1})\,D(AA^{-1}) = (AC^{-1})\,F(CA^{-1}) \qquad ...(1.35)$$

or

$$D = (AC^{-1})\,F(CA^{-1})$$

But $(AC^{-1})$ and $CA^{-1}$ are inverse to each other because $(AC^{-1})\,(CA^{-1}) = AEA^{-1} = E$. We believe that $AC^{-1}$ is one of the elements of the group. Let us assume that it be $H$. So,

$$H\,F\,H^{-1} = D \qquad ...(1.36)$$

Thus $(B, D, F)$ are conjugate to each other. We can make different sets of conjugate elements out of the elements of group. The elements of each set are conjugate to each other. As an example, consider the multiplication table of the group $\{E, P, Q, R, S, T)$ given below). (Table 1.12)

Let us collect the inverse of each element from this table. These are $P^{-1} = Q$, $Q^{-1} = P$, $R^{-1} = R$, $S^{-1} = S$, $T^{-1} = T$. The conjugate elements are $(E)$, $(PQ)$, $(RST)$ as calculated below.

| | | | |
|---|---|---|---|
| 1. $EPE = P$ | 2. $EQE = Q$ | 3. $ERE = R$ | 4. $ESE = S$ |
| $PPQ = P$ | $PQQ = Q$ | $PRQ = S$ | $PSQ = T$ |
| $QPP = P$ | $QQP = Q$ | $QRP = T$ | $QRP = T$ |
| $RPR = Q$ | $RQR = P$ | $RRR = R$ | $RRR = R$ |
| $SPS = Q$ | $SQS = P$ | $SRS = T$ | $SSS = S$ |
| $TPT = Q$ | $TQT = P$ | $TRT = S$ | $TST = R$ |

**Table 1.12:** Multiplication table

| | *E* | *P* | *Q* | *R* | *S* | *T* |
|---|---|---|---|---|---|---|
| *E* | *E* | *P* | *Q* | *R* | *S* | *T* |
| *P* | *P* | *Q* | *E* | *T* | *R* | *S* |
| *Q* | *Q* | *E* | *P* | *S* | *T* | *R* |
| *R* | *R* | *S* | *T* | *E* | *P* | *Q* |
| *S* | *S* | *T* | *R* | *Q* | *E* | *P* |
| *T* | *T* | *R* | *S* | *P* | *Q* | *E* |

It is thus concluded that in the group {*E*, *P*, *Q*, *R*, *S*, *T*}, we have three classes, [*viz.* (*E*) (*P*, *Q*) and (*R*, *S*, *T*)]. Let us designate them as $C_1$, $C_2$ and $C_3$ respectively. It is also seen that *E* transforms into itself by all the elements of the group. The set (*P*, *Q*) and (*R*, *S*, *T*) are each transformed into themselves by all the elements of the group. It is important to know that the sets which are transformed into themselves by all the elements of the group are called *conjugate classes* or *classes*.

The rigorous approach for determining which elements or operators are in the same class is by "Similarity Transformation". Since this process is rather tedious for groups of large orders, a simpler approach is however useful. This is given below.

**2. Simpler Approach:** (1) First we find out if the group is Abelian. In Abelian groups (*e.g.*, $C_n$, *etc.*) every operation forms its own class. (2) The useful distinctions between various operations should be made. The following are in the same or separate classes—(*a*) all vertical planes (*b*) $C_n$ and $C_n^{-1}$, (*c*) $S_n$ and $S_n^{-1}$. A rule that works is that the symmetry elements that are not interchanged by any other element (except cyclic group) are in separate classes. It is always true that all $\sigma_d$ planes are interchanged by $C_n$ but $\sigma_v$ and $\sigma_d$ are always in the different classes because these planes are not interchanged by any operation of the group. (3) For non-Abelian groups containing $C_n$ axes, $C_n^1$ and $C_n^{n-1}$ are in the same class; but for group having $C_3$-axis (*e.g.*, $C_{3h}$, $C_{3v}$ *etc.*), ($C_3^1$, $C_3^2$) and ($S_3^1$, $S_3^5$) are in separate classes. Rotation through the same angle about two different axes or reflections in two distinct planes belong to the same class if and only if the two axes or two planes can be brought into each other by some element of the group. Some of these rules are helpful in deciding the classes of the group elements.

A few of the important properties of classes are the following:

1. No element of the group transforms into two different classes.
2. No element occurs in more than one class.
3. The number of elements in a class is a factor of the order of the group, thus if *g* is the order of group (*e.g.*, in 6 order group {*E*, *P*, *Q*, *R*, *S*, *T*} and

the group has a class with $h$ elements (*e.g.*, 3), then each element of the class is transformed into itself exactly $g/h$ times.

4. In an Abelian group each element is a class by itself.
5. In all groups Identity element $E$ and $\sigma_h$ form class by itself because any similarity transform of the identity or $\sigma_h$ will always generate the identity element or $\sigma_{h.}$.
6. All elements in the conjugacy class containing elements $x_i$ have the same order.

## 1.7 CLASS ALGEBRA

Let us find out how one multiplies two classes of a group. Let us take the group of symmetry transformations of an equilateral triangle. Its multiplication table is given in Table 1.12. We can express this group in terms of classes as {(*E*), (*PQ*), (*RST*}. We will learn later that all groups in their Character Tables are represented in terms of classes. There is a reason of doing so which we will learn at a later stage. In this way, we have three classes in the symmetry group, *viz.* (*E*) (*P*, *Q*) and (*R*, *S*, *T*). Let us name them as $C_1$, $C_2$ and $C_3$ respectively. However, there is another fundamental reason of writing a group in the terms of classes. The reason is that there exists a class algebra. An example of which is given below. Thus we have,

Class 1: $C_1$ contains $E$; Class 2: $C_2$ contains $P$ and $Q$ and Class 3: $C_3$ contains (*R*, *S*, *T*). We present these classes as

$$C_1' = E$$
$$C_2' = 1/2\ (P + Q) \qquad \text{...(1.37)}$$
$$C_3' = 1/3\ (R + S + T)$$

We multiply these classes together in following way:

$$C_1' \circ C_1' = E \circ E = E = C_1';\ C_1' \circ C_2' = E \circ (1/2)(P + Q)$$
$$= (1/2)\ (P + Q) = C_2'$$
$$C_1' \circ C_3' = E \circ (1/3)\ (R + S + T) = (1/3)\ (R + S + T) = C_3'$$
$$C_2' \circ C_2' = [(1/2)(P + Q)] \circ [(1/2)(P + Q)]$$
$$= (1/4)(P \circ P + P \circ Q + Q \circ P + Q \circ Q).$$

Using multiplication table, we get it as

$$(1/4)\ [Q + E + E + P] = (1/2)E + (1/4)(P + Q) = (1/2)\ C_1' + (1/2)\ C_2'$$
$$= (1/2)\ (C_1' + C_2')$$
$$C_3' \circ C_2' = (1/3)\ (R + S + T) \circ [(1/2)\ (P + Q)]$$
$$= (1/6)\ [R \circ P + R \circ Q + S \circ P + S \circ Q + T \circ P + T \circ Q].$$

From the multiplication table, we get

$$= (1/6)\ [S + T + T + R + R + S]$$
$$= (1/3)\ [R + S + T] = C_3'$$

and
$$C_3' \circ C_3' = (1/9)\ [R + S + T) \circ (R + S + T)$$
$$= (1/9)\ [R \circ R + R \circ S + R \circ T + S \circ R + S \circ S + S \circ T + T \circ R + T \circ S + T \circ T].$$

Using multiplication table, this equals to

$$(1/9)\ [E + P + Q + Q + E + P + P + Q + E] = (1/3)\ [E + P + Q] = (1/3)[C_1' + 2C_2'].$$

In this manner, we complete the following class multiplication table (Table No. 1.13). From this table it is obvious that, in general, the product of multiplying two classes together is of the form

$$C_i' \circ C_j' = C_j' \circ C_i' = \sum_k \in_k C_k' \qquad ...(1.38)$$

where the sum is over all the classes and $\in_k$ is a coefficient. A rather good question now arises "Is it possible to obtain a linear sum of the classes of the form given in eq. (1.39).

$$\xi = \sum_j p_j C_j' \qquad ...(1.39)$$

**Table 1.13:** Class multiplication table

| $\circ$ | $C_1$ | $C_2$ | $C_3$ |
|---|---|---|---|
| $C_1$ | $C_1$ | $C_2$ | $C_3$ |
| $C_2$ | $P_2$ | ½ $(C_1 + C_2)$ | $C_3$ |
| $C_3$ | $C_3$ | $C_3$ | 1/3 $(C_1 + 2C_2)$ |

and which has the property that when multiplied by any other class $C_i$ (say) it gives an equation of the form given by (1.40) (Eigenvalue equation),

$$C_i'\xi = C_j'\left[\sum_j p_j C_j'\right] = \lambda\left[\sum_j p_j C_j'\right] \qquad ...(1.40)$$
$$= \lambda\xi \text{ (Eigenvalue equation).}$$

The eigenvalue λ, when determined leads directly to the character of the class in the character Table. However, the values of λ are actually not the characters; but are related to the characters of the classes. (For its proof *see* 'Applied Group Theory' by G.G. Hall, Longman Harlow, U.K. 1967).

In the case of Abelian group, every element is a class by itself. Therefore, the classes of Abelian groups form groups under class multiplication; but this is trivial because the classes will be isomorphic to the elements of the group itself.

2

# Introducing Symmetries
## (Molecular and Non-Molecular)

*Whenever we can apply the theory of groups, we are able to ignore the bewildering variety to see among similarities, differences and among differences, similarities. The changelessness in a changing world.*

**—C. Reid**

We usually talk of symmetry in our daily life because we instinctively assume that we understand its notion. It is natural as many objects around us like buildings, flowers, *etc.*, have external features which are related to each other. For instance a flower with five petals, when viewed by a stationary observer would appear the same by rotating it through an angle of 72° around its stem and a cube when rotated by an angle of 90° about an axis perpendicular to one of its faces will appear to be the same as that before rotation indicating the relationship of the faces of the cube with each 90° rotation. The relationship is not limited to rotation operation only; but can be extended to other operations also, like reflection in a plane, a point, or a line. These operations bring the object, like cube, after the operation to a state indistinguishable from that prior to operation. Any object which possesses this property is said to possess symmetry. It implies that the object has a certain regularity or balance in shape. If we look around us, we will notice that nature has provided us a very large number of examples of objects which possess these kinds of regularities, *e.g.*, leaves of plants, crystals, *etc.* This kind of regularity is observed not only in nature, but in man-made objects too. This may possibly be due to his feelings of happiness that he gets in regularity. We would like to emphasize here that we should not think that the ultimate limit of relationship concept lies between symmetry and regularity of geometrical objects alone. It is much deeper and widespread in its meaning. It is applicable to abstract ideas, like scientific laws, mathematical equations *etc.* But we will however, be concerned for the text largely with geometrical symmetries and geometrical operations because these symmetry operations give an insight in essentially, chemical problems, and especially related to structures of molecules. In fact, the theory underlying many powerful tools like I.R., and U.V. spectroscopy, dipole moments *etc.*, involves considerations of symmetry properties. Determination of crystal structures depends heavily on symmetry. Since this book is primarily concerned with the properties of molecules, it has been thought that our approach to develop the concept of symmetry should be geometric in nature. Because "Group measures Symmetry", a brief discussion on group theory is also warranted prior to application of symmetry properties of any system to measure its properties. Therefore, the concept of group (chapter one) and space groups and their applications to study the properties of molecules will also form part of this chapter.

## 2.1 DEFINITION OF SYMMETRY

Hermann Weyl defined symmetry applicable to symmetrical objects (geometrical symmetry), symmetrical systems and symmetrical laws, as follows:

"A thing, an object, a system or a law is symmetric under a specific transformation (or operation) provided its appearance is not altered by that operation". For instance, the operation of rotating a sphere upon any axis passing through its centre does not change the appearance of the sphere so the sphere is symmetric with respect to rotation about any axis through its centre.

In the above definition, the words transformation and operation are used interchangeably. However, there is a subtle difference between these two terms while defining the symmetry (*See J. Rosen's Book "Symmetry"*).

Although the definition appears to be simple but a number of mathematical concepts are involved in it. Let us examine it without involving ourselves too much into its mathematical details. The definition consists of the words,

(*a*) Thing (or system)

(*b*) Transformation (operation)

(*c*) Specific transformation which does not alter the appearance of the thing.

The "Thing" (or system) is something which is of interest to us with regard to its symmetry properties. It can be either a concrete object or a physical entity like a molecule, a plant, an animal, earth, solar system, galaxy, the whole universe or a process taking place in time, *e.g.*, scattering of elementary particles, a chemical reaction, biological growth, a piece of music, an idea or a concept like the laws of physics, a mathematical relationship or a feeling, *etc.* Thus, one can study the symmetry properties of any system which can be concrete or abstract in nature.

The word transformation (or operation) has a little mathematical implication (not discussed here). We will define it tentatively as an action that alters an object, from its initial to some final state (Image state).

In the above statement we have

$$\underset{\text{(I)}}{\text{Initial State of a thing}} \xrightarrow[\text{or Operation}]{\text{Transformations}} \underset{\text{(II)}}{\text{Final State of the thing (image state)}}$$

The final state of thing which is known as the "Image State" of the thing is derived from its initial state by performing a transformation.

The study of the symmetry properties of the system in which we are interested involves only those *specific transformations* which affect the initial state of the system in such a manner that the images obtained by the transformations are indistinguishable from the corresponding initial state of system. These transformations signify the "concept of Invariance" whenever the initial state and the image state of any system after the transformations remains invariant. The thing is then said to be "symmetric" with respect to those transformations and these specific transformations or operations are known as "Invariant" or "Symmetry Transformations". Therefore, "*Symmetry means invariance under one or more transformations*". The larger the number of symmetry transformations under which the system or object remains invariant, the higher is the degree of symmetry of the object. Lack of symmetry is known as "Asymmetry".

Earlier we have stated that the object or system whose symmetry properties we consider can be anything (concrete or abstract). In the light of the diversity of the systems, the symmetry transformations can also be equally diverse and imaginative. Thus, they may be geometric transformations (change of space, size, displacements, rotation, reflections), related to time (Displacement by equal increment on time scale, reversal of chronological order of events *i.e.*, reflection in time coordinate (Temporal symmetry), interchange of positive and negative charges, change of velocity *etc.* So the symmetry transformations can be anything. The only condition is the invariance of image state of the object or system from that of the initial state. The book will deal with the geometric transformations *i.e.*, those transformations which will not change the appearance of the objects like the shape or size (*i.e.*, Invariance in the geometric properties of the system).

## 2.2 SYMMETRY ELEMENTS AND SYMMETRY TRANSFORMATIONS: EXAMPLES—TRIANGLE, SQUARE AND TETRAHEDRON

**Example 1:** We take the example of an equilateral triangle to understand the basic concept of geometric symmetry (Fig. 2.1). We assume that the triangle has a definite size and located at a definite position in space with definite orientation (Important point). So any operation on it which changes its shape, position, size and/or its orientation in space will not be considered in the type of symmetry operations which we would like to consider.

We wish to study and collect only those transformations which produce its image state indistinguishable from its original state, keeping its shape position or size the same as those of initial state.

Let us consider an axis *D* perpendicular to the plane of the triangle and intersecting at its geometric centre *O* (Geometric entity) (Fig. 2.1). If we rotate the triangle by 2π/3 about this axis clockwise (operation), the triangle will come to a state (image state) which is indistinguishable from its initial state as a result of 2π/3 rotation operation about the axis. Thus, this operation is a symmetry operation for the equilateral triangle. Similarly rotation about the same axis by 4π/3 and 6π/3 (=2π) are two more symmetry operations of the triangle.

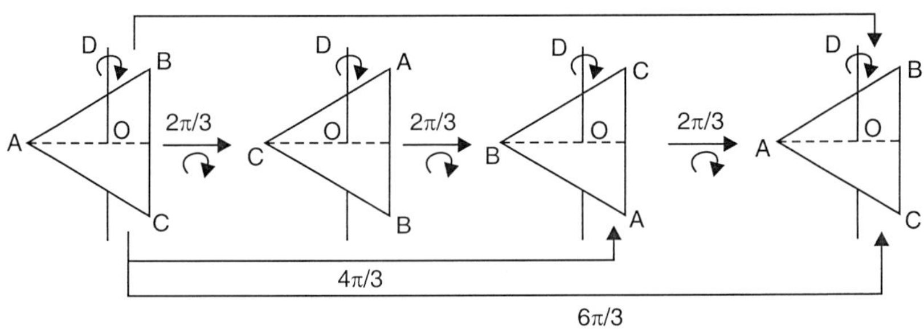


**Fig. 2.1** Equilateral triangle and its rotations about axis *D* clockwise by 2π/3, 4π/3, and 6π/3

We have rotated the triangle clockwise. We could have rotated the triangle anticlockwise also with the same results. To be consistent, we will follow temporarily the clockwise convention for the rotation operation.

We have taken in the above example a line (geometrical entity) with respect to which these symmetry rotation operations have been carried out. The geometrical entity *e.g.*, a line, a plane or a point is called symmetry element, while the action of rotation (in the above example) is called the symmetry operation or transformation. By now you must have realised that a symmetry operation or transformation is always defined with respect to a *symmetry element* (geometric entity). At the same time, the presence of symmetry element in any geometric system can be demonstrated only by showing the presence of an appropriate *symmetry operation*.

Besides the axis *D*, the triangle also possesses three more rotational axes (symmetry elements) which lie perpendicular to the axis *D* and in the plane of the triangle. Let us name them as *E*, *F* and *G* (Fig. 2.2). The clockwise rotation operations about these axes by 180° bring the triangle into configurations (Image states) indistinguishable from the initial states of the triangle (Fig. 2.2).

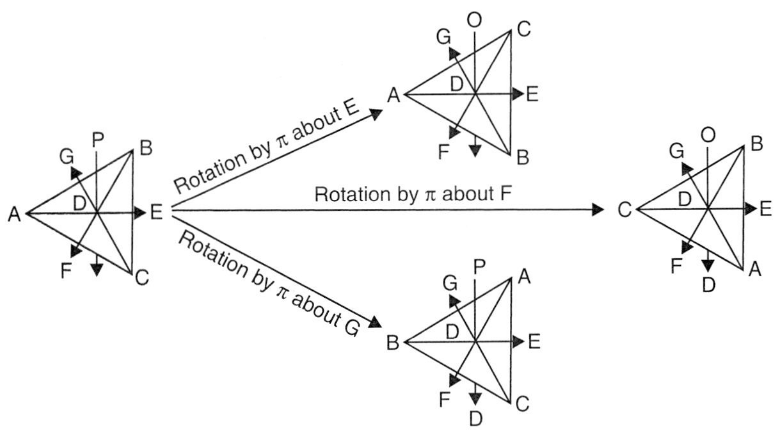


**Fig. 2.2** Clockwise rotations of equilateral triangle *ABC* about the axes *E*, *F*, *G*.

The triangle besides the rotation axes has another kind of symmetry element *viz.* mirror planes (geometric entity). They lie perpendicular to the plane of triangle and passing through one of vertices and through the mid-point of the side opposite to the vertex through which the plane is passing [Fig. 2.3(*a*) and 2.3(*b*)]. There are three such planes *X*, *Y* and *Z*.

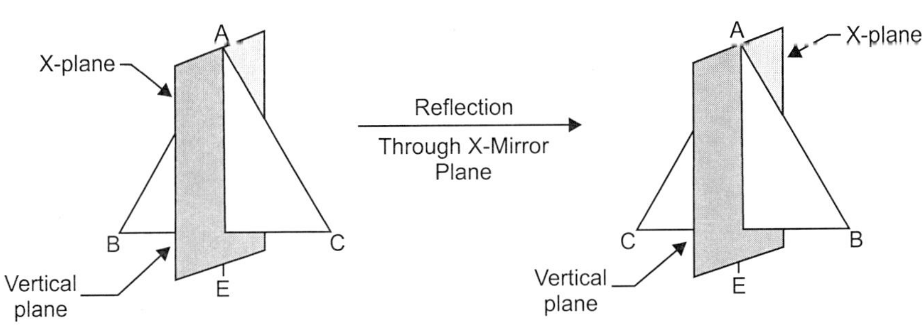


**Fig. 2.3** (*a*) Reflection plane *X* passing through vertex *A* and mid-point of side *BC* and through the centre *O*.

Reflections of triangle through these planes (Reflection operation = Action) bring it into "Self-coincidence" *i.e.*, the reflection action through these planes produces images indistinguishable from the original state of the triangle. A reflection Plane *X* passing through vertex *A* and mid-point of side *BC* and through the centre *O* is shown in [Fig 2.3(*a*)]. Similarly reflection through the reflection plane *Y* passing through vertex *B* and passing through the mid-point *AC* produces identical image state [Fig 2.3(*b*)], and through *Z* plane passing through vertex *C* and midpoint of side *AB* and centre *O* produces identical image state.

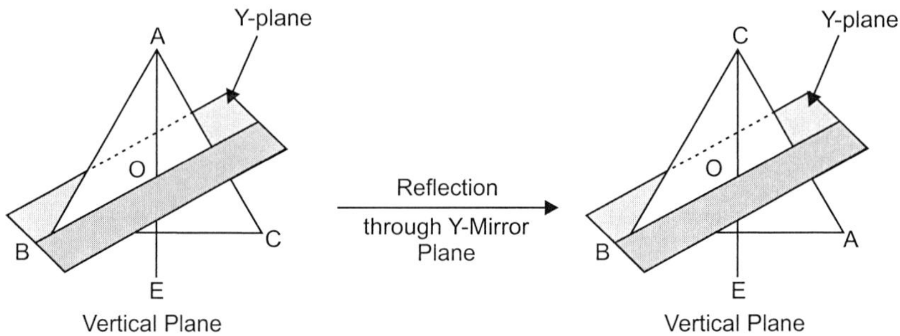


**Fig. 2.3** (*b*) A Reflection plane *Y* passing through vertex *B* and mid-point of side *AC* and through the centre *O*.

These three reflection operations through the mirror planes *AOE*, *BOF* and *COG* are examples of symmetry operations of the equilateral triangle. These three planes are vertical to the plane of the triangle. (It is assumed that the mirror is two sided and images are formed on both sides of the mirror.)

Another reflection plane *ABC*, present in the equilateral triangle is the plane of the triangle itself (one can imagine the triangle having a infinitesimal thickness and the mirror is lying in the middle of that thickness whereby the triangle is reflected in that mirror, to give its image state which is indistinguishable to the original state of the triangle. Fig. 2.4 (Horizontal plane-horizontal to the plane of the triangle *i.e.*, the plane of the triangle itself).

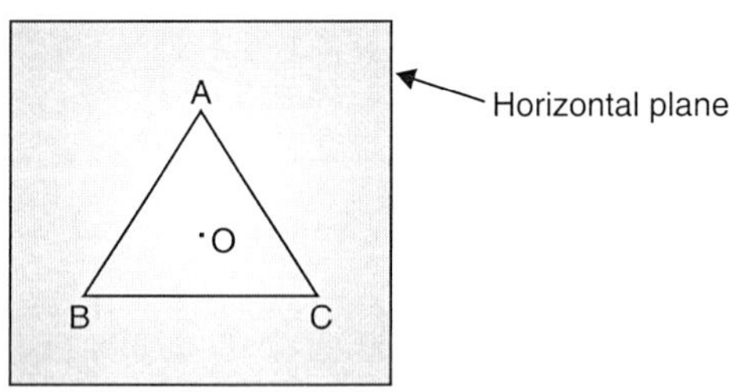


**Fig. 2.4** Reflection in the mirror plane *ABC* (plane of the triangle).

The preceding discussion thus indicates that an equilateral triangle or any molecule having equilateral triangular shape has the following symmetry transformations (Operating by these operations on the triangle brings it into "self-coincidence"). These operations are:

(*a*) Three rotation operations about the axis *D*, passing through the centre '*O*' of the triangle (clockwise rotations by 2π/3, 4π/3 and 6π/3).

(*b*) Six clockwise rotation-operations about the axes *AOE*, *BOF* and *COG* (clockwise rotations by π and 2π).

(*c*) Three reflection operations through mirror planes *X*, *Y* and *Z* passing through the centre *O* and vertex *A*, *B* and *C* lying perpendicular to the plane of the triangle.

(*d*) One reflection operation through the horizontal mirror plane *ABC* lying in the plane of the triangle *ABC*.

The triangle is symmetric with respect to three more symmetry operations which we will discuss later.

**Example 2:** As a second example, let us consider the symmetry of the square and find the symmetry operations with respect to which it is symmetrical (*See* Chapter 1). We have shown that the square has the following symmetry elements and operations (Fig. 2.5).

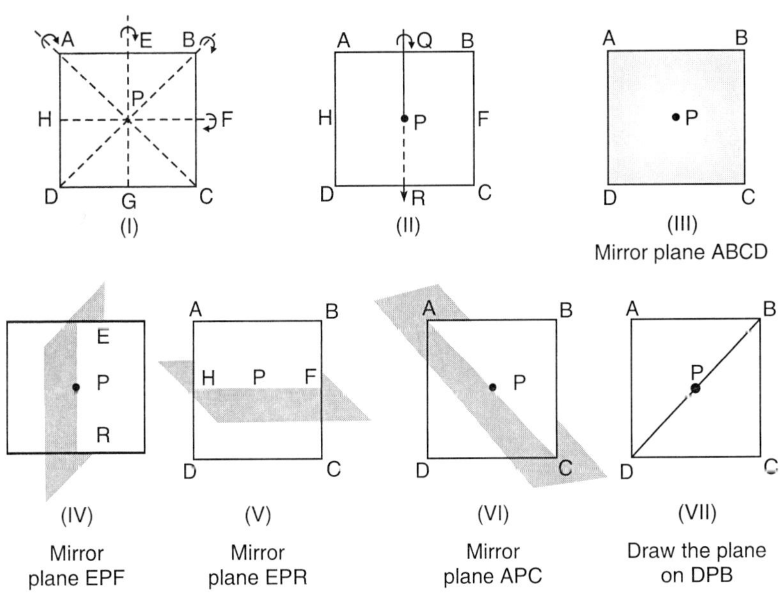


**Fig. 2.5** Symmetry elements and operations of square.

(*a*) Four rotation axes in the plane of the square *APC*, *BPD*, *EPG* and *FPH*. Clockwise rotation about these axes by 180° will send the square into "self-coincidence"

(*b*) Rotation axis *QPR* which is perpendicular to the plane of the square and passing through the centre *P* of square. Clockwise rotations about this axis by 2π/4, 2π/2, 3π/2 and 2π/1 (=*E*) send the square into "self coincidence".

(*c*) Mirror plane *ABCD* which is the plane of the square. This generates one reflection operation.

(*d*) Four mirror planes *APC*, *BPD*, *FPH* and *EPG* which are perpendicular to the plane of the square and pass through the mid-point *P* of the square. These planes generate four reflection operations (One operation per reflection plane).

Accordingly the square is symmetric with respect to these thirteen operations. Besides these, the square is also symmetric to another operation called "inversion" operation. It implies that the square has a centre of symmetry (or centre of inversion). Let us examine the centre *P* of the square (Fig. 2.6).

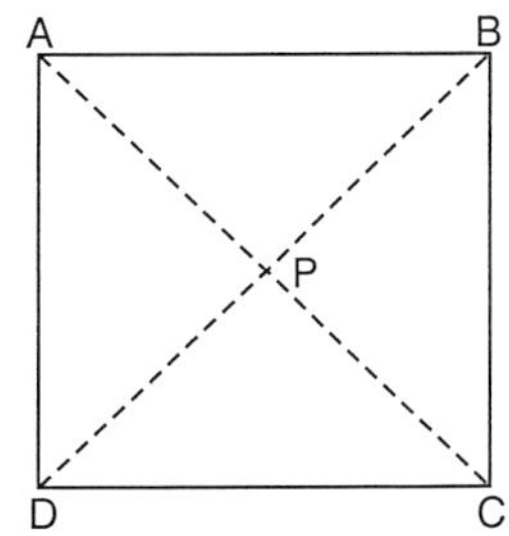


**Fig 2.6** Centre of inversion (*P*) in a square.

If we reflect the square through the centre *P*, we obtain an image state of the square which is indistinguishable from its original state. If the system has such a point (centre of mass), it is called as the centre of symmetry (Point geometrical Entity) and the reflection operation (or Inversion operation) is the symmetry operation. (Note that the equilateral triangle does not have such a point.)

The square besides these operations also has a few more symmetry operations which will be obvious in the next section of the chapter.

**Example 3:** Let us examine a regular tetrahedron $AB_4$ and find out the symmetry operations with respect to which it is symmetrical (Fig. 2.7).

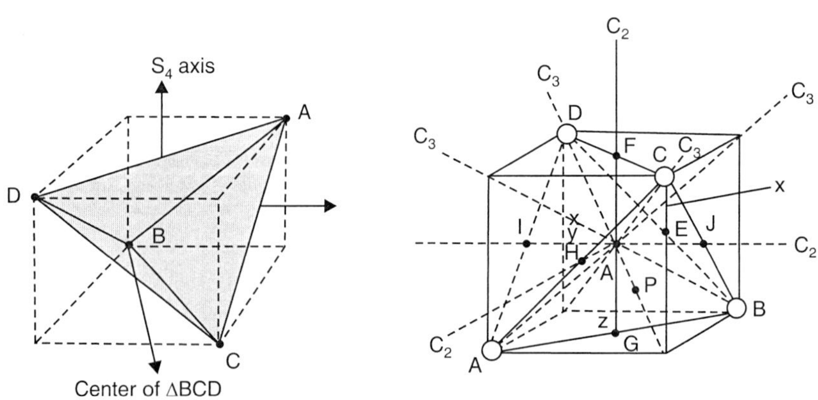


**Fig. 2.7** Symmetry elements of tetrahedron (Point out the symmetry elements of tetrahedron).

The following reflection and rotation operations are present:

1. Six reflection planes (*AEC*, *ABF*, *AJD*, *BHD*, *BCF*, *CDG*). Each plane generates a reflection symmetry operation.
2. Four rotation axes (*A* and mid point of $\Delta BCD$, *B* and mid point of $\Delta ACD$, *C* and the mid point of $\Delta ABD$, *D* and mid point of $\Delta ABC$). Clockwise rotation about these axes by $2\pi/3$, $4\pi/3$, $6\pi/3$ (= *E*) will generate $3 \times 4 = 12$ operations. Out of these 12 operations, 4 operations will be $(6\pi/3 = 2\pi) = E$ identity operations. Three rotation axes *FG*, *HE* and *IJ*, clockwise rotations about these axes by 180° will generate three symmetry operations.

## Tetrahedron–ABCD

1. **$C_3$ axes:** *AX*, *BY*, *CZ*, *DP*. Here tetrahedron *ABCD* which contain *A*, *B*, *C* and *D* as vertices and *X*, *Y*, *Z* and *P* which are mid points of $\Delta BCD$, $\Delta ACD$, $\Delta ABD$, and $\Delta ABC$, respectively.
2. **Reflection planes:** *ACE*, *ABF*, *ADJ*, *BDH*, *BCF*, *CDG* here *E*, *F*, *J*, *H*, *I*, *G*, are the mid point of sides *BD*, *CD*, *BC*, *AC*, *AD* and *AB*, respectively.
3. **$C_2$ axes:** passing through mid points of edges *AB*, *BC*, *CD*, *AC*, *AD*, *BD* and the mid point of the whole tetrahedron.
4. ***E*:** Identity.
5. **$S_4$ rotation-reflection axes:** Mid point of *AC* passing through *BC* and *CD* diagonals.

Besides these 17 operations, there will be six more operations which send the tetrahedron into "self-coincidence". These are called "Rotation-Reflection" operations. The latter operations are also called improper rotations while the one discussed earlier is known as proper rotation. The axis (geometric entity) about which the improper rotation takes place is called the improper axis. The improper rotation may be thought of taking place in two steps: first a proper rotation about an axis and then a reflection through a plane perpendicular to the rotation axis or *vice versa*. Now let us consider the tetrahedron *ABCD* as given in the Fig. 2.8.

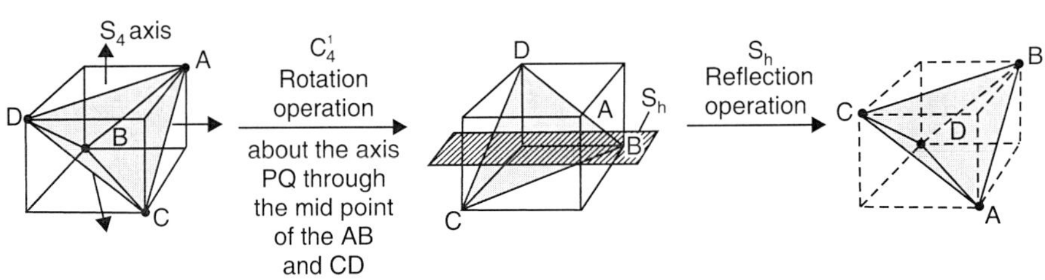


**Fig. 2.8** An improper-Rotation operation (Two operations).

The two operations together are equivalent to one improper rotation operation and is represented as $S_n^1$ as shown in Fig. 2.8. How many operations on $S_n$ axis generate is shown in Fig. 2.9.

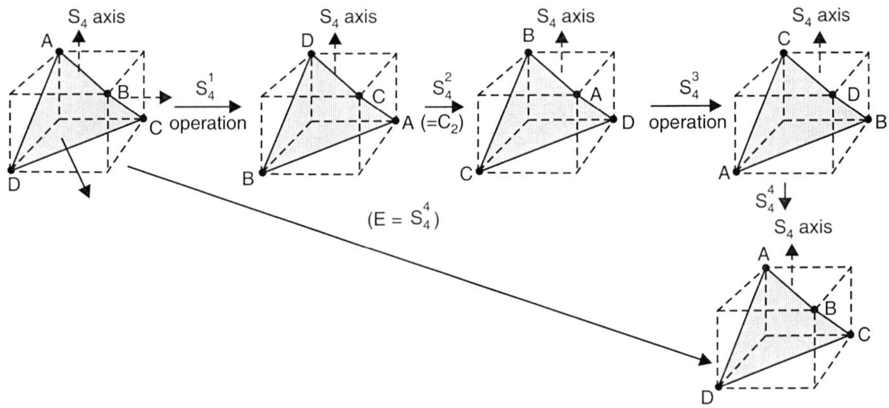


**Fig. 2.9** $S_4^1$, $S_4^2$ (=$C_2$), $S_4^3$ and $S_4^4$ (=$E$) operations of a tetrahedron.

As shown in Fig. 2.9, each improper $S_4$ axis generates two independent operations (*viz.* $S_4^1$, $S_4^3$). It is because $S_4^2$, $S_4^4$ are equivalent to $C_2$ and $E$ operation respectively. We have already counted these latter operations among the $C_2$ and $E$ operations of tetrahedron. Now there are three such improper axes along the $X$, $Y$ and $Z$ coordinate axes of the tetrahedron. These three $S_4$ axes will therefore generate six operations.

Thus, the total number of symmetry operations with respect to which a regular tetrahedron is symmetrical are,

(*i*) *Three improper axes:* Each generates two independent operations (total 6 operations).

(*ii*) *Three proper $C_2$ rotation axes about which one rotates the tetrahedron by an angle of 180°:* Each generates one operation (total of three operations).

(*iii*) *Four proper $C_3$ rotation axes about which one rotates the tetrahedron by an angle of $2\pi/3$, $4\pi/3$ and $6\pi/3$:* Each generates three operations (total 8 operations) because rotation operation by $6\pi/3$ (=$2\pi$) is equivalent to $E$.

(*iv*) *Six mirror planes:* Each generates one operation (total 6 operations).

In this manner, the tetrahedron is symmetrical with respect to the above calculated 24 operations.

In the above examples, we have so far considered symmetries in terms of transformations generated by symmetry elements (geometric entities) which leave a particular geometric figure, invariant. We would like to clarify here that the word invariant may be used in any one of the following two senses:

(*i*) *A geometric figure is mapped onto itself:* It means that both the distance between any two points and the angle between three points in the figure remain exactly the same before and after the transformations. These transformations are known as *Isometric Transformations* (we will be considering only these types of transformations in the book).

(*ii*) *Transformations which preserve the geometric properties of the object or the figure:* For instance, one can carry transformations which preserve the squaredness of a square. By some transformations a square $P$ goes into square $Q$ whose size may differ (similarity transformations). We will not consider these types of transformations here.

Only the following types of isometric transformations are used to map a geometric figure onto itself (for point group molecular symmetries).

(*a*) Identity (to be discussed later)

(*b*) Rotation (Proper)
(*c*) Inversion
(*d*) Reflection
(*e*) Rotation-Reflection (or improper) rotation

Category I type

The above transformations are used to determine the point group of the molecules or of any geometric figure. Under all these operations one point in the molecule or the figure remains fixed. Besides these transformations there are other transformations which are also known as Isometric transformations. These are given below.

1. Translation
2. Screw Rotation
3. Glide Reflection

Category II type (for Non-molecular symmetries)

The transformations needed to know the invariance of an object or figure in the sense of category I have already been discussed in the various examples given above. In translation symmetry operations (category II) the object is repeated in space at regular intervals by moving it in a given direction through a constant distance '*t*'. This transformation is symmetry transformation only if the object of the system is of infinite dimension and is not supposed to have an end. For example, a letter *b* is translated along the line by an interval '*t*' gives one-dimensional pattern (Fig. 2.10).

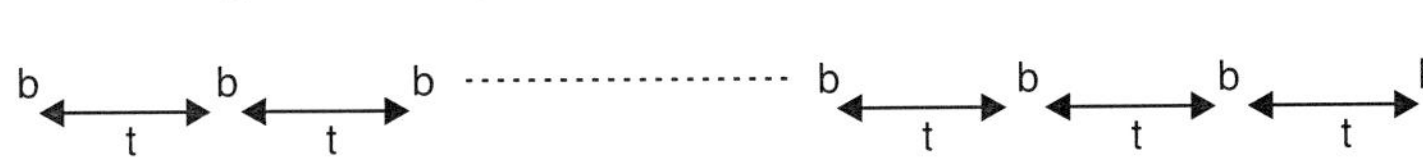


**Fig. 2.10** An example of Translational Symmetry.

If the same letter is repeated in two dimensions by two translational intervals (*viz.* $t_1$, $t_2$), it gives two-dimensional pattern and if it is repeated in three-dimensions by three translational intervals (*viz.* $t_1$, $t_2$, $t_3$) it will give a three-dimensional pattern. This type of translational symmetry is observed in crystals. Such kind of symmetry and their other combinations (category II) produce a displacement of the object or the system (as a whole) in space. This implies that unlike (category I) symmetries there is no point in the system or crystal which remains unmoved in space as a result of the application of category II symmetry operations. This kind of symmetry operations are unlike those of category I symmetry operations in which at least one point in the system (centre of gravity) remains unshifted. Category II symmetry transformations will be discussed in the later sections of the chapter.

In the category I transformations, at least one point remains fixed and does not change its position during the transformations. Thus during rotations all the points on the axis of rotation remain invariant as opposed to translation transformations (category II). If the system has two or more symmetry axes, (*e.g.*, in the case of tetrahedron) intersecting at a point, the point of intersection remains invariant during rotation symmetry transformations about the two axes. Similarly if the system has a mirror plane, during reflection operation, all the points lying on the mirror plane will remain invariant. If two such mirror planes intersect, all the points lying on the line of intersection remain invariant. In the case of inversion transformation the point of inversion through which one reflects the system (*e.g.*, square) remains invariant. In this manner category I, transformations are such that at least one point (generally centre of mass) in the system remains invariant during all the transformations. Presently we plan to discuss category I transformations and various point groups derived from them. *Point group is the set of those symmetry transformations carried out on a system in which at least one point remains invariant during all the transformations. That is why the name is "Point Group".*

### 2.2.1 Identity Element and Identity Operation

Besides the symmetry elements and symmetry operations discussed so far, there is one more symmetry element which is present in all objects or systems. It is named as the identity element and the operation associated with it is termed as identity operation. This operation in reality is not an operation. It involves no change in the system and basically is a pseudo operation. It implies an operation "of doing nothing" to the system. One picks up a system and leaves it as it is (it is like adding 0 to a number). The important concept of 'doing nothing' to the system (identity operation) is mathematically very important operation when group theory concept is applied to a set of symmetry operations, associated with any system. Thus, in example 1 above, rotation about three-fold rotation axis $D$ by $6\pi/3$ restores the system to its original or identical state. Hence, this operation (rotation by $6\pi/3$ about the axis $D$) is equivalent to doing nothing to the system. In this fashion one can identify all those operations which are equivalent to 'doing nothing' and these are grouped together and termed as $E$. All the remaining independent non-equivalent operations are also grouped together and these along with the identity element form a separate set and this set forms a point group. Identity element and the operation associated with it is denoted by the letter $E$. This symmetry operation is present in every system.

## 2.3 NOTATIONS AND PROPERTIES OF SYMMETRY ELEMENTS AND OPERATIONS (OR TRANSFORMATIONS OF MOLECULAR SYMMETRY)

**(*a*) Rotations:** If an object or a system (*e.g.*, an equilateral triangle) has a $p$-fold rotation symmetry axis, the smallest rotation about this axis by an angle $2\pi/p$ brings the object or system into itself-coincidence (*e.g.*, rotation by $2\pi/3$ about the axis $D$ in the case of equilateral triangle). This axis is a $p$-fold rotation axis and is of order $p$. It is denoted as $C_p$ ($C$ = cyclic, $p$ = order of the axis and refers to the number of symmetry rotation operations associated with the axis during a 360° rotation of the object). Thus, an equilateral triangle has a three-fold rotation axis of symmetry ($D$) passing through the centre of triangle and is normal to the plane of the triangle. A square has a four-fold rotation axis passing through the mid point of square and is normal to the plane of the square.

The total number of independent symmetry rotation operations associated with a $C_p$ axis is $(p - 1)$ and a particular operation is represented by writing a superscript on the R.H.S. of $C_p$. Thus, $C_p^n$ is the $n^{\text{th}}$ rotation operation associated with $C_p$ rotation axis. The first rotation (*i.e.*, rotation by $2\pi/p$) is written as $C_p^1$, (or simply $C_p$), Second rotation operation (*e.g.*, rotation by $2 \times 2\pi/p$) as $C_p^2$, third rotation operation (*i.e.*, rotation by $3 \times 2\pi/p$) as $C_p^3$ and so on until we carry a $p^{\text{th}}$ rotation operation (*i.e.*, rotation by $p \times 2\pi/p = 2\pi$) and which is written as $C_p^p$. However, $C_p^p$ operation produces the configuration of the object that is identical to the initial state of the object and hence, we write $C_p^p = E$ (or identity operation $E$). For example, in the case of an equilateral triangle, rotation by $3 \times 2\pi/3$ ($C_3^3$) is equivalent to the identity operation $E$. (Symbolically $C_p^{p+1}$ is equivalent = $C_p^1$). It must be emphasized here that $C_p^m$ which is a rotation operation of the body by an angle $(m \times 2\pi/p)$ about the $C_p$ axis is also equivalent to a rotation of the system by an angle $(2\pi/p)$ carried successively $m$ times. Further we also have $C_p^p = E$, $C_p^{p+1} = C_p^1$, $C_p^{p+2} = C_p^2$ and so on.

A body or a system may have more than one axis of rotation. For instance, a square (Example II) has one $C_4$ and four $C_2$ axes. An equilateral triangle has a

$C_3$ axis perpendicular to the plane of the triangle and passing through its centre. Besides this, it also has three 2-fold rotation axes ($C_2$) lying in the plane of the triangle bisecting the angle $A$, $B$, and $C$. Linear molecules have an $\infty$-fold rotation axis ($C_\infty$) which is the same as the molecular axis.

In case of molecules or objects having more than one axis of rotation and of different orders, the axis of the highest order is conventionally known as the principal axis of rotation. Thus $C_4$ rotation axis of a square is its "Principal" axis. An object or a molecule having more than one rotation axis of the same order, the axis on which the maximum number of atoms lies will be the "Principal axis". This is merely a convention.

The important properties of the rotation axes and rotation operations are the following:

(*i*) No molecule or object can possess just two 2-fold axis (either it should have, one, three or more than three 2-fold rotation axes). Thus, ethylene, methane (tetrahedron), *etc.*, have three 2-fold rotation axes.

(*ii*) Three-fold axes are quite common for *e.g.*, triangular pyramid or ($XY_3$) planar molecules, or equilateral triangle *etc.*

(*iii*) If an object or a molecule has a $C_n$ axis ($n$ = odd ) and also a two-fold axis perpendicular to $C_n$ axis, it automatically means that the other ($n - 1$) two-fold axes at an angle $[2 \times (2\pi/n)]$, $[3 \times (2\pi/n)]$, $[4 \times (2\pi/n)]$, ... $[(n - 1) \times (2\pi/n)]$ to the first 2-fold axis must exist. This is because once we carry out $C_n^1$ operation, we automatically generate a second $C_2$ axis from the first and on carrying $C_n^2$ rotation, we generate the third $C_2$-axis from the first and on carrying $C_n^3$ rotation, we generate the fourth $C_2$ axis and so on. In this manner, ($n - 1$) $C_2$-axes will be generated and thereby a total of number of $n$ two-fold axis must be present. This principle is not only applicable for $C_2$ axes but also applies to other symmetry elements *e.g.*, perpendicular planes containing the $C_n$ axis ($n$ = odd) because $n$ rotations about $C_n$-axis will generate $n$-symmetry elements (planes) lying perpendicular to and containing the $n$-fold axis.

The case when the number $n$ in $C_n$ is even, is a little different from that when $n$ is odd. Let us consider when $n = 4$ (for example a square), It has a $C_4$-axis and a $C_2$-axis which is perpendicular to $C_4$-axis (Fig. 2.11). On carrying out a $C_4^1$ operation, it is rotated by $\pi/2$ and generates a new $C_2$ axis. Let us name it as $C_2$(II). On carrying out a second $C_4^2$ operation, we rotate the $C_2$(I) axis by 180° ($2 \times 2\pi/2$). It means $C_2$(I) goes into $C_2$(III) which is identical to $C_2$(I) (That is, $C_2$(III) coincides with $C_2$(I) and no new $C_2$ axis is generated. Now we carry out a $C_4{}^3$ operation *i.e.*, rotation by 270° ($3 \times 4\pi/3$). This operation takes $C_2$(I) into a new $C_2$ axis. Let us name it as $C_2$(IV). This newly generated $C_2$(IV) axis is the same as $C_2$(II). Thus, we have only two $C_2$ axis, $C_2$(I) and $C_2$(II) (*See* Fig. 2.11). Similarly a system which has $C_6$ axis and a $C_2$ axis perpendicular to $C_6$ axis or a plane containing $C_6$-axis will generate new $C_2$ axis or three new planes.

You should realize here that in case II ($n$ = even), another two $C_2$-axes are present. (1) between $C_2$(I) and $C_2$(II) and (2) between $C_2$(II) and $C_2$(III). We name them as $C_2'$ and $C_2''$. Thus a total of four $C_2$ axes (*viz.* $C_2$(I), $C_2$(II), $C_2'$ and $C_2''$) must be present in the system.

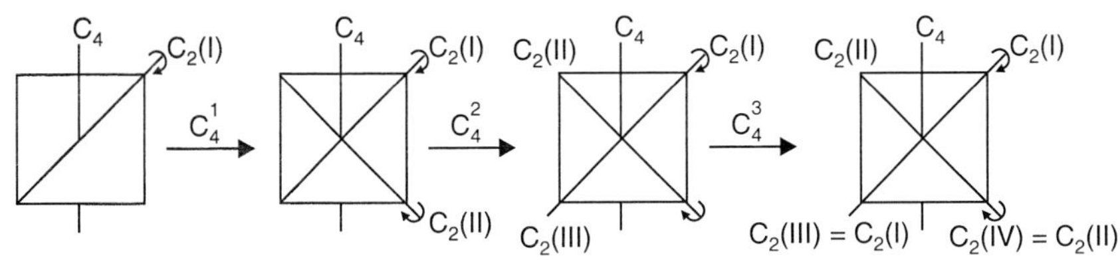


**Fig. 2.11** Generation of new $C_2$ axes by the operations of principal axis [*viz.* ($C_4$)].

If a system has a $C_n$ axis and if there is plane perpendicular to the axis not containing the $C_n$-axis, the plane is not replicated by the $C_n^m$ operation because by carrying out $C_n$ operations the whole plane goes into itself.

**(*b*) Reflection plane:** A symmetry plane (geometric entity) is indicated by the symbol ($\sigma$). If the symmetry plane contains the principal axis of rotation (proper or improper) which is also present in the object or the system then the plane is represented by ($\sigma_v$). If the reflection plane is perpendicular to the principal rotation axis, it is denoted by ($\sigma_h$). If the system has two ($\sigma_v$) planes, then there exists a third vertical plane, $D$ also, bisecting two ($\sigma_v$) planes. The plane $D$ is designated as $\sigma_d$ (Fig. 2.12). The subscripts $v$, $h$ and $d$ stand for vertical, horizontal and diagonal respectively. They refer to the orientation of the plane with respect to the principal rotation axis ($C_p$) which conventionally is taken as vertical to the plane of molecule.

The same symbols *viz.* $\sigma_v$, $\sigma_h$ $\sigma_d$ are also used to denote the reflection operations through the plane. This is so because the existence of one symmetry plane generates only one reflection operation and the effect of applying the same reflection operation consecutively two times ($\sigma_m^1 \,.\, \sigma_m^1$) is to bring the system back to the original state and so ($\sigma_m^1 \,.\, \sigma_m^1$) becomes effectively equal to $E$ (identity operation *i.e.*, $\sigma_m^1 . \sigma_m^1 = \sigma_m^2 = E$ ). Here superscript denotes the number of operations. In general, one can write $\sigma_m^n = E$ or $\sigma_m$ depending upon the value of $n$. If $n$ is even $\sigma_m^n = E$, if odd, $\sigma_m^n = \sigma_m$ ($E$ = identity operation).

There are systems that do not contain any plane of symmetry (*e.g.*, in SFOICl, tetrahedral structure). One general class of such systems is non-planar molecules in which the number of atoms of each type, present therein is odd. On the other hand there may be systems which have infinite number of planes of symmetry, whose general class consists of linear molecules because any plane containing the molecular axis is a symmetry plane. In general, systems may contain one or more than one symmetry planes (*e.g.*, $F_2SCl_2$ tetrahedral) has one symmetry plane which bisects the distance between F and Cl atoms and passes through the atom *S*. The $H_2O$ molecule has two symmetry planes ($\sigma_h$ and $\sigma_v$). A tetrahedron has six mirror planes (cf. example IV).

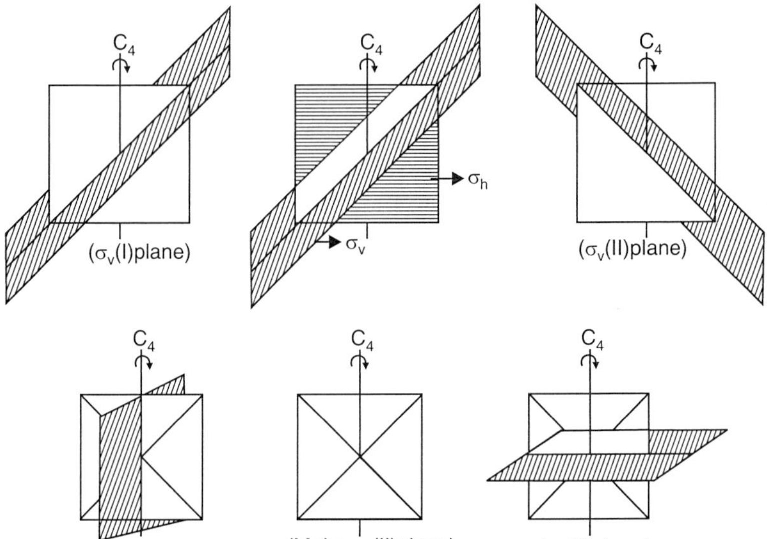


**Fig. 2.12** $\sigma_v$, $\sigma_d$ and $\sigma_h$ Planes.

## EXERCISE

1. Find the symmetry planes in $NH_3$, $CH_2Cl_2$, $CHCl_3$, $CH_2Cl_2$, $[PtCl_4]^{2-}$.
   If a systems contain more than one symmetry planes of the same type they are designated ($p\ \sigma_m$) where $p$ is the number of symmetry planes of the same type *i.e.*, $\sigma_m$ type ($m = v$ or $d$).

**(*c*) Reflection through centre of mass of the system (Inversion):** If the system possesses a centre of symmetry (geometrical entity, a point) it is designated as '*i*'. '*i*' stands for inversion. The same symbol '*i*' is used for inversion operation because the existence of a centre of symmetry generates only one symmetry operation. This is so because analogous to reflection-plane, the effect of applying the inversion operation twice to a system is to bring the system back to its initial state. One can therefore, write *i.i* (system) = *E* (identity operation). Obviously $i^p = E$ when $p$ is even and $i^p = i$ when $p$ is odd.

If a centre of inversion is present in a molecular system, then all the atoms except one (unique atom) in general occur in pairs. The odd atom (named as unique atom) must be present at the centre of symmetry because it is only the unique atom in the molecule that remains fixed when the inversion operation is performed. All the other atoms must occur in pairs because every atom is exchanged with its twin atom when inversion operation is performed. It implies that all molecules that contain an odd number of atoms of one type of species will lack a centre of symmetry, for example $CHCl_3$, $C_5H_5$ *etc.*, but the reverse statement "the molecule in which all the atoms except one occur in pairs must contain centre of symmetry" is not true *e.g.*, $CH_4$, $CH_2Cl_2$ do not have a centre of symmetry.

Example of systems having a centre of symmetry are octahedron, cylinder, circle, square, rectangle, $[PtCl_4]^{2-}$, staggered ethane, chair form of cyclohexane, benzene, homonuclear diatomic molecules, ethylene, *etc.*

## EXERCISE

**1.** Find out if the centre of symmetry is present in the following molecules – cyclopentane ($C_5H_5$), trans-($MA_2B_2$) molecules, tetrahedron, cone, tetrahedral molecules of the type $AB_4$ ]

**Answer:** none except trans-$[MA_2B_2]$

---

***(d)* Improper rotation axis and improper rotation operation:** An improper axis is denoted by a symbol $S_n$. The operation generated by this axis (Symmetry element) is a combination of two operations which individually may or may not be symmetry operations of the system by themselves. As discussed earlier the first operation is a rotation operation around the improper axis by an angle $2\pi/n$. (This operation may not generate the image state of the system indistinguishable to the initial state of the system). The second operation is the reflection operation through a plane perpendicular to the improper rotation axis which by itself may not be a plane of symmetry in the molecule. The final image thus obtained of the system after the above two successive operations is indistinguishable to its initial state. (These two operations are commutative). This combined operation is written as $S_n^1$. The second operation of improper rotation is written as $S_n^2$ which is expressed as given below and is equivalent to,

$$S_n^1.S_n^1 = C_n^1.\sigma_h\ C_n^1.\sigma_h = C_n^2.\sigma_h.\sigma_h = C_n^2.E = C_n^2$$

In general,

$$S_n^m = m^{\text{th}}\text{operation} = S_n^1.S_n^1.S_n^1 \cdots S_n^1\ (m\,\text{times})$$

$$= C_n^1.C_n^1.C_n^1.C_n^1\ (m\,\text{times})\ \sigma_h.\sigma_h.\sigma_h.\sigma_h\ (m\,\text{times})$$

$$= C_n^m.\sigma_h.\sigma_h.\sigma_h.\sigma_h \cdots (m\,\text{times})$$

$$C_n^m \cdot \begin{cases} (E) \text{ if } m = \text{even} \\ (\sigma_h) \text{if } m = \text{odd} \end{cases}$$

It should be noted that if a system has a centre of inversion then it must also posses a $S_2$ axis (*i.e.*, $i = S_2$) (Fig. 2.13).

Also $\qquad S_2^1 . S_2^1 = C_2^1 . \sigma_h . C_2^1 . \sigma_h = C_2^2 \, \sigma_h . \sigma_h = E$

It is obvious from the figure that $S_2$ operation corresponding to rotation round the principal axis through 180° followed by reflection through $\sigma_h$ is equal to the inversion operation (Fig. 2.13). Similarly $S_1$ operation corresponding to first 360° rotation around the axis, followed by reflection in the plane perpendicular to the axis is equivalent to $\sigma_h$ ($S_1 = \sigma_h$).

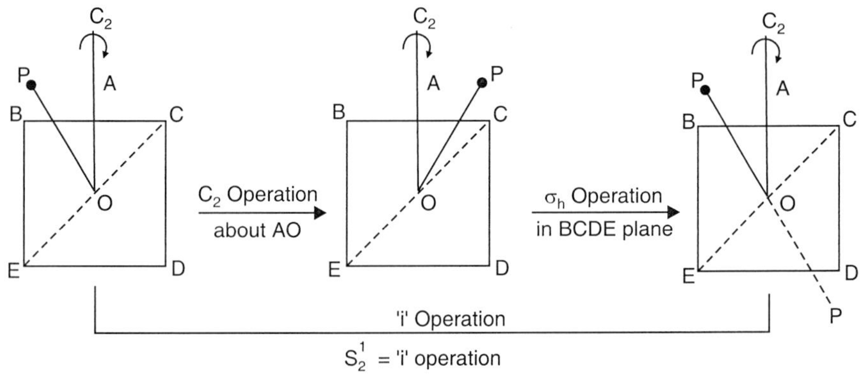


**Fig. 2.13** $S_2^1 = I$ operation and $S_2^2 = S_n^1 . S_n^1 = C_n^1 . C_n^1 . \sigma_h . \sigma_h = C_n^2$ operation.

Let us now consider a few more properties of $S_n$ operations. These are given below.

**(d$_1$) Operations generated by $S_n$ axis when $n$ = even (*e.g.*, $S_2$, $S_4$, $S_6$, ..., $S_n$):** $S_n$ axis generates $S_n^1, S_n^2, S_n^3, \cdots, S_n^n$ operations. For even values of $n$ we can write the $S_n^n$ operations as,

$$1. \quad \begin{cases} = (C_n^1 . C_n^1 . C_n^1 \cdots C_n^1 \,(n\,\text{times})\,(\sigma_h . \sigma_h . \sigma_h \cdots \sigma_h \; n\,\text{times}) \\ = C_n^n . (\sigma_h . \sigma_h . \cdots n\,\text{times}) \\ = C_n^n . E \\ = E . E \,(\text{When } n = \text{even} \equiv E \text{ operation}) \\ = E \end{cases}$$

$$2. \quad \begin{cases} S_n^{n+1} = S_n^n . S_n^1 = E . S_n^1 = S_n^1 \\ S_n^{n+2} = S_n^n . (C_n^1 . C_n^1)(\sigma_h . \sigma_h) \\ \qquad = E . C_n^2 . E = C_n^2 \end{cases}$$

3. $$\begin{cases} S_n^p = C_n^1 . C_n^1 \cdots C_n^1 \,(p \text{ times})\, \sigma_h . \sigma_h \cdots \sigma_h \,(p \text{ times}) \\ \text{Now, for even } p, \\ \text{we have,} \\ S_n^p = C_n^p . (\sigma_h . \sigma_h \cdots p \text{ times}) \\ \quad = C_n^p \\ \text{for odd } p \text{ we have,} \\ S_n^p = C_n^1 . C_n^1 \cdots C_n^1 \,(p \text{ times})\, \sigma_h . \sigma_h \cdots \sigma_h \,(p \text{ times}) \\ \quad = C_n^p . \sigma_h = S_n^p \end{cases}$$

4. $$\begin{cases} \text{when } p = n/2, \text{we have} \\ S_n^p = (C_n^1 . C_n^1 \cdots p \text{ times})\, (\sigma_h . \sigma_h \cdots p \text{ times}) \end{cases}$$

Now we will have two cases

**Case 1:** Let us consider $S_n^{n/2}$ operations ($n$ = even)

$$\begin{aligned} S_n^{n/2} &= (C_n^1 . C_n^1 . C_n^1 \ldots n/2 \text{ times})\, (\sigma_h . \sigma_h . \sigma_h \ldots n/2 \text{ times}) \\ &\equiv C_n^{n/2} . \sigma_h \,(\text{when } n/2 = \text{odd and } n = \text{even}) \\ &= (\text{rotation through } 180°) . \sigma_h \\ &= i(\text{inversion}) \end{aligned}$$

**Generalisation 1:** Thus in general, the presence of an $S_n$ axis ($n$ = even) requires the presence of inversion centre ($i$).

Let us consider again the operations generated by an $S_n$ axis when $n$ = even

$$\begin{aligned} S_n^1 &= C_n^1 . \sigma_h = S_n^1 \\ S_n^2 &= (C_n^1 . \sigma_h)\, (C_n^1 . \sigma_h) = (C_n^1 . C_n^1)\, (\sigma_h . \sigma_h) = C_n^2 \\ S_n^3 &= (C_n^1 . \sigma_h)\, (C_n^1 . \sigma_h)\, (C_n^1 . \sigma_h) = S_n^3 \\ S_n^4 &= (C_n^1 . \sigma_h)\, (C_n^1 . \sigma_h)\, (C_n^1 . \sigma_h)\, (C_n^1 . \sigma_h) = C_n^4 \\ S_n^5 &= S_n^5 \\ S_n^6 &= C_n^6 \\ S_n^{n/2} &= i \ldots\ldots \text{ so on} \end{aligned}$$

In this manner, for even $n$ we have generated,

$$(S_n^1, C_n^2, S_n^3, C_n^4, S_n^5, C_n^6 \ldots\ldots S_n^{n/2} = i \ldots\ldots)$$

These can grouped as $(S_n^1, S_n^3, S_n^5, i, S_n^{n/2+2} \ldots)\ (C_n^2\ C_n^4\ C_n^6 \ldots\ldots C_n^n)$

$(S_n^1, S_n^3, S_n^{n/2} = i \ldots\ldots)$ and (operations generated by $C_{n/2}$ axis)

**Generalisation 2:** We can thus generalise that the presence of $S_n$ axis of *even order* requires the presence of $C_{n/2}$ axis in the system.

(check the above generalisation by taking the examples of $S_2$, $S_4$, $S_6$, $S_8$ … *etc.*)

**(d$_2$) Operations generated by $S_n$ axis when $n$ = odd:** Let us take a system having an $S_n$ axis when $n$ is odd and find out the operations generated by the axis. We will have the following operations.

$$S_n^1 = C_n^1 . \sigma_h = S_n^1$$
$$S_n^2 = (C_n^1 . C_n^1)(\sigma_h . \sigma_h) = C_n^2$$
$$S_n^3 = C_n^3 . \sigma_h = S_n^3$$
$$S_n^4 = C_n^4 (\sigma_h . \sigma_h . \sigma_h . \sigma_h) = C_n^4 . E = C_n^4$$
$$S_n^5 = C_n^5 . \sigma_h = S_n^5$$
$$S_n^6 = C_n^6 E = C_n^6$$
$$S_n^7 = C_n^7 . \sigma_h = S_n^7$$

..............................

..............................

$$S_n^n = (C_n^1 . C_n^2 . C_n^3 \text{ ... } n \text{ times}).(\sigma_h . \sigma_h . \sigma_h \text{ ..... } n \text{ times})$$
$$= E . \sigma_h$$
$$= \sigma_h$$

**Generalisation 3:** The presence of $S_n$ axis with $n$ = odd requires the presence of $\sigma_h$ since $S_n^n$ operation does not generate $E$ (when $n$ = odd), one gets more than $n$-operations. Let us find out what the operations $S_n^{n+1}, S_n^{n+2}, S_n^{n+3} \ldots S_n^{2n}$ stand for. We have,

$$S_n^{n+1} = (C_n^{n+1})(\sigma_h.\sigma_h \ldots n+1 \text{ times}) \equiv C_n^{n+1}.E = C_n^1$$

($n$ + 1 is even, *i.e.*, when $n$ = odd) $\qquad C_n^{n+1} = C_n^1$

$$S_n^{n+2} = (C_n^{n+2})(\sigma_h.\sigma_h \ldots n+2 \text{ times}) \equiv C_n^2.\sigma_h = S_n^2$$

$$S_n^{n+3} = (C_n^{n+3})(\sigma_h.\sigma_h \ldots n+3 \text{ times}) \equiv C_n^3.E = C_n^3$$

because $n$ + 3 = even

$$S_n^{2n} = (C_n^{2n})(\sigma_h.\sigma_h \ldots 2n \text{ times}) \equiv E.E = E$$

because, $(C_n^{2n} = E), (\sigma_h \ldots . 2n \text{ times} = E)$

**Generalisation 4:** Thus, $S_n$ axis when $n$ is odd generates $2n$ operations *viz.*

$$[(S_n^1, C_n^2, S_n^3, C_n^4, \ldots, S_n^n (\equiv \sigma_h) \ldots)] \; [(C_n^1, S_n^2, C_n^3, S_n^4, \ldots, S_n^{2n} (\equiv E))]$$

We can group these operations as,

$$[(S_n^1, S_n^2, S_n^3, S_n^4, \ldots, S_n^n (\sigma_h \ldots)] + [(C_n^1, C_n^2, C_n^3, C_n^4, \ldots, C_n^{n-1}, E]$$

These $2n$ operations are all different and if one starts again from $S_n^{2n+1}$, the repetition starts. Out of these $2n$ operations $(n - 1)$ operations are $S_n^p$ type, whereas the other $n$-operation are $(C_n^1, C_n^2, \ldots, E)$ type and one operation is $\sigma_h$.

**Generalisation 5:** One then generalises that if $n$ = odd, we have $2n$ operations and the system has independently a $C_n$-axis and a horizontal plane, $\sigma_h$ to generate $C_n^p$ and $\sigma_h$ operations respectively in their own right besides having $S_n$ axis, which generates $S_n^p$ operations.

## EXERCISE

1. Take the systems having $\sigma_h$ plane along with having $C_3$, $C_5$, $C_7$ ... axis and enumerate the symmetry operations present therein.
2. Find out the $S_n$ axis and the operations generated from them in the following molecule.

   (*a*) Chloroethylene-1, 1 (*b*) Dibromo 2,2 dichlorocyclopropane,

   (*c*) Ethane (eclipsed), (*d*) Methane

   (*e*) Cyclopentane (*f*) Benzene

   (*g*) Cyclopentane (*h*) Octahedron (regular)

   **Answer:** $S_1$, $S_1$, $S_3$, $S_4$, $S_5$, $S_6$, $S_7$; $S_4$, $S_n^3$.

(*e*) **Inverse transformation (Inverse operation):** We have considered a transformation as some operation which takes a body from its initial state to its image state. If the image state is indistinguishable from the initial state of the object, we designate such a transformation (or operation) as symmetry or invariant transformation. But one can also take the image state of an object '*P*' obtained after carrying out the transformation $T_1$, on its initial state $O$ and carry out another transformation $T_2$ on this to get back its original initial state $O$.

Thus, the effect of $T_2$ is to cancel the effect of $T_1$ on the initial state of the object $O$. If such a transformation exists, then $T_2$ is called the "INVERSE" transformation of $T_1$. A common example of such transformation can the given as follows. Suppose you have the system consisting of two buckets $A$ and $B$ (Initial state of the system is bucket $A$ full and bucket $B$ empty). Now you transfer the water of bucket $A$ into bucket $B$. This may be assigned as transformation $T_1$ that produces the image state of the system. Let it be $Q$ state (Bucket $A$ empty and bucket $B$ full). The latter image state $O$ can now be considered as the new initial state of the system (*viz.* Bucket $B$ full + Bucket $A$ empty). We now carry out transformation $T_2$ on $Q$ state. This can be carried out by transferring the water from bucket $B$ to the bucket $A$. We obtain the initial state $P$ *i.e.*, transformation $T_2$. Thus $T_2$ becomes inverse transformation of $T_1$ and *vice versa*.

In general if the initial state $P$ and the image state of $Q$ of the object obtained after $T_1$ operation are subjected to $T_2$ operation and the image state obtained by

$T_2$ is indistinguishable from $P$ then the transformation $T_2$ is the inverse symmetry transformation of the symmetry transformation $T_1$.

Let us exemplify the above transformation by taking an example of a molecule. Take a square *ABCD* in the state obtained after rotation clockwise through an angle $2\pi/4$ about the principle axis (Fig. 2.14) to get the image state of square (*viz. S* state) by operation $C_4^1$. To get the square back in its original state *R* from states *S*, one rotates it clockwise about the same axis by an angle of $6\pi/4$ (*i.e.*, carrying out operation $C_4^3$). Thus $C_4^3$ operation is inverse of operation $C_4^1$ and vice versa. Similarly one can prove that the inverse of operation $C_4^2$ is $C_4^2$. You must have realised by now that the transformation $T_1$ and its inverse transformation $T_2$ are related through the operation *E* by the following relationship.

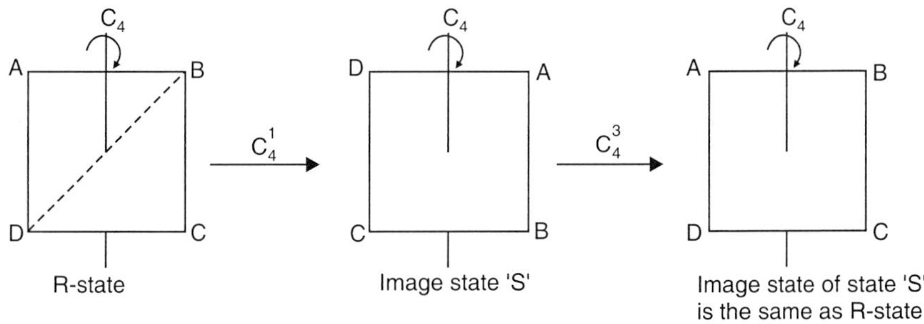


**Fig. 2.14** Inverse operation.

(The operation $T_1$) (Operation $T_2$) = operation *E* on the system.

or $$C_4^1 \cdot C_4^3 = E$$

or $$C_2^1 \cdot C_2^1 = E$$

In general, $\sigma_m \cdot \sigma_m = E$ [$\sigma_m$ is inverse of $\sigma_m$ ($m = v, d, h$)]

$i.i = E$ ($I$ is inverse of $i$)

[$C_n^m$ is the inverse of $C_n^{n-m}$ because about the same axis

$C_n^m \cdot C_n^{n-m} = C_n^n = E$ and when $n$ = even]

$S_n^m \cdot S_n^{n-m} = E$ [*i.e.*, $S_n^m$ is inverse of $S_n^{n-m}$ and vice versa].

(Prove the above relationships by taking concrete examples).

**(*f*) Successive transformations and combinations (or product) of transformations:** In this section, we extend our discussion regarding the effect of two successive transformations that has already been introduced while dealing improper axis (Sec. 2.3).

Let us take a system in its initial state *P* and a transformation $T_1$ takes it into its image state *Q*. We now carry out a second transformation $T_2$ on its image state $Q$ which takes it into its second image state, say $R$. Thus, the combined transformations $T_2 . T_1$ bring the initial state ($P$) of the system to a state ($R$). The combined transformation of the two transformations is written as $T_2 . T_1 = (T)$ (notice the order of two transformations). Here $T$ is the transformation which brings initial state $P$ of the system to the image state $R$. The convention to write the combination of two or more transformations is to write the first transformation ($T_1$), then the second transformation $T_2$ on the L.H.S. of the first transformation (*viz.* $T_1$). If there is a third transformation $T_3$ which carries the image state $R$ to another image state $S$ of the system, this transformation is written on the L.H.S. of $T_2 . T_1$ (*i.e.*, $T_3 . T_2 . T_1$).

In general if one carries out $n$-successive transformations (*viz.* $T_1 . T_2 . T_3$ ...... $T_n$) on the initial state $P$ of the system then the combined transformations say $X$,

$$(T_{n'} . T_{n-1} \ldots\ldots T_3 . T_2 . T_1) = (\text{combined transformation } i.e., \text{ say } X)$$

or, $T_{n'} . T_{n-1} . T_{n-2} \ldots\ldots T_3 . T_2 . T_1) = X$

Let us exemplify it by taking the case of a regular hexagon. (Fig. 2.15). Suppose we start with state (I) of regular hexagon and carry out a transformation $C_6^1$ ($T_1$ operation) which carries it to a state (II). We next carry out the second transformation $\sigma_v$ on the image state (II) of the hexagon. This carries the hexagon to its next image state (III) by $[\sigma_v . C_6^1]$ from the image state (II). It can be seen that the state (III) of the hexagon can be brought to the image state (IV) by a third independent transformation (*viz.* $\sigma_d$). One can then write in general, the net transformation as $[\sigma_d = \sigma_v . C_6^1]$.

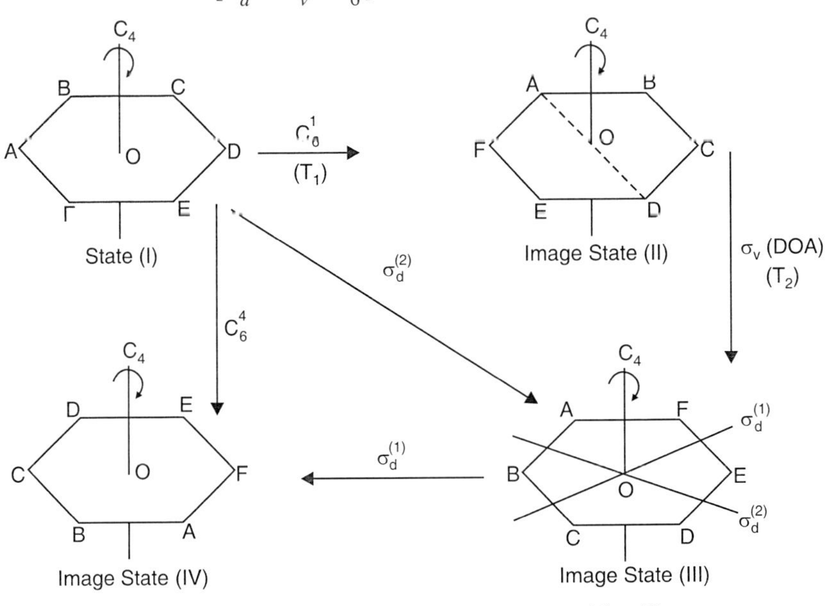


**Fig. 2.15** Successive transformations of $C_6H_6$ ($\sigma_d^{(1)}$, $\sigma_d^{(2)}$ are two diagonal planes of $C_6H_6$).

These transformations may in general be symmetry (invariant) transformations associated with the system. We will later show that the combination of any two or more than two symmetry transformations is the same as that produced by a single symmetry transformation associated with the system. For example,

$$C_6^1 = \sigma_v \cdot \sigma_d \ (2)$$

An important property about the combined transformations, in general, is that the effect of combining of two consecutive transformations depends on their order of successive applications. Therefore, the final image state of the system after the successive application of first $T_1$ and then $T_2$ may or may not be the same as that obtained after the successive application of first $T_2$ and then $T_1$. That is, the effect of

$T_1 \cdot T_2$ may or may not be the same as $T_2 \cdot T_1$ in all cases. We write this as

$$T_1 \cdot T_2 = \text{or} \neq T_2 \cdot T_1$$

(This depends on the nature of the operation of the system.)

In case $T_1 \cdot T_2 = T_2 \cdot T_1$ (*i.e.*, final image does not depend upon the order of the combined transformations) then the transformation $T_1$ is said to commute with $T_2$ (or both transformations are commuting transformations) but if $T_1 \cdot T_2 \neq T_2 \cdot T_1$ then $T_1$ does not commute with $T_2$. In that case $T_1$ and $T_2$ are said to be as non-commuting transformations. As an example it is obvious from Fig. 2.16 that the net effect of transformation $T_1$ (*i.e.*, rotation about an axis by an angle $\alpha$), followed by rotation transformation $T_2$ by an angle $\beta$ about the same axis is identical to the net effect of rotation transformation, $T_2$ by an angle $\alpha$, followed by rotation transformation $T_1$ by an angle $\beta$ about the same axis. The net effect is the combined transformation $T$ which is a rotation transformation by an angle $(\beta + \alpha)$ about the same axis. Therefore, two rotation transformations ($T_1$ and $T_2$) about the same axis commute ($T_1 \cdot T_2 = T_2 \cdot T_1$).

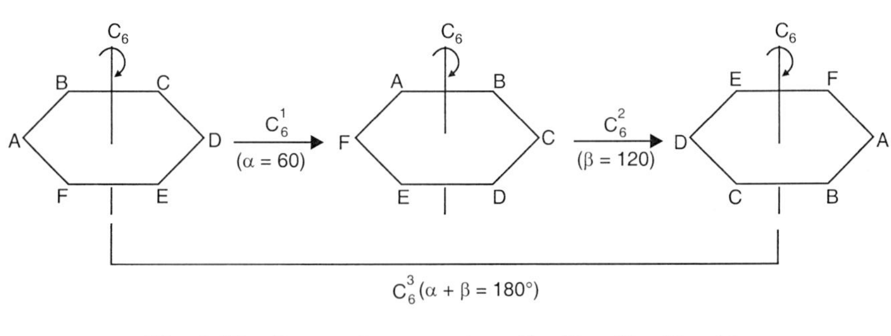


**Fig. 2.16** Commuting operators $T_1 \cdot T_2 = T_2 \cdot T_1 = T_3$.

One can prove that the following pairs of symmetry operations always commute:

1. Reflections through planes perpendicular to each other (*e.g.*, $\sigma_h$ and $\sigma_v$).

2. The inversion symmetry operation and either any symmetry reflection or symmetry rotation operation.
3. Two $C_2$ symmetry rotations about the axes perpendicular to each other.
4. Rotation operation about an axis and reflection operation in a plane, perpendicular to the rotation axis.

(Prove the above identities by taking suitable examples.)

Now let us take an example where two operations do not commute. Take the case of an equilateral triangle.

Let us combine $C_3^1$ and $\sigma_v^{(2)}$ operation. It is obvious from the Fig. 2.17 that the effect of transformation $C_3^1$ (first) followed by reflection in the plane $\sigma_v^{(2)}$ brings the triangle into state $A$. The latter state $A$ can be obtained from the original state by the transformation $\sigma_v^{(3)}$ [$\sigma_v^{(2)} . C_3^1 = \sigma_v^{(3)}$]. But, if one carries out the transformation $\sigma_v^{(2)}$ first, followed by rotation transformation $C_3^1$, the triangle is brought to the state $B$. The state $B$ can be obtained from the original state by a single transformation $\sigma_v^{(1)}$ [$C_3^1 \sigma_v^{(2)} = \sigma_v^{(1)}$]. Thus, $C_3^1$ and $\sigma_v^{(2)}$ are non-commuting operators (or transformations). In general, a rotation operation about an axis

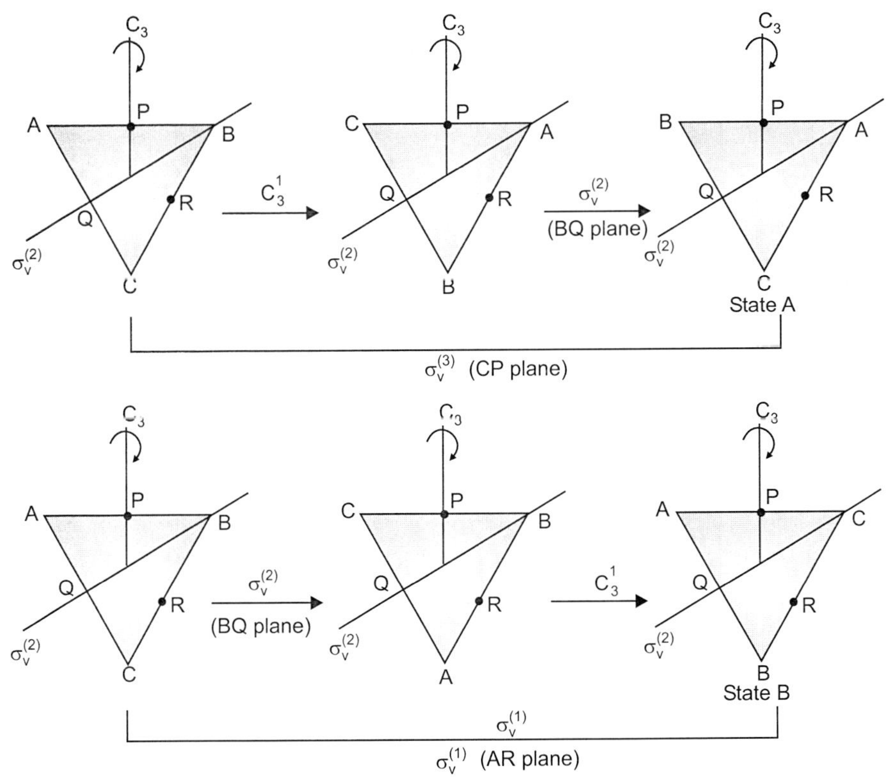


**Fig. 2.17** Non-commuting transformations $\sigma_v^{(2)} . C_3^1 \neq C_3^1 . \sigma_v^{(2)}$.

(*viz.* $C_n^x$) will not commute with a reflection operation in a vertical plane $\sigma_v$ containing the axis of rotation. [Justify this statement further by taking the example of a square or a regular pentagon or hexagon.]

**(*g*) Inverse of a combined transformation:** Suppose we have a transformation $T$ which is equal to a combined transformations. That is

$$T = [T_n \,.\, T_{n-1} \ldots\ldots T_3 T_2 T_1]$$

The inverse transformation of $T$ will be $T^{-1}$ and $T^{-1}$ can be written as

$$[T]^{-1} = [T_n \,.\, T_{n-1} T_{n-2} \ldots\ldots T_3 T_2 T_1]^{-1}$$

It can be easily proved that the expression in the square bracket is equal to,

$$[T_n \,.\, T_{n-1} T_{n-2} \ldots\ldots T_3 T_2 T_1]^{-1} = T_1^{-1} T_2^{-1} T_3^{-1} \ldots\ldots T_{n-1}^{-1} \,.\, T_n^{-1}$$

Thus, the inverse of a transformation T (*i.e.*, $T^{-1}$) which is obtained by the successive application of non-commuting transformations ($T_1, T_2, T_3, \ldots, T_{n-1}$ and $T_n$) is the transformation obtained by the consecutive application of the inverse of the non-commutating transformations, but in the reverse order. This is obvious from the (Fig. 2.18).

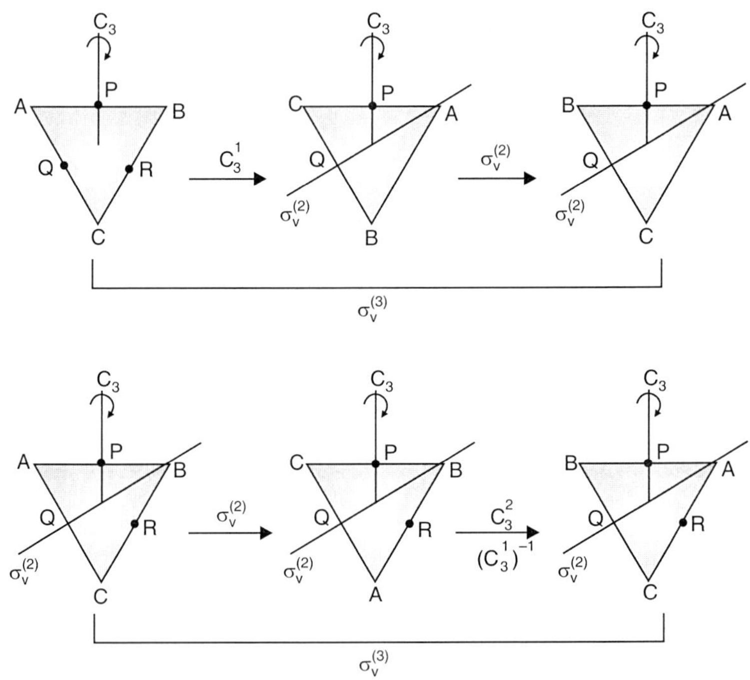


**Fig. 2.18** Inverse of combined operations.

In the case of symmetry transformation, the inverse of the symmetry transformation is also a symmetry transformation. Therefore, the inverse of $X^{-1}$ of a symmetry transformation, $X$ obtained by the consecutive application of non-commutating symmetry transformations, $X_1\ X_2\ X_3\ \ldots\ldots\ X_p$ is also a symmetry transformation obtained by the consecutive application of the inverses of the non-commutating transformations *viz.* $X_p^{-1}\ \ldots\ldots\ X_3^{-1}\ X_2^{-1}\ X_1^{-1}$, but in the reverse order. Thus, if

$$X = X_1\ X_2\ X_3\ \ldots\ldots\ X_{p-1}X_p$$

Then $$X^{-1} = X_p^{-1}\ X_{p-1}^{-1} \ldots\ldots\ X^{3-1X2\ -1}X_1^{-1}$$

In the case of commutating operators, the order of application of transformations does not make any difference.

**(*h*) Degree of symmetry of a system:** In order to know the degree of symmetry of any system, we make a list of all these symmetry transformations (or operations) which are associated with the system and which bring the system into its "self coincidence". We list by inspection all the symmetry elements possessed in it and then list all the symmetry transformations (or operations) generated by each of these elements. We will exemplify it by examining a few systems.

**Example 4:** The system $(C_2A_2B_2)$ that we have chosen is geometric planar (Fig. 2.19). It can be seen that the system has four symmetry elements (*viz.* $E$, $C_2$, $\sigma_h$ and $i$ ) as discussed earlier. Identity element generates only one operation ($E$). $C_2$ (which is two – fold axis passing through $P$ and perpendicular to the plane of the figure also generates one operation, ($C_2^1$). Mirror plane which is the plane of the system will generate one reflection operation, ($\sigma_h$). The centre of symmetry ($i$) at $P$ will generate one inversion operation ($i$) All these four symmetry transformations when applied to the system restore it to a configuration which is indistinguishable from the original state.

It should be noted that no other symmetry operation is possible in this case. We therefore have a set of four symmetry operations for our system (*viz.* $E$, $C_2^z$, $\sigma_h$, . $i$ ) In all these operations one point (*viz.* $P$) (centre of gravity of the molecule) is left unshifted. The molecules belonging to this class of the system are trans-dichloroethylene, *trans-*$PA_2\,B_2$ molecules [Collect some molecules of this shape], 2, 5-dichloro 3, 6-dibromobenzene, *trans* $[N_2F_2]$ *etc.* The order of symmetry of $[C_2\,A_2\,B_2]$ is 4.

**Example 5:** An equilateral triangle [Fig 2.19(*II*)] . It can be easily seen that it has the ten different symmetry element (*viz.* $E$, $C_3$, $C_2$, $C_2'$, $C_2''$, $\sigma_v$, $\sigma_v'$, $\sigma_v''$, $\sigma_h$, and $S_3$ axis) . These symmetry elements generate the operations $E$, $C_3^1$, $C_3^2$, $C_2$, $C_2'$, $C_2''$, $\sigma_v$, $\sigma_v'$, $\sigma_v''$, $\sigma_h$, and $S_3^1$, $S_3^5$. Here the symmetry elements $E$, $C_2$, $C_2'$, $C_2''$, $\sigma_v$, $\sigma_v'$, $\sigma_v''$, $\sigma_h$, generate only one operation while $C_3$ and $S_3$ generate two operations each [*viz.* ($C_3^1\ C_3^2$)] and [$S_3^1$, $S_3^5$]. These are the only symmetry operations possible in this system, with respect to which an equilateral triangle is symmetric. We have a set {$E$, $C_3^1$, $C_3^2$, $C_2$, $C_2'$, $C_2''$, $\sigma_v$ $\sigma_v'$, $\sigma_v''$, $\sigma_h$ and $S_3^1\ S_3^5$} and its order is 12.

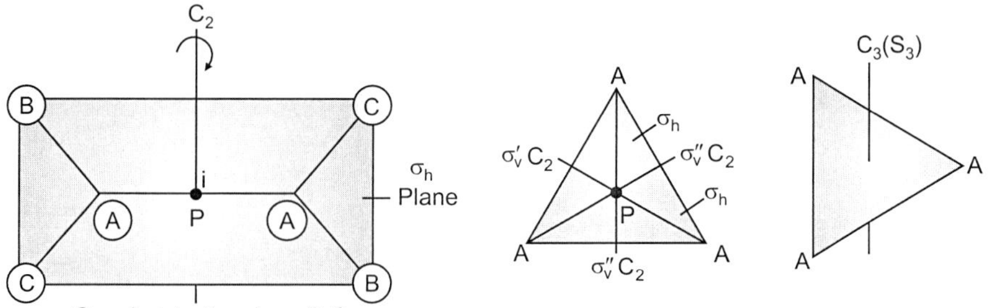


$C_2$-axis ⊥ to the plane $B_2C_2$
P = Centre of Inversion

Example (4)

Example (5)

**Fig. 2.19** (I and II) Shapes of molecules given in examples (4 and 5).

Some of examples of molecules having these operations are planar $AB_3$ *e.g.*, $BF_3$, $BCl_3$, $AlCl_3$, $CO_3^{-2}$, $NO_3^{-1}$, cyclopropane, 1, 3, 5-trichlorobenzene, trigonal bipyramidal molecules like $PCl_5$. [Collect some more molecules having these symmetry operations.]

**Example 6:** Triangular pyramid shaped Molecules ($AB_3$-type) Triangular pyramid shaped molecules have five symmetry elements. These are

(1) Identity $E$,

(2) $C_3$ rotation axis passing through vertex $A$ and the mid-point of the triangular base $B_3$, opposite to the vertex $A$,

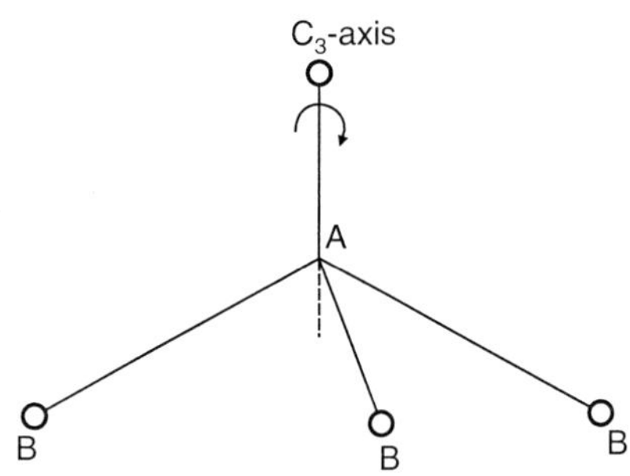


(3) three vertical planes containing the $C_3$ axis and passing through the vertex $A$ and the mid-point of the sides of $B_3$ equilateral triangle. The element $E$ generates one operation $E$. $C_3$ generates two operations $C_3^1$ and $C_3^2$, the third operation $C_3^3$ equals $E$. Each $\sigma_v$ plane generates only one operation and so three planes ($\sigma_v^{(1)}$ $\sigma_v^{(2)}$ $\sigma_v^{(3)}$) generate three operations. So the total number of operations are six (*viz.* $E$, $C_3^1$, $C_3^2$, $\sigma_v^{(1)}$, $\sigma_v^{(2)}$, $\sigma_v^{(3)}$). Examples of pyramidal $AB_3$ type molecules are $NH_3$, $PCl_3$, $CHCl_3$, $CBrCl_3$ *etc.* In all these systems one point remains invariant during transformations. (Fig. 2.20).

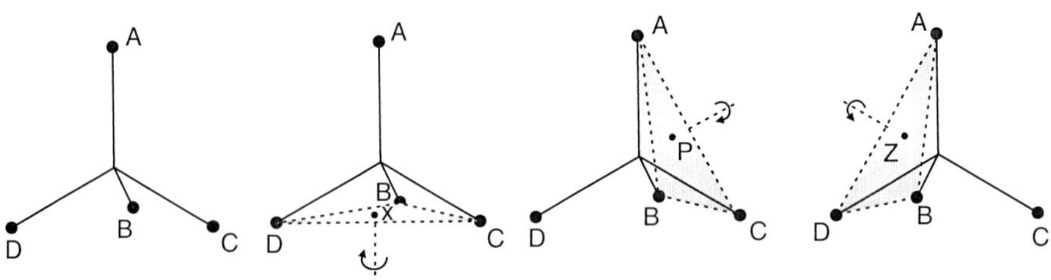


**Fig. 2.20** $C_3$ symmetry operations in $AB_3$ molecules.

**Example 7:** (Square) The symmetry elements associated with a square are the following: $E$, $C_4$ axis, $C_2'$ axis, $C_2''$ axis $i$ (centre of inversion), $S_4$ axis, $\sigma_h$ (horizontal plane), $\sigma_v$ (vertical mirror plane) and $\sigma_d$ (diagonal mirror plane).

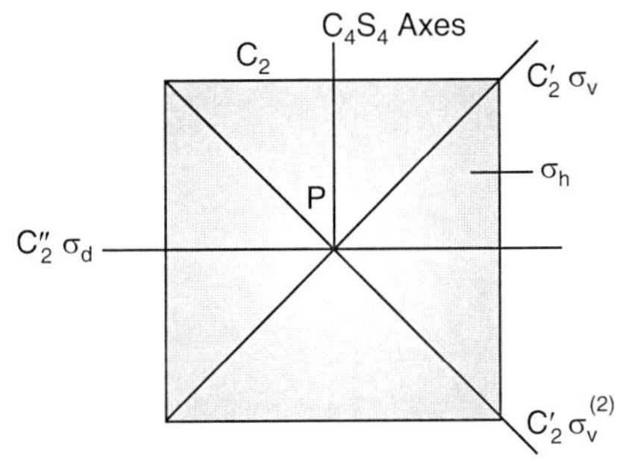


**Fig. 2.21** Symmetry elements of square shaped molecules.

All these symmetry elements intersect at one common point $P$ which remains invariant during all the operations generated by them. The operations generated by them are following:

$E$ (Identity) generates only one symmetry operation, $E$

$C_4$ (Vertical axis) generates three operations $C_4^1$, $C_4^2$ ($\equiv C_2$), $C_4^3$

$C_4^4$ operation ($\equiv E$) already taken care of by $E$.

$C_2^1$ (axis in the plane) generates only one operation, $C_2^1$ [$C_2^2 = E$ which is already taken care of. There are two such axes and hence two such operations.

$i$ (centre of inversion) generates only one operation $i$ [$i.i. = E$].

$S_4$ (Improper four-fold axis) generates two operations, $S_4^1$ and $S_4^3$. [$S_4^2 = C_2$] and [$S_4^4 = E$] which we have already taken care of.]

$\sigma_h$, $\sigma_v$ and $\sigma_d$ (mirror planes) generates only one operation each ($\sigma_h^2$ $\sigma_v^2$ $\sigma_d^2 = E$) (which has already been taken care of by the operation $E$.)

Hence, the total number of symmetry operations associated with square shaped molecules are $E$, $C_4^1$, $C_2$, $C_4^3$, $2C_2'$, $2C_2''$, $2\sigma_v$, $2\sigma_d$, $\sigma_h$, $i$ and $S_4^1$, $S_4^3$ which are written in the set form as {$E$, $2C_4$, $C_2 2C_2'$, $2C_2'' i$, $\sigma_h$ $2\sigma_v$ and $2S_4$}. It can be verified that there are no other symmetry operations associated with the square.

Some of the molecules having this symmetry are {$AB_4$} type of planar molecules, *e.g.*, $[PtCl_4]^{-2}$, $XeF_4$, trans [$AB_4C_2$ (Octahedral)] type, *e.g.*, $[Co(NH_3)_4Cl_2]^+$ *etc.*

**Example 8: (Regular Tetrahedron)** The symmetry elements associated with regular tetrahedron are $E$, (Identity), [$C_3^a\ C_3^b\ C_3^c\ C_3^d$] (four different axis) [$S_4^a\ S_4^b\ S_4^c$] (three different $S_4$ axis). All these symmetry elements cross at one point (centre of mass of tetrahedron) which remains invariant during the symmetry transformations (or operations) generated by the symmetry elements. (*See* Fig. 2.22).

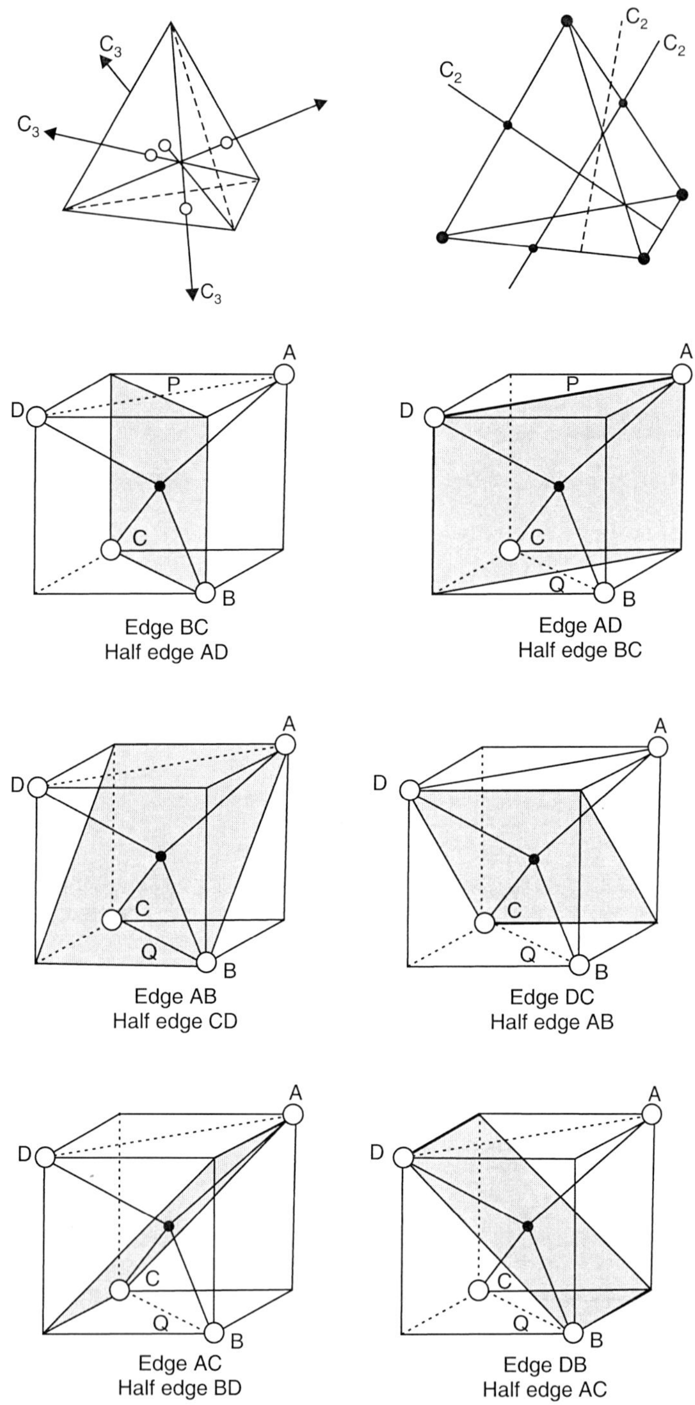


**Fig. 2.22** Symmetry operations and elements of tetrahedral molecules.

This three-fold axis (axis a, $B^1AB^1$) passes through vertex 1 $C = B$ centre of mass $A$ centre $B^1$ of equilateral triangle (2 3 4) and generates two rotation operations ($C_3^1$ $C_3^2$). Let us designate this axis as $C_3^a$ and operations as [$C_3^{a1}$, $C_3^{a2}$] Fig. 2.22. Similarly other three $C_3$ axis are:

$C_3^b$ [= axis $b$ $B_2AB_2'$] here 2 is vertex $A$ (centre of mass $B_2'$ = centre of equilateral triangle 134). This generates operations $C_3^{1b}$ $C_3^{2b}$.

$C_3^c$ [= axis $c$, $B^3AB^3$] here 3 is vertex $A$ (centre of mass) $B_3'$ = Centre of equilateral triangle (124). This generated operations $C_3^{c1}$ $C_3^{c2}$.

$C_3^d$ [= axis $d$, passing through $4B^4AB^{4\prime}$] here 4 is vertex; $A$ (centre of mass) $B_4'$ = Centre of equilateral triangle (123). This generated operations $C_3^{d1}$ $C_3^{d2}$.

Axis [$C_2^a$(1, 2) $A$ (3, 4)] = $C_2^a$ (axis passing through 1, 2 and 3, 4), the mid points of edges (1, 2) and (3, 4), respectively and $A$ (Centre of mass ) it generates only one operations $C_2^{a1}$.

Axis [$C_2^b$, 3) $A$ (2, 4)] = $C_2^b$ (axis passing through 2, 3 and 1, 4), the mid points of edges (2, 3) and (1, 4), respectively and through $A$ (Centre of mass ) it generates only one operations $C_2^{b1}$.

$\sigma_d^a$ [plane (1, 2 $A$ (3, 4))] = $\sigma_d^a$ passing through the edge (1, 2) and mid point of the edge (3, 4) and centre of mass $A$ . It will generate only one operation $\sigma_d^a$. Other five reflection planes are: (1, 3 $A$ (2, 4))] = $\sigma_d^b$, (1, 4 $A$ (2, 3))] = $\sigma_d^c$, (2, 3 $A$ (1, 4))] = $\sigma_d^d$, (3, 4 $A$ (1,2))] = $\sigma_d$ and (2, 4 $A$ (1, 3))] = $\sigma_d^f$. Thus these generate six operations.

$S_4^x$ axis generates two operations $S_4^{1x}$ and $S_4^{3x}$. Other $S_4^y$ and $S_4^z$ these generates four operations $S_4^{1y}$ $S_4^{3y}$ $S_4^{1z}$ and $S_4^{3z}$.

Thus the total symmetry transformations associated with a regular tetrahedron are {$E$, $C_3^{(a1)}$ $C_3^{(a2)}$ $C_3^{(b1)}$ $C_3^{(b2)}$ $C_3^{(c1)}$ $C_3^{(c2)}$ $C_3^{(d1)}$ $C_3^{(d2)}$ $C_2^{(a1)}$ $C_2^{(b1)}$ $C_2^{(c1)}$ $\sigma_d^a \sigma_d^b \sigma_d^c \sigma_d^d \sigma_d^e \sigma_d^f$ $S_4^{x1}$ $S_4^{x3}$ $S_4^{y1}$ $S_4^{y3}$ $S_4^{z1}$ $S_4^{z3}$}, which we write in the from {$E$, $8C_3$, $3C_2$, $6\sigma_d$, $6S_4$}. There are no more symmetry transformations with a tetrahedron.

Some of the molecules having this symmetry are tetrahedral shaped molecules having the formula $AB_4$ *e.g.*, $CH_4$, $CCl_4$, $MnO_4^-$, $Ni(CO)_4$.

In all the above examples, only those symmetry transformations have been considered by which one point of the system (centre of mass) remained invariant in all the operations belonging to a particular set. These sets of operations have very special significance which we will point out in the next section. We have other sets also in which the operations shift the positions of all the points, *e.g.*, the translation operations present in many systems like wall paper design, crystallographic lattice, *etc.* However, the emphasis in the present book is on those symmetry operations which are isometric and keep one point invariant by the transformations under consideration.

## 2.4 PROPERTIES OF THE SYMMETRY TRANSFORMATIONS PRESENT IN A SET

In the preceding paragraphs, we have seen that a system may have either one or more than one symmetry operations (hereafter we will write the word, operation in place of transformation/operation) with respect to which the system is symmetrical. There is only a fixed number of symmetry operations for a particular system. We thus have a set of symmetry operations for a particular system. We have also noticed while dealing with the combination (or product) of two or more than two symmetry operations that the result of two or more successive operations belonging to a system is itself a symmetry operation of the system. This forms the basis of application of the group theory (Group theory is the mathematical language of symmetry). Thus, if we wish to apply group theoretical principles on these sets of symmetry operations, we will have to prove that the operations belonging to a set also obey the four criteria of the group *viz*. closure, presence of an identity, presence of inverse of every element and the associative law. It is easy to prove that the members of every set of symmetry operations of a system obey all the group criteria and they do from a group. Further, if there is a set whose members also satisfy an additional property of commutativity, besides the four group properties, that is for every two of its members ($A$ and $B$), if

$$A.B = B.A,$$

then the set forms an Abelian group.

A finite group is a group having a definite number of operations in the set while an infinite group is the one having infinite number of operations in the set. Thus, a finite group is completely specified when all possible products ($A.B$) are known. The symmetry groups with which we are concerned here are mostly finite, except groups formed by linear molecules. The number of elements in a set forming a group is called as "Order of group" and is conventionally designated by a letter "$h$". Only things which the group is concerned with, in general, are its four axioms and theorems derived therefrom. It is not concerned with the nature of elements or the law of combination of the set. The symmetry and the group theory are related in the sense that the symmetry operators of any molecular system form a set whose members obey group axioms. When the rule of combination is the successive applications of these operators on a system, one can then apply group theory theorems on them.

## 2.5 MULTIPLICATION TABLES

The structure of the group of symmetry transformations can be determined by the product of every two transformations. If the order of group of symmetry transformations is '$h$' there will be '$h^2$' binary products and the group (at least in the abstract sense) can be completely defined. This information (*i.e.*, $h^2$ binary products can be arranged in the form of a table which is called "Multiplication Table" (cf. chapter one). In this table the members of the set forming the group are arranged in the top row and in the first column of the table and the binary products are written as shown in the Table 2.1.

The members of set forming the group are $\{E, a, b, c, d, f, \ldots\}$. The table exhibits a certain structure which is sufficient to specify an abstract group.

The entries in the table are the binary products of two elements, (say $a$, $b$). The order of the members of binary products is very specific. The order is (column member) $\circ$ (row member). Thus, $a \circ b$, is the binary product of (column member $b$) $\circ$ (row member $a$). The binary product ($a \circ b$) is written at the intersection of column labelled $a$ and the row labelled $b$. In this manner, all the binary products ($h^2$ in number) have been given in the Table 2.1.

**Table 2.1:** Multiplication Table

| ($\circ$) | **E** | ***a*** | ***b*** | ***c*** | ***d*** | ***f*** |
|---|---|---|---|---|---|---|
| $E$ | $E$ | $E \circ a$ | $E \circ b$ | $E \circ c$ | $E \circ d$ | $E \circ f$ |
| $a$ | $a$ | $a \circ a$ | $a \circ b$ | $a \circ c$ | $a \circ d$ | $a \circ f$ |
| $b$ | $b$ | $b \circ a$ | $b \circ b$ | $b \circ c$ | $b \circ d$ | $b \circ f$ |
| $c$ | $c$ | $c \circ a$ | $c \circ b$ | $c \circ c$ | $c \circ d$ | $c \circ f$ |
| $d$ | $d$ | $d \circ a$ | $d \circ b$ | $d \circ c$ | $d \circ d$ | $d \circ f$ |
| $f$ | $f$ | $f \circ a$ | $f \circ b$ | $f \circ c$ | $f \circ d$ | $f \circ f$ |

If the members of the set obey group axioms, then the following properties of the multiplication table will be observed:

1. Every entry in the table will be a member of the set because of the closure axiom of the group.
2. Each member of the set will occur once and only once in every row and column of the table (uniqueness of the product of two elements).
3. Table has a member $E$ (identity operation) once and only once in each row and column. So that $E.x = x.E$, where $x$ is any member of the set.
4. No two rows or columns can be identical.
5. Every element will have an inverse which occurs only once in each row and column (*i.e.*, if $x_i \,.\, x_j = E$), then $x_i = (x_j)^{-1}$ or $x_j = (x_i)^{-1}$].
6. Associative law will hold good.
7. In case, the group formed by the members of the set is Abelian then the multiplication table will be symmetrical with respect to its main diagonal with $E$ as its diagonal elements.

We have mentioned earlier that the members of the set of symmetry operations of any molecule form a group and it will give a specific multiplication table which will define the structure of that group. We then say that the set of operations of the molecule is the "Realisation" of the abstract group with a particular multiplication table. (Chapter one).

In physics and chemistry, one is very often concerned with such realisations of abstract groups formed by a set of symmetry operations, or by a set of vectors, or by a set of permutations, *etc.* These realisations are very significant in their studies. We will not discuss them any more. We however give below a few examples of the groups with their specific multiplication tables formed by a set of symmetry operations.

**Example 9:** As a first example, we take all the molecules which are either triangular bipyramid or equilateral triangle in shape. These belong to the same group. We take the molecule $PF_5$ as a representative. These systems are symmetrical with respect to a set of symmetry operations (*viz.* $E$, $C_3^1$, $C_3^2$, $S_3^1$, $S_3^5$, $C_2$, $C_2'$, $C_2''$, $\sigma_h$, $\sigma_v$ $\sigma_v'$ $\sigma_v''$) generated by the symmetry elements ($E$, $C_3$, $S_3$, $C_2$, $C_2'$, $C_2''$, $\sigma_h$, $\sigma_v$ $\sigma_v'$, $\sigma_v''$ ). These are shown in $PF_5$ molecule in Fig. 2.23.

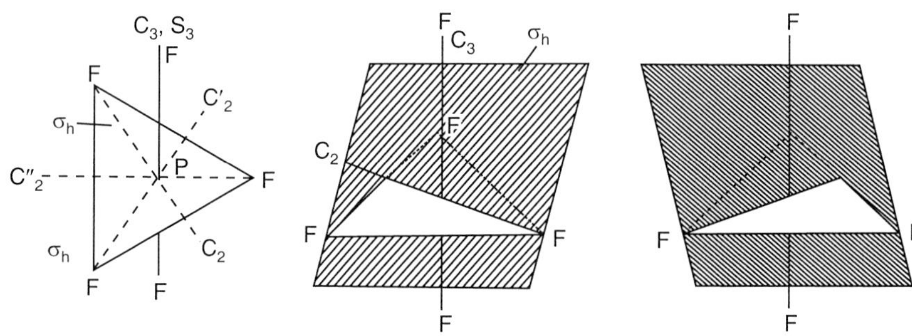


**Fig. 2.23** Symmetry elements of $PF_5$.

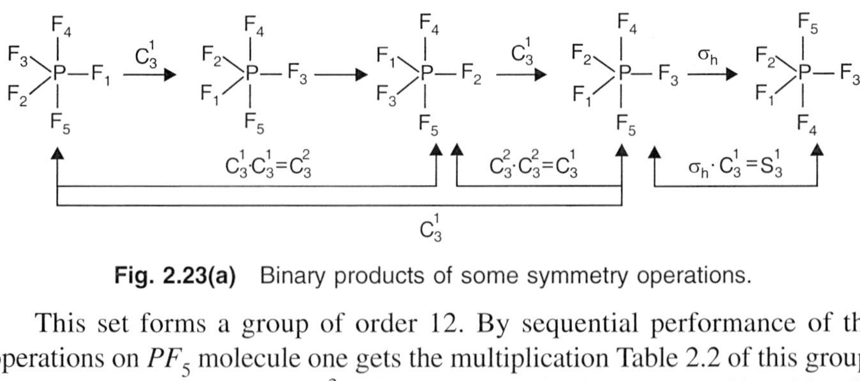


**Fig. 2.23(a)** Binary products of some symmetry operations.

This set forms a group of order 12. By sequential performance of the operations on $PF_5$ molecule one gets the multiplication Table 2.2 of this group, containing 144 entries (*viz.* $h^2$) of binary combinations. From this table, the following are obvious conclusions.

1. The binary product of any two members in the set is also a member of the set (closure property).
2. The set contains an element $E$ (Identity element).
3. Every member of the set has an inverse (Inverse property).
4. Associative law holds good (for example, $[C_3^1 \, . \, S_3^1 \, . \, C_2] = [S_3^5 \, . \, C_2 = \sigma_v'$ and $C_3^1]$.

$$(S_3^1 \, . \, C_2) = C_3^1 \, . \sigma_v'' = \sigma_v'.$$

5. Commutative law (*viz. A.B = B.A*) does not hold good in every case (*e.g.,*) $C_3^1 \sigma_v \neq \sigma_v \, . \, C_3^1$ because $C_3^1 \sigma_v = \sigma_v''$ and $\sigma_v \, . \, C_3^1 = \sigma_v'$.
6. Every row and column of the table is a particular rearrangement of the operations (members of the set). (Each member appears only once in each row and column). This particular rearrangement decides the structure of the abstract group (given below) of which the group formed by the symmetry operations of $PF_5$ is a realisation of the abstract group. The structure of the abstract group is given in Table 2.3.
7. All symmetry elements associated with $PF_5$ intersect at one point (*viz. P*), which remains unshifted in all the symmetry operations of $PF_5$.

**Table 2.2:** Multiplication Table for the set of operations associated with an equilateral triangle or a trigonal pyramid

**First Operation**

SECOND OPERATION

| $PF_5$ | $E$ | $C_3^1$ | $C_3^2$ | $\sigma_v$ | $\sigma_v'$ | $\sigma_v''$ | $\sigma_h$ | $S_3^1$ | $S_3^5$ | $C_2$ | $C_2'$ | $C_2''$ |
|---|---|---|---|---|---|---|---|---|---|---|---|---|
| $E$ | $E$ | $C_3^1$ | $C_3^2$ | $\sigma_v$ | $\sigma_v'$ | $\sigma_v''$ | $\sigma_h$ | $S_3^1$ | $S_3^5$ | $C_2$ | $C_2'$ | $C_2''$ |
| $C_3^1$ | ${C_3}^1$ | $C_3^2$ | $E$ | $\sigma_v''$ | $\sigma_v$ | $\sigma_v'$ | ${S_3}^1$ | $S_3^5$ | $\sigma_h$ | $C_2''$ | $C_2$ | $C_2'$ |
| $C_3^2$ | ${C_3}^2$ | $E$ | $C_3^1$ | $\sigma_v'$ | $\sigma_v''$ | $\sigma_v$ | ${S_3}^5$ | $\sigma_h$ | ${S_3}^1$ | $C_2'$ | $C_2''$ | $C_2$ |
| $\sigma_v$ | $\sigma_v$ | $\sigma_v'$ | $\sigma_v''$ | $E$ | $C_3^1$ | $C_3^2$ | $C_2$ | $C_2$ | $C_2$ | $\sigma_h$ | $S_3^1$ | $S_3^5$ |
| $\sigma_v'$ | $\sigma_v'$ | $\sigma_v''$ | $\sigma_v$ | $C_3^2$ | $E$ | $C_3^1$ | $C_2'$ | $C_2''$ | $C_2$ | $S_3^5$ | $\sigma_h$ | $S_3^1$ |
| $\sigma_v''$ | $\sigma_v''$ | $\sigma_v$ | $\sigma_v'$ | $C_3^1$ | $C_3^2$ | $E$ | $C_2''$ | $C_2$ | $C_2'$ | $S_3^1$ | $S_3^5$ | $\sigma_h$ |
| $\sigma_h$ | $\sigma_h$ | $S_3^1$ | ${S_3}^5$ | $C_2$ | $C_2'$ | $C_2''$ | $E$ | $C_3^1$ | $C_3^2$ | $\sigma_v$ | $\sigma_v'$ | $\sigma_v''$ |
| $S_3^1$ | ${S_3}^1$ | $S_3^5$ | $\sigma_h$ | $C_2''$ | $C_2$ | $C_2'$ | $C_3^1$ | $C_3^2$ | $E$ | $\sigma_v''$ | $\sigma_v$ | $\sigma_v'$ |
| $S_3^5$ | ${S_3}^5$ | $\sigma_h$ | $S_3^1$ | $C_2'$ | $C_2''$ | $C_2$ | $C_3^2$ | $E$ | $C_3^1$ | $\sigma_v'$ | $\sigma_v''$ | $\sigma_v$ |
| $C_2$ | $C_2$ | $C_2'$ | $C_2''$ | $\sigma_h$ | $S_3^1$ | $S_3^5$ | $\sigma_v$ | $\sigma_v'$ | $\sigma_v''$ | $E$ | $C_3^1$ | $C_3^2$ |
| $C_2'$ | $C_2'$ | $C_2''$ | $C_2$ | $S_3^5$ | $\sigma_h$ | $S_3^1$ | $\sigma_v'$ | $\sigma_v''$ | $\sigma_v$ | $C_3^2$ | $E$ | $C_3^1$ |
| $C_2''$ | $C_2''$ | $C_2$ | $C_2'$ | $S_3^1$ | $S_3^5$ | $\sigma_h$ | $\sigma_v''$ | $\sigma_v$ | $\sigma_v'$ | $C_3^1$ | $C_3^2$ | $E$ |

**Table 2.3:** Abstract group of which the group of symmetry operators of $PF_5$ is a realisation

| *GROUP* $G_1$ | $E$ | $C_3^1$ $a$ | $C_3^2$ $b$ | $\sigma_v$ $c$ | $\sigma_v'$ $d$ | $\sigma_v''$ $f$ | $\sigma_h$ $g$ | $S_3^1$ $h$ | $S_3^5$ $j$ | $C_2$ $k$ | $C_2'$ $l$ | $C_2''$ $m$ |
|---|---|---|---|---|---|---|---|---|---|---|---|---|
| $E$ | $E$ | $a$ | $b$ | $c$ | $d$ | $f$ | $g$ | $h$ | $j$ | $k$ | $L$ | $m$ |
| $a$ | $a$ | $b$ | $E$ | $f$ | $c$ | $d$ | $h$ | $j$ | $g$ | $m$ | $k$ | $l$ |
| $b$ | $b$ | $E$ | $a$ | $d$ | $f$ | $c$ | $j$ | $g$ | $h$ | $l$ | $m$ | $k$ |
| $c$ | $c$ | $d$ | $f$ | $E$ | $a$ | $b$ | $k$ | $l$ | $m$ | $g$ | $h$ | $j$ |
| $d$ | $d$ | $f$ | $c$ | $b$ | $E$ | $a$ | $l$ | $m$ | $k$ | $j$ | $g$ | $h$ |
| $f$ | $f$ | $c$ | $d$ | $a$ | $b$ | $E$ | $m$ | $k$ | $l$ | $h$ | $j$ | $g$ |
| $g$ | $g$ | $h$ | $j$ | $k$ | $l$ | $m$ | $E$ | $a$ | $b$ | $c$ | $d$ | $f$ |
| $h$ | $h$ | $j$ | $g$ | $m$ | $k$ | $l$ | $a$ | $b$ | $e$ | $f$ | $c$ | $d$ |
| $j$ | $j$ | $g$ | $h$ | $l$ | $m$ | $k$ | $b$ | $e$ | $a$ | $d$ | $f$ | $E$ |
| $k$ | $k$ | $l$ | $m$ | $g$ | $h$ | $j$ | $c$ | $d$ | $f$ | $E$ | $a$ | $b$ |
| $l$ | $l$ | $m$ | $k$ | $j$ | $g$ | $h$ | $d$ | $f$ | $c$ | $b$ | $E$ | $a$ |
| $m$ | $m$ | $k$ | $l$ | $h$ | $j$ | $g$ | $f$ | $c$ | $d$ | $a$ | $b$ | $E$ |

Since the members of set of operations associated with $PF_5$ obey all the axioms of being a group, the set is a group. Let us name it as $G_1$ which is the same as abstract group $G_1$ (Table 2.3). This is not an Abelian group.

In the same fashion, if one collects all the symmetry operations of $BF_3$, $CO_3^{2-}$, $NO_3^-$ $(C_2H_6)$ (eclipsed), *etc.*, one finds that these molecules also have exactly the same symmetry operations and the same multiplication table as those of $PF_5$. Therefore, the set of operations of these molecules form the same group as that of $PF_5$ (*viz.* $G_1$).

**Example 10:** As a second example let us take the case of planar $C_2H_4$ molecule. Its structure is shown below.

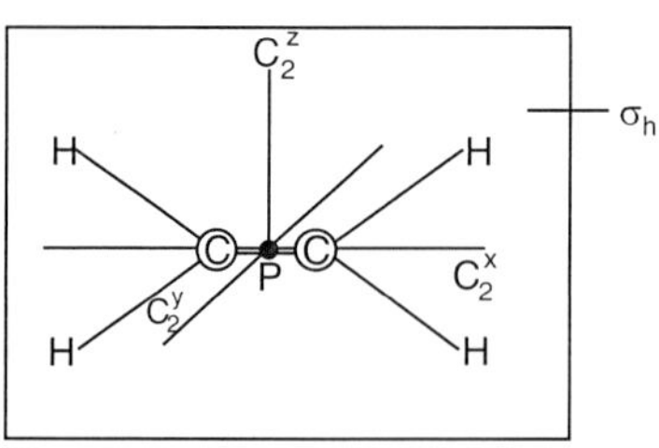


(a)

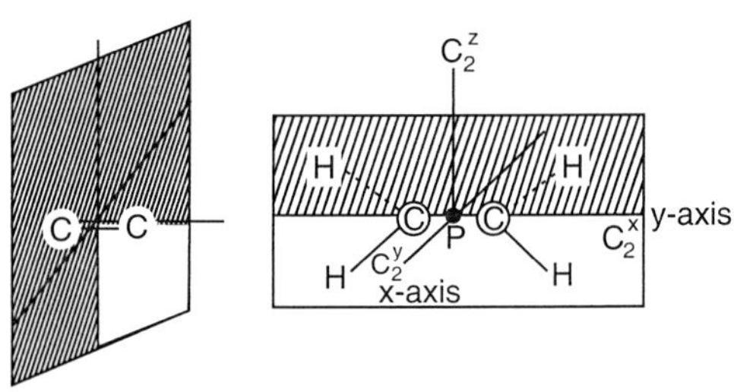


(b)

**Fig. 2.24** Structure of $C_2H_4$.

One can easily notice that the symmetry elements associated with $C_2H_4$ (planar in the *XY* plane) are $\{E, C_2^x, C_2^y, C_2^z, i\ (= S_2), \sigma_v^{xy}, \sigma_v^{yz}$ and $\sigma_v^{xz}\}$ which generate symmetry operations $\{E, C_2\, C_2'\, C_2'', i, \sigma_v, \sigma_v', \sigma_v''\}$. These will generate 64 entries in its multiplication Table 2.4.

**Table 2.4:** Multiplication Table of $C_2H_4$ symmetry operations

| $G_2$ | $E$ | $C_2^x$ | $C_2^y$ | $C_2^z$ | $i$ | $\sigma_v^{xy}$ | $\sigma_h$ | $\sigma_v^{xz}$ |
|---|---|---|---|---|---|---|---|---|
| $E$ | $E$ | $C_2^x$ | $C_2^y$ | $C_2^z$ | $i$ | $\sigma_v^{xy}$ | $\sigma_h$ | $\sigma_v^{xz}$ |
| $C_2^x$ | $C_2^x$ | $E$ | $C_2^z$ | $C_2^y$ | $\sigma_h$ | $\sigma_v^{xz}$ | $i$ | $\sigma_v^{xy}$ |
| $C_2^y$ | $C_2^y$ | $C_2^z$ | $E$ | $C_2^x$ | $\sigma_v^{xz}$ | $\sigma_h$ | $\sigma_v^{xy}$ | $i$ |
| $C_2^z$ | $C_2^z$ | $C_2^y$ | $C_2^x$ | $E$ | $\sigma_v^{xy}$ | $i$ | $\sigma_v^{xz}$ | $\sigma_h$ |
| $i$ | $i$ | $\sigma_h$ | $\sigma_v^{xz}$ | $\sigma_v^{xy}$ | $E$ | $C_2^z$ | $C_2^x$ | $C_2^y$ |
| $\sigma_v^{xy}$ | $\sigma_v^{xy}$ | $\sigma_v^{xz}$ | $\sigma_h$ | $i$ | $C_2^z$ | $E$ | $C_2^y$ | $C_2^x$ |
| $\sigma_h$ | $\sigma_h$ | $i$ | $\sigma_v^{xy}$ | $\sigma_v^{xz}$ | $C_2^x$ | $C_2^y$ | $E$ | $C_2^z$ |
| $\sigma_v^{xz}$ | $\sigma_v^{xz}$ | $\sigma_v^{xy}$ | $i$ | $\sigma_h$ | $C_2^y$ | $C_2^x$ | $C_2^z$ | $E$ |

It can be seen from Table 2.4 that every row and every column is one of the rearrangements of all the operations of the group. Each operation appears only once in each row and column. This particular arrangement of the operations is one of the large numbers of other possible arrangements of the operations of the set which form the structure of the group. Let us call it $G_2$ group. Thus, the group of the symmetry operations associated with $C_4H_4$ type of molecules as the realisation of group $G_2$ (say). It is an Abelian group. You will further notice that all the symmetry elements (geometric entities) intersect at one point (centre of mass or $\iota$ element) which remains unshifted during all the symmetry operations. Other molecules belonging to this group ($G_2$) are biphenyl, $B_2H_6$, pyrazine ($C_4H_4N_2$), $Al_2Cl_6$, $[Ni(C_7H_8)_2]$, *etc.* These have the same set of symmetry elements and symmetry operations as $C_2H_4$. The structure of the multiplication table is the same as that of $G_2$ (Table 2.4). Hence, these molecules belong to $G_2$ group.

The symmetry elements, symmetry operation, the multiplication tables along with a representative molecule belonging to the respective groups for a few molecules are given below as examples:

(A) Multiplication Table of $AB_3$ type molecules (Table 2.5)

(B) Symmetry elements $E$, $C_3$, $\sigma_v$

Symmetry operations $\{E, C_3^1, C_3^2, \sigma_v^{(1)}, \sigma_v^{(2)}, \sigma_v^{(3)}\}$

**Table 2.5:** Multiplication Table of $AB_3$ (Pyramidal) symmetry operations

| $G_3$ | $E$ | $C_3^1$ | $C_3^2$ | $\sigma_v^{(1)}$ | $\sigma_v^{(2)}$ | $\sigma_v^{(3)}$ | ***Examples*** |
|---|---|---|---|---|---|---|---|
| $E$ | $E$ | $C_3^1$ | $C_3^2$ | $\sigma_v^{(1)}$ | $\sigma_v^{(2)}$ | $\sigma_v^{(3)}$ | $AB_3$ type of pyramidal molecules |
| $C_3^1$ | $C_3^1$ | $C_3^2$ | $E$ | $\sigma_v^{(3)}$ | $\sigma_v^{(1)}$ | $\sigma_v^{(2)}$ | $CHCl_3$, $NH_3$, $CBrCl_3$, $PH_3$ |
| $C_3^2$ | $C_3^2$ | $E$ | $C_3^1$ | $\sigma_v^{(2)}$ | $\sigma_v^{(3)}$ | $\sigma_v^{(1)}$ | 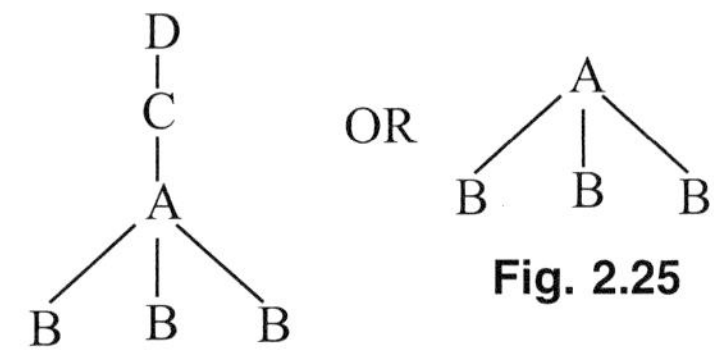 |
| $\sigma_v^{(1)}$ | $\sigma_v^{(1)}$ | $\sigma_v^{(3)}$ | $\sigma_v^{(2)}$ | $E$ | $C_3^1$ | $C_3^2$ | |
| $\sigma_v^{(2)}$ | $\sigma_v^{(2)}$ | $\sigma_v^{(1)}$ | $\sigma_v^{(3)}$ | $C_3^2$ | $E$ | $C_3^1$ | **Fig. 2.25** |
| $\sigma_v^{(3)}$ | $\sigma_v^{(3)}$ | $\sigma_v^{(2)}$ | $\sigma_v^{(1)}$ | $C_3^1$ | $C_3^2$ | $E$ | |

You will notice that if the set of elements and operations are $\{E, C_3\ C_2\}$ and $\{E, C_3^1\ C_3^2\ C_2, C_2'\ C_2''\}$ respectively, the multiplication table of this will be the same as that of $G_3$. In this group $C_2$-axes are perpendicular to the principal $C_3$-axis. The examples are (staggered) ethane or staggered $C_2Cl_6$. The shape of the molecule is given below.

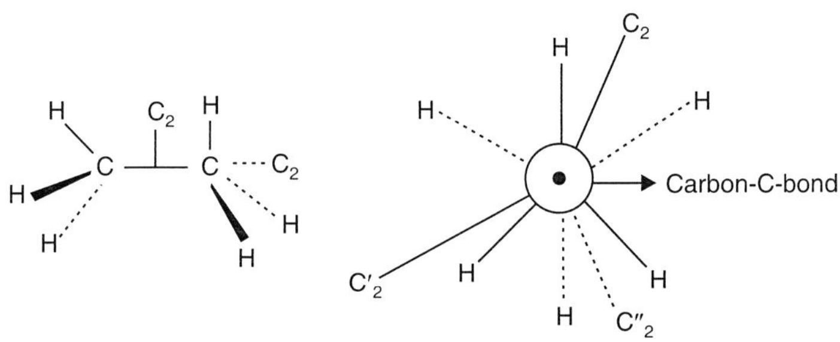


**Fig. 2.26** Structure $A_2B_6$ molecular (staggared).

**Table 2.6:** Multiplication Table of $G_4$
Symmetry Elements $\{E, S_3, C_3, \sigma_h\}$
Symmetry Operations $\{E, S_3^1, S_3^5, C_3^1, C_3^2, \sigma_h\}$

| $G_4$ | $E$ | $S_3^1$ | $S_3^5$ | $C_3^1$ | $C_3^2$ | $\sigma_h$ | ***Examples*** | |
|---|---|---|---|---|---|---|---|---|
| $E$ | $E$ | $S_3^1$ | $S_3^5$ | $C_3^1$ | $C_3^2$ | $\sigma_h$ | $M(OH)_3$ | 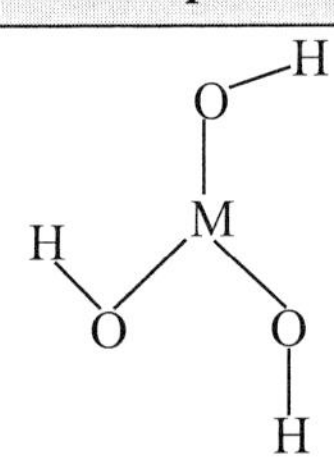 |
| $S_3^1$ | $S_3^1$ | $C_3^2$ | $E$ | $S_3^5$ | $\sigma_h$ | $C_3^1$ | Planar | |
| $S_3^5$ | $S_3^5$ | $E$ | $C_3^1$ | $\sigma_h$ | $S_3^1$ | $C_3^2$ | $A-(B-X)_3$ | |
| $C_3^1$ | $C_3^1$ | $S_3^5$ | $\sigma_h$ | $C_3^2$ | $E$ | $S_3^1$ | | |
| $C_3^2$ | $C_3^2$ | $\sigma_h$ | $S_3^1$ | $E$ | $C_3^1$ | $S_3^5$ | | |
| $\sigma_h$ | $\sigma_h$ | $C_3^1$ | $C_3^2$ | $S_3^1$ | $S_3^5$ | $E$ | | Shape of $A(Bx)_3$ (Planar) **Fig. 2.27** |

**Table 2.7:** Multiplication Table of $G_5$
Symmetry Elements $\{E, S_4\}$
Symmetry Operations $\{E, S_4^1, S_4^3, C_2^1\}$

| $G_5$ | $E$ | $S_4^1$ | $S_4^3$ | $C_2^1$ | *Example* | *Structure* |
|---|---|---|---|---|---|---|
| $E$ | $E$ | $S_4^1$ | $S_4^3$ | $C_2^1$ | Boat form of | 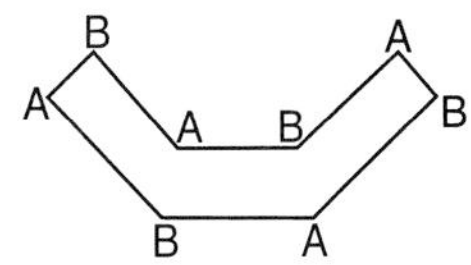 |
| $S_4^1$ | $S_4^1$ | $C_2^1$ | $E$ | $S_4^3$ | $A_4B_4$ of shape | |
| $S_4^3$ | $S_4^3$ | $E$ | $C_2^1$ | $S_4^1$ | as shown | |
| $C_2^1$ | $C_2^1$ | $S_4^3$ | $S_4^1$ | $E$ | | **Fig. 2.28** Boat form of $A_4B_4$ molecule. |

**Table 2.8:** Multiplication Table of $G_6$
Symmetry Elements $\{E, C_4, \sigma_v, \sigma_d\}$
Symmetry Operations $\{E, C_4^1, C_4^3, C_2^1, \sigma_v^1, \sigma_v^2, \sigma_d^1, \sigma_d^2\}$

| $G_6$ | $E$ | $C_4^1$ | $C_4^3$ | $C_4^2 (= C_2^1)$ | $\sigma_v^{(1)}$ | $\sigma_v^{(2)}$ | $\sigma_d^{(1)}$ | $\sigma_d^{(2)}$ | *Examples* |
|---|---|---|---|---|---|---|---|---|---|
| $E$ | $E$ | $C_4^1$ | $C_4^3$ | $C_4^2$ | $\sigma_v^{(1)}$ | $\sigma_v^{(2)}$ | $\sigma_d^{(1)}$ | $\sigma_d^{(2)}$ | $AB_4CD$ type |
| $C_4^1$ | $C_4^1$ | $C_4^2$ | $E$ | $C_4^3$ | $\sigma_d^{(1)}$ | $\sigma_d^{(2)}$ | $\sigma_v^{(2)}$ | $\sigma_v^{(1)}$ | $[Cu(NH_3)_4Cl\ Br]$ |
| $C_4^3$ | $C_4^3$ | $E$ | $C_4^2$ | $C_4^1$ | $\sigma_d^{(2)}$ | $\sigma_d^{(1)}$ | $\sigma_v^{(1)}$ | $\sigma_v^{(2)}$ | |
| $C_4^2$ | $C_4^2$ | $C_4^3$ | $C_4^1$ | $E$ | $\sigma_v^{(2)}$ | $\sigma_v^{(1)}$ | $\sigma_d^{(2)}$ | $\sigma_d^{(1)}$ | 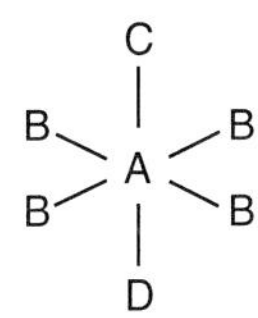 |
| $\sigma_v^{(1)}$ | $\sigma_v^{(1)}$ | $\sigma_d^{(2)}$ | $\sigma_d^{(1)}$ | $\sigma_v^{(2)}$ | $E$ | $C_4^2$ | $C_4^3$ | $C_4^1$ | |
| $\sigma_v^{(2)}$ | $\sigma_v^{(2)}$ | $\sigma_d^{(1)}$ | $\sigma_d^{(2)}$ | $\sigma_v^{(1)}$ | $C_4^2$ | $E$ | $C_4^1$ | $C_4^3$ | |
| $\sigma_d^{(1)}$ | $\sigma_d^{(1)}$ | $\sigma_v^{(1)}$ | $\sigma_v^{(2)}$ | $\sigma_d^{(2)}$ | $C_4^1$ | $C_4^3$ | $E$ | $C_4^2$ | |
| $\sigma_d^{(2)}$ | $\sigma_d^{(2)}$ | $\sigma_v^{(2)}$ | $\sigma_v^{(1)}$ | $\sigma_d^{(1)}$ | $C_4^3$ | $C_4^1$ | $C_4^2$ | $E$ | **Fig. 2.29** |

**Table 2.9:** Multiplication Table of $G_7$
Symmetry Elements $\{E, C_2^z, \sigma_h^{(x,y)}, i\}$
Symmetry Operations $\{E, C_2^z, \sigma_h^{(x,y)}, i\}$

| $G_7$ | $E$ | $C_2^z$ | $\sigma_h^{(x,y)}$ | $i$ | *Example* | *Structure* |
|---|---|---|---|---|---|---|
| $E$ | $E$ | $C_2^z$ | $\sigma_h^{(x,y)}$ | $i$ | Trans $A_2B_2C_2$ | 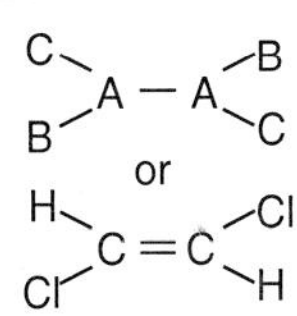 |
| $C_2^z$ | $C_2^z$ | $E$ | $i$ | $\sigma_h^{(x,y)}$ | type *e.g.*, | |
| $\sigma_h^{(x,y)}$ | $\sigma_h^{(x,y)}$ | $i$ | $E$ | $C_2^z$ | trans dichloro | |
| $i$ | $i$ | $\sigma_h^{(x,y)}$ | $C_2^z$ | $E$ | ethylene | **Fig. 2.30** |

**Table 2.10:** Multiplication Table of $G_8$
Symmetry Elements $\{E, C_4\}$
Symmetry Operations $\{E, C_4^1, C_4^2, C_4^3\}$

| $G_8$ | $E$ | $C_4^1$ | $C_4^2$ | $C_4^3$ | *Example* | *Structure* |
|---|---|---|---|---|---|---|
| $E$ | $E$ | $C_4^1$ | $C_4^2$ | $C_4^3$ | 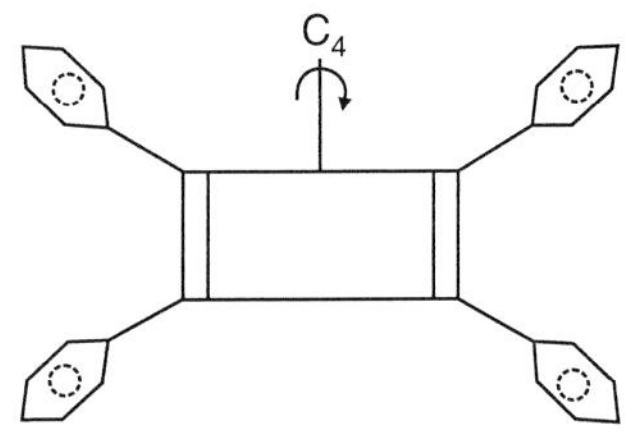 | |
| $C_4^1$ | $C_4^1$ | $C_4^2$ | $C_4^3$ | $E$ | | |
| $C_4^2$ | $C_4^2$ | $C_4^3$ | $E$ | $C_4^1$ | | |
| $C_4^3$ | $C_4^3$ | $E$ | $C_4^1$ | $C_4^2$ | Fig. 2.31 | |

We have given the symmetry operations associated with molecules having tetrahedral shapes. Examples of $AB_4$ type molecules are $CH_4$, $CCl_4$, $CBr_4$. One can thus form multiplication tables of their symmetry operations and obtain a group. Let us call it as $G_9$. These contain a total of twenty-four symmetry operations with multiplication table having $(24)^2$ entries of $(AB)$ types of binary combinations. One should attempt to write this multiplication table.

One can also write down all the symmetry elements and their generated symmetry operations of octahedral $AB_6$ type molecules. There are 48 symmetry operations for such molecules having a general formula $AB_6$. Its multiplication table has not been given. It will have $(48)^2$ number of entries of binary combinations of $A \circ B$ type. The examples of such molecules are in large numbers, *e.g.*, $SF_6$, $[M(H_2O)_6]$ $[XeF_6]$. Let the group of 48 symmetry operations be $G_{10}$.

## 2.6 POINT GROUPS

In the preceding paragraphs, we have given the symmetry operations of a number of molecules having different shapes. It implies and be generalised that any molecule or a geometrical object can be described by a set of symmetry operations associated with it. The members of this set satisfy all the conditions of forming a group. Therefore, one can associate a particular group to a molecule to which the molecule belongs. We have also seen that all the symmetry elements of the group to which the molecule belongs intersect at one point whose position remains invariant by all the operations associated with the molecule. This unshifted point lies inside the molecule. Because of the presence of this fixed point during all symmetry operations of groups of the groups associated with such molecules are known as *Point Groups* or *Molecular Point Groups*. Thus in the preceding section, all the groups ($G_1$, $G_2$, $G_3$,.....,$G$ *etc.*) are the different point groups to which various molecules belong.

One must not think that the groups are limited only to point groups. There are groups called the space groups which contain operations involving "Translatory motions" because of which the symmetry elements associated with the system do not intersect at one unshifted common point. They are not called as "Point groups" but Space groups. The elementary idea of such groups will be in the coming sections of this chapter.

Every molecule that we come across can be assigned a definite number of symmetry operations forming a group which will be one of the infinite numbers of theoretical possible *point groups*. However, one must also realise that there are only a limited number of points groups which are significant. We will discuss them in the following paragraphs. There are two systems for their notations,

(1) Schonfliess notations based on C-σ-*i*-*S* notations of symmetry elements. We will use them in our discussion.

(2) Hermann-Manguin notation which is generally used by crystallographers. It is also called the International system adopted and recommended by the International Union of Crystallographers. It is based on enumerating the minimum number of symmetry elements needed to define a point group.

We list below the different common point groups.

**Type I Point groups having no rotational symmetry axis:** These are designated as $C_1$, $C_s$, and $C_i$. In $C_1$ type there is no symmetry element except $E$ (Identity element). It is of order one. An example of molecule belonging to $C_1$ point group is [$HC(CH_3)(Br)(Cl)$].

In $C_s$ point group, there are only two symmetry elements, *viz.* $E$ and a mirror plane (σ) or an $S_1$ axis. This is a point group of order two and has two symmetry operations *viz.* $\{E, \sigma\}$. Examples are [HOCl], [β-chloronaphthalene].

In $C_i$ point group, there are two elements $E$ and $i$ (or $S_2$ axis). It is of order 2 and consists of two operations $\{E, i\}$. An example is [HClFC-CHClF] (staggered configuration).

**Type II Point groups which has only one rotation axis:** These are designated as either as $C_p$ or $S_p$.

**(a) $C_p$ point groups:** In these point groups, only one proper rotation axis of order $C_p$ is present. It generates a set of rotation operations (*viz.* $C_p^1$, $C_p^2$, $C_p^3$......$C_p^{p-1}$, $E$). It is of order $p$. These are known as "Cyclic Groups" and are Abelian Groups. Thus, we have:

1. $C_2$ point group-Example - [$CH_2Cl–CH_2Cl$] (staggered) or $H_2O_2$ (staggered)
2. $C_3$ point group-Example - [$CH_3–CBr_3$] (staggered) (Fig. 2.26)
3. $C_4$ point group-Example: See Fig. 2.30
4. $C_6$ point group-$C_6(C_6H_5)_6$

**(b) $S_p$ point groups:** These types of point groups contain only one improper axis ($S_p$). Here we should consider $p$ to be odd or even separately. As we have considered earlier when $p$ is even, then $S_p$ element generates $P$ symmetry operations, *viz.* $S_p^1\ S_p^2\ S_p^{p/2}$ (= $C_2$), $S_p^{p/2+1}$... $S_p^{p-1}$ and $S_p^p$ (= $E$). These are known as $S_p$ point group. If $p$ is odd, then $S_p$ axis generates $2p$ operations {*viz.* $S_p^1\ S_p^{p+1}\ S_p^{p+2}$...... $C_p^1\ C_p^2\ C_p^3$...... $C_p^n$ and $S_p^{2p}$ (= $E$) and $S_p^p$ (= $\sigma_h$)}. These groups are also designated as $C_{nh}$ ($C_{nh}$ implies that there is a $C_n$ axis and a $\sigma_h$ plane perpendicular to $C_n$ axis. The combination of $C_n$ axis and $\sigma_h$ mirror plane means that there will automatically be an $S_n$-axis, $S_p$ ($p$ = odd) implies the existence of $C_p$ and $\sigma_h$. An example for $S_4$ point group is [$C_8(CH_3)_4\ H_4$] 1, 3, 5, 7 tetramethyl cyclo octatetraene.

**Type III Point group $C_{pv}$:** These point groups arise by adding a vertical plane ($\sigma_v$) to the $C_p$ axis. Let us recall that that the operations generated by $C_p$-axis are ($C_p^1\ C_p^2$......$C_p^{p-1}$). Now, if we add a vertical plane ($\sigma_v$) and if $p$ of $C_p$ is odd, then $C_p$ axis will generate $p$ number of ($\sigma_v$) planes. Example: $NH_3$, which contain three ($\sigma_v$) planes because it has $C_3$-axis (3 = odd number).

If $p$ in $C_p$ -axis is even, the ($\sigma_v$) planes generated by $C_p$ will be of two types *viz.* ($n$/2) ($\sigma_v$) and ($n$/2) ($\sigma_d$) types. Example is [$PtCl_4$] with $C_4$-axis with $2\sigma_v$ and $2\sigma_d$. However in both cases ($p$ = even or odd), the set of operations generated by $C_p$ and all $\sigma_v$ or $\sigma_d$ constitutes a complete set. Such point groups are designated as $C_{nv}$. Thus we have,

(*i*) Point group $C_{2v}$: ($C_2$-axis + $2\sigma_v$). An example is 2, 6 dichloroquinone.

(*ii*) Point group $C_{3v}$: ($C_3$-axis + $3\sigma_v$). An examples are $AB_3$ pyramidal and $CXY_3$.

(*iii*) Point group $C_{4v}$ ($C_4$-axis + $2\sigma_v$ + $2\sigma_d$ planes). Examples are pyramidal $AB_4$ molecules.

**Type IV Point groups $C_{ph}$:** These types of point groups contain one $C_p$-axis (principal rotational axis) and one $\sigma_h$ mirror plane perpendicular to $C_p$-axis. We have discussed earlier while dealing with improper axis ($S_n$) the number of operations generated in this case depends upon the value of the $p$ (*i.e.*, $p$ is even or odd). Thus, if $p$ is an even number, then the operations are {$C_p^1\ C_p^2$ ...... $C_p^{p/2}$ (= $C_2$) ...... $C_p^{p-1}$, $S_p^1$, $S_p^5$, $S_p^{p-2}$, $\sigma_h$, $S_p^{p/2} = i$}.

If $p$ = odd then the number of operations containing the set for $C_{ph}$ are ($C_p^1\ C_p^2\ C_p^3$ ...... $C_p^{p-1}\ C_p^p$ (= $E$), $S_p^1\ S_p^3\ S_p^5$, $S_p^{p+2}\ S_p^{p+4}$...... $S_p^p$ (= $\sigma_h$)}.

Examples are:

For $C_{2h}$ {$E\ C_2\ \sigma_h$ }, -trans (1, 2-dichloroethylene)

For $C_{3h}${$E\ C_3^1\ C_3^2\ S_3^1\ S_3^5\ \sigma_h$ }-[$B(OH)_3$]

For $C_{4h}$ {$E\ C_4^1\ C_2^1\ C_4^3\ S_4^1\ S_4^3\ i\ \sigma_h$ }-[$C_4(CH_3)_4$]

**Type V Point groups $D_p$:** These point groups are obtained by adding one 2-fold axis ($C_2$-axis) at right angles, to the principal axis $C_p$. Now adding a $C_2$-

axis at right angle to $C_p$-axis means that there must be $p$ such 2-fold axes separated by an angle of $(360°/p)$. Now $C_p$ axis generates $\{C_p^1\ C_p^2\ ......C_p^p\ (= E)\}$ operations. Further one $C_2$-axis will generate one $C_2$ operation and so $p$ such $C_2$-axes will generate $p$ $C_2$ operations. The $n$ operations generated by $C_p$-axis plus $n$ operations generated by $n$ $C_2$ will give a total of $2n$ operations. This set of $2n$ operations will give $D_n$ point group. Now if $p$ is,

(*a*) **Even,** then $C_2$-axis will generate two types of operations *viz.* $p/2$ operations will be of $C_2'$ and remaining $p/2$ operations will be of $C_2''$.

(*b*) **Odd,** then $C_2$-axis will generate all $p$ number of $C_2$-operations.

So, for *even p* we will have a set of $\{E\ C_p^1\ C_p^2......C_p^{p-1}\ C_p^{p/2}(= C_2)\ C_2'{}_{(1)}\ C_2'{}_{(2)}\ C_2'{}_{(p/2)}, C_2''{}_{(1)}\ C_2''{}_{(2)}\ ......C_2''{}_{(p/2)}\}$ and for odd $p$, the set will contain $\{E = C_p^1\ C_p^2\ ......\ C_p^{p-1}\ C_{2\,(1)}\ C_{2\,(2)}\ C_{2\,(p)}\}$

Examples are rare.

**Type VI Point groups $D_{ph}$:** These point groups are obtained by adding a $\sigma_h$ plane to a system having a $C_p$-axis (principal axis) and $p$ number of 2-fold axis at right angles to $C_p$-axis. It implies that these $p$ number of $C_2$-axis lie in the $\sigma_h$ plane. The symmetry operations of these point groups are obtained by combining $\sigma_h$ operations with each operation of $D_n$ point group. It can be proved that the combination of $C_2'$-axis and a $\sigma_h$ plane generates $\sigma_v$ planes. For even $p$, we have two types of $C_2$-axis $C_2'$ and $C_2''$. So combining $C_2'$ and $C_2''$ axis with $\sigma_h$ will generate two different types of $\sigma_v$ planes ($\sigma_v$ and $\sigma_d$). For odd value of $p$, we will have only one type of $\sigma_v$ planes. For even $p$ values, we will also have an inversion centre. While for odd values of $p$, we will not be having any centre of inversion.

Besides the above operations, one should remember that $C_p$ axis with a $\sigma_h$ plane will also generate $S_n$-axis. Therefore, we will also have $S_p^{\text{th}}$ operations. Thus if $p$ is even, one obtains new operations $S_p^1$, $S_{p/2}(= i)$ ...... $S_p^{(p-2)/2}$, $S_p^{p-1}$ If $p$ is odd, one obtains new $(p - 1)$ improper rotation operations of the type $S_p^1$ $S_p^3\ S_p^5$ ...... $S_p^{2p-3}\ S_p^{2p-1}$ and $\sigma_h$. If we collect all these operations, $D_{pb}$ will have the following operations.

(*a*) $p$ = even, $(E, 2C_p, 2C_p^2......2C_p^{p/2}(= C_2)\ (1/2p)C_2'', \sigma_h, 2S_p\ (2C_p, \sigma_h)$ $(1/2p)\ \sigma_v', (1/2p)\ \sigma_v''$

(*b*) $p$ = odd, $\{E, 2C_p, 2C_p^2$ ...... $2C_p^{(p-1/2)}, pC_2, \sigma_h, 2\ S_p^1, 2S_p^3$ ...... $2\ S_p^{p-2}$, $\sigma_v)$

Examples are:

(*i*) $D_{2h}$ $\{E, C_2\ C_2', C_2''\ \sigma_h\ \sigma_v, \sigma_v'\}$ $C_2H_4$, naphthalene, pyrazine, $Al_2Br_6$, *etc.*

(*ii*) $D_{3h}$ $\{E, C_3^1\ C_3^2\ 3C_2'\ \sigma_h, 3\sigma_v\ 3S_p^1\}$ cyclopropane, ethane (eclipsed), $PF_5$, $BF_3$, $AlCl_3$.

(*iii*) $D_{4h}$ $\{E, 2C_4, C_2, 2C_2', 2C_2'', i, \sigma_h, 2S_4^1\ 2\sigma_v\ 2\sigma_v'\}$ $[PtCl_4]^{2-}$ trans- $[MA_4B_2]$, octahedral, $S_4$, *etc.*

**Type VII Point group $D_{pd}$:** It was shown in the preceding paragraph that the addition of $\sigma_h$ to $D_p$ automatically generates $\sigma_v$ planes. So one should not expect the formation of new type of point groups by the addition of $\sigma_v$ planes containing $C_2$ axis to $D_p$ groups. But the addition of $\sigma_d$ planes which bisects the angle between two adjacent $C_2$-axis ($p$ = odd) or between $C_2'$ and $C_2''$ axes ($p$ = even) to the $D_p$ point groups generate new point groups. For odd values of $p$, the angle between two $C_2'$-axes is (360°/4$p$) and so the angle between a $\sigma_d$ and a $C_2$-axes is also (360°/4$p$). A right angle will be formed by $p$ such angles and so in $Dp_d$ point group when ($p$ = odd), it is possible to find a $C_2'$ axis located at right angle to a $\sigma_d$ plane. The product of these two operation ($\sigma_d$ and $C_2'$) is an inversion operation $I$ and so $D_{pd}$ ($p$ = odd) groups have a centre of symmetry. The operations for $D_{pd}$ point group are

(***a***) $\boldsymbol{p}$ **= even** $\{E,\ C_p^1,\ C_p^{p-1},\ C_p^2,\ C_p^{p-2} \ldots\ldots C_p^{p/2}\ (= C_2),\ pC_2,\ p\sigma_d,$ $S_{2p}\ S_{2p}^{2p-2}\ S_{2p}^3\ S_{2p}^{2p-3}\ \ldots\ldots\ S_p^{p-1}, S_p^{p+1}\}$

(***b***) $\boldsymbol{p}$ **= odd** $\{E, C_p^1, C_p^{p-1}, C_p^2, C_p^{p-2} \ldots\ldots\ C_p^{p-1/2}, C_p^{p+1/2}, pC_2, p\sigma_d, S_{2p}^1,$ $S_{2p}^{2p-2}, S_{2p}^p (= i)\}$

Examples are:

(*i*) $D_{2d}$......$\{E, C_2\ C_2', C_2''\ \sigma_d, \sigma_d'\ S_4^1, S_4^3\}$ $C_3H_2$ (allene) $C_9H_8$, *etc.*

(*ii*) $D_{3d}$......$\{E, C_3^1\ C_3^2\ C_2\ C_2', C_2'', i, S_6^1\ S_6^4\ \sigma_d\ \sigma_d'\ \sigma_d''\}$ ethane (staggered)

(*iii*) $D_{5d}$......$\{E, C_5^1\ C_5^4\ C_5^2\ C_5^3\ 5C_2\ i\ S_{10}^1\ S_{10}^8\}$ ferrocene (staggered)

**Type VIII Point groups ($\boldsymbol{T, T_h, T_d, O, O_h, I, I_h}$ and $\boldsymbol{K_h}$):** In all the above point groups which we have discussed it has been tactfully assumed that the systems or the molecules have only one principal axis $C_p$ ($p \geq 2$). The point groups with one principal axis are then generated by adding $C_2$-axis, at right angles to $C_p$-axis . ($p \geq 2$) and/or by adding $\sigma_d$, $\sigma_v$ planes containing the $C_p$ axis or a $\sigma_h$ plane at right angles to it. We have however not investigated the consequences of adding $C_n$-axis /axes ($n > 3$) at right angles to $C_p$-axis. It can be shown that one can generate a number of new point groups by adding such axes. Some of the systems belonging to these points groups are present in nature. These are very highly symmetric systems. We will not investigate them in any great detail. We will only write the symmetry operations forming the point groups $T$ $T_h$, $T_d$, $O$, $O_h$, $I$, $I_h$ and $K$. The symmetry operations associated with these point groups are shown below.

**1. T point group:** $\{E, 4C_3^1, 4C_3^2, 3C_2^1\}$-order of the group = 12

**2. $\boldsymbol{T_d}$ point group:** $\{E, 4C_3^1, 4C_3^2, 3C_2, 6\sigma_d, 3S_4^1, 3S_4^3\}$-order of the group = 24

Examples: All $AB_4$ (tetrahedral molecules)

**3. $T_h$ point group:** $\{E, 4C_3^1\ 4C_3^2\ 3C_2, i\ (= S_6^3)\ 4S_6^1\ 4S_6^5, 3\sigma_h\}$-order of the group = 24

Examples: Octahedral molecule $[Co(NO_2)_6]$ and $[M(H_2O)_6]$ in which all the atoms of trans-$H_2O$ ligands lie in $\sigma_h$ plane.

**4. Point group $O$:** $\{E, 3C_4^1\ 3C_4^3\ 3C_4^2\ 6C_2'\ 4C_3^1\ 4C_3^2\}$ order = 24. This group is similar to $T_d$ point group except that $6\sigma_d$ of $T_d$ point group are replaced by $6C_4$ axes operations and the $6C_2'$ axes are the axes which bisect diagonally opposite edges of the cube.

**5. Point group $O_h$:** This group is generated by $O$ by adding three $\sigma_h$ planes at right angles to each 4-fold axes. Since $C_2.\sigma_h = i$, it must necessarily contain a centre of inversion $i$. Other operations are,

$\{E, 4C_3^1\ 4C_3^2\ 3C_4^1, 3C_4^3, 3C_4^2\ (= C_2)\ 6C_2'\ i\ 3S_4^1\ 3S_4^3\ 4S_6^1, 4S_6^5, 3\sigma_h, 6\sigma_d\}$.
Order = 48.

Examples: All octahedral molecules ($AB_6$). Furthermore, *Cube* also has the same set of symmetry operations (Ref. Space or lattice point group in Sec. 2.10) and belongs to $O_h$ point groups. All molecules of the type $AB_8$ (cubic with $A$ at the centre ) also belong to $O_h$ point group (Ref. body centered-cubic lattice).

**6. Point group $I$:** This point group is the sub group of $I_h$ point group to which the dodecahedron and icosahedron belong. It has the symmetry operations $\{E, 6C_5^1\ 6C_5^4, 6C_5^2\ 6C_5^3, 10C_3^1, 10C_3^2, 15C_2\}$. Order = 60.

**7. Point group $I_h$(or $I_i$ or $P_i$ or $P_h$):** The addition of reflection planes $\sigma_h$ perpendicular to the 2-fold axis (not 5-fold) to the I point group generate new symmetry elements. Thus $C_2.\sigma_h = i$, implies the presence of centre of inversion $i$. Other symmetry operations are $\{E, 6C_5^1, 6C_5^4, 6C_5^2\ 6C_5^3, 10C_3^1, 10C_3^2, 15C_2\ i\ 6S_{10}^1, 6S_{10}^9, 6S_{10}^3, 6S_{10}^7, 10S_{10}^1, 10S_{10}^5\ 15C_2^2\}$ Order = 120.

Examples: All icosahedral or dodecahedral molecules like $(B_{12}H_{12})$, $(B_{12})$ $(C_{20}H_{20})$.

**8. Point group $K_h$:** This is the Centro-symmetric point group of infinite order. Sphere belongs to this point group. All possible symmetry elements and operations belong to this group.

The point groups $T\ T_h$, $T_d$, $O$, $O_h$, $I$, $I_h$ are the only possibilities in which one expects to generate new point groups by adding symmetry rotation axes of order higher than 2 at right angles or only at any other angles to the principal axis $C_n$. These generate the following platonic solids (Polyhedra):

(*a*) Polyhedra whose faces are regular polygons *e.g.*, equilateral triangles, squares, pentagons or hexagons,

(*b*) Polyhedra whose vertices are all symmetrically equivalent,

(*c*) Polyhedra whose edges are equivalent.

These polyhedra are tetrahedron ($4F, 4V, 6E$), cube ($6F, 8V, 12E$), octahedron ($8F, 6V, 12E$), dodecahedron ($12F, 20V, 30E$) and icosahedron ($20F, 12V, 30E$). Here $F$ = face, $V$ = vertex and $E$ = edge.

One can prove that the above solids polyhedra represent all the possibilities of solids structures.

## 2.7 PROCEDURE TO FIND THE POINT GROUP OF A MOLECULE OR SOLID

How do we assign point groups to molecules? It is a practical problem. We do it in the following steps:

**Step 1:** Find out if the molecules has any axis of rotation, if you find "none", then search for a mirror plane if it has then it belongs to $c_s$. If not, search if it has any centre of inversion. If it has, the point group is $C_i$. If not, then it belongs to $C_{1.}$

**Step 2:** If the molecule has an axis of rotation, then look for the highest order rotational axis. This will be the principal axis. If the axis is of infinite order, then examine if the molecule is linear. In case it is, then it belongs to $D_{\infty h}$ or $C_{\infty v}$ point group. In case it has a centre of symmetry and a $\sigma_h$ plane, the molecule belongs to $D_h$ otherwise to $C_{\infty v}$. In case the molecule is spherical shape, then it belongs to $K_h$ point group.

**Step 3:** If the principal axis is 3-, 4-, or 5-fold. (And not of $\infty$ order) then look for another axis or axes of the *same order*. If you find one then one of the following cases arise:

If there are six 5-fold axes, then the molecule belongs to $I$ or $I_h$ point group.

(*a*) In case of yes, then look for the presence of any reflection plane of symmetry. If you find one then the molecule belongs to $I_h$ and in case no such reflection plane exists, then the molecule belongs to $I$.

(*b*) In case you do not find another axis of the same order, but you find the presence of three 4-fold axes and if these are mutually perpendicular to each other, the molecule belongs to either $O$, or $O_h$ point group. In that case look for the presence of any reflection plane, if such plane exists, the molecule belongs to $O_h$ otherwise to $O$.

(*c*) If you do not find three 4-fold or 5-fold axes at right angles to each other but you find four 3-fold axes at an angle of 109° 28′ from each other, then the molecule belongs to $T$-point group. If the molecule also has a centre of inversion then molecule belongs to $T_h$-point group, otherwise to $T$. In case of the *absence* of $i$, the molecule has $6\sigma_d$ planes, then it belongs to $T_d$ point group.

**Step 4:** In the *absence* of more than one principal axis of order higher than 3 if it has *one* principal axis of any order and no other rotation axis of any order (even of 2-fold), then the point group of molecule will be either $C_p$ or $S_p$. In case the principal axis is also a $2p$ fold improper rotation axis, the molecule belongs to $S_p$ point group, otherwise to $C_p$ point group. One should also look for reflection planes. There will be three alternatives.

(*a*) If there exist no planes, then the molecular point group will be $C_p$ or $S_p$.

(*b*) If there exist $p$ mirror planes containing the principal axis also, ($\sigma_v$ planes) then the molecule belongs to either $C_{pv}$ or $S_{2pv}$ (= $D_{pd}$).

(*c*) If there exists a mirror plane perpendicular to the principal axis ($\sigma_h$ plane), the point group to which the molecule belongs will be $C_{ph}$ and the molecule does not belong to $S_{2ph}$ point groups.

**Step 5:** If there exists one $p$-fold principal axis and there also exist $p$ number of other 2-fold axes perpendicular to the principal axis, the following alternative point groups may arise.

(*a*) Besides $p$ number of 2-fold axes perpendicular to the principal axis of order $p$, no other reflection plane ($\sigma_v$ or $\sigma_h$) are present, then the point group of the molecule will be $D_p$.

(*b*) Besides, $p$ number of 2-fold axes of (*a*), a mirror plane ($\sigma_h$ type) is present. Then the point group of the molecule is $D_{ph}$. These point groups will also have $\sigma_v$ planes.

(*c*) Besides $p$ number of 2-fold axes as in (*a*), there is no $\sigma_h$ plane (cf. *b*), there are mirror planes which contain the principal axis (*viz.* $\sigma_v$ planes) and which also bisect the angle between any two 2-fold axis that are perpendicular to the principal axis (*i.e.*, $\sigma_d$ planes) then the point group of the molecule is $D_{pd}$.

**Examples:** Ruthenocene (Eclipsed) [$Ru(C_5H_5)_2$]. The structure of ruthenocene is given in Fig. 2.32. In order to find the point group of this molecule, let us investigate stepwise as given in the preceding paragraphs.

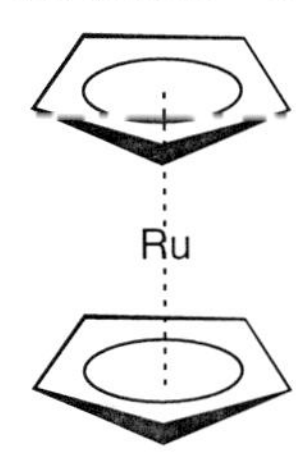


**Fig. 2.32**

**Step 1:** The molecule has a 5-fold principal rotation axis. So does not belong to any one of the following point groups $C_1$, $C_s$, $C_i$, $C_{\infty v}$, $D_{\infty h}$.

**Step 2:** The system has a 5-fold rotation axis and no other axis of the same order. So it does not belong to any of the following point groups $T$, $T_h$, $T_d$, $O$, $O_h$, $I$, $I_h$.

**Step 3:** It has one 5-fold axis and five 2-fold axes perpendicular to the principal $C_5$ axis. So it does not belong to $C_p$ or $S_{2p}$, $C_{pv}$, $S_{pv}$ or $C_{ph}$ point groups. It should belong to $D$-type point groups.

**Step 4:** It has only one 5-fold axis and five 2-fold axes perpendicular to $C_5$ principal axis. It therefore belongs to $D_{5h}$.

**Examples:**

(*a*) Prove that $[Cr\ (OX)_3]^{-3}$, $[Co(en)_3]^{+3}$ belong to $D_3$ point group.

(*b*) Prove that $S_8$ molecule belong to $D_{4d}$.

(*c*) Prove that trans $[Co\ (NH_3)_3\ (H_2O)_3]^{3+}$ belongs to $C_{2v}$ point group.

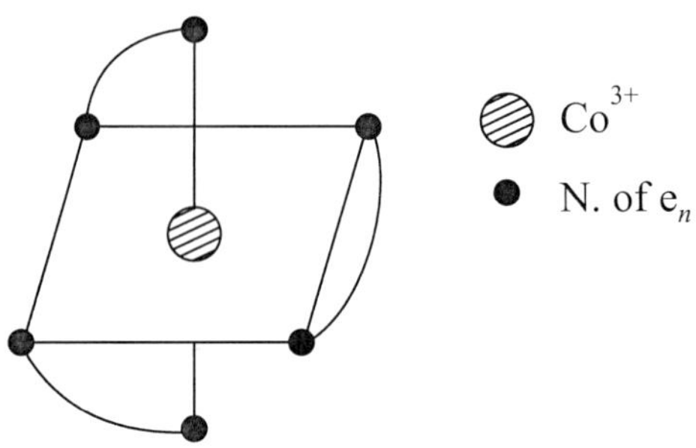


**Fig. 2.33** Structure of $[Co(en)_3]^{3+}$.

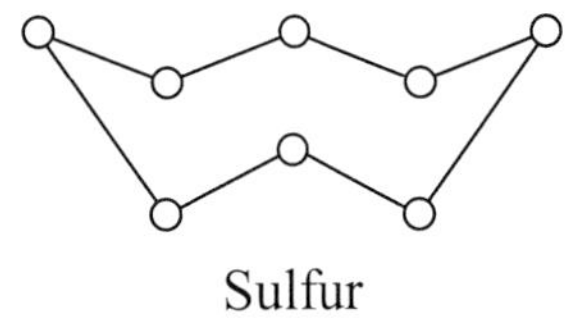


**Fig. 2.34** Structure of $S_8$ molecule.

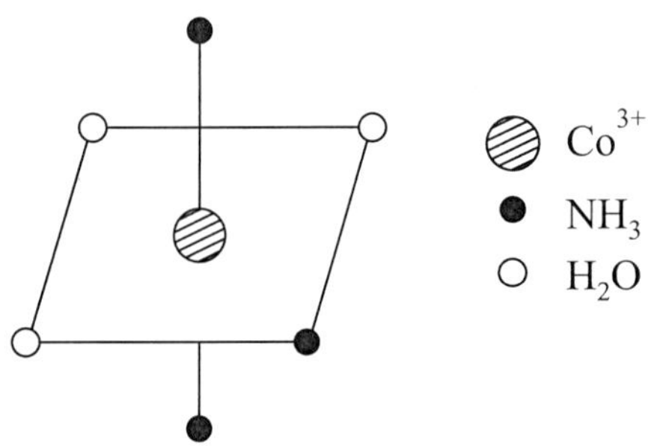


**Fig. 2.35** Structure of *trans* – $[CO(NH_3)_3\ (H_2O)_3]^{3+}$.

## 2.8 CLASSES OF SYMMETRY OPERATIONS

In the earlier sections of the chapter, we have defined a symmetry group, It is a set of symmetry operations with a definite law of binary combination of two symmetry operations, The symmetry operations of the set forming the group obey all the four axioms of a group (*viz.* closure, associative law, presence of identity element and presence of inverse of every element operation). It has been noted in chapter one that the members of the set forming a group can be further subgrouped into smaller sets termed as classes by the application of an operation which is termed as "Similarity Transformation" defined as given below.

$$B = X^{-1}AX$$

Here $A$, $B$, $X$, $B^{-1}$ are the members of the set. We express this relationship by saying that $B$ is the similarity transform of $A$. $A$ and $B$ are conjugate to each other.

Thus by the application of similarity transformation on all the symmetry operations of a group, we can collect all those members which are conjugate to each other. These smaller subsets of mutually conjugate elements will form a class. In this manner we can classify all the members (symmetry operations) of the set forming a group, into different "conjugacy classes". We will illustrate this process of classifying the members of set into classes by taking symmetry operations of $NH_3$ molecule.

The set of symmetry operations of $NH_3$ (Point group $C_{3v}$) is $\{E, C_3^1\ C_3^2, \sigma_v^{(1)}\ \sigma_v^{(2)}\ \sigma_v^{(3)}\}$. We will first write its multiplication table as shown in Table 2.11.

**Table 2.11:** Multiplication table for the $NH_3$ molecule

| $C_{3v}$ | $E$ | $C_3^1$ | $C_3^2$ | $\sigma_v^{(1)}$ | $\sigma_v^{(2)}$ | $\sigma_v^{(3)}$ |
|---|---|---|---|---|---|---|
| $E$ | $E$ | $C_3^1$ | $C_3^2$ | $\sigma_v^{(1)}$ | $\sigma_v^{(2)}$ | $\sigma_v^{(3)}$ |
| $C_3^1$ | $C_3^1$ | $C_3^2$ | $E$ | $\sigma_v^{(3)}$ | $\sigma_v^{(2)}$ | $\sigma_v^{(1)}$ |
| $C_3^2$ | $C_3^2$ | $E$ | $C_3^1$ | $\sigma_v^{(2)}$ | $\sigma_v^{(3)}$ | $\sigma_v^{(1)}$ |
| $\sigma_v^{(1)}$ | $\sigma_v^{(1)}$ | $\sigma_v^{(2)}$ | $\sigma_v^3$ | $E$ | $C_3^1$ | $C_3^2$ |
| $\sigma_v^{(2)}$ | $\sigma_v^{(2)}$ | $\sigma_v^{(3)}$ | $\sigma_v^1$ | $C_3^2$ | $E$ | $C_3^1$ |
| $\sigma_v^{(3)}$ | $\sigma_v^{(3)}$ | $\sigma_v^{(1)}$ | $\sigma_v^2$ | $C_3^1$ | $C_3^2$ | $E$ |

We construct another table showing the combinations $X^{-1}AX$ with the help of the above multiplication table.

Table 2.12 shows the combinations (or product $X^{-1}AX$ formed by using the multiplication table of $NH_3$. From Table 2.12, it is obvious that

1. $E$ is conjugate to itself. It implies that it forms one class containing $E$.
2. $C_3^1$ and $C_3^2$ are conjugate to one another, so a smaller subset $\{C_3^1\ C_3^2\}$ forms another class.
3. $\{\sigma_v^{(1)}\ \sigma_v^{(2)}\ \sigma_v^{(3)}\}$ are conjugate with each other, so the subset $(\sigma_v^{(1)}\ \sigma_v^{(2)}\ \sigma_v^{(3)})$ forms another class.

**Table 2.12:** Conjugate Elements

| $X$ | $X^{-1}EX$ | $X^{-1}C_3^1X$ | $X^{-1}C_3^2X$ | $X^{-1}\sigma_v^1X$ | $X^{-1}\sigma_v^2X$ | $X^{-1}\sigma_v^3X$ |
|---|---|---|---|---|---|---|
| $E$ | $E$ | $C_3^1$ | $C_3^2$ | $\sigma_v^{(1)}$ | $\sigma_v^{(2)}$ | $\sigma_v^{(3)}$ |
| $C_3^1$ | $E$ | $C_3^1$ | $C_3^2$ | $\sigma_v^{(3)}$ | $\sigma_v^{(1)}$ | $\sigma_v^{(2)}$ |
| $C_3^2$ | $E$ | $C_3^1$ | $C_3^1$ | $\sigma_v^{(2)}$ | $\sigma_v^{(3)}$ | $\sigma_v^{(1)}$ |
| $\sigma_v^{(1)}$ | $E$ | $C_3^2$ | $C_3^1$ | $\sigma_v^{(1)}$ | $\sigma_v^{(3)}$ | $\sigma_v^{(2)}$ |
| $\sigma_v^{(2)}$ | $E$ | $C_3^2$ | $C_3^1$ | $\sigma_v^{(3)}$ | $\sigma_v^{(2)}$ | $\sigma_v^{(1)}$ |
| $\sigma_v^{(1)}$ | $E$ | $C_3^2$ | $C_3^1$ | $\sigma_v^{(2)}$ | $\sigma_v^{(1)}$ | $\sigma_v^{(3)}$ |

Thus, the set of six operations forms three classes, *viz.* $(E)$, $(C_3^1\ C_3^2)$ and $(\sigma_v^1\ \sigma_v^2\ \sigma_v^3)$. Since each element in a subset class behaves in a similar fashion, so we know the properties of all the elements of a class by knowing the properties of one element of the class. The number of members in each class (subset) is written before the symbol representing the class. Thus, the list of symmetry operations of $NH_3$ *viz.* $\{\{E, (C_3^1\ C_3^2), (\sigma_v^1\ \sigma_v^2\ \sigma_v^3)\}$ is written as $\{E, 2C_3\ 3\sigma_v\}$. In the same way if one calculates the classes for $C_{4v}$ point group *viz.* $[E, C_4^1, C_4^3, C_2, \sigma_v^{(1)}, \sigma_v^{(2)}, \sigma_d^{(1)}, \sigma_d^{(2)}]$, one gets classes $\{E\}$ $\{C_4^1\ C_4^3\}$, $\{C_2\}$,$\{\sigma_v^{(1)}\ \sigma_v^{(2)}\}$,$\{\sigma_d^{(1)}\ \sigma_d^{(2)}\}$. The symmetry operations of $C_{4v}$ are written in terms of classes as $\{[E, 2C_4, C_2, 2\sigma_v, 2\sigma_d)$. Thus the symmetry operations of all point groups can be divided into classes and the operations are written in terms of classes. This division of operations of point group into classes of conjugate operations (or member) of the set is a natural subdivision of group using the abstract definition of class. This subdivision implies subdivisions having geometrically equivalent operations. For bigger group, this method of subdividing into conjugated classes is a little too long and cumbersome. But there are simpler rules for dividing the operations of a point group into its classes. Some of these rules are as follows:

1. A rotation group $C_p^m$ and its inverse $(C_p^{p-m})$ belong to the same class (and different class for each value of $m$ ($m = 1, 2, 3......m - 1$)) under the following conditions. Either the molecule has ($a$) $\sigma_v$ containing $C_p$ axis or ($b$) there is a 2-fold rotation axis perpendicular the $C_p$- axis. If these conditions are not satisfied then $C_p^m$, $C_p^{p-m}$ are in class by themselves. This rule is also true for $S_p$-axis.
2. Two $\sigma_v$ and $\sigma_d$ planes belong to the same class if there exists in the group a symmetry operation which moves all the points on one plane of symmetry to the corresponding point on the second symmetry plane. Thus, in example I, all the three $\sigma_v$ operations belong to the same class, because all the points on a $\sigma_v$ plane can be moved into corresponding points on $\sigma_v^{(2)}$ by $C_3^1$ operation and on $\sigma_v^{(3)}$ by $C_3^2$ operation.
3. Two rotations $C_p^k$ and $C_q^k$ about two different symmetry axes ($C_p$ and $C_q$) belong to the same class if the point group has a symmetry operation which moves all the points on one of the axis of rotation onto the corresponding points on the other.
4. The rule (3) is also satisfied by the improper rotation operation ($S_n^a$ and $S_q^a$).
5. If the point group contains $i$ ((inversion operation), it always forms a class by itself, since it commutes with every other operation of the group.

6. If the point group contains $\sigma_h$ it is always a class by itself.
7. In an Abelian group each operation forms a class by itself because each operation is conjugate only with itself *i.e.*, $X^{-1} AX = A$ in every case. As a practice, the students should take the operations of a $T_d$ point group $\{E, 4C_3^1, 4C_3^2, C_2', C_2'' \; 3S_4^1 \; 3S_4^2 \, \sigma_d^{(1)} \sigma_d^{(2)} \sigma_d^{(3)} \sigma_d^{(4)} \sigma_d^{(5)} \sigma_d^{(6)}\}$ and use the above rules to classify these operations into the following classes. $\{E, 8C_3, 3C_2, 6S_4, 6\sigma_d\}$.

From the above rules, one may say that the operations constituting a class are EQUIVALENT operations. By the words equivalent operations, we mean that the two operations *A* and *B* are said to be equivalent if one *e.g.*, *A* is converted into the other (*viz*. *B*) by applying operation *C* and *vice versa*. The latter statement provides a geometrical significance to the operations belonging to a class. This can be stated as follows. Let there be a vector *P* in a coordinate system (*m*) and we carry operations of *A* and *B* of the point group *G* on the vector *P*. These operators (*A* and *B*) of group *G* convert the vector *P* into vectors $P^A$ and $P^B$ in the coordinate system *m*. In this case coordinate system *m* is fixed and we move the vector, *P*). Now, if there is a third operation *C* belonging to the point group *G*, and if we apply this operator on the coordinate system (*m*) to obtain a new coordinate system *n*, keeping the vectors, *P*, $P^A$ and $P^B$ fixed, then the vector *P* in the coordinate system (*m*) will become *P′* in the new coordinate system *n*. Now in the new coordinate system if we apply operations *A* and *B* on vector *P′* then operator *A* will transform *P* into $P^b$ while *B* will transform *P′* into $P^A$. It implies that the roles of *A* and *B* on a vector are interchanged in the new coordinate system (*n*) from what they were in the old coordinate system (*m*). Thus, the operation *C* belonging to the point group *G* which converts coordinate system (*m*) into (*n*) and if *A* and *B* are also the members of the same point group *G* then *A* and *B* belong to the same class.

In a nutshell, one can generalise that the arrangement of operations of a point group into classes is to arrange them into sets of equivalent operations. These sets are the classes of the group.

One can also prove that " The order *k* of any class belonging to a group *G* is an integral factor of the order *h* of the point group *G* (*i.e.*, $h/k$ = whole number) (cf. Chapter 1).

## EXERCISE

**1.** Prove that if a group contains an element *Q* which commutes with every other elements *R*, then *Q* is in a class by itself.

**2.** Prove that every element of an Abelian group (*e.g.*, $C_n$) is in a class by itself *i.e.*, every class of an Abelian group is of order one.

## 2.9 SUBGROUPS

Until now we have discussed the concepts of a group (*i.e.*, a set with algebraic structure whose law of binary combination satisfies the group postulates). Let us now introduce the concept of a "subgroup" *i.e.*, groups within a group (for mathematical details *see* Chapter 1).

Let us again examine the multiplication table for the point group $C_{3v}\{E, C_3^1$ $C_3^2, \sigma_v^{(1)}\ \sigma_v^{(2)}\ \sigma_v^{(3)}\}\}$ given in Table 2.11. Let us consider the following subsets of the group. $\{E, C_3^1\ C_3^2\}$ $(\{\{E, \sigma_v^{(1)}\}$ $\{\{E, \sigma_v^{(2)}\}$ $\{E, \sigma_v^{(3)}\}$, $\{E\}$ which are of order 3, 2, 2, 2 and 1, respectively. These smaller subsets also satisfy the axioms of a group. Their multiplication tables are written as

| $C_3$ | $E$ | $C_3^1$ | $C_3^2$ |
|---|---|---|---|
| $E$ | $E$ | $C_3^1$ | $C_3^2$ |
| $C_3^1$ | $C_3^1$ | $C_3^2$ | $E$ |
| $C_3^2$ | $C_3^2$ | $E$ | $C_3^1$ |

| $C_s$ | $E$ | $\sigma_v^1$ |
|---|---|---|
| $E$ | $E$ | $\sigma_v^1$ |
| $\sigma_v^1$ | $\sigma_v^1$ | $E$ |

| $C_s$ | $E$ | $\sigma_v^2$ |
|---|---|---|
| $E$ | $E$ | $\sigma_v^2$ |
| $\sigma_v^2$ | $\sigma_v^2$ | $E$ |

| $C_s$ | $E$ | $\sigma_v^3$ |
|---|---|---|
| $E$ | $E$ | $\sigma_v^3$ |
| $\sigma_v^3$ | $\sigma_v^3$ | $E$ |

The symmetry elements of $C_{3v}$ are $\{\{\{E, C_3, \sigma_v^{(1)}\ \sigma_v^{(2)}\ \sigma_v^{(3)}\}$ while the symmetry elements of the subgroup $\{E, C_3^1\ C_3^2\}$ are $E$ and $C_3$ and the symmetry elements of the other three subgroups, (*viz.*) $\{E, \sigma_v^{(1)}\}\{\{E, \sigma_v^{(2)}\}$ and $\{E, \sigma_v^{(3)}\}$ are $E$ and mirror planes.

By our rules for defining a group, the sets $\{E\}$,$\{C_3^1, C_3^2\}$and $\{E, \sigma_v^1\}$, $\{E, \sigma_v^2\}$ $\{E, \sigma_v^3\}$ also form groups of order smaller than 6 which is the order of $C_{3v}$ point group. These are called the subgroups of the entire group. The first subgroup $\{E, C_3^1\ C_3^2\}$ is formed by removing the symmetry reflection planes from the parent group $C_{3v}$. The degree of symmetry is lowered. Similarly the subgroup $\{E, \sigma_v^1\}$ of order 2 is formed by eliminating $C_3$ axis, $\sigma_v^{(2)}$ and $\sigma_v^{(3)}$ planes from the point group $C_{3v}$. Similarly other subgroups are obtained by removing $C_3$ axis and $\sigma_v^{(1)}\ \sigma_v^{(3)}$ planes (Fig. 2.36).

There are a number of subgroups of the point group $D_{4h}$. Thus, $(E, C_2)$, $(E, C_2')$, $(E, C_2'')$ $(E, \sigma_v^1)(E, \sigma_d^1)$ *etc.*, are the subgroups of order 2 (Refer to Multiplication table of $D_{4h}$) point group that is given earlier.

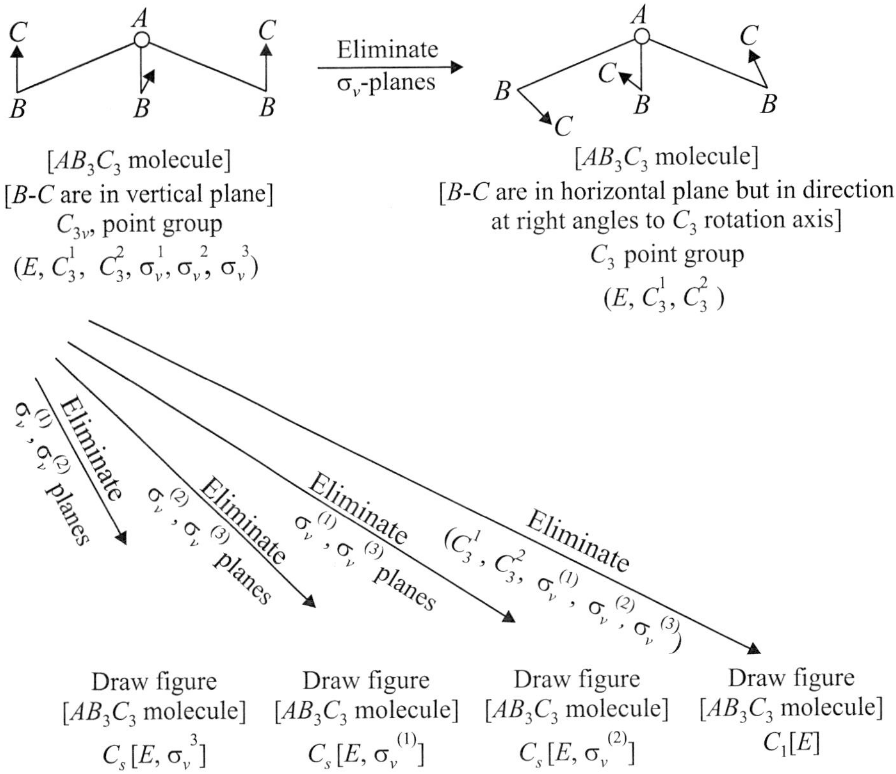


**Fig. 2.36** Lowering of symmetry from $C_{3V}$ to $C_3$, $C_s$ and $C_1$.

## EXERCISE

1. Find all the subgroups of the point group of $D_{4h}$.

It should be realised that every group contains two trivial subgroups-$\{E\}$ and the whole group itself. These two trivial subgroups are called "Improper" subgroups while the other subgroups except $\{E\}$ and (whole group) are said to be "Proper" subgroup.

## 2.10 EQUIVALENT ATOMS

Another concept which is of great utility in the application of symmetry to solve physical problems, is concerned with the symmetrically equivalence of atoms in a molecule. As a matter of fact, the word "Equivalence" has a mathematical meaning and is an extension of the word "Equality". The equality relation in the theory of sets has the property of reflexive, symmetric, and transitive. We will not, however, discuss these terms here. We would like to know simply the answer to the question. "How many atoms in a molecule belonging to a given point group G are equivalent *i.e.*, those atoms which are interchanged into one another by symmetry operations of the point group?" Naturally equivalent atoms must belong to the same chemical species. Thus, in $NH_3$ the three hydrogen atoms are equivalent because they are interchanged by the operations $C_3^1$ $C_3^2$, *etc.* Other examples of equivalent atoms include

(1) hydrogen atoms of $CH_4$, $C_6H_6$,

(2) fluorine atoms of $SF_6$ *etc.*

Examine $PF_5$ molecule. Though all the three atoms of the fluorine in the equatorial region can be interchanged by the symmetry operations of $D_{3h}$ point group, but the fluorine atom in the axial positions can not be interchanged with the flourine atoms present in the equatorial plane by any symmetry operation of the $D_{3h}$. Hence $F$ atoms in the equatorial plane of $PF_5$ are not equivalent to those in the axial positions. Similarly, hydrogen atoms in the $\alpha$ and $\beta$ positions of naphthalene are not equivalent.

It must be realised that in NQR and solid state physics, conformations of molecules in solids are generally fixed. Therein the molecular symmetry may not necessarily be the same as the crystal symmetry. Consequently a given pair of atoms may be equivalent in a single molecule, but they need not be equivalent in the solid.

While calculating rotational partition functions, one generally encounters with the terms "symmetry number". Symmetry number simply measures the number of equivalent positions which are attained by a complete rotation of the molecule about a symmetry rotational axis. Consequently it is just the order of the rotation axis.

## 2.11 EXAMPLE OF POINT GROUPS AND THE MOLECULES BELONGING TO THEM

A few examples of molecules with the point group assigned to them are given in Table 2.13.

**Table 2.13**

| *No.* | *Group* | *Pt GR* | *Examples* | *Symmetry operations* |
|---|---|---|---|---|
| 1. | Cyclic($C_n$) | $C_1$ | ($CH_3COC_2H_5$), $BF(C_6H_5)_2$ $Co(en)_2Cl_2$, CHFClBr | $E$ |
| 2. | | $C_2$ | $BF(C_6H_5)_2$, $NH_2NH_2$, $H_2O_2$ | [$E$, $C_2$] |
| 3. | | $C_3$ | $As(C_6H_5)_3$ | $E$, $C_3^1$, $C_3^2$ |
| 4. | | $C_4$ | Cyclo[$C_4(C_6H_5)_4$] | $E$, $C_2$, $C_4^1$, $C_4^3$ |
| 5. | | $C_6$ | $C_6(C_6H_5)_6$ | $E$, $C_6$, $C_3$, $C_2$, $C_3^2$ and $C_6^5$ |
| 6. | Cyclic($Cn_v$) | $C_{2v}$ | HCHO, 2-butene, Zeise's Salt, phenanthroline | $E$, $C_2^z$, $\sigma_v^{23}$, $\sigma_v^{yz}$ |

| | | | | |
|---|---|---|---|---|
| 7. | | $C_{3v}$ | $NH_3$, $CHCl_3$ $Cr(CO)_3(n^6\text{-}C_6H_6)$ | $E, 2C_3, 3\sigma_v$ |
| 8. | | $C_{4v}$ | $[(Cu(NH_3)_4ClBr]$ | $E, 2C_4, C_2, 2\sigma_v, 2\sigma_d$ |
| 9. | | $C_{5v}$ | $(C_5H_5)A$.(A on 5-fold axis) | $E, 2C_5^z, 2C_3^z, 5\sigma_v$ |
| 10. | | $C_{6v}$ | $(C_6H_6)Ag$(Ag on 6-fold axis) | $E, 2C_5^1, 2C_3^z, C_2^z, 3\sigma_v, 3\sigma_d$ |
| 11. | $C_{nh}$ | $C_{1h}$ | Metachlorobromobenzene | $E, \sigma_h$ |
| 12. | | $C_{2h}$ | Trans-dichloroethylene | $E, C_2^i, \sigma_h, (\equiv i.C_2)$ |
| 13. | | $C_{3h}$ | $M(OH)_3$ | $E, C_3, C_3^2, \sigma_h, S_3, S_3^2$ |
| 14. | | $C_{4h}$ | $(Cu(NH_3)_4)$ | $E, C_4, C_2, C_4^3$, i, $S_4^3, \sigma_h, S_4^1$ |
| 15. | | $C_{5h}$ | – | $E, C_5, C_5^2, C_5^3, C_5^4,$ $\sigma_h, S_5, S_5^7, S_5^9, S_5^3$ |
| 16. | | $C_{6h}$ | Hexamethylbenzene (6 H of $CH_3$ in one plane) | $E, C_6, C_3, C_2, C_3^2, C_6^5,$ $i, S_3^5, S_6^5, \sigma_h, S_6, S_3$ |
| 17. | $S_n$ | $S_2 \equiv C_i$ | 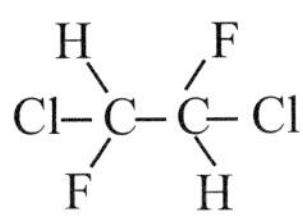 | $E, i$ |
| 18. | | $S_4$ | Boat form $A_4B_4$ (see text of the book) | $E, C_2, S_4, S_4^3$ |
| 19. | | $S_6$ | Chair form $C_6H_{12}$ | $E, C_3, C_3^2, i, S_6^1, S_6^5$ |
| 20. | $D_n$ (Dihedral) | $D_1 \equiv C_2$ | | $(E, C_2)$ |
| 21. | | $D_2$ | $C_6H_5C_6H_5$ Two $C_6H_5$ planes at right angles | $E, C_2^x, C_2^y, C_2^z$ |
| 22. | | $D_3$ | $CH_3\text{-}CH_3$(staggered), $Co(en)_3$ | $E, 2C_3, 3C_2$ |
| 23. | | $D_4$ | | $E, 2C_4, C_2, 2C_2', 2C_2''$ |
| 24. | | $D_5$ | | $E, 2C_5, 2C_5^2, 5C_2$ |
| 25. | | $D_6$ | | $E, 2C_6, 3C_2, 3C_2'',$ $2C_3, C_2$ |
| 26. | $D_{nd}$ | $D_{2d}$ | Allene two H-C-H planes (at right angles) | $E, 2S_4, C_2, 2C_2', 2\sigma_d$ |
| 27. | | $D_{3d}$ | $CH_3\text{-}CH_3$(staggered) | $E, 2C_3, 3C_2, i, 2S_6, 3\sigma_d$ |
| 28. | | $D_{4d}$ | $Mn_2(CO)_{10}$ | $E, 2S_8, 2C_4, 2S_8^3,$ $C_2, 4C_2', 4\sigma_d$ |
| 29. | | $D_{5d}$ | Staggered $Fe(n^5\text{-}C_5H_5)_2$ | $E, 2C_5, 2C_5^2, 5C_2,$ $i, 2S_{10}^3, 2S_{10}, 5\sigma_d$ |

| | | | | |
|---|---|---|---|---|
| 30. | | $D_{6d}$ | Staggered $Cr(n^6\text{-}C_6H_6)_2$ | $E, 2S_{12}, 2C_6, 2S_4,$ $2C_3, 2S_{12}^5, C_2\ 6C_2', 6\sigma d$ |
| 31. | $D_{nh}$ | $D_{2h}$ | $C_2H_4$, $B_2H_6$, $C_6H_5\text{-}C_6H_5$ | $E, C_2^x, C_2^y, C_2^z, i, \sigma^{xy},$ $\sigma^{yz}, \sigma^{xz}$ |
| 32. | | $D_{3h}$ | $NO_3^-$, $BF_3$, $PF_5$, Anthracene | $E, 2C_3^z\ 3C_2^z\ 3C_2', \sigma_h^{xy},$ $2S_{5,}\ 3\sigma_v$ |
| 33. | | $D_{4h}$ | Cyclobutadiene, $Cu(NH_3)_4\text{-}Cl_2$ Cu(pthalocyanine) | $E, 2C_4^z, C_2, 2C_2', 2C_2'',$ $i, 2S_4, \sigma_h, 2\sigma_v, 2\sigma_d$ |
| 34. | | $D_{5h}$ | Eclipsed $Fe(n^5\text{-}C_5H_5)_2$, Cyclopentadiene | $E, 2C_5, 2C_5^2, 5C_2, \sigma_h,$ $2S_5, 2S_5^3, 5\sigma_v$ |
| 35. | | $D_{6h}$ | Benzene Eclipsed $Cr(n^6\text{-}C_6H_6)_2$ | $E, 2C_6^z, 2C_3, C_2, 3C_2',$ $3C_2'', i, 2S_3, 2S_{6,}\sigma_h,$ $3\sigma_v, 3\sigma_d$ |
| 36. | | $T_d$ | $CH_4$,$CCl_4$ | $E, 8C_3, 3C_2, 6S_4, 6\sigma_d$ |
| 37. | | $O_h$ | $MA_6$, $S_6$, $SF_6$ | $E, 8C_3, 6C_2, 6C_4, 3C_2,$ $i, 6S_4, 8S_6, 3\sigma_h$ |
| 38. | | $I_h$ | $B_{12}H_{12}$ | $E, 12C_5, 12C_5^2, 20C_3,$ $15C_2, i, 12S_{10},$ $12S_{10}3, 20S_6, 15\sigma_h$ |
| 39. | | $C_s$ | $SOCl_2$, HOCl, 1-Chloronaphthalene, $CH_2ClF$ | $E, \sigma_v$ |

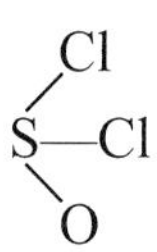


($C_s$)

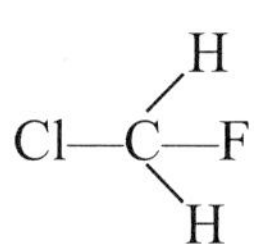


($C_s$)

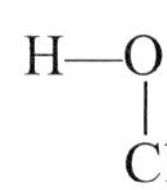


($C_s$)

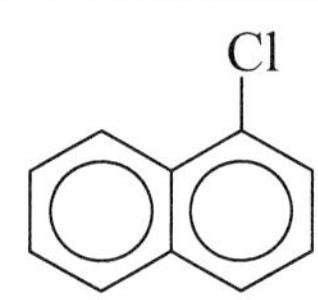


($C_s$)

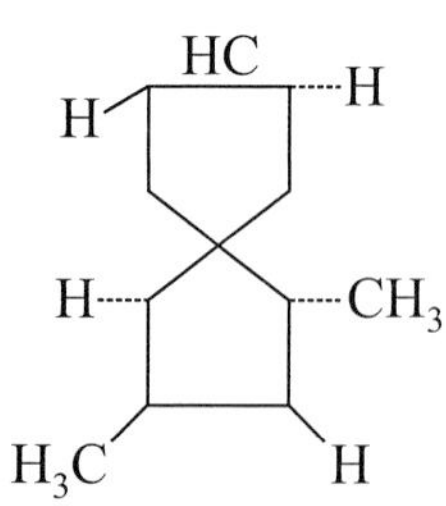


($S_4$)

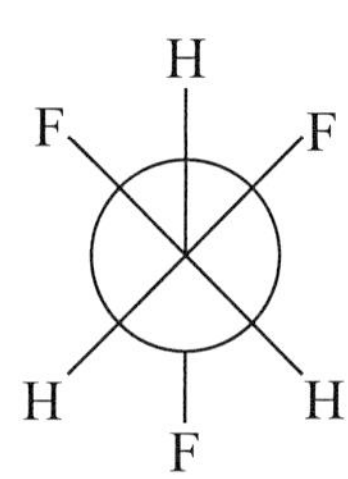


($C_3$)

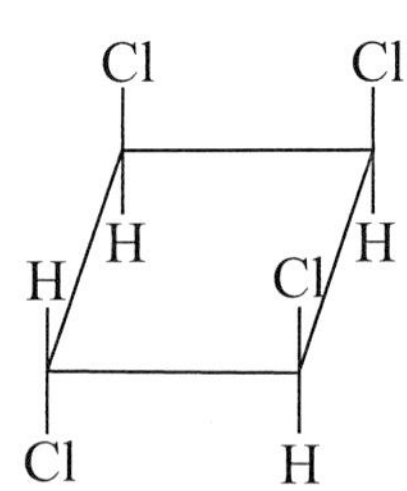


($C_4$)

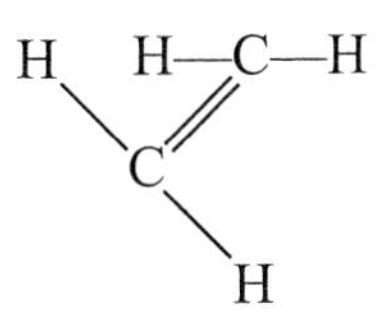

($D_2$)

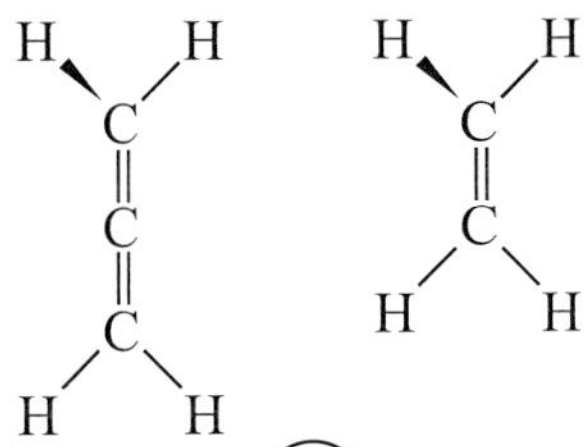

($D_{2d}$)

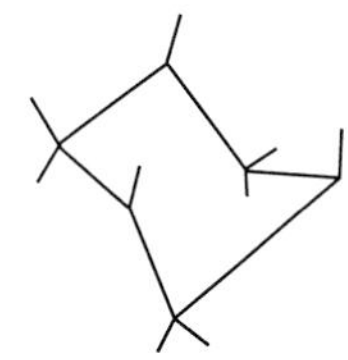

($D_{3d}$)

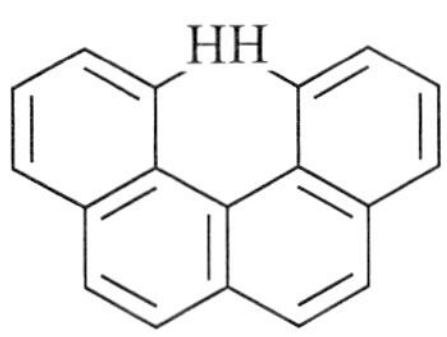

($C_2$)

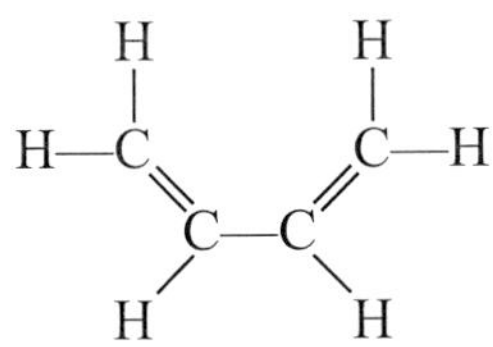

($C_{2v}$)

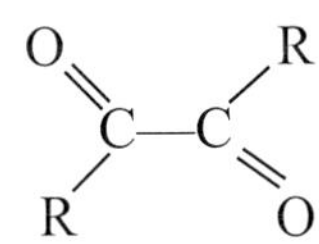

($C_{2h}$)

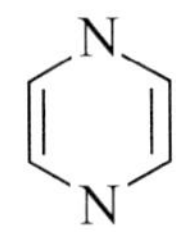

($D_{2h}$)

## 2.12 THE CRYSTALLINE STATE (SPACE GROUPS)

**Lattices:** In the preceding a sections, we have been mainly concerned with the geometrical symmetry elements and symmetry operations of molecules which leave a point in the system (centre of gravity) invariant and hence with the point groups. When we extend our study to the symmetry properties of the crystals, we add new operations to the point group operations in the treatment of solids. These include those which repeat the basic unit throughout the crystal. Such operations are "Translation Operations". To illustrate them, let us take some arbitrary molecule $AB_2$ (*e.g.*, $H_2O$, or HOCl *etc.*) and repeat it as many times as you wish after moving a distance $x$ in a straight line in one or two directions. This is shown in Fig. 2.37.

If we now displace the centre of gravity of each molecule [or O-atoms of $H_2O$ or any other point of $H_2O$], we obtain what we term as *Lattice* of $AB_2$ (*e.g.*, $H_2O$). This is shown in Fig. 2.37. In this lattice system ($Q$ of Fig. 2.37) we can choose a certain area and consider the area as unit of repetition, we call this area as "*Unit Cell*". This area (unit cell) can be chosen in a number of ways (Fig. 2.38). In the figure, we have a choice of choosing area *a*, *b*, *c*, *d*, *e*, *etc*. In the unit cell *a*, *b*, *c*, each vertex of the unit cell has one point, but this point is shared by four unit cells, and each point belong only ¼ of the point to one cell. So the total point in one cell is only one (*i.e.*, $4 \times \frac{1}{4}$). This type of unit cell is called the *Primitive cell* (one point per unit cell). But in the unit cell *d*, the cell contains two points, or we may choose other cells, which contain three, four or more points. Such unit cells are termed as multiple (double or triple *etc.*) cells. The two translations $x_i$ and $x_j$ along which the lattice repeats are known as conjugate translations. The choice of unit cell is totally arbitrary. It is purely a matter of convenience.

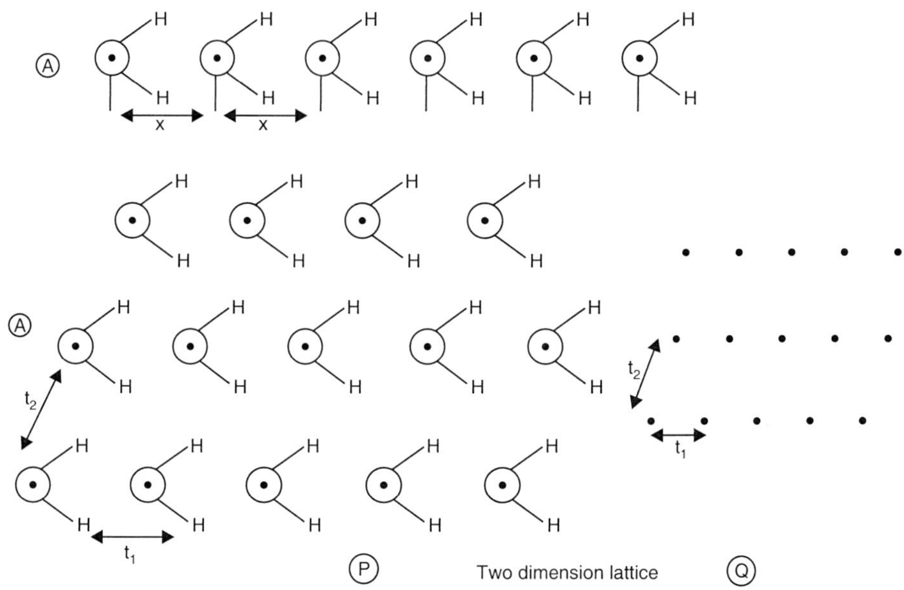


**Fig. 2.37** Lattices in one and two dimensions.

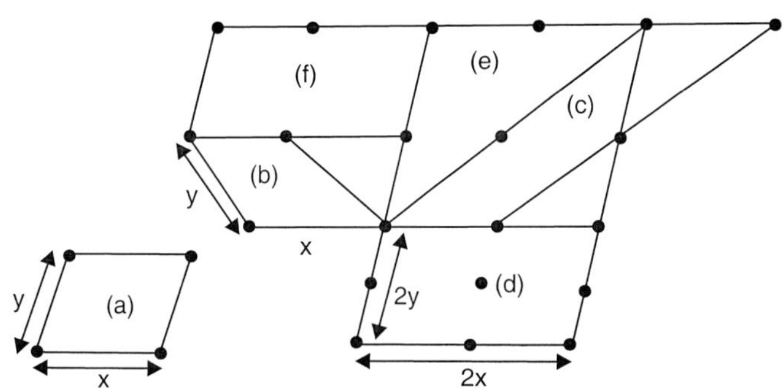


**Fig. 2.38** Unit cells in two dimensions.

This lattice can of course, be extended in the three dimension by a third translation $x_k$, which is not coplanar with $x_i$ and $x_j$. Similarly a three-dimensional lattice array of real molecule can also be obtained *e.g.*, for $H_2O$, HOCl, *etc.* In three dimensions, there are fourteen such arrangements of points. Analogous to two-dimensional lattices we assign a small volume and a lattice point which we term as unit cell. This repeats in the three dimension and fill up the whole space. Similarly the lattice array of real object or molecules repeated by translations which can be distinguished from the lattice of imaginary points extended into a third direction.

**Symmetry properties of unit cells (Point groups):** Suppose we take the case of three-dimensional lattice with HOCl molecule placed at each lattice point (Fig. 2.39). Since HOCl has no symmetry in the two dimension, and absence of any element of symmetry within the plane allows us to make a *general lattice* with no symmetry with different unit lengths *a* and *b* and any angle α in between *a* and *b*. But, if we choose a parallelogram unit cell and at such vertex we keep a molecule having a centre of symmetry (*e.g.*, *trans*-dichloroethylene), then the unit cell will be having a centre of symmetry (Fig. 2.40).

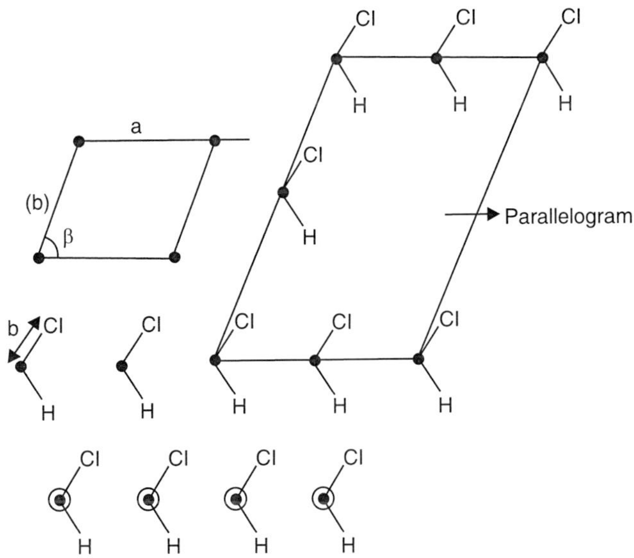


**Fig. 2.39** Two-dimensional Lattices.

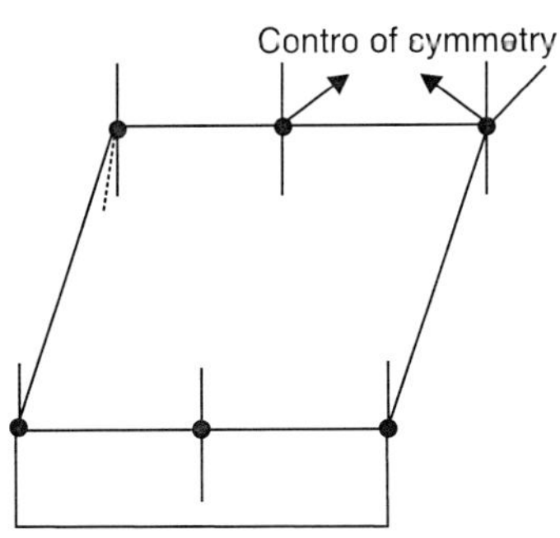


**Fig. 2.40** Lattice *P* [having a centre of symmetry] parallelogram.

If the lattice has a plane of symmetry (*m*) it will require the lattice points to lie on rows both parallel and perpendicular to *m*, resulting in either a rectangular

($a = b$, $\alpha = 90°$) or a diamond shape lattice *i.e.*, ($a = b$). Both these lattices also accommodate second plane and 2-fold axis. [Fig. 2.41 (*a*, *b*)]. Generally diamond shape lattice is replaced by a double rectangular lattice [Fig. 2.41 (*b*)] having a lattice point at the centre, body centered or *C*-centered lattice. A special case of diamond shaped lattice is when the angle ($\alpha$) 60° or 120° which is known as rhombic lattice, which can accommodate 3-fold or 6-fold axis, and three mirror planes. Similarly, one can also have a square lattice ($a = b$, $\alpha = 90°$). This lattice can accommodate all symmetry operations of a square. These five lattices (*viz.* lattice with no symmetry, lattice with 2-fold axis, lattice with 3-or 6-fold axis, lattice with 4-fold axis) with a specific number of mirror planes are the only lattices allowed in two-dimentional lattices.

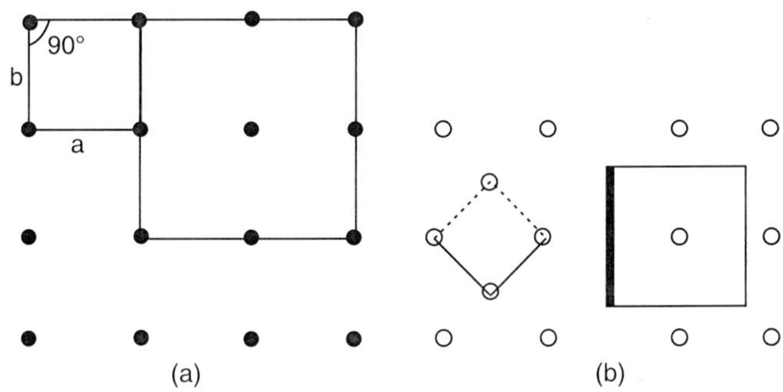


**Fig. 2.41** (*a*) Rectangular lattices (*b*) Diamond shaped and rectangular lattices.

It can be proved that the only possible rotation axes allowed in two-dimensional lattices are 1-, 2-, 3-, 4- and 6-fold, and no other axes are possible. These generate a specific number of mirror planes. A particular number of mirror planes is associated with a particular rotation axis.

The rotational operations along with mirror reflection operations form a point group. For a two-dimentional lattice one can have only a certain number of planar point groups (10 in number). Thus, every point of a particular two-dimentional lattice is associated with a particular point group with lattice points which remain invariant during symmetry operations.

Let us now consider the symmetry of crystals with the lattice as an imaginary construct in three dimensions consisting of an infinite number of geometric points in space. These lattice points are arranged in such a manner that the environment of a point (*i.e.*, the arrangement of points about this point) is the same for every point of the lattice. It also implies that the environment about a point has symmetry and that symmetry is the same around all the points. All the symmetry operations associated with the environment around a lattice point form a point group. As described above in the two-dimensional cases, it can be shown that in the case of three-dimensional lattices also, only a certain number of point groups are compatible with a particular lattice structure.

In addition to the above rotational symmetry of lattice points, we also have translational symmetry. Since all lattice points are equivalent, the lattice points can be translated in the three directions (*a*, *b* and *c*) by an amount of (*ma, nb*, and *pc* respectively), (*n*, *m* and *p* are whole numbers) in such a way that the displaced lattice and the lattice prior to displacement are invariant. Such lattices are called "Primitive translations". Since primitive translations can be "binary combined" in a way as to satisfy the group postulates, thus the set of such translations form a translation groups, However, there are some non-primitive translations (*viz. m*, *n* and *p* are fractional) which by themselves do not form a group; but the rotation about this axis in combination with translation operation along this axis form a symmetry operation. These operations are known as *Screw axis operations* [Fig. 2.42(*a*)]. If the non-primitive translations are combined with reflection (*m*), then the combined operations are known as "Glide planes" [Fig. 2.42(*b*)].

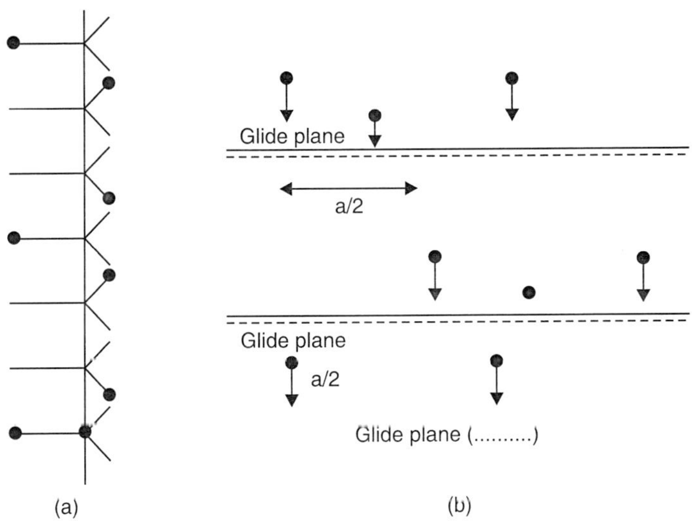


**Fig. 2.42** (*a*) Screw axis 3-fold (*b*) Screw axis and glide plane.

Using the rotational symmetry alone, there are a total of seven different lattice systems (Rotations are restricted to 1-, 2-, 3-, 4- and 6-rotational axis.) Similar to two-dimensional lattice there are seven systems (cubic, orthogonal, trigonal, rhombic hexagonal, tetragonal, monoclinic, triclinic). These seven crystal systems are classified according to the *point group symmetry* of the lattice points. Some of these crystal systems are compatible with more than one lattice (Table 2.14). These are called *Bravais Lattices* and are a total of 14 in number.

**Table 2.14:** Bravais Lattices

| *Crystal System* | *Their Symmetries* | | |
|---|---|---|---|
| | *Crystal Classes* | *Bravais Lattice* | *Lattice Group* |
| Cubic | $O_h$ $T_h$ $T_d$ $T$ | $P\,I\,F$ | $O_h$, $T_h$ |
| Tetragonal | $D_{4h}$ $C_{4h}$ $D_2$ $C_{4v}$ $D_4$ $C_4$ $S_4$ | $P\,I$ | $D_{4h}$, $C_{4h}$ |
| Orthorhombic | $D_{2h}$ $C_{2v}$ $D_2$ | $P\,I\,C\,F$ | $D_{2h}$ |
| Trigonal | $D_{3d}$ $D_3$ $C_{3v}$ $S_6$ $C_3$ | $P$ | $D_{3d}$, $S_6$ |
| Hexagonal | $D_{6h}$, $C_{6h}$, $D_{3h}$, $C_{6v}$, $D_6$, $C_{3h}$, $C_6$ | $P$ | $D_{6h}$, $C_{6h}$ |
| Rhombic | | | |
| Monoclinic | $C_{2h}$, $i$, $C_2$, $C_s$ | $P\,C$ | $C_{2h}$ |
| Triclinic | $C_1$ $C_i$ | $P$ | $C_i$ |

There are three points which are not clear from the table:

1. The point group symmetries associated with a particular system are many (*e.g.*, for cubic system, the point groups associated with it are $O_h$, $T_d$, $T_h$ $O$ and $T$). How are the different point groups associated with one system?
2. Generally one of the point groups out of a set of different point groups associated with a particular class of crystal are the subgroups of one main point group. Thus, $T_h$, $T_d$, $O$ and $T$ are the subgroups of $O_h$. How the $C_{4v}$ which is also a subgroup of $O_h$ does not belong to cubic system?
3. Lattices of crystals are the network of imaginary points. But in real crystals, the lattice points are occupied with molecules which have either lower or higher symmetries of the lattice point. How are the two point groups associated with the same lattice point?

To answer the first question, let us take the cubic system which can have three different lattices, [*viz. P* (primitive)], *I* (body centred cubic), and *F* [(face-centred cubic)]. The point group of a cube is $O_h$. In the primitive lattice, each corner of the cubic has a point which has $O_h$ point group. However, you must also realise that the body centre of each cube forming the lattice also has $O_n$ symmetry and therefore by adding a lattice point, we do not change the symmetry of the lattice and so we may also have a body centred cubic lattice. According to the same arguments, the centre of each face of a cubic lattice has also $O_n$ symmetry, and therefore, face centred cubic lattice (*F*) is also possible. However, we can not have lattice points on the corner, body centre as well as face centre simultaneously, because by placing the points both on body centre and also on face centre we do not create a new lattice. We only end up with a smaller unit cubic cell (primitive type) by joining the body centre, face centres and the corner points that is why we get the different types of lattices for the same point group of the lattice point.

To answer the third question prior to the second, we know that our lattice points are purely imaginary points and we have considered imaginary lattices formed by these geometrically equivalent imaginary points. In reality we have a crystal and on each imaginary point we keep some objects (*e.g.*, atoms molecules, or ions). We term these objects as the *basis* of the crystal. Now these basis (objects molecules, or atoms) have a *certain symmetry e.g.*, symmetry of atoms as basis has a higher symmetry (sphere) than $O_h$ or the basis may be an ion take $CO_3^{2-}$ which has lower symmetry than the site (lattice point) symmetry. Therefore, the site of the basis may be lower or higher than the site (lattice point) symmetry. There is a rule "The symmetry of basis in a crystal cannot be higher than the site symmetry". Thus, if the symmetry of the basis is higher, then the crystal forces reduce the symmetry of basis until it is not higher than the site symmetry. Now how much the symmetry of the basis is lowered depends upon the second rule and which answers our second question. The second rule is based on the following principles:

(*a*) Only certain basis symmetry are compatible with the symmetry of a given crystal system (like cubic, orthogonal, *etc*.).

(*b*) These *basis symmetries* which are *compatible* with a particular crystal system are termed as *crystal classes*.

(*c*) The symmetry of a crystal class is given by a *subgroup* (or the whole group) of the symmetry (or point group) of the *crystal system* to which it belongs. Thus, the symmetry of crystal class is lower than *crystal system*, and then the former must be a subgroup of the whole group of the system.

(*d*) It is also possible that the crystal class (or a subgroup of the parent group) may belong (or be a sub group of the group) to a *crystal system* having lower symmetry for example $C_{4v}$ point group is subgroup of both the symmetry of orthogonal as well as to a *cubic* system. In such cases the rule is that the crystal class (*i.e.*, $C_{4v}$) will belong to the lower symmetry *i.e.*, the one which has lower symmetry *i.e.*, orthogonal system. In a case where the crystal class belongs to many crystal systems, then that crystal class will belong to a crystal system having LOWEST symmetry. That explains why $C_{4v}$ is the crystal class of orthorhombic system and does not belong to cubic system although it is subgroup of both systems.

(*e*) The total number of crystal classes that are compatible with the symmetry of all the crystal system is thirty-two (Table 2.14). These are known as thirty two crystallographic point groups.

We now consider those crystals that contain more than one type of atoms or molecules in the unit cell and these occupy sites other than the points of the Bravais lattice. What is the symmetry of such sites? In these cases, the symmetry of the site depends upon its position in the unit cell. For example, a general site on the edge of a cubic lattice (*e.g.*, along *a*-axis) is invariant under the operations of $C_{4v}$. It will, therefore, have $C_{4v}$ symmetry. A general point of the diagonal of a cube is invariant under the operations of $C_{3v}$ symmetry. An object (or basis) which is spherical, when free if present on diagonal cannot take the symmetry greater than the site. It will, therefore, have the symmetry of $C_{3v}$ which is the symmetry of the site in the crystal.

Until now we have not considered translational glide plane and screw axis symmetry and used rotational (proper and improper) reflections and inversion operations only. Now the space group in which we consider only rotations (proper and improper), mirror planes, and translations only (no glide plane and screw axes), we call such space groups as "Symmorphic". In those space groups, where we consider glide planes and screw axes along with other symmetry operations (*viz*. rotations, reflections, and translations are called non-symmorphic space group. There are 73 symmorphic and 157 non-symmorphic space groups making a total of 230 space groups. (We will not discuss them in at this point as the subject matter of the book is limited to molecular symmetry). The discussion of these space groups can be found in any book of crystallography.

Another important point worth mentioning is that the *X*-ray diffraction patterns of most crystal are centro-symmetric whether or not the crystals have a centre of symmetry or not. Those crystallographic point groups which have a centre of symmetry are of great significance. Those space groups having the centre of symmetry are known as *Laue groups*.

In the following sections, we will describe some of the important symmetries found in nature. The description will be very brief and sketchy.

## 2.13 SOME IMPORTANT SYMMETRIES

For a long time during the development of symmetry principles, discrete geometric symmetries, described in the preceding paragraphs and their application in arts *etc*., were the only ones which have been extensively employed. The applications of continuous symmetries in three dimensions began at the beginning of the 20$^{th}$ century which has brought a breakthrough in the thinking of physicists. Besides the use of continuous symmetries, there are a number of other discrete and continuous symmetries which are not geometrical in nature. We will not discuss these types of symmetries here and will make a simple passing reference of some of these just to familiarise the way of thinking of the scientists. These are:

**(*a*) Space-time (or Fundamental) Symmetries:** This type of symmetry may be thought of a kind of geometrical symmetry in four-dimensional space. One considers space and time as one entity and thus enlarging the three-dimensional world of the space coordinates $(x, y, z)$ with a fourth coordinate, *viz.* $t$ ( time). We have discussed the geometrical symmetries in three dimensions by various transformations (*viz.* rotational, translational and reflections *etc.*). In space-time space, we consider symmetry transformations in four dimensions *viz.* $(x, y, z, t)$. These transformations are termed as *Lorentz transformations.* These relate the transformed coordinates $(x', t')$ to the original coordinates $(x, t)$ for a body moving in the $x$-direction with velocity $v$ by the following equations:

$$x' = \frac{x - vt}{\sqrt{1 - v^2/c^2}}$$

and

$$t' = \frac{t - (v/c^2)x}{\sqrt{1 - v^2/c^2}}$$

Here $c$ is speed of light.

These space-time transformations are very important and played a major role in the development of special theory of relativity. Because of their immense importance, these are also known as Fundamental Symmetries.

Although one can graphically display the space coordinate transformations $(x, y, z)$ but the space time coordinate transformations cannot be easily displayed. The only way to recognise them is by investigating the consequences that the $(x, y, z, t)$ symmetry transformations have on physical phenomena. For example, the length of a body moving with velocity $v$ in the $x$-direction appears to be contracted in the reference frame moving with the body. Thus, one studies the various consequences to study the symmetry of system in four dimensions.

**(*b*) Permutation Symmetry:** This symmetry is relevant to the description of a set of identical objects. For example, the physical or any other property of two identical objects remains invariant by the exchange of these objects. This symmetry played a very significant role in the development of quantum mechanics. Thus, the properties of a body like, electrons, atoms, protons or nuclei are described by a wave functions $\psi$, which depend upon coordinates of the particles $(x, y, z)$ forming the set *i.e.*, $\psi$ (1, 2, 3...). If the members of the set are identical like electrons, then although the function, $\psi$, remains invariant if we exchange the coordinates of two particles; but the sign of the wave function may either remain the same or the sign may change. Thus, we have two types of symmetries

(*i*) $\psi$ (1, 2......) = $\psi$ (2, 1......) ......(Bosons )

or (*ii*) $\psi$ (1, 2......) = (–) $\psi$ (2, 1......)...... (Fermions)

Thus, these particles for which $\psi$ (1, 2......) remains invariant are known as Bosons and those for which $\psi$ (1, 2......) change sign are known as Fermions. The sets of permutation transformations are denoted by $S$ and these follow the group postulates. Thus, the set of these transformations form a group which are known as permutation groups, and denoted by $S_n$. These are termed as symmetric groups. (See chapter one for further discussion).

**(*c*) Dynamic Symmetries:** The notion of this type of symmetry is rather complex. We can see its consequences only by which we investigate. Thus, let us examine the symmetry of a plane circle (Fig. 2.43) which belongs to point group $O(2)$. The set of this point group operations has all the rotation operations around an axis perpendicular to the plane of the circle and which passes through to centre $C$. Now it so happens that by some reasons (usually of dynamic origin) the symmetry is lowered or broken. But the condition is that the breaking or lowering of symmetry is not at random but is done in such an ordered way that the figure (*e.g.*, circle) remains invariant under a set having a smaller number of rotation operations. For example, if we tie the opposite ends of the rim of circle at four places as shown in Fig. 2.43 at right angles to each other or a wheel (circle) with parts of one metal (white) and with parts of another metal (white) (Fig. 2.43). This will cause the circle or the wheel to remain invariant only under the set of rotations by angles multiple of 90°. It will lower the symmetry of the circle to $C_n$. This kind of ordered breaking of symmetry caused *usually by dynamic reasons*, are termed as *Dynamic symmetry*. It implies that the symmetry of the system in static state is the higher which become lower when in dynamic state (Fig. 2.43).

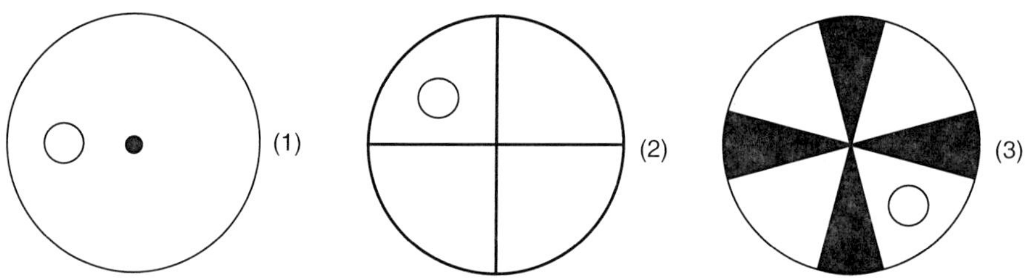


**Fig. 2.43** Colour symmetry.

These types of symmetries are very useful because one can get detailed understanding of a complicated physical system. This symmetry became very popular after Gell-Mann and Nem independently in 1961 used it to describe the properties of *Hadrons*.

Geometric symmetries act on the familiar three-dimensional space. It implies that operators of the geometric symmetry group are those under which the potential energy of the particles remains invariant. However, there are operations which involve simultaneous transformations of coordinates (static) and momenta (dynamic) and which leave the Hamiltonian invariant as a *whole*. This dynamic symmetry act usually on some abstract space and therefore, one can display visually only by the effects of these symmetries. Thus those systems which are described by quantum mechanics (dynamic state), the effect of dynamic system is displayed on the eigenstates of the system. For example, for H-atom, the energy of the quantum states are classified as (1*s*), (2*s*, 2*p*) (3*s* 3*p* 3*d*)) *etc.*, these are labeled by the letter *n*, (2*l* + 1). The important point regarding the effect of dynamic symmetry is the regular pattern of these energy levels. The similar effects have been observed in the elementary particle physics and its example is provided by the energy level diagram of these particles where also a regular pattern has been observed which is caused by the dynamic symmetry. It is to be noted that the dynamic symmetries of baryons led to the development of the quark model in which baryons themselves are formed out of more fundamental particle, called quarks.

Dynamic symmetries are related to the combination of objects rather than to single objects. For example, the constituents of nuclei *viz.* protons and neutrons, bind together in pairs (*viz. s* and *d*). Thus, the nuclei can be treated as an assembly of paired and unpaired particles, the dynamic properties of pairs provides classification scheme for the energy levels of nuclei.

One should note here that most symmetries encountered in science are not exact but broken so one generally observes the dynamic symmetries in physics as broken, rather then exact.

***(d)* Gauge Symmetry:** Gauge Symmetry is another kind of symmetry which is commonly used in physics present in the laws of electrodynamics (Maxwell's laws). In terms of (Maxwell's Laws), the most of the basic interactions of nature possess this symmetry. It is described by a scalar $\phi$ and a vector $A$ potential. The laws of electrodynamics are invariant under the set of following transformations.

$$\bar{A} \rightarrow A + \bar{V}A$$

$$\phi \rightarrow \phi - \frac{1}{c}\frac{\partial A}{\partial t}$$

This invariance is called gauge invariance. When this invariance is applied to quantum mechanics the electrodynamics becomes quantum electrodynamics. It has been further seen that the week and strong interactions are also governed by this symmetry. In the latter case, the theory is known as quantum chromo dynamics.

***(e)* Time Reversal Symmetries:** This is another type of symmetry which has been very extensively used by physicists.

***(f)* Super Symmetries:** The seed for the development of this type of symmetry was sown in 1971 in an attempt to further unify the laws of physics. As we have seen earlier (permutation symmetry), that there exists two types of particles (Fermions and Bosons) in nature. All the theories of symmetries deal either Bosons or Fermions. There was no theory which deals with both types of particles. Now one may start with an enquiry as to what happens if bosons and fermions are exchanged in terms of permutation symmetry. The symmetry which studies such exchanges between bosons and fermions is termed as *Super Symmetry*. There are two applications of the super symmetry until now. The first application is in the study of space-time symmetry (*viz.* space-time super symmetry). The second application is applicable to dynamic symmetry, (termed as Dynamic Super Symmetry) in which this super symmetry (dynamics) is applied in study of energy levels of nuclei. In such a study, the pairs of particles in the nuclei (see dynamic symmetry applies to nuclear particles) behave like bosons while others that remained as singles behave like fermions. While applying super symmetry to such pairs and single particle, the invariance of the theory under some set of transformations of boson into bosons, fermions into fermions and *bosons into fermions* produce certain regular patterns in the energy levels spectra.

Super symmetry concept is highly mysterious as compared to normal symmetry we will not discuss it anymore.

**(*g*) Quasi-Symmetries:** This type of symmetry is present where there is no symmetry but there remains still some regularity. The majority of examples where quasi symmetry is applied are *geometric quasi symmetry*. The latter being geometric in nature can be visually displayed.

**Conclusion:** In this chapter on symmetry, we have discussed mainly geometric symmetry, though there are other kinds of symmetries that have been extensively used in the development of science, particularly physics as per Weyl's statement, "One can construct very elaborate symmetries starting from very simple symmetry like Left/Right symmetry. The abstractness of the concept of symmetry is obvious, especially those symmetries which do not act on three-dimentional space, now the basic motivation for the study of the symmetry". None-the-less the basic motivation for the study of symmetry in science is the ancient motivation that "*The Beauty is bound up with Symmetry*". This motivation has been further extended following P.A.M. Dirac statement that "*If the theory of Nature is beautiful, it must be true*".

# 3

# Vector Spaces and Representations

In chapter two, we have defined symmetries of stationary molecules of solid objects by a set of symmetry operators which transform them into their self-coincidence. The set of operations belonging to a molecule form a point group assigned to it. The principles of group theory are then applied. This way of classifying molecules or systems into point groups is of little use to us. To understand the behaviour or the properties of the systems under the action of symmetry operators, we require the algebraic formulations of symmetry operators. Therefore, a method has to be developed so that the abstract nature of the point group could be reduced to manipulatable numbers while solving physical problems. The theory of representation, a branch of abstract linear algebra—provides us a powerful tool for the purpose. This chapter is concerned with such a study.

## 3.1 MEANING OF REPRESENTATION

One may define in all its generality a representation of a point group as "*Any Realisation of an Abstract Group*" by certain matrices which in turn, reflect the structure of the point group. It implies that the multiplication table formed by the set of matrices associated with the operators of the group must be identical with that formed by the operators themselves. The abstract group is then said to be represented by the set of these matrices. The group formed by the set of matrices is homomorphic or isomorphic to the abstract group. The preference of matrix representation over any other realisation of a group is because of the following reasons.

(*a*) Many groups have their natural representations in terms of matrices.

(*b*) Matrices are handy mathematical structures because we know their relatively easier algebraic manipulations.

(*c*) The product of two matrices is generally not commutative and so they can serve to represent faithfully non-Abelian groups.

By this procedure, one constructs a matrix vector space with matrices as vectors and whose base vectors are formed by a maximal set of linearly independent matrices.

Any system (*e.g.*, a molecular framework) can be defined as a collection of points which may be represented by fixed point vectors with respect to an arbitrarily chosen coordinate system with origin at a point which remains unshifted by the symmetry operations. In that coordinate system, any symmetry operation in the form of an operator acting on points (or point vector) of the system will transform them into new vectors in the same vector space (*e.g.*, $\Gamma \overline{P} = \overline{P}'$ ) where $\Gamma$ is a symmetry operator, $\overline{P}$, any vector and $\overline{P}'$, transformed vector).

**Note:** We expect the students to be familiar with elementary knowledge of various phase spaces and matrix manipulations.

One can extrapolate this concept to all the point vectors of the system forming the concerned space. Mathematically one can express it as,

$$\Gamma\{\overline{P}_1, \overline{P}_2, \ldots, \overline{P}_n\} = \{\overline{P}_1', \overline{P}_2', \ldots, \overline{P}_n'\}$$

where $\{\overline{P}_1, \overline{P}_2, \ldots\}$ are the vectors forming the system and $\{\overline{P}_1', \overline{P}_2', \ldots\}$ are the transformed vectors after applying the symmetry operations and belonging to the same vector space. These transformed vectors $\{\overline{P}_1', \overline{P}_2', \ldots\}$ can be represented by linear combinations of the original basis vectors (*viz.* $\overline{P}_1, \overline{P}_2, \ldots, \overline{P}_n$ ). It implies a relationship between the vectors of a vector space belonging to a system (*e.g.*, a molecule) and the symmetry operations applied on the vectors of the space.

The vector space may not only be defined in terms of point vectors in three-dimensional space but also to spaces which are more than three–dimensional (*e.g.*, a configuration space or/and a function space where the vectors of the space are functions of variable coordinates). At the heart of much of the theory of group representation, there lies a union of a group concept with that of the linear vector space (in general). The above classification and characterisation of all possible point groups by a set of matrices apparently seem to be an impossible task. Firstly because in constructing the representation, we may use vector spaces of any dimension and secondly the choice of the basis vectors in a particular vector space on which the effects of operators are to be seen is purely arbitrary. Because of these latter factors, it will be possible to construct many numerically different sets of matrices for the same point group which multiply together according to the same rules as applied to construct the multiplication table of the point group under consideration (cf. Chapter 2).

However, this problem of classification may be solved firstly by regarding all the choices of basis as essentially equivalent and secondly by proving that all equivalent representations can be described in terms of smaller irreducible representations that not only depend on the choice of basis vectors and also on the size of matrices that represent a particular group. Therefore, the end result is that only a set of simple rules can be formed for using these irreducible representations. In this chapter and the following one, we will discuss the formulation of reducible and irreducible representations[1].

## 3.2 VECTOR SPACES (RECAPITULATIONS)

**(A) Physical Space:** We have seen that the geometry of the framework of any molecule describes its symmetry. If we wish to discuss the effect of symmetry[2] of molecules on their various properties, like I.R., electronic spectra, *etc.*, we have to talk about such things as their wave functions in the language of vector spaces (mathematical aspect of which is a little too involved so as to discuss it here in detail). Its basic concept however arises from the consideration of the properties of ordinary three-dimensional space. (You have already gone through in your elementary classes a few of its related properties). We will recapitulate them with the aim of continuity of topic under discussion, (*viz* representations of groups).

For fixing a point in space, we choose any one arbitrary point which we use it as the origin of the coordinate system. Related to this origin, a fixed frame of reference (coordinate system) is arbitrarily chosen and then the position of any point $P$ is known by the length of the point $P$ from the origin $O$ and its direction by the values of its coordinates $(x, y, z)$ from the three axes. We then call the directed line $OP$ as the position vector of $P$ and denote it by $\overline{P}$. Thus, we can have an infinite number of vectors $\overline{P}, \overline{Q}, \overline{R}, \overline{S}, \ldots$ *etc.* in space (one vector for each point of space) which describes their positions in space. The geometric relations between their position vectors can be represented by the following algebraic equations.

---

1. Hereafter Irreducible representation will be designated by I.R. and infrared spectra by *i.r.*
2. Symmetry is an operator.

**1. Equality:**

(*a*) The position vectors of two points *P* and *Q* are equal in both length and direction (*e.g.*, $\overline{P} = \overline{Q}$), when two points coincide with each other.

(*b*) In any vector space (*e.g.*, our three-dimensional space), we can add any two vectors $\overline{P}$ and $\overline{Q}$ to get a third vector $\overline{R}$ using parallelogram law. $\left[\overline{R} = \overline{P} + \overline{Q}\right]$.

(*c*) The order of writing the two vectors [*viz.* ($\overline{P} + \overline{Q}$ or $\overline{Q} + \overline{P}$)] for addition is totally irrelevant.

(*d*) The addition of vectors can be extended to any number of vectors and the resultant after addition of all the vectors $\overline{X}$ can be written as,

$$\overline{P} + \overline{Q} + \overline{R} + \overline{S} + \ldots = \overline{X}$$

(*e*) The addition of vectors observes the associative law.

$$(\overline{P} + \overline{Q}) + \overline{S} = \overline{S} = \overline{P} + (\overline{Q} + \overline{S}) = (\overline{P} + \overline{Q} + \overline{S})$$

(*f*) The space has a zero vector. It implies that the position vector of the origin (0, 0, 0) is itself a zero vector such that $\overline{0} + \overline{P} = \overline{P}$ or $\overline{0} + \overline{Q} = \overline{Q}$ *etc.* Its operation is irrelevant.

(*g*) In view of the case (*f*) (*viz.* the presence of zero vectors), we can define the inverse of any vector. Thus, we can define a vector $(-\overline{P})$ such that $\overline{P} + (-\overline{P}) = \overline{0}$.

$(-\overline{P})$ is a vector whose length is equal to that of $\overline{P}$ but whose direction is opposite to that of $\overline{P}$.

(*h*) The above properties indicate that the set of all the vectors of the three-dimensional space form a group with respect to addition of two vectors.

(*i*) Multiplication of any vector $\overline{P}$ by any, scalar *a* (number *a*, where *a* is a real number) we get the vector $(a\overline{P})$ which is also a vector of the space. The number *a* can be positive, negative or zero. In brief, such operation will be called scalar multiplication (of a vector by a scalar).

Besides above properties, the following laws are also fulfilled.

$$b(a\overline{P}) = ab\overline{P}$$

$$(a+b)\overline{P} = a\overline{P} + b\overline{P} \quad \ldots \text{(distribution law)}$$

Prior to defining a vector space, let us define a field. Let *F* be a set of scalars (*e.g.*, numbers *a*, *b*, *c*, …) and we define two binary operations (*viz.* addition and multiplication). The set *F* forms a field (*i*) if *F* is an Abelian group under addition with identity 0; and (*ii*) non-zero elements of *F* also constitute an Abelian group under multiplication with 1 as identity element. Now, using the definition of field and the cases ($a$ to $i$) above, we define a vector space ($L$), A set of element { $\bar{P}, \bar{Q}, \bar{R},...$ } defined over a field $F$ forms a vector space if the following conditions are satisfied:

- $L$ forms an Abelian group with respect to vector addition.
- Any scalar of the field $F$ and any element of $L$ { $\bar{P}, \bar{Q}, \bar{R}...,$ } can be combined by an operation called "scalar multiplication" to give an element of $L$ such that for every $\bar{P}, \bar{Q}, \bar{R}...$ belonging to $L$, we also have,

$$\left.\begin{aligned} & a(\bar{P}+\bar{Q}) = a\bar{P}+a\bar{Q} \\ \text{and}\quad & (a+b)\bar{P} = a\bar{P}+b\bar{P} \\ & a(b\bar{P}) = ab\bar{P} \\ \text{and}\quad & (1).\bar{P} = \bar{P} \text{ and} \\ & 0\bar{P} = (\bar{0}) \end{aligned}\right\} \text{which belong to } L$$

Here $a$, $b$, 0 and 1 are scalars and $\bar{0}$ is a vector. The elements of vector space are called vectors.

There are different vector spaces two of them are described below.

**(B) Function Space:** One of the many types of vector spaces is a function space. The vectors of this space are functions. The condition for the set of functions to form a vector space is that these functions should have some properties common to all members of the set (*i.e.*, functions). These function spaces have a large number of applications in quantum mechanics.

Let there be a set of functions $f$, $g$, $h$, ... of a variable $x$. Thus, if the set forms a function (*e.g.*, vector) space then we should have the following:

1. The addition of two functions (*e.g.*, $f$ and $g$) produces a third function which is also a member of the set *i.e.*,

$$h(x) = f(x) + g(x) \qquad ...(3.1)$$

2. The function $h$ should be function of $x$ and eq. (3.1) is satisfied for every value of $x$,

$$h(x) = f(x) + g(x) \qquad ...(3.2)$$

3. The set should have *zero function* which means, that the function takes the value "Zero" for every value of $x$.
4. If the function space has zero function in the set of functions, we can use it to define $(-f)$ which should also be a member of the set. It implies that $[(f) + (-f)](x) = 0\,(x) = 0$. That is, it takes the value zero for every value of $x$ (zero function).

5. We can define the multiplication of the function by a scalar $a$, to get a new function $(af)$, such that $(af)$ is a function which for each value of $x$ has the value $[a.f(x)]$.
6. For any two arbitrary scalars, $a$ and $b$, the functions follow the following rules:

$$\begin{aligned}(a + b) f(x) &= af(x) + bf(x) \\ a(f(x) + g(x)) &= af(x) + ag(x) \\ a(bf(x)) &= abf(x)\end{aligned} \qquad ...(3.3)$$

The resulting functions after the operations with scalars $a$ and $b$, are also the members of the set.

Thus, the set of all such functions over a field of scalars forms a function (vector) space. We give below an example of function space.

**Example 1:** We know that the wave functions $\psi_i$s representing the states of an atom, an electron or a molecule in a state of energy $E$ are obtained by solving Schrödinger equation.

$$\hat{H}\psi = E\psi, \qquad ...(3.4)$$

where

$$\hat{H} = \sum_{k=1}^{n}(h^2/8\pi^2 me)\nabla_k^2 + V$$

Here $\hat{H}$ is a linear operator.

If $\psi$ and $\phi$ are two functions of the same variables defined over $\hat{H}$ with the energy $E$ and $a$ and $b$ are two scalars, then the following relations hold:

$$\left.\begin{aligned}\hat{H}(a\phi) &= a(\hat{H}\phi) \\ \hat{H}(\phi + \psi) &= (\hat{H}\phi + \hat{H}\psi) \\ \text{and} \quad a(b\hat{H}\phi) &= ab(\hat{H}\phi)\end{aligned}\right\} \qquad ...(3.5)$$

(Here, the common property which remains invariant is the energy.)

We therefore conclude that the set of all possible wave functions $\{\psi, \phi, a\psi, a\phi, b\psi, b\phi, (a + b)\psi, \phi\,(a + b)...\}$ of the system with the *same energy E* forms a function (vector) space over a scalar field.

**(C) Configuration Space:** To understand the concept of configuration space, we start with a molecule consisting of $N$ nuclei and $n$-electrons. This is a very important space as far as the applications of symmetry to the molecular problems are concerned. In order to understand its concept, let us first take an orthonormal coordinate system with unit vectors $e_1, e_2, e_3,..., e_n$ into the three-dimensional physical space of molecule in which the $n + N$ particles move. Let us take the origin of the coordinate system at a point which remains unshifted during the symmetry operations of the molecule. With respect to this coordinate system, the position of $k^{\text{th}}$ particle is described by ($x_1^k$, $x_2^k$, $x_3^k$ ) ($k$ = number of the $k^{\text{th}}$ particle = 1, 2, 3, …, $n + N$). Thus, the coordinates of all the particles can be written as $\{x_1^1, x_2^1, x_3^1, x_1^2, x_2^2, x_3^2, \ldots, x_1^{n+N}, x_2^{n+N}, x_3^{n+N}\}$. Thus, any particular set of the $x$ ($x_i^k$) values describes a particular configuration of the molecule at a particular instant during their motion.

We assume that as these ($n + N$) particles move, the coordinates of all the particles change, and we get a different configuration. In this way, the configuration of the molecule goes on changing with the time. One can have a relationship between the time and the configurations of the molecules. It therefore implies that there must exist a relationship between time and the coordinates ($x_1^i$, $x_2^i$, $x_3^i$) ($i$ = 1,…, $n + N$), of the particles. Thus, a set of ($n + N$) numbers describes a configuration of the molecule.

Instead of regarding *these* 3($n + N$) *numbers (i.e., coordinates) of* ($n + N$) *particles moving in a common three-dimensional space, we may also regard them as components or coordinates of a single vector in a vector space of* 3($n + N$) *dimensional space.* Thus, in 3($n + N$) dimensional space, any vector will be represented by 3($n + N$) numbers which become the coordinates of a vector in a 3($n + N$) dimensional space. We are thus able to find a method to express a point in 3($n + N$) dimensional space by 3($n + N$) coordinates with respect to 3 ($n + N$) basis unit vectors ($e_1$, $e_2$,…, $e_{3n+3N}$). Therefore, each point in the 3($n + N$) dimensional space represents one configuration of the molecule. Such a 3($n + N$) dimensional space is known as configuration space of the molecule.

In order to determine these ($n + N$) coordinates in configuration space, we assume,

(*i*) That there are ($n + N$) coordinate systems located at each ($n + N$) particles separately,

(*ii*) That there are ($n + N$) number of origins (one origin at each atom) with 3($n + N$) coordinate axes and with 3($n + N$) basis unit vectors $(e_1^k, e_2^k, e_3^k)$.

(*iii*) There will be 3($n + N$) coordinate for ($n + N$) particles measured with respect to their *own* coordinate systems. We will therefore be having a set of 3($n + N$) coordinates to locate the position of ($n + N$) particles at a particular moment (or instant) of time. We will thus be able to describe a state of the molecule at a particular moment by ONE point in the configuration space.

We know that particular state of motion of the ($n + N$) particles of the molecule moving simultaneously will be described by a wave function which will be a function of all the coordinates of the molecules as variables. Thus, wave function will give one numerical value of $\psi(x_i^k)$, $i$ = 1, 2, 3, $k$ = 1, 2, 3,…, ($n + N$) at each point *i.e.*, a numerical value at each point of the configuration space of the molecule.

Thus, in dealing with molecular problems, we are concerned with vector space of different types. These are,

(*i*) Physical space in which the particles forming the molecules are moving,

(*ii*) The configuration spaces of molecules as a whole,

(*iii*) The function spaces associated with the various energy levels of the molecule.

Although to solve Schrodinger equation with $(n + N)$ particles is impossible but the Born-Openheimer approximation allows us to discuss their different motions separately

(*i*) The translational and the rotational *motions* of the *particles* in the framework of the molecule as a whole and thus we require three coordinates only.

(*ii*) The vibrational motions of the nuclei forming the framework of the molecule, about the equilibrium positions of each nuclei.

(*iii*) The motions of the electrons moving in the fixed framework of the nuclei in which we use $3n$-dimensional configuration space for the $n$ electrons with the coordinates $x_1^1, x_2^1, x_3^1, x_1^2, x_2^2, x_3^3, x_1^n, x_2^n, x_3^n$ as variables and keeping the nuclei $3N$ coordinates as fixed. In solving the vibrational problems, we use $3N$-dimensional coordinate space $x_1^1$, $x_2^1$, $x_3^1$, $x_1^2$, $x_2^2$, $x_3^2$, $x_1^N$, $x_2^N$, $x_3^N$ as the variables which correspond to the displacements of the nuclei from their equilibrium positions.

Thus, in all the above problems we are concerned with the framework formed by the nuclei of the molecule in their equilibrium position. The symmetry of the framework can therefore be helpful in classifying the vectors of these vector spaces. The method of doing it is described in the following chapters.

### 3.2.1 Base Vector and Coordinate System Defined in Vector Space

Let us start with a three-dimensional space. We can choose any three non-coplanar vectors $(\bar{P}, \bar{Q}, \bar{R})$ in this space. Because of a unique correspondence between the position vector $\bar{A}$ for any point of space and the set of three numbers $(a_1, a_2, a_3)$ which will characterise the position vector $\bar{A}$, any vector of the three-dimensional space can be written as a linear sum of three vectors

$$\bar{A} = a_1 \bar{P} + a_2 \bar{Q} + a_3 \bar{R} \qquad \ldots(3.6)$$

Thus, by choosing three definite non-coplanar unit vectors $\bar{P}, \bar{Q}, \bar{R}$ we have set up a coordinate system. In this coordinate system, the vectors $\bar{P}, \bar{Q}, \bar{R}$ are known as "Base Vectors" and $a_1, a_2$ and $a_3$ are the coordinates of a point.

In order to make sure that $\bar{P}$, $\bar{Q}$ and $\bar{R}$ are non-coplanar vectors, one must see that the only possible solution of the equation

$$\alpha \bar{P} + \beta \bar{Q} + \gamma \bar{R} = 0 \text{ is } \alpha = \beta = \gamma = 0, \qquad \ldots(3.7)$$

Conclusively, non-zero values of these coefficients can be found satisfying (eq. 3.7). It implies that the zero vector ( $\overline{0}$ ) can be obtained when $\alpha = \beta = \gamma = 0$. If it does not, let us assume that $\alpha \neq 0$, then we can divide R.H.S. and L.H.S. of the equation by $\alpha$ and get the,

$$-\overline{P} = \left(\frac{\beta}{\alpha}\right)\overline{Q} + \left(\frac{\gamma}{\alpha}\right)\overline{R} \qquad \ldots(3.8)$$

which causes $\overline{P}$ to lie in the plane, formed by $\overline{Q}$ and $\overline{R}$ but that contradicts the hypothesis that $\overline{P}, \overline{Q}$ and $\overline{R}$ are non-coplanar vectors. When the conditions imposed on eqs. (3.7) and (3.8) are satisfied, we say that the vectors $\overline{P}, \overline{Q}$ and $\overline{R}$ are linearly independent or these vectors are not the functions of each other. It also implies that among the infinite number of choices of these vectors ( $\overline{P}, \overline{Q}, \overline{R}$), we can choose only three vectors in the three-dimensional space. We express it by saying that our space of vectors has three dimensions.

The above concept can be generalised for any vector space. You will see that spaces with which we are concerned will have a finite number of dimensions say, $n$. In any $n$-dimensional space, we will find only $n$ linearly independent vectors (*viz.* $\overline{T}_1$, $\overline{T}_2$, $\overline{T}_3, \ldots, \overline{T}_n$) and these $n$-linearly independent vectors will form the basis for the coordinates and any vector $\overline{B}$ in that space can be written as a linear combination of these $n$ vectors;

$$\overline{B} = \alpha_1\overline{T}_1 + \alpha_2\overline{T}_2 + \alpha_3\overline{T}_3 + \ldots \alpha_n\overline{T}_n \qquad \ldots(3.9)$$

For some numbers ($\alpha_1$, $\alpha_2$, $\alpha_3$, ..., $\alpha_n$), these numbers are called the components of the vector $\overline{B}$ with respect to above chosen coordinate system and with vectors $T_1$, $T_2$, $T_3, \ldots, T_n$ as unit vectors along the coordinate axes.

**Example 2:** Consider the function space of the all the solutions of the differential equation.

$$\frac{d^2y}{dx^2} = -y \qquad \ldots(3.10)$$

The eq. (3.10) has two solutions (1) $f_1 = \cos x$ and $f_2 = \sin x$. These solutions ($f_1$, $f_2$) are linearly independent since it is impossible to satisfy the equation $\alpha_1 \sin x + \alpha_2 \cos x = 0$ with any fixed non-zero values of $\alpha_1$ and $\alpha_2$ for every value of $x$. Thus, if we choose $\sin x$ and $\cos x$ as basis coordinate system in our function space, then any vector $\overline{V} = \alpha_1 \sin x + \alpha_2 \cos x$ will be represented by $\alpha_1$ and $\alpha_2$. The vector $\overline{V}$ is also a solution of the differential equation. Thus, every solution of (eq. 3.10) written as $\alpha_1 \sin x + \alpha_2 \cos x$ will behave like a vector in two-dimensional vector space with $\sin x$ and $\cos x$ as basis of chosen coordination system. The vector will be written as $\alpha_1 \sin x + \alpha_2 \cos x$ and $\alpha_1$ and $\alpha_2$ as components of the vector.

In general, we have chosen a set of linearly independent base vectors in a vector space. The operations of vector addition and multiplication of vectors by scalars are expressed in terms of components ($\alpha_1, \alpha_2, \alpha_3, \ldots, \alpha_n$) and components ($\beta_1, \beta_2, \beta_3 \ldots, \beta_n$). Thus, if a vector $\bar{R}_1$ is written as ($\alpha_1, \alpha_2, \alpha_3, \ldots, \alpha_n$) and another vector $\bar{R}_2$ as components ($\beta_1, \beta_2, \beta_3 \ldots, \beta_n$). Then,

$$\bar{R}_1 + \bar{R}_2 = (\alpha_1 + \beta_1)(\bar{T}_1) + (\alpha_2 + \beta_2)(\bar{T}_2) + \ldots + (\alpha_n + \beta_n)(\bar{T}_n),$$

where $\bar{T}_1, \bar{T}_2, \ldots, \bar{T}_n$ are the base vectors and $(\alpha_1 + \beta_1), (\alpha_2 + \beta_2), \ldots, (\alpha_n + \beta_n)$ of vector $\bar{R}_1 + \bar{R}_2$ and the components of $\bar{R}_1\bar{R}_2$ $(= (\alpha_1\beta_1)\bar{T}1 + (\alpha_2\beta_2)\bar{T}2 + \ldots$ $+ (\alpha_n\beta_n)\bar{T}n)$ as $\alpha_1\beta_1$, $\alpha_1\beta_2$.

### 3.2.2 Orthogonal Coordinate System

We have seen in the preceding paragraph that in a three-dimensionl physical space, the three non-coplanar vectors will serve as basis for a coordinate system and the directions of basis vectors define the directions of three coordinate axes. So whenever we say that a vector $\bar{R}$ has the components ($\alpha_1, \alpha_2, \alpha_3$) we assume that the length of the base vectors are of unit length. But how do we define a unit length and an angle between two vectors. For this purpose, we have to make our vector space as a *metric space* or an *inner product* space. A vector space over a given field $F$ is called a product space, if we impose one more condition (namely, the concept of inner product) on the vector space. The condition is the inner product that is defined on the space *i.e.,* with every pair of elements (vectors) $\bar{P}$ and $\bar{Q}$ of the space, there is associated a unique number belonging to the field $F$ which is denoted by $(P \cdot Q)$ (inner or scalar product) of $\bar{P}$ and $\bar{Q}$ satisfying the following properties:

(*i*) $(\bar{P} \cdot \bar{Q}) = |\bar{P}||\bar{Q}| \cos\theta = (\bar{Q} \cdot \bar{P})$

(*ii*) $(\alpha\bar{P} \cdot \beta\bar{Q}) = (\alpha\beta)\ |P|Q|\cos\theta = (\alpha\beta)(\bar{P} \cdot \bar{Q})\infty$ ...(3.11)

and (*iii*) $\bar{R}.(\alpha_1\bar{P} + \alpha_2\bar{Q}) = \alpha_1(\bar{R}.\bar{P}) + \alpha_2(\bar{R}.\bar{Q})$

Now, eq. (3.11) can define the length and the angle between any two vectors $\bar{P}$ and $\bar{Q}$. Thus taking the inner product of $\bar{B}$ with itself, we get from eq. (3.9).

$$(\bar{B} \cdot \bar{B}) = \sum_{i=1}^{n} \alpha_i^2 (\bar{T}_i, \bar{T}_i) + \sum_{i>j} \alpha_i \alpha_j (\bar{T}_i \bar{T}_j) = \| B \|^2 \cos\theta = \| B \|^2 \quad \ldots(3.12)$$

If we assume $\bar{T}_i$'s as unit vectors with angle between $T_i$ and $T_j$ as 90°, then eq. (3.12) becomes,

$$\| B \|^2 = \sum_{i=1}^{n} (\alpha_i)^2 \quad \ldots(3.13)$$

Thus, according to eq. (3.13) the length of $B$ can be either +ve or (–ve). The positive value of the root (*viz.* $|B|$) is called the 'norm' of the vector. We have also defined the angle between the two vectors *e.g.*, $\overline{P}$ and $\overline{Q}$ as

$$\cos\theta = \frac{(\overline{P}.\overline{Q})}{|P||Q|} \qquad ...(3.14)$$

Using the above conditions and applying them in our three-dimensional space, we assume, that

1. Three non-planar vectors serve as basis vectors.
2. The directions of the base vectors define the directions of three coordinate axes.
3. The lengths of the base vectors define a unit of length for measurement along each axis *i.e.*, $\left(\left|\overline{T}_q\right|\right) = +1$.
4. If we choose (not a condition) the angle between two base vectors of unit length to be equal to 90° (*i.e.*, they are mutually perpendicular to each other), then $(T_i \cdot T_j)$ becomes zero. Such vectors are known as *Orthogonal vectors*.
5. For $\overline{B} = \alpha_1 T_1 + \alpha_2 T_2 + \alpha_3 T_3$

$$\|\overline{B}\| = \left\|\sum_{i=1}^{n}(\alpha_i)^2\right\|^{\frac{1}{2}}$$

6. We denote such base vectors by $\overline{e}_1, \overline{e}_2$, and $\overline{e}_3$. Here $\overline{e}_i$ are vectors of unit length and are mutually perpendicular to each other.

We have, thus, defined the unit vectors with 90° angle between them in three-dimensional space. But this concept may be extended to any dimensional vector space. Only way to extend the concept in more than three-dimensional space, function space *etc.*, is to redefine our concept of inner product. Any definition of scalar product that we choose, must satisfy the following conditions.

$$\left.\begin{aligned} &(i)\ (b\overline{P}\cdot b\overline{S}) = b(\overline{P}\cdot\overline{S}) \\ &(ii)\ [\overline{P}\cdot(\overline{Q}+\overline{R}) = (\overline{P}\cdot\overline{Q}) + (\overline{P}\cdot\overline{R})] \\ &(iii)\ (\overline{P}\cdot\overline{S}) = 0 (\text{If } \overline{P} = 0, \text{or } \overline{S} = 0 \\ &(iv)\ (\overline{P}\cdot\overline{P}) \geq 0 \end{aligned}\right\} \qquad ...(3.15)$$

### 3.2.3 Orthonormal Vectors

Let us see how we can apply the above condition (eq. 3.15) to our function space. Let us first define normalisation of a vector. Suppose for $(\overline{P} \neq 0)$, we have,

$$(\overline{P}\cdot\overline{P}) = N^2$$

then $$\overline{P}' = \frac{1}{N}(\overline{P}) = 1 = \left[\frac{1}{N^2}(\overline{P}\cdot\overline{P})\right]^{\frac{1}{2}}$$

Thus, $\overline{P}'$ becomes a unit vector. It is called the normal vector of unit length. (Normalised vector)

In case, $(\overline{P}\cdot\overline{Q}) = 0$, and $|\overline{P}| = 1$, $|\overline{Q}| = 1$, then such vectors are known as *orthonormal vectors.*

*It can be proved* (1) *that the two orthonormal vectors are linearly independent.* (2) *that it is always possible to find a set of n orthogonal unit vectors in an n-dimensional space*. Thus, a set of $n$ linearly independent base unit ($\overline{e}_1$, $\overline{e}_2$, $\overline{e}_3$, $\overline{e}_4,\ldots,\overline{e}_n$) can be obtained. These can serve as a basis for a coordinate system in our vector space. We call such a coordinate system of $n$ unit vectors as orthogonal coordinate system and say that its base vectors ($\overline{e}_1$, $\overline{e}_2$, $\overline{e}_3$, $\overline{e}_4,\ldots,\overline{e}_n$) form an *orthonormal basis for the coordinate system.*

We now find how to apply the concept of scalar product and orthonormality to function space. Let us assume that the functions are real functions. (*i.e.*, the value at a point $x$, of the function is real and not imaginary. Suppose there are two functions, $f_1$ and $f_2$ and we wish to find a scalar product *viz.* $(f_1 \cdot f_2)$. This product must give a number which must be associated with both functions as a whole. The (scalar) product should not depend on a particular set of values of the variables of which they are functions. The only way to obtain such a number independent of values of variables is to integrate the product of the functions over a range of values of their variables *i.e.*, over a domain $[a, b]$. Thus, we define the scalar product over a domain $[a, b]$ as,

$$(f_1 \cdot f_2) = \int_a^b f_1 \cdot f_2 \partial\tau \qquad \ldots(3.16)$$

the integration is carried out over all the variables within a certain range (domain $[a, b]$) of which $f_1$ and $f_2$ are functions. It thus gives a number associated with the product of two functions $f_1$ and $f_2$. It can be easily seen that this number has the following properties.

$$\left.\begin{aligned} &(i)\ (f_1 \cdot af_2) = a(f_1 \cdot f_2) \\ &(ii)\ [f_1 \cdot (f_2 + f_3)] = (f_1 \cdot f_2) + (f_1 \cdot f_3) \end{aligned}\right\} \qquad \ldots(3.17)$$

Here $f_1, f_2$ and $f_3$ are three functions of function space. We also get

$$(iii)\ (f_1 \cdot f_1) = \int_a^b f_1^{\,2} \partial\tau \geq 0, \text{ and} \qquad \ldots(3.18)$$

(*iv*) $(f_1 \cdot f_1) = 0$ only when $f_1 = 0$ for all the values of variables ($x$) of which $f_1$ is a function.

(*v*) Now, if $(f_1 \cdot f_1) = N_f^2$, then $N_f$ is termed as the normalisation of $f_1$ and is analogous to the length of position vector. Thus, when $N_f = 1$, we say that the function is normalised and we can get a normalised function $f_1$ by dividing the function $f_1$ by $\frac{1}{N_{f_i}}$, which is the unit vector in the function space like $e_1, e_2, e_3,\ldots$ in physical vector space. If two functions are orthogonal, then we must have

$$\int_a^b (f_1 \cdot f_2)\partial\tau = 0 \qquad \ldots(3.19)$$

As stated earlier in any $n$-dimensional space whether a physical space or function space or else it is always possible to find $n$ number of normalised and orthogonal functions (or vectors) as basis functions $(f_1, f_2, f_3,\ldots, f_n)$ which satisfy the relation.

$$\int_a^b (f_1 \cdot f_2).\partial\tau = \delta_{ij} \qquad \ldots(3.20)$$

In terms of such orthonormal (*i.e.*, orthogonal as well as normalized) functions, functions $g$ and $h$ of the function space can be defined as

$$\left.\begin{aligned} g &= \alpha_1 f_1 + \alpha_2 f_2 + \alpha_3 f_3 \ldots \alpha_n f_n \\ \text{and} \quad h &= \beta_1 f_1 + \beta_2 f_2 + \beta_3 f_3 \ldots \beta_n f_n \end{aligned}\right\} \qquad \ldots(3.21)$$

If $f_1, f_2 \ldots f_n$ are the set of orthonormal functions, then we have

$$(g \cdot g) = \alpha_1^2 + \alpha_2^2 + \alpha_3^2 + \ldots \alpha_n^2 = N_\alpha^2 \text{(say)} \qquad \ldots(3.22)$$

$$(h \cdot h) = \beta_1^2 + \beta_2^2 + \beta_3^2 + \ldots \beta_n^2 = N_\beta^2 \text{(say)} \qquad \ldots(3.23)$$

Further if $g$ and $h$ are orthogonal, we have

$$\alpha_1\beta_1 + \alpha_2\beta_2 + \alpha_3\beta_3 \ldots \alpha_n\beta_n = 0 \qquad \ldots(3.24)$$

**Example 3:** We reconsider (Example 2). The set of functions $(\alpha_1 \sin x + \alpha_2 \cos x)$ forms a function space formed by the solution of the differential equation

$$\frac{d^2y}{dx^2} = -y$$

This function space consists of functions of the type

$$g = \alpha_1 f_1 + \alpha_2 f_2 \qquad \ldots(3.25)$$

Here, $f_1$ and $f_2$ are functions $\sin x$ and $\cos x$, respectively. We take the values of $x$ in the range $-\pi$ to $+\pi$. The scalar product of $g$ and $h$ is

$$f(g, h) = \int_{-\pi}^{+\pi} g(x)h(x)\partial\tau \qquad \ldots(3.26)$$

With this definition of scalar product, we get

$$(f_2 \cdot f_2) = \int_{-\pi}^{+\pi} \cos^2 x\partial x = \pi$$

$$(f_1 \cdot f_1) = \int_{-\pi}^{+\pi} \sin^2 x\partial x = \pi$$

and
$$(f_1 \cdot f_2) = \int_{-\pi}^{+\pi} \cos x \sin x\partial x = 0$$

Thus, the normalised $f_1$ and $f_2$ functions over the domain [–1, +1] are

$$f_1' = \frac{1}{\sqrt{\pi}} f \text{ and } f_2' = \frac{1}{\sqrt{\pi}} f_2 \qquad \ldots(3.27)$$

Also, $f_1$ and $f_2$ are orthogonal to each other for

$$(f_1 \cdot f_2) = 0$$

Therefore, the functions $f_1'$ and $f_2'$ are orthonormal functions.

## 3.3 REPRESENTATIONS

**(A) Symmetry Operations and Operators:** In Section 3.2, we have defined three different types of vector spaces which we will be discussing in the book. These are

(1) Physical vector spaces

(2) Function vector spaces

(3) Configuration spaces. In chapter two, we have also defined different transformations associated with molecules that define the point groups of the molecules. We regarded these transformations as symmetry operations *viz.* (*i*) Rotation, (*ii*) Reflection, (*iii*) Inversion, (*iv*) Rotation-reflection and (*v*) Identity.

The connection between these symmetry operations and the vector spaces is that we regard the operations as operators which act on the vectors of a vector space to produce other vectors of the same vector space according to some definite rules.

For example, in an $n$-dimensional physical space a vector $\bar{P}$ with components $(x_1^1, x_2^1, x_3^1, \ldots, x_1^n, x_2^n, x_3^n)$ is transformed to another vector $\bar{P}'$ with components $\{(x_1^1)', (x_2^1)', (x_3^1)', (x_1^2)', (x_2^2)', (x_3^2)', \ldots, (x_1^n), (x_2^n)', (x_3^n)'$ as a result of symmetry transformation $T$. That is,

$$\hat{T}(\overline{P}) \rightarrow \overline{P}' \qquad ...(3.28)$$

The equation (3.28) implies that when the operator $\hat{T}$ operates on $\overline{P}$ and produces $\overline{P}'$, then there will be a definite relation characteristic of $\hat{T}$ between the coordinate of $\overline{P}$ and $\overline{P}'$. The same concept is applicable to the vectors of other vector spaces. In this chapter, we would like to quantify this concept and to get a quantitative relationship which gives rise to the concept of representation of a group by means of matrices. We will develop the theory of groups using these group representations. Once we are able to develop a theory of groups in terms of these representations, we will be in a position to apply the theory of groups in solving quantum mechanical problems in any allied areas.

**(B) Operators in Physical Vector Space:** Let us start to find matrices representing transformations in physical vector space. We start a three-dimensional space and consider the effect of symmetry operation on a point vector $\overline{P}$. Let us take a point $O$ in the space which remained unmoved by the symmetry operations as origin of an orthonormal coordinate system with unit base vectors $(\overline{e}_1, \overline{e}_2 \text{ and } \overline{e}_3)$. With respect to the chosen arbitrary set, let the coordinates of the point vector $\overline{P}$ be $(x_1, x_2, x_3)$. [Fig 3.2($a$)].

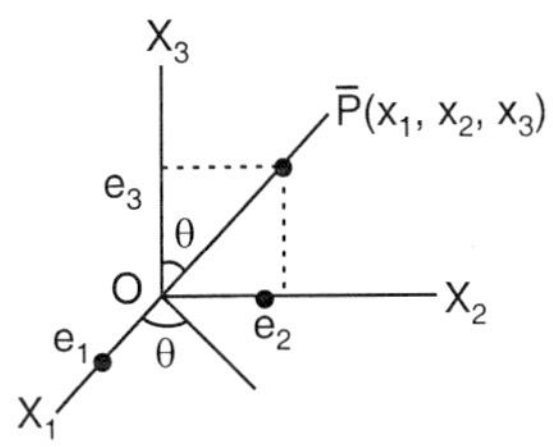


**Fig. 3.1** Coordinats of a Point $\overline{P}$.

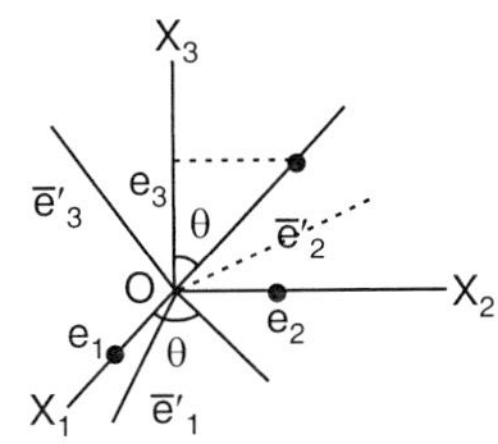


**Fig. 3.2 (a)** Rotation about $x_3$-axis with point $\overline{P}$ constant and $\overline{e}_1, \overline{e}_2, \overline{e}_3$ changed to $\overline{e}_1', \overline{e}_2', \overline{e}_3'$.

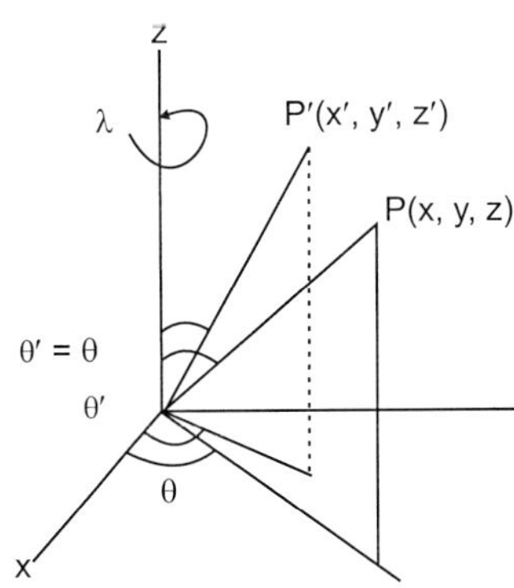


**Fig. 3.2 (b)** Rotation about $Z$-axis with $e_1$, $e_2$, $e_3$ and $P$ changed to point $P'$ is $\hat{R}\overline{P} = \overline{P}'$.

We find the effect of the symmetry operator $\hat{T}$ (Rotation or reflection *etc.*) on a position vector $\overline{P}$ in the three-dimensional vector space. It results a shift of the vector $\overline{P}$ to a new vector $\overline{P}'$ under the operator $\hat{T}$. That is,

$$\hat{T}(\overline{P}) \rightarrow \overline{P}' \qquad \ldots(3.29)$$

This operator $\hat{T}$ (in general, all symmetry operations of the type we are considering) will not change the length of the vector $\overline{P}$, but it will change the direction of the vector. That is,

$$\hat{T}(a\overline{P}) \rightarrow a\overline{P}' \qquad \ldots(3.30)$$

and
$$\hat{T}(\overline{P}+\overline{Q}) = \hat{T}\overline{P}+\hat{T}\overline{Q} \qquad \ldots(3.31)$$

Eqs. (3.30) and (3.31) indicate that the operation $\hat{T}$ is a linear operator. This is the most important property of symmetry operators.

In our arbitrarily chosen coordinate system, a vector $\overline{P}$ is expressed as

$$\overline{P} = x_1\overline{e}_1 + x_2\overline{e}_2 + x_3\overline{e}_3 \qquad \ldots(3.32)$$

Thus, in the same vector space, we will have, (cf. eq. 3.30)

$$\hat{T}\overline{P} = \overline{P}' = x_1(\hat{T}\overline{e}_1)+x_2(\hat{T}\overline{e}_2)+x_3(\hat{T}\overline{e}_3) \qquad \ldots(3.33)$$

$$= x_1(\overline{e}_1')+x_2(\overline{e}_2')+x_3(\overline{e}_3') \qquad \ldots(3.34)$$

We will now express the base vectors $(\overline{e}_1, \overline{e}_2 \text{ and } \overline{e}_3)$ in terms of the transformed vectors $(\overline{e}_1', \overline{e}_2', \overline{e}_3')$ as follows: [(Fig 3.2 (*b*)] and eq. (3.32)

$$\left.\begin{aligned} \hat{T}\overline{e}_1 &= \overline{e}_1' = r_{11}\overline{e}_1 + r_{21}\overline{e}_2 + r_{31}\overline{e}_3 \\ \hat{T}\overline{e}_2 &= \overline{e}_2' = r_{12}\overline{e}_1 + r_{22}\overline{e}_2 + r_{32}\overline{e}_3 \\ \text{and} \quad \hat{T}\overline{e}_3 &= \overline{e}_3' = r_{13}\overline{e}_1 + r_{23}\overline{e}_2 + r_{33}\overline{e}_3 \end{aligned}\right\} \qquad \ldots(3.35)$$

note the order of *ij* in $r_{ij}$ (*i* increases along the column *j* increases along the row)

or
$$\hat{T}(\overline{e}_1, \overline{e}_2, \overline{e}_3) = (\overline{e}_1', \overline{e}_2', \overline{e}_3') = \underline{|e_1, e_2, e_3|} \underbrace{\begin{vmatrix} r_{11} & r_{12} & r_{13} \\ r_{21} & r_{22} & r_{23} \\ r_{31} & r_{32} & r_{33} \end{vmatrix}}_{\text{matrix assigned to operator } \hat{T}} \qquad \ldots(3.36)$$

In general,

$$\hat{T}\overline{e}_k = \sum_{j=1}^{3} r_{jk}\overline{e}_j (j=1, 2, 3) \qquad \ldots(3.37)$$

In eqs. (3.35–3.37) $r_{jk}$ are the components of the vector $\hat{T}\overline{e}_k$ $(= e_j')$ with respect to the original base vectors $(\overline{e}_1, \overline{e}_2, \overline{e}_3)$. Thus, eq. (3.33) becomes

$$\begin{aligned}\bar{P}' &= \hat{T}\bar{P} = x_1(r_{11}\bar{e}_1 + r_{21}\bar{e}_2 + r_{31}\bar{e}_3) + x_2(r_{12}\bar{e}_1 + r_{22}\bar{e}_2 + r_{32}\bar{e}_3) \\ &\qquad + x_3(r_{13}\bar{e}_1 + r_{23}\bar{e}_2 + r_{33}\bar{e}_3) \\ &= (x_1r_{11} + x_2r_{12} + x_3r_{13})\bar{e}_1 + (x_1r_{21} + x_2r_{22} + x_3r_{23})\bar{e}_2 \\ &\qquad + (x_1r_{31} + x_2r_{32} + x_3r_{33})\bar{e}_3 \ldots \\ &= x_1'\bar{e}_1 + x_2'\bar{e}_2, x_3'\bar{e}_3 \end{aligned} \qquad \ldots(3.38)$$

Where $\left.\begin{aligned}(x_1') &= (x_1r_{11} + x_2r_{12} + x_3r_{13}) \\ (x_2') &= (x_1r_{21} + x_2r_{22} + x_3r_{23}) \\ (x_3') &= (x_1r_{31} + x_2r_{32} + x_3r_{33})\end{aligned}\right\}$

or
$$x_k' = \sum_{i=1}^{3} x_i r_{ki} \quad (k = 1, 2, 3) \qquad \ldots(3.39)$$

$$(x_1', x_2', x_3') = (x_1, x_2, x_3)\begin{vmatrix} r_{11} & r_{21} & r_{31} \\ r_{12} & r_{22} & r_{32} \\ r_{13} & r_{23} & r_{33} \end{vmatrix}$$

$$= (x_1x_2x_3) \text{ (matrix)}$$

(because $\bar{e}_1', \bar{e}_2', \bar{e}_3'$ are orthonormal).

Thus, we have

$$\bar{e}_k' = \hat{T}\bar{e}_k = \sum_{j=1}^{3} r_{jk}\bar{e}_j \qquad \ldots(3.40)$$

and
$$x_k' = \sum_{j=1}^{3} r_{kj}x_k \qquad \ldots(3.41)$$

Note carefully the coefficients of $\bar{e}_k'$ and $x_k'$ in eqs. (3.40) and (3.41). This example shows how one can regard a symmetry operation in terms of operator which acts on a general vector of a vector space or on a point of a configuration space or on a function space because in each of these spaces, a vector is defined in terms of a basis (orthonormal) arbitrarily chosen in the space and which can be expressed in terms of matrix, that is specific for a particular operation. Thus, once we know the numerical values of $r_{ki}$ in the above defined matrices, one can find a relationship in terms of numbers between the coordinates $(x_1, x_2, x_3)$ of a vector $P$ and the coordinates $(x_1', x_2', x_3')$ of the $\bar{P}'$ (*viz.* $\hat{T}\bar{P} = \bar{P}'$.) In this manner, the set of nine numbers $(r_{ik})$ gives a numerical representation of the symmetry operator $\hat{T}$ with respect to the orthonormal basis $(\bar{e}_1, \bar{e}_2, \bar{e}_3)$ in a three-dimentional space.

At this point, we can extrapolate the concept of representing the symmetry operators in terms of matrices. Thus, if we have a $n$-dimensional space (physical, function, or configuration) defined by $n$-orthonormal bases $(\overline{e}_1, \overline{e}_2, \ldots, \overline{e}_n)$, then the matrices for the operator $\hat{T}$ will be $(n \times n)$. Or the symmetry operator $\hat{T}$ will be represented by $n^2$ numbers or $(n \times n)$ matrix. Thus, when the $T$ matrix is applied on the base vectors, it is written as

$$(\overline{e}_1', \overline{e}_2', \ldots, \overline{e}_n') = (\overline{e}_1, \overline{e}_2, \ldots, \overline{e}_n)\,\hat{T} = (\overline{e}_1 + \overline{e}_2, \ldots, \overline{e}_n)\begin{vmatrix} r_{11} & r_{12} & r_{13}\cdots & r_{1n} \\ r_{21} & r_{22} & r_{23}\cdots & r_{2n} \\ r_{n1} & r_{n2} & r_{n3}\cdots & r_{nn} \end{vmatrix}$$

Here in $r_{ji}$, the first suffix indicates the row and the second suffix indicates the number of column. The way in which $\hat{T}$ is related to eqs. (3.39) and (3.40) should be noted. When the operator $\hat{T}$ is applied on the basis vectors $(\overline{e}_1, \overline{e}_2, \overline{e}_3, \ldots, \overline{e}_n)$, then the components $r_{k1}, r_{k2} \ldots r_{kn}$ of the vector $\hat{T}\, e_k$ are written in order down the $\hat{T}\overline{e}_k$ column while the coefficients ( $r_{k1}, r_{k2}, \ldots, r_{kn}$ ) when appear in eq. (3.39) for $x_k'$ are written across the $k^{th}$ row of the matrices. In the following paragraphs, we write down and derive different matrices which represent different types of symmetry operators (Rotations, Reflection, Inversion, Rotation-reflection).

**(C) Matrix Notations for Symmetry Operators:** We now derive various representations of all types of symmetry operators defining point groups by their corresponding matrices. Symmetry operators which we have used to define various point groups in chapter two are the following:

(*i*) Identity operator ($E$)

(*ii*) Proper rotation operators about an axis ($C_n^{\alpha}$)

(*iii*) Reflection operators in a plane ($\sigma_v$, $\sigma_h$, $\sigma_d$)

(*iv*) Inversion operators ($I$ or $i$)

(*v*) Rotation-reflection (or Improper rotation) operator ($S_n^{\alpha}$)

We first derive matrices for these operators in three-dimensional physical space, followed by examining if these matrices follow the rules of multiplication table of a point group. If they do, then we will be sure that a set of matrices of operators form a faithful representation of the point group (Isomorphic or homomorphic to the point groups).

**1. Identity operation:** Let us take a vector $\overline{P}$ in the space whose coordinates are ($\xi_1$, $\xi_2$, $\xi_3$) with respect to the basis of unit orthonormal vectors $(\overline{e}_1, \overline{e}_2, \overline{e}_3)$

$$\overline{P} = \xi_1\overline{e}_1 + \xi_2\overline{e}_2 + \xi_3\overline{e}_3 \qquad \ldots(3.42)$$

Since the operation, $E$ leaves all the vectors unchanged. so,

$$\hat{T}_{(E)}\overline{P} = \hat{T}_{(E)}(\xi_1\overline{e}_1 + \xi_2\overline{e}_2 + \xi_3\overline{e}_3)$$
$$= \xi_1\overline{e}_1 + \xi_2\overline{e}_2 + \xi_3\overline{e}_3 \qquad ...(3.43)$$

In the matrix notation, eq. (3.43) can be written as:

$$\hat{T}_{(E)}\begin{vmatrix}\xi_1\overline{e}_1\\ \xi_2\overline{e}_2\\ \xi_3\overline{e}_3\end{vmatrix} = |\xi_1\overline{e}_1, \xi_2\overline{e}_2, \xi_3\overline{e}_3|\begin{vmatrix}1 & 0 & 0\\ 0 & 1 & 0\\ 0 & 0 & 1\end{vmatrix} \qquad ...(3.44)$$

The matrix representing $E$ is, therefore, given as

$$E = \hat{T}_{(E)} = \begin{vmatrix}1 & 0 & 0\\ 0 & 1 & 0\\ 0 & 0 & 1\end{vmatrix} \qquad ...(3.45)$$

**2. Rotation about $z$-axis (or $e_3$-axis):** Let us rotate the unit bases vectors $(\overline{e}_1, \overline{e}_2, \overline{e}_3)$ about the $\overline{e}_3$ axis by an angle $\alpha$ (counterclockwise). In this case, $\overline{e}_3$ will be left unchanged by the operation $C_n^{\alpha}$ about $\overline{e}_3$ but $\overline{e}_1$ and $\overline{e}_2$ will be rotated to the position $C_n^{\alpha}\overline{e}_1$ and $C_n^{\alpha}\overline{e}_2$ as shown in Fig. 3.3.

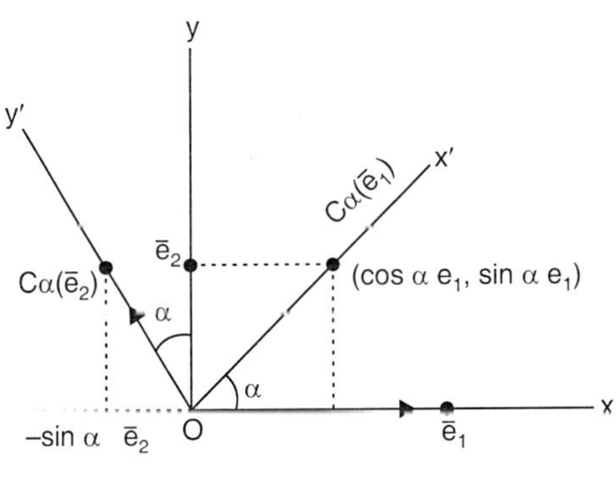


**Fig. 3.3** Rotation operations $C_n^{\alpha}$ about $\overline{e}_3$ -axis.

These are unit vectors with no (*i.e.*, zero) components along $\overline{e}_3$ vector. The $x_1$ and $x_2$ components of $C_n^{\alpha}(\overline{e}_1)$ and $C_n^{\alpha}(\overline{e}_2)$ vectors will be (cos $\alpha$, sin $\alpha$) and (–sin $\alpha$, cos $\alpha$) respectively. Thus,

$$\left.\begin{aligned} C_n^{\alpha}(\overline{e}_1) &= (\cos\alpha)\overline{e}_1 + (\sin\alpha)\,\overline{e}_2 + 0\,\overline{e}_3 \\ C_n^{\alpha}(\overline{e}_2) &= -(\sin\alpha)\overline{e}_1 + (\cos\alpha)\overline{e}_2 + 0\,\overline{e}_3 \\ C_n^{\alpha}(\overline{e}_3) &= 0\overline{e}_1 + 0\overline{e}_2 + 1\,\overline{e}_3 \end{aligned}\right\} \qquad ...(3.46)$$

the matrix of $C_n^{\alpha}$ with respect to $(\bar{e}_1, \bar{e}_2, \bar{e}_3)$ is

$$(i) \text{ about } z\text{-axis } C_{\alpha}(z) = \begin{vmatrix} \cos\alpha & \sin\alpha & 0 \\ -\sin\alpha & \cos\alpha & 0 \\ 0 & 0 & 1 \end{vmatrix} \quad ...(3.47)$$

Similarly, the effects of rotations about $y$-axis and $x$-axis may be expressed with respect to the base unit vectors $(\bar{e}_1, \bar{e}_2, \bar{e}_3)$ as follows.

$$(ii) \text{ about } y\text{-axis } C_{\alpha}(y) = \begin{vmatrix} \cos\alpha & 0 & \sin\alpha \\ 0 & 1 & 0 \\ -\sin\alpha & 0 & \cos\alpha \end{vmatrix} \quad ...(3.48)$$

and

$$(iii) \text{ about } x\text{-axis } C_{\alpha}(x) = \begin{vmatrix} 1 & 0 & 0 \\ 0 & \cos\alpha & \sin\alpha \\ 0 & -\sin\alpha & \cos\alpha \end{vmatrix} \quad ...(3.49)$$

In case the effect of rotation is to be observed around any arbitrary axis, the matrix for rotation about $C_{\psi}$ (any axis) is given below,

$$C_{\psi} \text{ (any axis)} = \begin{vmatrix} [\cos\psi + 1^2(1-\cos\psi)] & [1(1-\cos\psi) + n\sin\psi] & [1(1-\cos\psi) + m\sin\psi] \\ [1(1-\cos\psi) + n\sin\psi] & [\cos\psi + m^2(1-\cos\psi)] & [m(1-\cos\psi) + 1\sin\psi] \\ [1(1-\cos\psi) + m\sin\psi] & [n(1-\cos\psi) - 1\sin\psi] & [\cos\psi + n^2(1-\cos\psi)] \end{vmatrix} \quad ...(3.50)$$

[In eq. (3.50), we have, l = sin θ cos ϕ; $m$ = sin θ sin ϕ and $n$ = cos θ]

The transformation matrices given by eqs. (3.48–3.50), are with respect to orthonormal basis vectors $(\bar{e}_1, \bar{e}_2, \bar{e}_3)$. In case one wishes to find the effects of rotation about any arbitray axis on the coordinates $x$, $y$, $z$ the matrices for such transformation will be given by the transpose of the matrices given in eqs. (3.48)–(3.50) because the basis vectors are orthonormal. (*i.e.*, $e_i e_j = \delta_{ij}$).

**3. Reflection in a plane containing $\bar{e}_3$, (*i.e.*, $\sigma_v$–plane):** If $\hat{P}_R = \hat{\sigma}_V$ a reflection $\sigma_v$ in a plane *containing* $\bar{e}_3$ *axis* making an angle β with $\bar{e}_1$ axis (angle β, with the plane containing $\bar{e}_1$, $\bar{e}_3$); we have situation as shown in (Fig. 3.3). From the figure, it is obvious that

$$\begin{aligned}\hat{\sigma}_V \, \bar{e}_1 &= (\cos 2\beta)\bar{e}_1 + (\sin 2\beta)\bar{e}_2 + 0\bar{e}_3 \\ \hat{\sigma}_V \, \bar{e}_2 &= (\sin 2\beta)\bar{e}_1 - (\cos 2\beta)\bar{e}_2 + 0\bar{e}_3 \\ \hat{\sigma}_V \, \bar{e}_3 &= 0\bar{e}_1 + 0\bar{e}_2 + 1\bar{e}_3\end{aligned} \quad \text{...(3.51)}$$

$$\begin{vmatrix} \bar{e}_1 \\ \bar{e}_2 \\ \bar{e}_3 \end{vmatrix} \equiv \hat{\sigma}_V \begin{vmatrix} \bar{e}_1 \\ \bar{e}_2 \\ \bar{e}_3 \end{vmatrix} = \begin{vmatrix} \cos 2\beta & \sin 2\beta & 0 \\ \sin 2\beta & -\cos 2\beta & 0 \\ 0 & 0 & 1 \end{vmatrix} \begin{vmatrix} \bar{e}_1 \\ \bar{e}_2 \\ \bar{e}_3 \end{vmatrix} \quad \text{...(3.52)}$$

Therefore, the matrix for $\hat{\sigma}_V$ operator which will show the effect on the coordinates $(\xi_1, \xi_2, \xi_3)$ of the point $P$ or also on $(\bar{e}_1, \bar{e}_2, \bar{e}_3)$ will be given as

$$\begin{vmatrix} \cos 2\beta & \sin 2\beta & 0 \\ \sin 2\beta & -\cos 2\beta & 0 \\ 0 & 0 & 1 \end{vmatrix} \quad \text{...(3.53)}$$

**4. Reflection in the plane perpendicular to $\bar{e}_3$ ($\sigma_h$ Plane):** The image of any vector $\bar{P}(\xi_1, \xi_2, \xi_3)$ when reflected in the plane $(e_1 e_2)$ (or *XY* plane) is shown in the Fig. 3.4(*b*). Here the vectors $\bar{e}_1$ and $\bar{e}_2$ remain unaltered in the plane; but the vectors along the $\bar{e}_3$ axis will be reversed. Therefore, the coordinate $\xi_3$ will become $(-\xi_3)$ while $\xi_1$ and $\xi_2$ remain the same. Therefore, the matrix for $\sigma_h$ will be

$$\hat{\sigma}_h = \begin{vmatrix} 1 & 0 & 0 \\ 0 & 1 & 0 \\ 0 & 0 & -1 \end{vmatrix} \quad \text{...(3.54)}$$

**5. An improper rotation about $\bar{e}_3$ (*i.e.*, $S_n^\alpha$):** The effect of the improper rotation $S_n^\alpha$ on $e_1$ and $\bar{e}_2$ will be the same as that of rotation around $\bar{e}_3$, and at the same time it also involves reflection in the plane $(e_1 e_2)$. The reflection in $(e_1 e_2)$ plane will change the coordinate $\xi_3$ to $(-\xi_3)$. Thus,

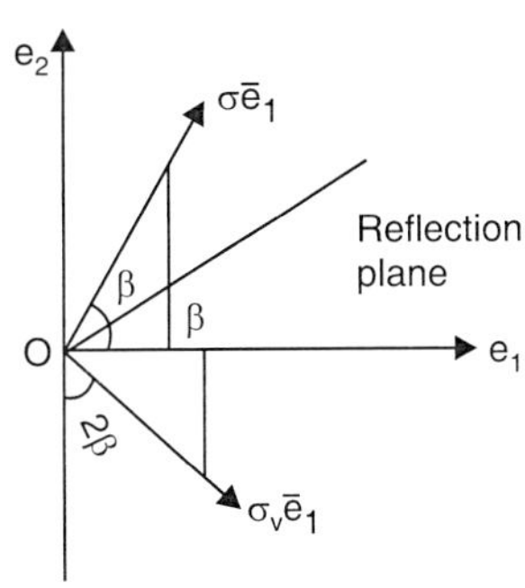


Fig 3.4 (*a*)

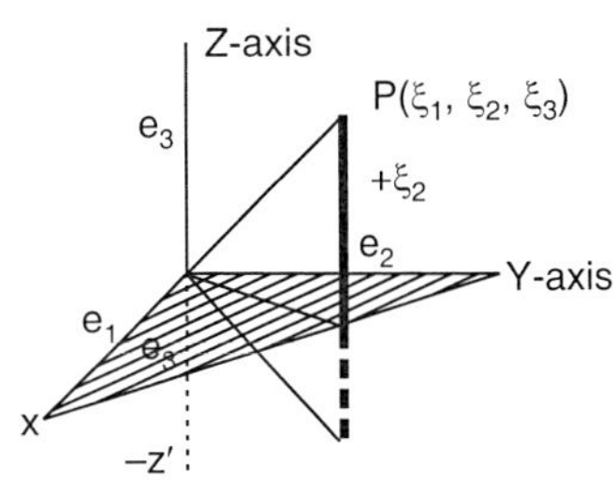


Fig 3.4 (*b*)

$$\left(S_n^\alpha\right)^2 = \hat{C}_n^\alpha \hat{\sigma}_h = \begin{vmatrix} \cos\theta & \sin\theta & 0 \\ -\sin\theta & \cos\theta & 0 \\ 0 & 0 & 1 \end{vmatrix} \begin{vmatrix} 1 & 0 & 0 \\ 0 & 1 & 0 \\ 0 & 0 & -1 \end{vmatrix} \qquad \text{...(3.55)}$$

$$= \begin{vmatrix} \cos\theta & \sin\theta & 0 \\ -\sin\theta & \cos\theta & 0 \\ 0 & 0 & -1 \end{vmatrix} \qquad \text{...(3.56)}$$

**6. Inversion in the origin (*I* or *i*):** The effect of inversion operator (*I* or *i*) is simply to reverse the base vectors $\bar{e}_1, \bar{e}_2, \bar{e}_3$ to $-\bar{e}_1, -\bar{e}_2, -\bar{e}_3$. Therefore, the corresponding matrix will be

$$(I) \text{ or } (\hat{i}) = \begin{vmatrix} -1 & 0 & 0 \\ 0 & -1 & 0 \\ 0 & 0 & -1 \end{vmatrix} \qquad \text{...(3.57)}$$

**Table 3.1:** The matrices which represent various symmetry operations used to define various point groups

| *S.No.* | *Symmetry operations* | *Matrices* |
|---|---|---|
| 1. | Identity *E* | $\begin{vmatrix} 1 & 0 & 0 \\ 0 & 1 & 0 \\ 0 & 0 & 1 \end{vmatrix}$ |
| 2.(*a*) | Rotation about $e_3$ axis by an angle θ | $\begin{vmatrix} \cos\theta & \sin\theta & 0 \\ -\sin\theta & \cos\theta & 0 \\ 0 & 0 & 1 \end{vmatrix}$ |
| 2.(*b*) | Rotation about $e_1$ axis by an angle θ | $\begin{vmatrix} 1 & 0 & 0 \\ 0 & \cos\theta & \sin\theta \\ 0 & -\sin\theta & \cos\theta \end{vmatrix}$ |
| 2.(*c*) | Rotation about $e_2$ axis by an angle θ | $\begin{vmatrix} \cos\theta & 0 & \sin\theta \\ 0 & 1 & 0 \\ -\sin\theta & 0 & \cos\theta \end{vmatrix}$ |
| 3. | Reflection in a plane containing $e_3$ axis and the plane make an angle β with $e_1$ axis $\hat{\sigma}_V$ | $\begin{vmatrix} \cos 2\beta & \sin 2\beta & 0 \\ \sin 2\beta & -\cos 2\beta & 0 \\ 0 & 0 & 1 \end{vmatrix}$ |

| | | |
|---|---|---|
| 4. | Reflection in $(\overline{e}_1\overline{e}_2)$ plane $\hat{\sigma}_h$ | $\begin{vmatrix} 1 & 0 & 0 \\ 0 & 1 & 0 \\ 0 & 0 & -1 \end{vmatrix}$ |
| 5. | (Rotation-reflection) about an axis $e_3$ and reflection plane $\overline{e}_1\overline{e}_2$ $(\hat{S}_n^{\theta})$ | $\begin{vmatrix} \cos\theta & \sin\theta & 0 \\ -\sin\theta & \cos\theta & 0 \\ 0 & 0 & -1 \end{vmatrix}$ |
| 6. | Inversion operation through the origin O, ($i$ or $I$) | $\begin{vmatrix} -1 & 0 & 0 \\ 0 & -1 & 0 \\ 0 & 0 & -1 \end{vmatrix}$ |

### 3.3.1 Relationship between Symmetry Operators, Multiplication Table and Their Matrices

We will now show that the matrices (derived above for various symmetry operators) form a group following the same multiplication table as that of the corresponding point group. It implies that the matrices must follow the same four axioms defining a group. Let us see, if these matrices follow the same axioms with matrix multiplications as binary operation.

**1. Closure axiom:** We take three operators $\hat{S}$, $\hat{T}$ and $\hat{R}$ from the set forming a group. These are related as

$$\hat{S}.\hat{R}. = \hat{T} \qquad \ldots(3.58)$$

Let us take a vector $\overline{P}$ of a vector space (for our convenience, we take a three-dimensional space). The application of operator $\hat{R}$ on $\overline{P}$ will take it to a vector $\overline{P}'$ and the application of $\hat{S}$ operator on $\overline{P}'$ will take it to a new vector $\overline{P}''$. For each operator, we have a corresponding matrix. That is, we will have a matrix $|R|$ for $R$ operator and a matrix $|S|$ for the $S$ operator. The combined application of $S.R$ will give a matrix $|SR|$ which will be obtained by multiplying matrices $|S|$ and $|R|$. In the same way, the application of operator $\hat{T}$ on the vector $\overline{P}$ will give a matrix $|T|$ for the $\hat{T}$ operator. If the closure axiom is to follow then we should have

$$|S||R| = |SR| = |T| \qquad \ldots(3.59)$$

To establish the eq. (3.59), let us choose the three orthonormal unit vectors $(\overline{e}_1, \overline{e}_2, \overline{e}_3)$ which will define the vector $\overline{P}$ as

$$\overline{P} = \xi_1\overline{e}_1 + \xi_2\overline{e}_2 + \xi_3\overline{e}_3 \qquad \ldots(3.60)$$

The application of $\hat{R}$ operator on vector $\overline{P}$ will give

$$\hat{R}\overline{P} = \hat{R}|P| = \xi_1 \hat{R}\overline{e}_1 + \xi_2 \hat{R}\overline{e}_2 + \hat{R}\xi_3 \overline{e}_3 \qquad \text{...(3.61)}$$

As shown in the preceding sub-section, $\hat{R}$ with respect to $(\overline{e}_1, \overline{e}_2, \overline{e}_3)$ is given as

$$\hat{R} = \begin{vmatrix} r_{11} & r_{21} & r_{31} \\ r_{12} & r_{22} & r_{32} \\ r_{13} & r_{23} & r_{33} \end{vmatrix} \qquad \text{...(3.62)}$$

or, by eq. (3.40) $(\hat{R}\hat{e}_k) = \sum_{j=1}^{3} r_{jk} e_j,$ ...(3.63)

and,

$$(\hat{S}\hat{R})\hat{e}_k = \hat{S}(\hat{R}e_k) = \hat{S}\left(\sum_{j=1}^{3} r_{jk} e_j\right) = \sum_{j=1}^{3} r_{jk}(\hat{S}\hat{e}_j) \qquad \text{...(3.64)}$$

analogous to eq. (3.63), we have

$$\hat{S}\overline{e}_j = \sum_{i=1}^{3} s_{ij} e_i \qquad \text{...(3.65)}$$

where $(s_{ij})$ matrix similar to eq. (3.62) is given as,

$$\hat{S} = \begin{vmatrix} s_{11} & s_{21} & s_{31} \\ s_{12} & s_{22} & s_{32} \\ s_{13} & s_{23} & s_{33} \end{vmatrix} \qquad \text{...(3.66)}$$

Thus, eq. (3.64) becomes

$$\begin{vmatrix} r_{11} & r_{21} & r_{31} \\ r_{12} & r_{22} & r_{32} \\ r_{13} & r_{23} & r_{33} \end{vmatrix} \begin{vmatrix} s_{11} & s_{21} & s_{31} \\ s_{12} & s_{22} & s_{32} \\ s_{13} & s_{23} & s_{33} \end{vmatrix} \begin{vmatrix} e_1 \\ e_2 \\ e_3 \end{vmatrix} \qquad \text{...(3.67)}$$

$$= \begin{vmatrix} r_{11}s_{11} + r_{21}s_{12} + r_{31}s_{13} & r_{11}s_{21} + r_{21}s_{22} + r_{31}s_{23} \cdots \\ r_{12}s_{11} + r_{22}s_{12} + r_{32}s_{13} & r_{12}s_{21} + r_{22}s_{22} + r_{32}s_{23} \cdots \\ \vdots & \vdots \quad \cdots \end{vmatrix} \begin{vmatrix} e_1 \\ e_2 \\ e_3 \end{vmatrix} \qquad \text{...(3.68)}$$

$$= \left| \left( \sum_{i=1}^{3} \sum_{j=1}^{3} s_{ij} r_{jk} \right) e_i \right| \qquad \text{...(3.69)}$$

Now, the application of $\hat{T}$ on $e_k$ by eq. (3.40) gives

$$\hat{T}\bar{e}_k = \sum_{i=1}^{3} t_{ik} e_i \qquad \text{...(3.70)}$$

In order to satisfy the closure axiom of group, we must then have

$$\hat{T}\bar{e}_k = (\hat{S}\hat{R})\bar{e}_k = \hat{S}(\hat{R}\hat{e}_k) = \hat{S}\left(\sum_{j=1}^{3} r_{jk} e_j\right) = \sum_{j=1}^{3} r_{jk}(\hat{S}e_j)$$

$$\left.\begin{aligned} &= \sum_{j=1}^{3} r_{jk}\left(\sum_{i=1}^{3} s_{ij} e_i\right) \\ &= \sum_{i=1}^{3}\left(\sum_{j=1}^{3} s_{ij} r_{jk}\right) e_i \end{aligned}\right\} \qquad \text{...(3.71)}$$

Thus, eqs. (3.70) and. (3.71) must be same in order to fulfill the closure condition,

$$t_{ik} = \sum_{j=1}^{3} s_{ij} r_{jk} \qquad \text{...(3.72)}$$

In case the relationship given in eq. (3.72) follows, then we have deduced that the matrix representing the product operation $SR = T$ is the matrix whose elements are the numbers $t_{ik}$ defined in terms of the elements of $R$ and $S$ by eq. (3.72). We can say now that

$$|T| = |SR| \qquad \text{...(3.73)}$$

Eq. (3.73) implies that the matrix representing the product of two symmetry operators is the product of their representative matrices whose elements are given by eq. (3.72). This definition of the product of $S$ and $R$ concludes that the element in the $i^{\text{th}}$ row and $k^{\text{th}}$ column $R$ of the product matrix is given as

$$t_{ik} = (s_{i1}r_{1k} + s_{i2}r_{2k} + s_{i3}r_{3k}) \qquad \text{...(3.74)}$$

Or, $$[sr]_{ik} = (\text{row } i \text{ of } \hat{S})\,(\text{column } k \text{ of } \hat{R})$$

This is the closure condition which has to be followed for forming the group by matrices. This product of two matrices can be extended to any number of matrices. It can then be easily shown that if there are three matrices $A$, $B$ and $C$ then we can form a product matrix $BC$ and get a matrix $X$ and multiply it by the matrix $A$ to form $A$ ($BC$) or $AX$. The resultant matrix will be equal to a matrix $Y$ obtained by first multiplying $A$ and $B$ to get matrix $AB$ (= $P$) and then multiply $P$ by the matrix $C$ to get a matrix $Y$. These two matrices $X$ and $Y$ will be identical. That is,

$$A\,(BC) = (AB)\,C \qquad \text{...(3.75)}$$

**2. Other axioms of group (Presence of identity, inverse and associativity law):** Similarly, the matrices corresponding to the symmetry operations will also fulfill the other conditions of the group (*viz.* $\overline{R}'$ will be a inverse member of $R$ present in the set) and it will also contain the operation $E$ [identity].

Thus, the inverse of rotation $C_n(\alpha)$ about $e_3$ axis is a rotation $C_n(-\alpha)$ about $e_3$ axis. The corresponding matrix is given by substituting $(-\alpha)$ in place of $(\alpha)$ in eq. (3.47).

That is,

$$C^{-1}(\alpha) = C(-\alpha) = \begin{vmatrix} \cos(-\alpha) & \sin(-\alpha) & 0 \\ -\sin(-\alpha) & \cos(-\alpha) & 0 \\ 0 & 0 & 1 \end{vmatrix} = \begin{vmatrix} \cos\alpha & -\sin\alpha & 0 \\ \sin\alpha & \cos\alpha & 0 \\ 0 & 0 & 1 \end{vmatrix} \quad \ldots(3.76)$$

Thus, by taking matrices corresponding to other operations, one can show that

$$\sigma_v^2 = E \quad \ldots(3.77)$$

That is, the reflection is its own inverse and

$$(i^2) = (I^2) = E \quad \ldots(3.78)$$

That is, inverse operation is its own inverse.

We have shown that we can effectively replace the abstract symmetry operations of a molecular point group by the matrices which represent their effects on the orthonormal base vectors of a coordinate system $(\overline{e}_1, \overline{e}_2, \overline{e}_3)$ or on the coordinate of a point $P$ $(x_1, x_2, x_3)$ in any three-dimensional physical space of the molecule. All the relations that may hold between symmetry operations are mirrored by the same relations between their matrices. We may say that matrices represent the symmetry operations of a point group or any group and the set of these matrices forms a representation of the group. The set of orthonormal basis vectors $(\overline{e}_1, \overline{e}_2, \overline{e}_3)$ of the coordinate system with respect to which the matrices are defined, form the basis of a representation.

**Example 4:** We give below an example of matrix representation of the point group $C_{2v}$ [symmetry operations: $E$, $C_2$, $\sigma_v(xz)$, $\sigma_v(yz)$]. According to Table 3.1, we get the following matrices for the symmetry operations of $C_{2v}$ point group.

$$E = \begin{vmatrix} 1 & 0 & 0 \\ 0 & 1 & 0 \\ 0 & 0 & 1 \end{vmatrix}; \quad C_2(z) = \begin{vmatrix} -1 & 0 & 0 \\ 0 & -1 & 0 \\ 0 & 0 & 1 \end{vmatrix}, \quad \sigma_V(xz) = \begin{vmatrix} 1 & 0 & 0 \\ 0 & -1 & 0 \\ 0 & 0 & 1 \end{vmatrix},$$

[here $\theta = 180°$] [here $\beta = 0$]

$$\sigma_V(yz) = \begin{vmatrix} -1 & 0 & 0 \\ 0 & 1 & 0 \\ 0 & 0 & 1 \end{vmatrix} \quad \ldots(3.79)$$

[here $\beta = 90°$]

These matrices obey the group multiplication Table of $C_{2V}$ point group (*See* Chapter 2). This is shown below.

| $C_{2V}$ | $E$ | $C_2(z)$ | $\sigma_V(xz)$ | $\sigma_V(yz)$ |
|---|---|---|---|---|
| $E$ | $E$ | $C_2(z)$ | $\sigma_V(xz)$ | $\sigma_V(yz)$ |
| $C_{2V}$ | $C_2(z)$ | $E$ | $\sigma_V(xz)$ | $\sigma_V(yz)$ |
| $\sigma_V(xz)$ | $\sigma_V(xz)$ | $\sigma_V(yz)$ | $E$ | $C_{2V}$ |
| $\sigma_V(yz)$ | $\sigma_V(xz)$ | $\sigma_V(yz)$ | $C_{2V}$ | $E$ |

≡

| $C_{2V}$ | $\begin{vmatrix}1&0&0\\0&1&0\\0&0&1\end{vmatrix}$ | $\begin{vmatrix}-1&0&0\\0&-1&0\\0&0&1\end{vmatrix}$ | $\begin{vmatrix}1&0&0\\0&-1&0\\0&0&1\end{vmatrix}$ | $\begin{vmatrix}-1&0&0\\0&1&0\\0&0&1\end{vmatrix}$ |
|---|---|---|---|---|
| $\begin{vmatrix}1&0&0\\0&1&0\\0&0&1\end{vmatrix}$ | $\begin{vmatrix}1&0&0\\0&1&0\\0&0&1\end{vmatrix}$ | $\begin{vmatrix}-1&0&0\\0&-1&0\\0&0&1\end{vmatrix}$ | $\begin{vmatrix}1&0&0\\0&-1&0\\0&0&1\end{vmatrix}$ | $\begin{vmatrix}-1&0&0\\0&1&0\\0&0&1\end{vmatrix}$ |
| $\begin{vmatrix}-1&0&0\\0&-1&0\\0&0&1\end{vmatrix}$ | $\begin{vmatrix}-1&0&0\\0&-1&0\\0&0&1\end{vmatrix}$ | $\begin{vmatrix}1&0&0\\0&1&0\\0&0&1\end{vmatrix}$ | $\begin{vmatrix}-1&0&0\\0&1&0\\0&0&1\end{vmatrix}$ | $\begin{vmatrix}1&0&0\\0&-1&0\\0&0&1\end{vmatrix}$ |
| $\begin{vmatrix}1&0&0\\0&-1&0\\0&0&1\end{vmatrix}$ | $\begin{vmatrix}1&0&0\\0&-1&0\\0&0&1\end{vmatrix}$ | $\begin{vmatrix}-1&0&0\\0&-1&0\\0&0&1\end{vmatrix}$ | $\begin{vmatrix}1&0&0\\0&1&0\\0&0&1\end{vmatrix}$ | $\begin{vmatrix}-1&0&0\\0&-1&0\\0&0&1\end{vmatrix}$ |
| $\begin{vmatrix}-1&0&0\\0&1&0\\0&0&1\end{vmatrix}$ | $\begin{vmatrix}-1&0&0\\0&1&0\\0&0&1\end{vmatrix}$ | $\begin{vmatrix}-1&0&0\\0&1&0\\0&0&1\end{vmatrix}$ | $\begin{vmatrix}-1&0&0\\0&-1&0\\0&0&1\end{vmatrix}$ | $\begin{vmatrix}1&0&0\\0&1&0\\0&0&1\end{vmatrix}$ |

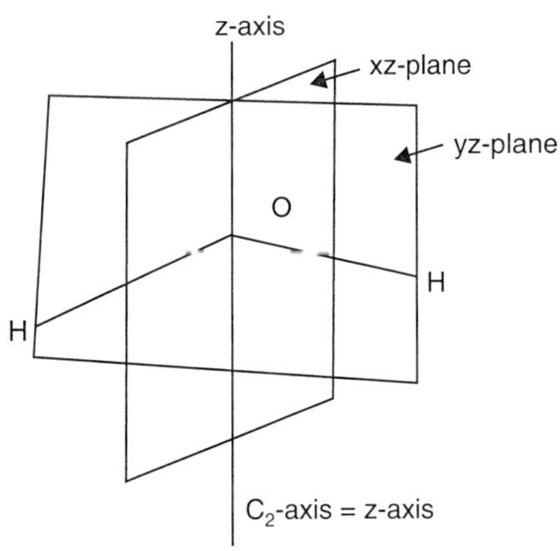


**Fig. 3.5** Planes of $H_2O$ molecules.

One can see one to one correspondence of all the elements in both the multiplication tables.

If we choose the coordinates of $O$, $H^1$ and $H^2$ as shown in Fig. 3.6, we get nine-dimensional representation. If we take the coordinates of $H^1$ and $H^2$, we get a six-dimensional representation with $(x_1, y_1, z_1)$ $(x_2, y_2, z_2)$as basis for the representation. Thus,

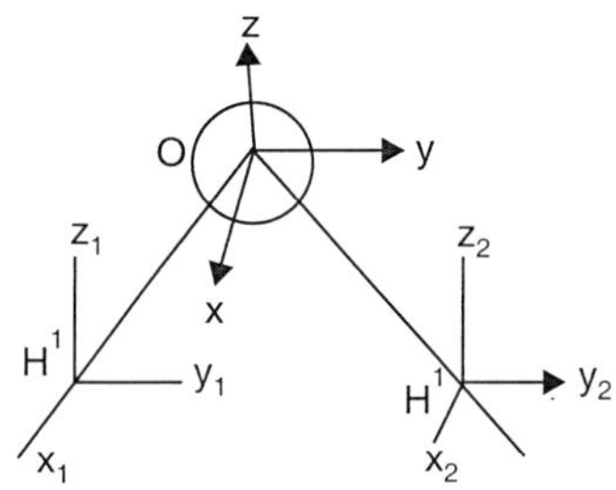


**Fig. 3.6** Coordinates of $H_2O$ molecules.

$$\hat{C}_2 \begin{vmatrix} x_1 \\ y_1 \\ z_1 \\ x_2 \\ y_2 \\ z_2 \end{vmatrix} = \begin{vmatrix} x_1' \\ y_1' \\ z_1' \\ x_2' \\ y_2' \\ z_2' \end{vmatrix} = \begin{vmatrix} -x_2 \\ -y_2 \\ z_2 \\ -x_1 \\ -y_1 \\ z_1 \end{vmatrix} = \begin{vmatrix} 0 & 0 & 0 & -1 & 0 & 0 \\ 0 & 0 & 0 & 0 & -1 & 0 \\ 0 & 0 & 0 & 0 & 0 & 1 \\ -1 & 0 & 0 & 0 & 0 & 0 \\ 0 & -1 & 0 & 0 & 0 & 0 \\ 0 & 0 & 1 & 0 & 0 & 0 \end{vmatrix} \begin{vmatrix} x_1 \\ y_1 \\ z_1 \\ x_2 \\ y_2 \\ z_2 \end{vmatrix}$$

$$\sigma_v(xz) \begin{vmatrix} x_1 \\ y_1 \\ z_1 \\ x_2 \\ y_2 \\ z_2 \end{vmatrix} = \begin{vmatrix} x_1'' \\ y_1'' \\ z_1'' \\ x_2'' \\ y_2'' \\ z_2'' \end{vmatrix} = \begin{vmatrix} x_2 \\ -y_2 \\ z_2 \\ x_1 \\ -y_1 \\ z_1 \end{vmatrix} = \begin{vmatrix} 0 & 0 & 0 & 1 & 0 & 0 \\ 0 & 0 & 0 & 0 & -1 & 0 \\ 0 & 0 & 0 & 0 & 0 & 1 \\ 1 & 0 & 0 & 0 & 0 & 0 \\ 0 & -1 & 0 & 0 & 0 & 0 \\ 0 & 0 & 1 & 0 & 0 & 0 \end{vmatrix} \begin{vmatrix} x_1 \\ y_1 \\ z_1 \\ x_2 \\ y_2 \\ z_2 \end{vmatrix} \qquad \ldots(3.80)$$

$$\text{and } \sigma_v(yz) \begin{vmatrix} x_1 \\ y_1 \\ z_1 \\ x_2 \\ y_2 \\ z_2 \end{vmatrix} = \begin{vmatrix} x_1^x \\ y_1^x \\ z_1^x \\ x_2^x \\ y_2^x \\ z_2^x \end{vmatrix} \begin{vmatrix} -x_1 \\ y_1 \\ z_1 \\ -x_2 \\ y_2 \\ z_2 \end{vmatrix} = \begin{vmatrix} -1 & 0 & 0 & 0 & 0 & 0 \\ 0 & 1 & 0 & 0 & 0 & 0 \\ 0 & 0 & 1 & 0 & 0 & 0 \\ 0 & 0 & 0 & -1 & 0 & 0 \\ 0 & 0 & 0 & 0 & 1 & 0 \\ 0 & 0 & 1 & 0 & 0 & 1 \end{vmatrix} \begin{vmatrix} x_1 \\ y_1 \\ z_1 \\ x_2 \\ y_2 \\ z_2 \end{vmatrix}$$

If we use above matrices for the symmetry operations of $C_{2v}$ point group to form the multiplication table for $C_{2v}$, one will find that these matrices also follow the multiplication table of $C_{2v}$. Therefore, these matrices also form a representation for $C_{2v}$ point group.

By now you must have realised that one can form infinite number of matrix representations for the same point group because we will obtain different matrices for different choices of coordinates. All these representations are known as "*Faithful Representations*" because in them each group operation has a different matrix and there is one to one correspondence between symmetry operation and the matrix that represents the operation.

### 3.3.2 Equivalent Representations

It was mentioned in the preceding paragraphs that one can obtain an infinite number of different matrix representations for any point group. This is because matrices forming a representation depend upon the choice of arbitrary coordinate system. Thus, we cannot have any one unique representation of a group that is independent of the choice of coordinate systems. Although one can however, have an infinite number of representations, but all these possible sets are essentially equivalent to each other because they all contain exactly the same information, *viz.* the effects of symmetry operations on a general vector of a vector space. All the matrices of one representation arc related to the matrices of another representation. The relationship can be derived easily. We will not give any derivation of this relationship but we will simply state it without proof.

Suppose we have chosen two sets of base vectors, *viz.* $(f_1, f_2, f_3, \ldots, f_n)$ and $(g_1, g_2, g_3, \ldots, g_n)$ which give the set of matrices $D^f(R)$ and $D^g(R)$, respectively. We thus have two representations. Here $D^f(R)$ is a set of matrices for operations $(R)$ with respect to base vectors $(f_1, f_2, f_3, \ldots, f_n)$ as basis and $D^g(R)$ is another set of matrices for the same operations $(R)$ with respect to $(g_1, g_2, g_3, \ldots g_n)$ as basis. These two bases $(f_1, f_2, f_3, \ldots, f_n)$ and $(g_1, g_2, g_3, \ldots, g_n)$ can be transformed into each other by the matrix $A$ or $B$ because each base vector of one basis can be expressed as

$$\left.\begin{aligned} f_k &= \sum_{j=1}^{n} a_{jk} g_j \\ &= \sum_{j=1}^{n} a_{jk} \sum_{i=1}^{n} b_{ij} f_i \end{aligned}\right\} \quad \ldots(3.81)$$

Thus, $$\left.\begin{aligned} \{f_1 f_2, \ldots, f_n\} &= |A| \begin{vmatrix} g_1 \\ g_2 \\ g_n \end{vmatrix} \\ \text{and} \quad \{g_1 g_2, \ldots, g_n\} &= |B| \begin{vmatrix} f_1 \\ f_2 \\ f_n \end{vmatrix} \end{aligned}\right\} \quad \ldots(3.82)$$

In can be proved that $|A||B|$ = unit matrix,

$$\left.\begin{aligned} &\text{or} && |A| = |B|^{-1} \\ &\text{and} && |B| = |A|^{-1} \end{aligned}\right\} \qquad \ldots(3.83)$$

With the results given in eqs. (3.82)–(3.83), one can easily show that the any operation.

$$\left.\begin{aligned} &\left[D^{g}(R_i)\right] = [A]\left[D^{f}(R_i)\right][A]^{-1} \\ \text{or} \quad &\left[D^{f}(R_i)\right] = [B]\left[D^{g}(R_i)\right][B]^{-1} \end{aligned}\right\} \qquad \ldots(3.84)$$

Thus, if we have a set of matrices, *e.g.*, $D^f(R)$ forming a representation of any point group with respect to a basis $(f_1, f_2, f_3, \ldots, f_n)$ eq. (3.84) tells us what the matrices representing the operators will become equivalent if we change to another basis whose vectors are $\{g_1, g_2, g_3, \ldots, g_n\}$. Only thing that we have to know is the matrices $A$ or $B$ which change one base vectors to another. These changes of basis do not affect the multiplication table of the group. Thus, we can define equivalent representations as,

**Two representations [$D^f(R)$ and $D^g(R)$] of a point groups formed by $|p \times p|$ matrices are equivalent if a matrix $|A|$ with its inverse $|A|^{-1}$ exists for which the relationship given by equation (3.84) holds for every operation $R$ of the point group.**

**Example 5:** Let us exemplify the above concept by taking the case of $NH_3$ molecule (point group $C_{3v}$) (Fig. 3.7) and (Fig. 3.8). Let us choose two base unit vectors. $(d_1, d_2, d_3)$ and $\overline{e}_1, \overline{e}_2, \overline{e}_3$. Point group is $C_{3v}[(E, C_3^1, C_3^2, \sigma_v^{(1)}, \sigma_v^{(2)}, \sigma_v^{(3)})]$. The three unit vectors $(d_1, d_2, d_3)$ point along the three $(N - H)$ or $(A - B)$ bonds.

Employing $(d_1, d_2, d_3)$ as unit base vectors we have the following matrix representation.

$$\begin{matrix} E & C_3^1 & C_3^2 & \sigma_v^1 & \sigma_v^2 & \sigma_v^3 \\ \begin{vmatrix} 1 & 0 & 0 \\ 0 & 1 & 0 \\ 0 & 0 & 1 \end{vmatrix}, & \begin{vmatrix} 0 & 1 & 0 \\ 0 & 0 & 1 \\ 1 & 0 & 0 \end{vmatrix}, & \begin{vmatrix} 0 & 0 & 1 \\ 1 & 0 & 0 \\ 0 & 1 & 0 \end{vmatrix}, & \begin{vmatrix} 1 & 0 & 0 \\ 0 & 0 & 1 \\ 0 & 1 & 0 \end{vmatrix}, & \begin{vmatrix} 0 & 0 & 1 \\ 0 & 1 & 0 \\ 1 & 0 & 0 \end{vmatrix}, & \begin{vmatrix} 0 & 1 & 0 \\ 1 & 0 & 0 \\ 0 & 0 & 1 \end{vmatrix} \end{matrix} \qquad \ldots(3.85)$$

(Prove the formation of above matrices for the six symmetry operations.)

Now we choose another orthonormal basis $(\overline{e}_1, \overline{e}_2, \overline{e}_3)$ for the representation of $C_{3v}$ point group. One can calculate the relationship between these two bases *viz.* $d_1, d_2, d_3$ and $\overline{e}_1, \overline{e}_2, \overline{e}_3$ as given as follows.

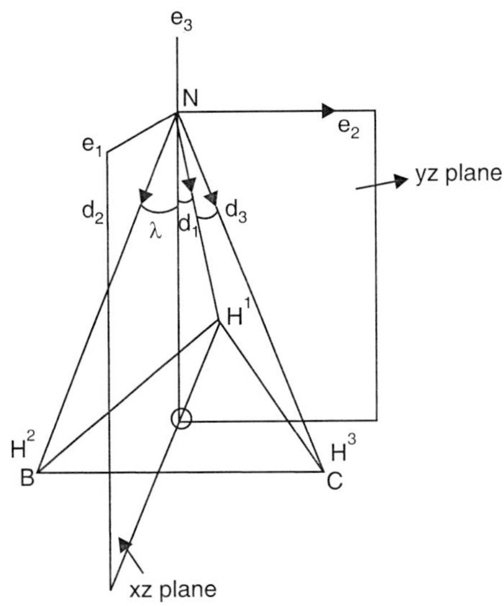


**Fig. 3.7** Coordinate Systems of $NH_3$ ($AB_3$).

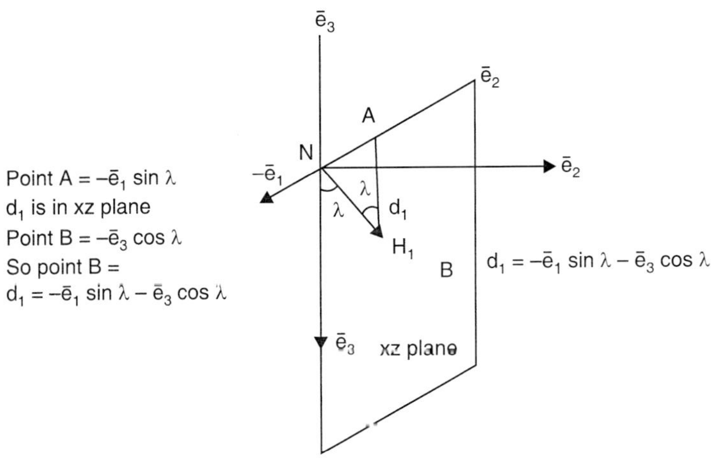


**Fig. 3.8** Showing $d_1 = -\bar{e}_1 \sin\lambda - \bar{e}_3 \cos\lambda$.

$$\left.\begin{aligned} d_1 &= -e_1 \sin\lambda - e_3 \cos\lambda \\ d_2 &= \frac{1}{2}(e_1 - \sqrt{3}e_2)\sin\lambda - e_3\cos\lambda \\ d_3 &= \frac{1}{2}(e_1 + \sqrt{3}e_2)\sin\lambda - e_3\cos\lambda \end{aligned}\right\} \qquad \ldots(3.86)$$

So the matrix | $A$ | of eq. (3.84) is given by

$$\begin{vmatrix} d_1 \\ d_2 \\ d_3 \end{vmatrix} = \underbrace{\begin{vmatrix} -\sin\lambda & 0 & -\cos\lambda \\ \frac{1}{2}\sin\lambda & -\frac{\sqrt{3}}{2}\sin\lambda & -\cos\lambda \\ \frac{1}{2}\sin\lambda & +\frac{\sqrt{3}}{2}\sin\lambda & -\cos\lambda \end{vmatrix}}_{|A|} \begin{vmatrix} e_1 \\ e_2 \\ e_3 \end{vmatrix} \quad \ldots(3.87)$$

From eq. (3.87), one can calculate $(\overline{e}_1, \overline{e}_2, \overline{e}_3)$ in terms of ($d_1$, $d_2$, $d_3$) as given below and it will be equal to $|A|^{-1}$

$$\begin{vmatrix} e_1 \\ e_2 \\ e_3 \end{vmatrix} = \begin{vmatrix} \underbrace{\begin{matrix} -\frac{2}{3\sin\lambda} & \frac{1}{3\sin\lambda} & \frac{1}{3\sin\lambda} \\ 0 & \frac{1}{\sqrt{3}\sin\lambda} & \frac{1}{\sqrt{3}\sin\lambda} \\ -\frac{1}{3\cos\lambda} & -\frac{1}{3\cos\lambda} & -\frac{1}{3\cos\lambda} \end{matrix}}_{|B|} \end{vmatrix} \begin{vmatrix} d_1 \\ d_2 \\ d_3 \end{vmatrix} \quad \ldots(3.88)$$

Since the base vectors ($d_1$, $d_2$, $d_3$) are not orthogonal, we have another relationship, which is given below.

$$d_1.d_2 = d_2.d_3 = d_3.d_1 = -\frac{1}{2}\sin^2\lambda + \cos^2\lambda \quad \ldots(3.89)$$

In order that ($d_1$, $d_2$, $d_3$) be orthogonal, their dot product must be zero *i.e.*,

$$-\frac{1}{2}\sin^2\lambda + \cos^2\lambda = 0$$

or

$$\frac{\sin\lambda}{\cos\lambda} = \tan\lambda = \sqrt{2} \quad \ldots(3.90)$$

That is,

$$\sin\lambda = \sqrt{\frac{2}{3}} \text{ and } \cos\lambda = \sqrt{\frac{1}{3}} \quad \ldots(3.91)$$

For such particular values of sin λ and cos λ, the matrix $B$ is inverse of $A$ or $|B| = |A|^{-1}$. Thus one can calculate | $B$ |, for the point group $C_{3v}$ with respect to the basis ($d_1$, $d_2$, $d_3$) by applying the similarity transformation as follows:

$$D^g(R)\ (g = d_1, d_2, d_3) = |A| \left| D^f(R) \right| |A|^{-1} \quad \ldots(3.92)$$

One can thus calculate all the matrices | $D^g(R)$ | from the matrices $|D^f(R)|$. These are given in (Table 3.2) below and these matrices will be the same as obtained in eq. (3.92). In the same table, we have also given another representation of $C_{2V}$ with respect to the basis $(p_1', p_2', p_3')$ in function space. Here $p_1'$, $p_2'$ and $p_3'$ are the $p$-orbitals $\left[viz.\ e^{\pm 2\pi i/3},\ p_3(=r\cos\theta)\right]$. This representation is denoted by $\Gamma_2$.

**Table 3.2:** (Equivalent Representations) $\Gamma_1(\bar{e}_1, \bar{e}_2, \bar{e}_3)$, $\Gamma_2(p_1', p_2', p_3')$ and $\Gamma_3(d_1, d_2, d_3)$ of $C_{3V}$ point group

| ***Operation*** | | $E$ | $C_3^1$ | $C_3^2$ | $\sigma_v^1$ | $\sigma_v^2$ | $\sigma_v^3$ |
|---|---|---|---|---|---|---|---|
| Representation | $\Gamma_1$ | $\begin{vmatrix} 1 & 0 & 0 \\ 0 & 1 & 0 \\ 0 & 0 & 1 \end{vmatrix}$ | $\begin{vmatrix} -c & -s & 0 \\ s & -c & 0 \\ 0 & 0 & 1 \end{vmatrix}$ | $\begin{vmatrix} -c & s & 0 \\ -s & -c & 0 \\ 0 & 0 & 1 \end{vmatrix}$ | $\begin{vmatrix} 1 & 0 & 0 \\ 0 & -1 & 0 \\ 0 & 0 & 1 \end{vmatrix}$ | $\begin{vmatrix} -c & -s & 0 \\ -s & c & 0 \\ 0 & 0 & 1 \end{vmatrix}$ | $\begin{vmatrix} -c & s & 0 \\ s & c & 0 \\ 0 & 0 & 1 \end{vmatrix}$ |
| | $\Gamma_2$ | $\begin{vmatrix} 1 & 0 & 0 \\ 0 & 1 & 0 \\ 0 & 0 & 1 \end{vmatrix}$ | $\begin{vmatrix} \in & 0 & 0 \\ 0 & \in^* & 0 \\ 0 & 0 & 1 \end{vmatrix}$ | $\begin{vmatrix} \in^* & 0 & 0 \\ 0 & \in & 0 \\ 0 & 0 & 1 \end{vmatrix}$ | $\begin{vmatrix} 0 & 1 & 0 \\ 1 & 0 & 0 \\ 0 & 0 & 1 \end{vmatrix}$ | $\begin{vmatrix} 0 & \in^* & 0 \\ \in & 0 & 0 \\ 0 & 0 & 1 \end{vmatrix}$ | $\begin{vmatrix} 0 & \in & 0 \\ \in^* & 0 & 0 \\ 0 & 0 & 1 \end{vmatrix}$ |
| | $\Gamma_3$ | $\begin{vmatrix} 1 & 0 & 0 \\ 0 & 1 & 0 \\ 0 & 0 & 1 \end{vmatrix}$ | $\begin{vmatrix} 0 & 0 & 1 \\ 1 & 0 & 0 \\ 0 & 1 & 0 \end{vmatrix}$ | $\begin{vmatrix} 0 & 1 & 0 \\ 0 & 0 & 1 \\ 1 & 0 & 0 \end{vmatrix}$ | $\begin{vmatrix} 1 & 0 & 0 \\ 0 & 0 & 1 \\ 0 & 1 & 0 \end{vmatrix}$ | $\begin{vmatrix} 0 & 0 & 1 \\ 0 & 1 & 0 \\ 1 & 0 & 0 \end{vmatrix}$ | $\begin{vmatrix} 0 & 1 & 0 \\ 1 & 0 & 0 \\ 0 & 0 & 1 \end{vmatrix}$ |

$\left[\text{Here } c = \cos\frac{\pi}{3} = \frac{1}{2},\ s = \sin\frac{\pi}{3} = \frac{\sqrt{3}}{2},\ \in = e^{\frac{2\pi i}{3}} = -c + i\right]$, $i$ being imaginary number $\sqrt{-1}$.

### 3.3.3 Characters (or) Trace of a Matrix

We have so far discussed:

1. Change of basis of a representation of a point group changes the matrices of representations.
2. Whatever are the chosen bases for the representations of the point group, the matrices of all the representations are equivalent.
3. The matrices of two equivalent representations are related to each other through a similarity transformation (*viz.* $A^{-1}D(R)A$), where the matrix $A$ with its inverse $A^{-1}$ are the transformation matrices from one basis to another.

The above points imply that the representations of any point group are basis dependent. This is problematic. We would like to discover some quantity or quantities associated with the representations which are (1) not basis dependent (2) remain invariant for all equivalent representations and (3) different for non-equivalent representations. Let this quantity be, $[\chi(D(R)]$. This property remains invariant (when one basis changes to another (or the representation $D(R)$ to $[A(D(R)A^{-1}]$ where $A$ is a matrix describing the change of basis). So the property $[\chi(D(R)]$ must be defined in such a way that

$$\chi[D(R)] = \chi[A\ D(R)A^{-1}] \qquad \ldots(3.93)$$

for any matrix $A$ with its inverse $A^{-1}$. It turns over that one such appropriate invariant quantity that we can easily associate with $[D(R)]$ is the TRACE of the matrix designated as $[Tr(D)]$. The trace of any matrix equals the sum of all the diagonal elements of a matrix. Thus, if we have a $n \times n$ matrix then

$$T_r(D) = \sum_{i=1}^{n} d_{ii} \qquad \ldots(3.94)$$

The trace of any matrix remains invariant when a change of basis is carried out on a representation. That is, for all equivalent representations, the value of $Tr(D)$ is the same. All equivalent representations of any point group can therefore be characterised by one specific set of traces of the matrices forming the representation. That is why $T_r[D(R)]$ is called the CHARACTER of the group element $R$ in the representation with matrix $[D(R)]$. We denote the character by the symbol $\chi(R)$. That is,

$$\chi(R) = T_r\ [D(R)] \qquad \ldots(3.95)$$

The complete set of characters, (one for each element of the group), is called the character of the representation. We say that the two representations are equivalent, if and only if they have the same characters.

The conjugate elements $P$ and $Q$ belong to the same class and these are obtained by using similarity transformations. This definition of the conjugacy relationship between $P$ and $Q$ elements in any class can be mirrored between the same $P$ and $Q$ in the representations formed by their respective matrices. Thus, if $P$ and $Q$ are two conjugate elements, there must be an element $R$ of the group for which,

$$D(P) = [D(R^{-1})][(DC(Q)\ [D(R)], \qquad \ldots(3.96)$$

where, $[D(R^{-1})] = [D(R)]^{-1}$.

Thus, according to properties of the trace of conjugate matrices, we will expect

$$T_r[D(P)] = T_r[D(Q)]. \qquad \ldots(3.97)$$

That is, the traces or the characters of two conjugate elements must be the same. Therefore, if we want to characterise a representation of a point group, it is sufficient to state the character of only one element from each class of the group.

As an example, we take the point group $C_{3v}$ which is divided into three classes $\{E, 2C_3, 3\sigma_v\}$. In its equivalent representations given in Table 3.3 below, the characters of $\{E, C_3, 3\sigma_v\}$ are given as

$$\chi(E = 3, \chi(C_3^1) = \chi(C_3^{-1}) = 0 \text{ and } \chi(\sigma_v^{(1)}) = \chi(\sigma_v^{(2)}) = \chi(\sigma_v^{(3)}) = 1 \quad ...(3.98)$$

**Table 3.3:** Characters of Classes of $C_{3v}$

| $C_{3v}$ | *Class I* <br> $E$ | *Class II* <br> $C_3^1, C_3^{-1}$ | *Class III* <br> $\sigma_v^{(1)}, \sigma_v^{(2)}, \sigma_v^{(3)}$ |
|---|---|---|---|
| Characters $\chi$ | 3 | 0 | 1 |

## 3.4 CLASSIFICATION OF THE REPRESENTATIONS OF GROUPS (REDUCIBLE AND IRREDUCIBLE)

### 3.4.1 Reducible

In the preceding paragraphs, we have shown as how to characterise a representation of a point group by the characters of the equivalent matrices which are independent of the choice of basis. We now introduce the concept of reducible and irreducible representations without which one can not precede even a step while applying the symmetry properties to solve a problem.

We again refer back to the three representations $\Gamma_1$, $\Gamma_2$ and $\Gamma_3$ of the group $C_{3v}$ given in Table 3.3. To form these equivalent representations, we have used three non-coplanar vectors with origin at a point that remained unmoved by the symmetry operations of the group (N-atom). The three sets of vectors [*viz*, $(\bar{d}_1, \bar{d}_2, \bar{d}_3)$, $(\bar{e}_1, \bar{e}_2, \bar{e}_3)$ . $(\bar{e}^{-2\pi i/3}, \bar{e}^{+2\pi i/3}, 1)$ ] served as the basis (unit vectors) to obtain three sets of 3 × 3 matrices representing $C_{3v}$. These are three matrix representations of $C_{3v}$. We have also shown that the matrices forming one representation of $C_{3v}$ can be transformed into the matrices of the other representations by a suitable choice of the transformation matrix $|A|$.

Out of these basis sets forming $C_{3v}$ group representations, the basis $(\bar{e}_1, \bar{e}_2, \bar{e}_3)$ is very simple. The operations of $C_{3v}$ are such as to leave $\bar{e}_3$ vector unaltered. This vector ($\bar{e}_3$) points along the axis of symmetry $C_3$ and the symmetry of $C_{3v}$ leaves all those position vectors pointing along $\bar{e}_3$ unaltered, and changes vectors which lie in the plane perpendicular to $\bar{e}_3$ *viz.* $\bar{e}_1$ and $\bar{e}_2$. The plane containing transformed vectors obtained after the transformation of $\bar{e}_1$ and $\bar{e}_2$ can be expressed as linear combinations of $\bar{e}_1$ and $\bar{e}_2$. That is,

$$\left.\begin{aligned} R\bar{e}_1 &= r_{11}e_1 + r_{21}e_2 + 0e_3 \\ R\bar{e}_2 &= r_{12}e_1 + r_{22}e_2 + 0e_3 \\ R\bar{e}_3 &= e_1 + e_2 + 1.e_3 \end{aligned}\right\} \quad ...(3.99)$$

In the basis of this kind, all the matrices of the representation can be expressed as,

$$D(R) = \left|\begin{array}{cc|c} r_{11} & r_{12} & 0 \\ r_{21} & r_{22} & 0 \\ \hline 0 & 0 & 1 \end{array}\right| \quad \ldots(3.100)$$

Thus, we can make each matrix $|D(R)|$ of the symmetry operations into two matrices, *viz.* $\begin{vmatrix} r_{11} & r_{12} \\ r_{21} & r_{22} \end{vmatrix}$ and $|1|$, One is two-dimensional and another one-dimensional. Let us designate $(2 \times 2)$ matrices by $|D'(R)|$ and $(1 \times 1)$ matrices by $|D''(R)|$. So our representation $[D(R)]$ becomes,

$$D(R) = \left|\begin{array}{c|c} D'R & 0 \\ \hline 0 & D''R \end{array}\right| \quad \ldots(3.101)$$

That is, $D(R)$ is the direct sum of the matrix representations, one $|D'(R)|$ with respect to $(e_1, e_2)$ as basis and $[D''(R)]$ with respect to $e_3$ as basis (*i.e.*, one-dimensional). The basis $e_1$ and $e_2$ can be chosen in an infinite number of ways but the representations formed by these matrices are equivalent. Let us symbolise the two-dimensional representation by $\Gamma'$ and one-dimensional one by $\Gamma''$ and the representation formed by $(e_1\ e_2\ e_3)$ as base vectors by $\Gamma$. We then have

$$\Gamma = \Gamma' + \Gamma'' \quad \ldots(3.102)$$

We can state from eq. (3.102) that $\Gamma$ is a reducible representation and is reduced to $\Gamma'$ and $\Gamma''$ representations. In other words $\Gamma$ representation, that is reducible, is reduced into two component representations $\Gamma'$ and $\Gamma''$.

We now attempt to reduce the $\Gamma'$ representation into two components of one-dimensional representations. If by a suitable choice of basis vectors (*viz.* $e_1'$ and $e_2'$) we would like to know "is it possible to reduce $\Gamma'$ into further reduced form and to express $\Gamma'$ as the direct sum of two one-dimensional representations". We can then express it as,

$$\left.\begin{aligned} Re_1'' &= r_1 e_1' \\ \text{and} \qquad Re_2'' &= r_2 e_2' \end{aligned}\right\} \quad \ldots(3.103)$$

All the matrices of $\Gamma'$ in the reduced representation with basis $(e_1', e_2')$ would then be of the form,

$$D'(R) = \begin{vmatrix} r_1 & 0 \\ 0 & r_2 \end{vmatrix} \qquad \ldots(3.104)$$

And so, $\Gamma' = \Gamma_1' + \Gamma_2''$

or, $$\Gamma = \Gamma_1' + \Gamma_2' + \Gamma'' \qquad \ldots(3.105)$$

This cannot be reduced any further. We call these ultimate reduced forms of representations as *Irreducible Representations*.

It can be shown that $\Gamma'$ for $C_{3v}$ point group cannot be further reduced to $\Gamma_1'$ and $\Gamma_2'$ representations by any change of basis coordinates and so it is an irreducible representation.

### 3.4.2 Reducible and Irreducible Representations (In General)

We have defined the reducibility of any representation in a three-dimensional space into irreducible ones by choosing an appropriate basis. This concept can be further generalised for an $n$-dimensional representation.

We start with an $n$-dimensional vector space. We can express any vector in this space as

$$\bar{V} = x_1\bar{e}_1 + x_2\bar{e}_2 + x_3\bar{e}_3 + \ldots + x_n\bar{e}_n \qquad \ldots(3.106)$$

Let us take a point group (say $C_{3v}$) with symmetry operations $\{E, C_3^1, C_3^2, \sigma_V^{(1)}, \sigma_V^{(2)}, \sigma_V^{(3)}\}$.

We choose $(e_1, e_2, e_3, \ldots, e_n)$ as our basis on which we will operate by the symmetry operators. Let us transform this basis to another orthonormal basis $(f_1, f_2, f_3, \ldots, f_n)$ by any matrix $|A|$. Let the second basis (*viz.* $f_1, f_2, f_3, \ldots, f_n$) be such that the effects of the symmetry operator $R$ can be expressed as:

$$R = \left[\begin{array}{cccc|ccccc} r_{11} & r_{12} & \cdots & r_{1m} & 0 & 0 & 0 & 0 & 0 \\ r_{21} & r_{22} & \cdots & r_{2m} & 0 & & & & \\ \vdots & & & & \vdots & & 0 & & \\ r_{m,1} & r_{m2} & \cdots & r_{mm} & 0 & 0 & 0 & 0 & 0 \\ \hline & & & & r_{m+1,m+1} & \cdots & & & r_{(m+1)n} \\ & & 0 & & r_{m+2,m+1} & \cdots & & & r_{m+2,n} \\ & & & & \vdots & & & & \\ & & & & r_{n,m+2} & \cdots & & & r_{nn} \end{array}\right] \begin{array}{l} m\text{-row} \\ \\ \\ \\ (n-m)\text{ row} \end{array} = \left|\begin{array}{c|c} D'R & \\ \hline & D''R \end{array}\right|$$

$$= R\begin{vmatrix} f_1 \\ f_2 \\ f_3 \\ \vdots \\ \vdots \\ f_n \end{vmatrix} = \begin{vmatrix} r_{11} & r_{12} & \ldots r_{1m} & 0 & 0 & \ldots 0(n) \\ r_{21} & \ldots r_{22} & \ldots r_{2m} & 0 & 0 & \ldots 0(n) \\ \ldots & \ldots & \ldots & \ldots & \ldots & \ldots 0(n) \\ r_{m1} & r_{m2} & \ldots r_{mn} & 0 & 0 & \ldots 0(n) \\ 0 & 0 & \ldots 0 & r_{m+1,m+1} & r_{m+1,m+2} & \ldots r_{m+1,n} \\ 0 & 0 & \ldots 0 & r_{m+2,m+1} & r_{m+2,m+2} & \ldots r_{m+2,n} \\ \ldots & \ldots & \ldots & \ldots & \ldots & \ldots \\ 0 & 0 & \ldots 0 & r_{n,m+1} & r_{n,m+2} & \ldots r_{n,n} \end{vmatrix} \begin{vmatrix} f_1 \\ f_2 \\ f_3 \\ \vdots \\ \vdots \\ f_n \end{vmatrix}$$

...(3.107)

In this representation with respect to $(f_1, f_2, f_3, \ldots, f_n)$, the first $m$ base vectors are transformed by the symmetry $R$ to other base vectors and the remaining base vectors $(f_{m+1}, f_{m+2}, \ldots, f_n)$ do not change by the symmetry operation $R$. Thus $m$-base vectors $(f_1, f_2, f_3, \ldots, f_m)$ by themselves form a vector space $V'$ of $m$-dimensions (a subspace of $V$). This vector space $V'$ has the property that any vector in the vector space is transformed into another vector of $V'$ by the symmetry operator $R$ (invariant subspace of $V$). Since, we have assumed orthonormal bases $(f_1, f_2, \ldots, f_n)$,then the base vectors $(f_{m+1}, f_{m+2}, \ldots, f_n)$ define another subspace $V''$ of $(n - m)$ dimension. $V''$ will have a property that any vector $f''$ in the subspace $V''$ will be orthogonal to the vector in $V'$ and the scalar products $(f' \cdot f'')$ and also $(Rf' \cdot Rf'')$ will be zero if $(f' \cdot f'') = 0$. Since vector $Rf'$ belong to $V'$ sub-space, so the vector $Rf''$ (belonging to $V''$) will be orthogonal to vector $Rf'$ and will be linear combinations of the vectors $(f_{m+1}, f_{m+2}, \ldots f_n)$, as shown in eq. (3.107).

So the operation $R$ can be represented as

$$R = \left[\begin{array}{cccc|ccccc} r_{11} & r_{12} & \cdots & r_{1m} & 0 & 0 & 0 & 0 & 0 \\ r_{21} & r_{22} & \cdots & r_{2m} & 0 & & & & \\ \vdots & & & & \vdots & & 0 & & \\ r_{m,1} & r_{m2} & \cdots & r_{mm} & 0 & 0 & 0 & 0 & 0 \\ \hline & & & & r_{m+1,m+1} & & \cdots & & r_{(m+1)n} \\ & & 0 & & r_{m+2,m+1} & & \cdots & & r_{m+2,n} \\ & & & & \vdots & & & & \\ & & & & r_{n,m+2} & & \cdots & & r_{nn} \end{array}\right] \begin{array}{l} m\text{-row} \\ \\ (n-m)\text{ row} \end{array} = \left[\begin{array}{c|c} D'R & \\ \hline & D''R \end{array}\right]$$

...(3.108)

From eq. (3.108), it can be concluded that the two invariant subspaces $V'$ and $V''$ give two independent representations $D'(R)$ and $D''(R)$ of the point group with respect to bases $(f_1, f_2, f_3, \ldots, f_m)$ and $(f_{m+1}, f_{m+2}, \ldots, f_n)$, respectively.

Let $V'$ produce the representation $\Gamma^1$ and $V''$ produce the representation $\Gamma^2$. We write it as

$$\Gamma = \Gamma^{(1)} + \Gamma^{(2)} \qquad \ldots(3.109)$$

It implies that $\Gamma$ is a reducible representation of the point group whose matrices give a suitable choice of basis vectors into reduced form (3.108) and can be expressed as the direct sum. The only condition is that the basis which has been arbitrarily chosen should be orthonormal which divides the $V$ space into two invariant subspaces $V'$ and $V''$.

We now continue this process of reducing the space into subspaces and assume that $V'$ and $V''$ can further be subdivided into smaller dimensional subspaces so that their matrices $D'(R)$ and $D''(R)$ can further be expressed as the direct sum of matrices belonging to representations of smaller dimensions by the change of basis and reduce $\Gamma^{(1)}$ and $\Gamma^{(2)}$ further to other representations. This process of reduction is continued until we eventually divide the space $V$ into a number of subspaces ($V_1, V_2, \ldots, V_k$) each of which gives independently a representation of the group of operators $R$. None of these subspaces can be further subdivided into invariant subspaces of smaller dimensions. When one reaches the final stage then all the matrices of the original $G$ will be transformed into a diagonal block form.

$$D(R) = \begin{vmatrix} D^1(R) & & & & 0 \\ & D^2(R) & & & \\ & & D^3(R) & & \\ & & & \ddots & \\ 0 & & & & D^p(R) \end{vmatrix} \qquad \ldots(3.110)$$

Here $D^1(R), D^2(R), \ldots, D^p(R)$ belong to different irreducible representations ($\Gamma^1.\Gamma^2\ldots \Gamma^p$) of the group which cannot be further reduced by any change of basis whatsoever.

We have thus, arrived at a stage where the reducible representation $\Gamma$ has been completely reduced into its component irreducible representations and we write.

$$\Gamma = \Gamma^1 + \Gamma^2 + \Gamma^3 + \Gamma^4 + \ldots + \Gamma^p \qquad \ldots(3.111)$$

In the reduction of $\Gamma$, it is also possible, that the same IR may appear more than once or, it might happen that $[D^i(R)]$ and $[D^j(R)]$, having the same dimension, are related by a similarity transformation (*viz.* $AD^i(R)A^{-1}$) with some particular matrix $|A|$ and $D^i(R)$ and $D^j(R)$ are equivalent representations. In that case, we should write the decomposition of $\Gamma$ as

$$\Gamma = \Gamma^1 + \Gamma^2 + \Gamma^g + \Gamma^m + \ldots + \Gamma^{(p-q)} \qquad \ldots(3.112)$$

The above concepts are fundamental for the application of group theory to various problems. In future chapters, we will see the problem of classifying the normal modes of vibrations of a molecule, electronic energy levels, finding selection rules *etc.* Whereby we will reduce a reducible representation into its components in which a representation of a point group of the molecule must be decomposed. Prior to the applications of group theory to various problems, we will describe IRs of various point groups, formation of symmetry adopted linear combinations, the basis vectors taken in a dynamic problem of **molecules** by projection operators.

## 3.5 RECAPITULATION

Since, this chapter deals with some of the very important concepts we would like to summarise the subject matter of representations.

(*a*) We have defined three important vector spaces, *viz.* physical, function and configuration (*n*-dimensional).

(*b*) We have arbitrarily chosen a coordinate system and defined orthonormal unit basis vectors, *viz.* ($\overline{e}_1, \overline{e}_2, \overline{e}_3, \ldots, \overline{e}_n$). We have used as an example a three-dimensional physical vector space with basis vectors $\overline{e}_1, \overline{e}_2, \overline{e}_3$.

(*c*) We have determined the effects of symmetry operators on the basis vectors and also on the coordinates of a vector with respect to ($\overline{e}_1, \overline{e}_2, \overline{e}_3$) It led us to the formation of matrices (one matrix for one symmetry operation).

(*d*) We have mentioned that the same sort of operations can be carried out on an *n*-dimensional physical, functional and configuration space. We did not prove it, but we have extrapolated the results of three-dimensional vector space to *n*-dimensional vector spaces.

(*e*) We have taken a point group $C_{3v}$ and determined the various sets of matrices corresponding to its symmetry operations forming the $C_{3v}$ point group. Thus, a set of matrices are formed determining the effects of symmetry operations of $C_{3v}$ with respect to a three-dimensional arbitrarily chosen orthonormal basis ($\overline{e}_1, \overline{e}_2, \overline{e}_3$). This formed a representation of $C_{3v}$.

(*f*) We have also chosen two other sets of basis vectors [*viz.* ($d_1$, $d_2$, $d_3$)] and [$e^{+2\pi i/3}$, $e_3$] and calculated two different sets of matrix representations with respect to these arbitrarily chosen bases.

(*g*) We have found that these representations are basis-dependent which are all equivalent representations in a vector space of the same dimensionality and are related through a similarity transformation [$AD(R)A^{-1}$] where $A$ matrix transforms one basis set into another. These representations are of no advantage as they all are basis dependent.

Secondly, when any symmetry operation is applied to the vectors of a basis, then the basis or the coordinates get mixed up among each other. Since, the dimensions of a representation are the same as that of the vector space, the dimension of the matrix goes on increasing with the increase of dimensionality of the vector space. The matrices of larger dimensionality are difficult to handle mathematically. These problems have to be solved.

(*h*) In search of any property of matrices which is basis independent, we found that the traces of all equivalent matrices are the same. So one can use the traces of the matrices formulating the point group representation in order to characterise the point groups. Thus, the traces of the matrices of matrix representation are used to characterise the point groups. These traces are known as characters representing the symmetry operators of the point groups. These characters are basis independent.

(*i*) The above said problem of the case (*g*) is taken care of for classifying the matrices or representations into reducible and irreducible representations. It has been carried as under.

(*j*) Suppose we have *n*-dimensional vector space with *n* number of orthogonal basis set $(f_1, f_2, \ldots, f_n)$. We then form *n*-dimensional matrices forming *n*-dimensional representations. In this representation (say it is non-diagonal and let us designate it as $G_{(p)}$. Here *p* varies from 1 to '*h*' where *h* is the order of the group so we have a representation of a point group of '*h*' order representation as

$$\{G_{(1)}, G_{(2)}, G_{(3)}, \ldots, G_{(p)} \ldots, G(h)\}$$

where $G_{(p)}$'*s* are different non-diagonal matrices, and represent it as $\Gamma_{red}$. We change the basis coordinates $\{(f_1, f_2, \ldots, f_n)\}$ to some other coordinate system $\{f_1', f_2', \ldots, f_n'\}$. Using this coordinate system we will obtain another equivalent representation. If we continue changing our basis coordinate system, one will get a coordinate system in which all matrices of the representation simultaneously in the blocked form in which all the blocks are on the diagonal of the matrix. Each block in all the matrices of all symmetry operations will be of the same dimension. These ultimately reduced form of matrices will form the IRs of the group.

$$G(p) = \begin{vmatrix} G_1(g) & 0 & 0 & 0 & 0 & 0 & 0 & 0 \\ 0 & G_2(p) & 0 & 0 & 0 & 0 & 0 & 0 \\ 0 & 0 & G_3(p) & 0 & 0 & 0 & 0 & 0 \\ \cdots & \cdots & \cdots & \cdots & \cdots & \cdots & \cdots & \cdots \\ 0 & 0 & 0 & 0 & 0 & 0 & 0 & G_h(p) \end{vmatrix}, \quad (p = 1, 2, 3, \ldots h) \qquad \ldots(3.113)$$

Here,

(*k*) Zero represents zero elements of the matrix.

(*l*) $G_1(1), G_1(2), \ldots, G_1(h)$ are of the same dimensions. Likewise $G_2(1)$, $G_2(2), \ldots, G_2(h)$ are also of the same dimension and so on.

(*m*) There will be '*m*' such set of matrices $\{G_1(p), G_2(p), G_3(p), \ldots, G_m(p)\}$, $(m < n)$, and $\{p = 1, 2, 3, \ldots, h\}$. From the rule of matrix multiplication in block form, it follows that the *m* sets $\{G_1(p), G_m(1), G_m(2), \ldots G_m(h)\}$, $(m = 1, 2, \ldots, m < n)$ will form *m* representations of the point group. If these blocked matrices cannot be further reduced to blocks of smaller dimensions by any change of orthogonal basis of unit vectors, then these *m* number of representations are called *Irreducible Representations*.

(*n*) Out of these irreducible representations (*m* in number), it is possible that, they can be transformed into each other by some similarity transformations.

(*o*) Each of these IRs will be characterised by the characters of the matrices.

(*p*) The characters of all the symmetry operations belonging to a class will be the same.

(*q*) The characters of symmetry operations belonging to different classes will be different in general.

Thus, the above discussion implies that we can characterise point groups by the set of characters of the matrices belonging to matrix representation.

Now, the following problems have to be solved:

1. For any point group, will there be a fixed number of IRs If so, how many IRs are there for a point group and what are their dimensionalities?
2. For any reducible representation formed by arbitrarily choosing a basis of the vector space, the characters of the matrices can be decomposed into characters of IRs of the point group. If so, which ones of IRs characters are present in the character of the reducible representation and how many of IRs are present in the reducible one.
3. Even if the problem given under above case (2) is solved then it is also very important as to how one goes about making the linear combinations of the basis vectors which has produced reducible representation so that there the binary combined basis vectors give the IRs of the point group.

If we can solve the above problems, we can very comfortably and easily apply symmetry properties of the molecules in solving many mathematical problems in the properties of the next two chapters. In the same chapters, we also discuss the character table of the point group and how one goes about using it?

# 4

# Properties of Irreducible Representations

In the previous chapter, we have discussed the formation of reducible representations of point groups by the sets of matrices. The matrices forming the representation multiply the same way as the symmetry operations of the group. Since for the same point group, one can form infinite number of matrix representations, it is impossible to classify and to characterise any point group by such representations. Besides, the matrices are dependent on the basis and also on the dimensionality of the vector space. All such matrix representations are termed as "Reducible Representations" and forbid us to characterise the point group uniquely. In the same chapter, we have also seen that the problem is not so bad. Because it is possible to choose a basis with respect to which a matrix representation can be formed which is independent of the chosen basis and dependent on the dimensionality of the vector space. In this representation, (obtained after applying similarity transformation to the reducible representation), matrices appear in the block-diagonal form with smaller dimensional matrices appearing along the diagonal. The form of these block diagonalised matrices is the same for all the matrices associated with every symmetry operations of the point group. The set of all the smaller blocks, (one from each blocked diagonal matrix, at one particular position of the blocked matrices corresponding to the symmetry operators) form one representation. This is found to be independent of the basis and depends upon the dimensions of the vector space. Such basis independent matrix-representations are known as "IRREDUCIBLE" representations. (Henceforth, these will be abbreviated as I.Rs). These I.Rs have very remarkable properties which we will be discussing in the present chapter. All the point groups can now be classified and characterised by the characters of the matrices, of these I.Rs once for all and are able to form the Character Tables for every finite point group.

## 4.1 PROPERTIES OF I.Rs OF THE POINT GROUPS

**Theorem I:** Let there be a group $G$ of order '$h$'. It possesses '$k$' different classes. The number of I.Rs $\Gamma_1, \Gamma_2, \ldots, \Gamma_k$ possessed by the group $G$ will precisely be equal to $k$. That is, the number of I.Rs is equal to the number of classes of the group.

**Theorem II:** Let the dimensions of I.Rs ($\Gamma_1, \Gamma_2, \Gamma_3, \ldots, \Gamma_k$) be equal to $n_1, n_2, n_3, \ldots, n_k$ respectively. The order of the group $G$ satisfies the following equation.

$$n_1^2 + n_2^2 + \ldots + n_k^2 = h \qquad \ldots(4.1)$$

Thus for a point group $G$ of order $h$ and classes $k$, there will be only one possible set of ($n_1, n_2, n_3, \ldots, n_k$) integers, whose sum of the squares will be equal to $h$. One can not have more than one set of integers $k$ satisfying eq. (4.1).

As a first example, we have $C_{3v}$ point group of order 6 (number of symmetry operations $\{E, (C_3^1, C_3^2), (\sigma_v^{(1)}, \sigma_v^{(2)}, \sigma_v^{(3)})\}$ with three classes, [*viz.* $\{k_1$, $k_2$ and $k_3$]. Only possible set of $n_1$, $n_2$ and $n_3$ integers satisfying eq. (4.1) is

$$1^2 + 1^2 + 2^2 = 6 \qquad \ldots(4.2)$$

We will therefore have two non-equivalent I.Rs each of dimensions one and one I.R of dimension two.

As a second example, we take point group $C_{2v} = \{E, C_2, \sigma_{xv}, \sigma_{xz}\}$ of order 4 with four classes ($k_1$, $k_2$, $k_3$, $k_4$). Since the point group contains four classes, Theorem II tells us that it will have four different I.Rs whose dimensions ($n_1, n_2, n_3, n_4$) satisfy eq. (4.1). It can be easily seen that the only one set of four integers, the sum of whose squares is four will be the set (1, 1, 1, 1). That is,

$$1^2 + 1^2 + 1^2 + 1^2 = 4 \qquad \ldots(4.3)$$

We conclude that $C_{2v}$ possesses four I.Rs each of dimension one.

As a third example, we take point group $T_d = \{E, 8C_3, 3C_2, 6S_4, 6\sigma_v\}$ of order 24 and having five classes. Thus according to theorem I, we will be having five different I.Rs ($\Gamma_1$, $\Gamma_2$, $\Gamma_3$, $\Gamma_4$ and $\Gamma_5$). Let their respective dimensions be ($n_1$, $n_2$, $n_3$, $n_4$ and $n_5$). The values of these $n_i$'s must satisfy eq. (4.1). That is,

$$n_1^2 + n_2^2 + n_3^2 + n_4^2 + n_5^2 = 24$$

It can easily be seen that the only set of five integers, the sum of whose squares is equal to 24 will be the set (1, 1, 2, 3, 3).

$$1^2 + 1^2 + 2^2 + 3^2 + 3^2 = 24 \qquad ...(4.4)$$

We conclude that $T_d$ possesses five I.Rs, two of which are one-dimensional, next one of dimension two, and the remaining two each of dimension three. No other set of five integers is possible whose sum of square is 24.

**Theorem III "The Great Orthogonality Theorem":** It is concerned about the elements of the matrices associated with I.Rs of a group satisfying eq. (4.1). In the equation, the dimensionality of $\Gamma_k$ is given by $n_k$ which also shows the dimensionality of the matrix of the irreducible representation $\Gamma_k$. When we square $(n_k)$, the number $(n_k^2)$ gives the number of elements in the $|n_k \times n_k|$ matrix. We will represent the elements of $|n_k \times n_k|$ matrix representing one symmetry operation of the group in the $\Gamma_k$ by the number $[\Gamma_k(R)_{ml}]$. The latter is the number present in the matrix at the cross of the $m^{th}$ row and $l^{th}$ column.

The great orthogonality theorem is

$$\sum_R [(\Gamma_i(R)_{mn})][(\Gamma_j(R)_{m'n'})]^* = \frac{h}{\sqrt{l_i l_j}} \delta_{ij}\delta_{mm'}\delta_{nn'} \qquad ...(4.5)$$

Here $l_i$, $l_j$ are dimensions of I.Rs. $i$ and $j$ respectively of the point group of order $h$.

We know that a group of order $h$ defines a $h$-dimensional space, the space of the group elements. In this group space, we would like to find $h$ number of linearly independent vectors of lowest possible dimensions which will span the whole group space. Let us assume $h$ number of orthonormal I.Rs. Each I.R. of the group space is a set of $h$-matrices (one matrix for one symmetry operation). The dimensionality of the matrices will be the same as that of I.R. of the group and will be represented by $(l_i \times l_i)$ matrices (one for each operation $R$ in the $i^{th}$ I.R.) These will contribute $(l_i \times l_i)$ components, one component for one vector and thus giving $(l_i \times l_i)$ vectors. For example, the components of one of the $l_i^2$ vectors will be $[(a_{11}^i)_E, (b_{11}^i)_{R_1}, (c_{11}^i)_{R_2}, ..., (p_{11}^i)_{R_p}]$, where $a_{11}, b_{11}, c_{11}, \ldots, p_{11}$ are the (1, 1) elements of the matrices representing I.R. Thus, the matrix of the one-dimensional I.R will be $|(x)|$, that of two-dimensional will be $|2 \times 2|$ giving four components and that of three dimensional will be $|3 \times 3|$ giving nine components. As an example, we have,

Symmetry operations of the group of order $h = E, R_1, R_2, R_3, \ldots, R_h$.

Matrices of two-dimensional irreducible representations are

$$\underset{E}{\begin{vmatrix} 1 & 0 \\ 0 & 1 \end{vmatrix}}, \underset{R_1}{\begin{vmatrix} a_1 & b_1 \\ c_1 & d_1 \end{vmatrix}}, \underset{R_2}{\begin{vmatrix} a_2 & b_2 \\ c_2 & d_2 \end{vmatrix}}, \underset{R_3}{\begin{vmatrix} a_3 & b_3 \\ c_3 & d_3 \end{vmatrix}}, \ldots, \underset{R_n}{\begin{vmatrix} a_n & b_n \\ c_n & d_n \end{vmatrix}}$$

Operation of the group =

Orthogonal vector no. one $= |1, a_1, a_2, a_3, \ldots, a_n|$

Orthogonal vector no. two $= |1, b_1, b_2, b_3, \ldots, b_n|$

Orthogonal vector no. three $= |1, c_1, c_2, c_3, \ldots, c_n|$

Orthogonal vector no. four $= |1, d_1, d_2, d_3, \ldots, d_n|$

} Here $a_i$, $b_i$, $c_i$ and $d_i$ are the components of the vectors, designated as $(R)_{mn}$

These vectors form the I.Rs of the group.

Eq. (4.5) expresses the fact that the square of the length of any such vector equals to $(h/l_k)$ and the total no. of such vectors will be equal to the order $h$ of the group (or the dimensionality of the group space).

We discuss the following rules or properties of I.Rs of the group and their characters with reference to above orthogonality theorem eq. (4.5).

$$\{(a_{11}^i)E, (b_{11}^i)R_1, (c_{11}^i)R_2, \ldots, (P_{11}^i)R_P$$

1. The sum of the squares of the dimensions of the I.Rs of the group is equal to the order of the group already given as eq. (4.1).

   (This can be derived from the orthogonality theorem, but the proof is lengthy and will not be given here). Since $\chi_i(E)$, the character of the representation of $E$ (identity element) in the I.Rs is equal to the dimensionality of the I.R. we can also write equation (4.1) as,

$$[\chi_1(E)]^2 + \{\chi_2(E)]^2 + [\chi_3(E)]^2 + \ldots [\chi_k(E)]^2 = h \qquad \ldots(4.6)$$

2. The sum of the squares of the characters in any I.R. equals to $h$. That is,

$$\sum_R [\chi_i(R)]^2 = h \qquad \ldots(4.7)$$

   (The proof of eq. (4.6) is given below.

   We know,

$$\sum_R [\Gamma_i(R)_{mm}] [\Gamma_i(R)_{m'm'}] = \frac{h}{l_i} d_{mm'}$$

   Summing over $m$ and $m'$, we get the (LHS).

$$\sum_m \sum_{m'} \sum_R [\Gamma_i(R)_{mm}] [\Gamma_i(R)_{m'm'}]^* = \sum_R \chi_i(R)\, \chi_i(R) = \sum_R [\chi_i(R)]^2 \qquad \ldots(4.8)$$

   Summing over $m$, $m'$ on the RHS, we get,

$$\frac{h}{l_i} \sum_m \sum_{m'} \delta_{mm'} = \frac{h}{l_i} . l_L = h \qquad \ldots(4.9)$$

Thus,

$$\sum_R [\chi_i(R)]^2 = h \qquad ...(4.10)$$

3. The vectors whose components are the characters of two different I.Rs are orthogonal. That is,

$$\sum_R \chi_i(R)\chi_j(R) = 0, \text{ when } i \neq j \qquad ...(4.11)$$

[It is easy to prove this relation from eq. (4.5)]

4. In a given representation (Reducible or Irreducible), the characters of all matrices belonging to operations in the same class are identical.

We have proved this earlier.

5. The number of irreducible representations of a group is equal to the number of classes in the group. This is Theorem No. 2 given earlier in the Chapter.
6. We would like to state as a corollary of property 1 that the $l_k^2$ vector corresponding to an I.R. of $h$-dimensional subspace. Since, the total space is the sum of its orthogonal subspaces formed by all the I.Rs of a group of order $h$, then eq. (4.5) follows immediately.
7. There will always be one I.R. which is termed as identity representation in which the matrices (and characters) for all the operations are of order (1 × 1) or Character 1.
8. Character of Identity operation for all I.Rs must be positive. It cannot have negative value.
9. The characters of Identity operation imply the dimensionality of the I.R. Thus for $\chi_E = 2$ the dimensionality of the I.R. is two, $\chi_E = 3$ shows the dimensionality of the I.R. three, *etc.*

**Example 1:** (Related to the above rules on properties of I.Rs)

Let us take the point group $C_{2v} = \{(E, C_2, \sigma_{xz}, \sigma_{yz})\}$. It contains four symmetry operations. Each operation is in separate class. Hence, according to the Property No. 5, we will be having four I.Rs for $C_{2v}$ and according to property number one, eq. (4.1), we should have,

$$n_1^2 + n_2^2 + n_3^2 + n_4^2 = 4.$$

The only solution for the above equation is

$$1^2 + 1^2 + 1^2 + 1^2 = 4.$$

So the dimensions of all I.Rs will be one.

## 4.2 DETERMINATION OF CHARACTERS

By the application of above rules or properties of I.Rs, we can calculate the characters of all the I.Rs of $C_{2v}$ as follows.

We write the classes and the I.Rs and calculate the characters of various I.Rs in the following steps:

**Step (*a*):** According to rule 7, there will always be one Identity I.R. in which the characters of all the operations will be 1 (identity representation). This representation is taken as unfaithful representation designated as $A_1$ or $\Gamma_{1.}$

So, we write the characters of $\Gamma_1$ as 1, for all the symmetry operations of point group $C_{2v}$ according to rule 6. (Thus according to rule 2, $1^2 + 1^2 + 1^2 + 1^2 = 4$. The characters of the remaining three representations will also be (±1) satisfying the rule 2.

**Step (*b*):** According to rule 3, the remaining three I.Rs (*viz.* $\Gamma_2$, $\Gamma_3$, $\Gamma_4$) must be orthogonal. That is,

$$\sum_R [\chi_i(R)][\chi_j(R)]^* = 0 \qquad \text{...(4.12)}$$

So in order that the I.R $\Gamma_2$ satisfy eq. (4.12), we must have the characters of two symmetry operations to be 1 and those of other two as (–1). That is, according to (4.12), we must have,

$$(1)(1) + (1)(-1) + (1)(-1) + (1)(1) = 0 \qquad \text{...(4.13)}$$

Therefore, we must have the characters as given in the Table 4.1. According to rule 6, the character of the operation $E$ can never be negative in our case.

In this manner we have calculated the characters of all the operations of $C_{2v}$ point group.

**Example 2:** Let us take the $C_{3v}$ point group ( $E, C_3^1, C_3^2, \sigma_v^{(1)}, \sigma_v^{(2)}, \sigma_v^{(3)}$ ) having three classes $\{E, 2C_3, 3\sigma_v\}$, of order 6. Thus according to rule 1, we will have three I.Rs of the group $\{\Gamma_1, \Gamma_2, \Gamma_3\}$. Let the dimension of the I.Rs be $n_1$, $n_2$ and $n_3$. by eq. (4.14), we get

$$n_1^2 + n_2^2 + n_3^2 = 6 \qquad \text{...(4.14)}$$

Eq. (4.14) can be satisfied only with the Set 1, 1 and 2. No other set of integers will satisfy eq. (4.14). We will therefore, be having two I.Rs of dimension-1 and the third one of dimension-2. We write it in the tabular form as given in Tables 4.1 and 4.2. As usual, there will be one *identity I.R* having the character 1 for all operations (Rule 7). The characters of identity operation for all the representations will be the same as those of the dimensions of the I.Rs (Rule 9). We now look for the characters for the second one-dimensional I.R ($\Gamma_2$). Its characters will be ±1 and it will be orthonormal to $\Gamma_1$. So, if we write the character of $3\sigma_v$ operations as (–1) and those of ($2C_3$) operations as 1, we will be having the following relationship according to rule 3.

**Table 4.1:** Characters of representation of $C_{2v}$ point group

| $C_{2v}$ | $E$ | $C_2$ | $\sigma_{xz}$ | $\sigma_{yz}$ |
|---|---|---|---|---|
| $\Gamma_1$ | 1 | 1 | 1 | 1 |
| $\Gamma_2$ | 1 | 1 | –1 | –1 |
| $\Gamma_3$ | 1 | –1 | 1 | –1 |
| $\Gamma_3$ | 1 | –1 | –1 | 1 |

**Table 4.2:** Characters of representation of $C_{3v}$ point group

| $C_{3v}$ | $E$ | $2C_3$ | $3\sigma_v$ |
|---|---|---|---|
| $\Gamma_1$ | 1 | 1 | 1 |
| $\Gamma_2$ | 1 | 1 | –1 |
| $\Gamma_3$ | 2 | –1 | 0 |

$$1 \,.\, 1 + 2 \,.\, 1 \,.\, 1 + 3 \,.\, (-1)\,(1) = 0 \qquad \text{...(4.15)}$$

Eq. (4.15) implies that $\Gamma_2$ is orthonormal to $\Gamma_1$. However, if we assign (–1) character to operation ($C_3$) and (1) to $\sigma_v$, then the orthogonality condition of $\Gamma_2$ will not be satisfied, because of the orthogonality condition of $\Gamma_1$ and $\Gamma_2$, we will have $\{1 \,.\, 1 + 2 \,.\, 1 \,.\, (-1) + 3 \,.\, 1 \,.\, 1 \neq 0\}$. So the first assignment of characters (1, 1, –1) is justified.

We now take the assignment of characters to the third I.R ($\Gamma_3$), which is of dimension 2 and therefore under the operation $E$, we write 2, since the character of the identify operation $E$, is given by the dimensionality of the I.R. In order to find the characters of $\Gamma_3$, we apply the orthogonal theorem. Let us assume that the characters of ($C_3$) and ($\sigma_v$) operations for $\Gamma_3$ I.R are $\chi_3(C_3)$ and $\chi_3(\sigma_v)$ respectively. Thus according to Rule (3), we have,

$$\sum_R \chi_1(R)\chi_3(R) = 1.2 + 2.1.\chi_3(C_3) + 3.1.\chi_3(\sigma_V) = 0 \qquad \text{...(4.16)}$$

and $$\sum_R \chi_2(R)\chi_3(R) = 1.2.1 + 1.2 + 2.1.\chi_3(C_3) - 3.1.\chi_3(\sigma_V) = 0 \qquad \text{...(4.17)}$$

Solving eqs. (4.16) and (4.17), we obtain,

$$\chi_3\,(\sigma_v) = 0,$$

and $$\chi_3\,(C_3) = (-1)$$

We may check our result by our rule 1 [*viz.* $(n_1^2 + n_2^2 + n_3^2 = h)$]

This gives,

$$(2)^2 + (-1)^2 \,.\, 2 + 3.\ (0)^2 = 6$$

The rule 1 is satisfied. Thus, we make the character table of $C_{3v}$ (Table 4.2).

## 4.3 REPRESENTATIONS FOR CYCLIC GROUPS

It has been stated earlier that all cyclic groups are Abelian. A group of order $h$ will have $h$ cyclic operations $(E, C_n^1, C_n^2, \ldots, C_n^{n-1})$ and each operation will form a class by itself. According to rule (1), there will be $h$ number of one-dimensional I.Rs. We have a general scheme for characters of such point groups. This is explained by taking the following example.

Let us consider the point group $C_5 = \{E, C_5^1, C_5^2, C_5^3, C_5^4\}$. We wish to know a set of five one-dimensional representations $[\Gamma_1, \Gamma_2, \Gamma_3, \Gamma_4, \Gamma_5]$. These I.Rs must be orthonormal. That is,

$$\sum_{m=1}^{5}[\Gamma_p(C_n^m)][\Gamma_q(C_n^m)]^* = h\delta_{pq} = 5\delta_{pq} \qquad ...(4.18)$$

We know that,

$$\text{Exp }(2\pi ip/5) = \cos\frac{2\pi ip}{5} + i\sin\frac{2\pi ip'}{5} \qquad ...(4.19)$$

We may abbreviate $(e^{2\pi i/5})$ as $\in$, so that

$$\text{Exp.}\frac{2\pi ip}{5} = \in^p, \text{ when } p = 5, 10, \text{ then Exp. } \frac{2\pi ip}{5} = 1 \qquad ...(4.20)$$

We can write the characters of $C_5$ point group as given below in the tabular form (Here for $\Gamma_p = \in^p$ for $C_5^1$, $\in^{2p} = C_5^2, \ldots, C_5^5$) (Table 4.3).

The characters of $C_5$ point group follow the rule of orthogonalisation.

Let us take $\Gamma_p$ and $\Gamma_q$ I.Rs and apply the orthonormality condition. Assuming $(q - p) = r$, we get

**Table 4.3**

| $C_5/\Gamma_p$ | $C_5^1$ | $C_5^2$ | $C_5^3$ | $C_5^4$ | $C_5^5 (=E)$ |
|---|---|---|---|---|---|
| $\Gamma_1$ | $\in$ | $\in^2$ | $\in^3$ | $\in^4$ | $\in^5 = E = 1$ |
| $\Gamma_2$ | $\in^2$ | $\in^4$ | $\in$ | $\in^3$ | $\in^{10} = E = 1$ |
| $\Gamma_3$ | $\in^3$ | $\in$ | $\in^4$ | $\in^2$ | $\in^{15} = E = 1$ |
| $\Gamma_4$ | $\in^4$ | $\in^3$ | $\in^2$ | $\in$ | $\in^{20} = E = 1$ |
| $\Gamma_5$ | $\in^5$ | $\in^{10}$ | $\in^{15}$ | $\in^{20}$ | $\in^{25} = E = 1$ |

=

**Table 4.4**

| $C_5/\Gamma_p$ | $C_5^1$ | $C_5^2$ | $C_5^3$ | $C_5^4$ | $C_5^5$ |
|---|---|---|---|---|---|
| $\Gamma_1$ | $\in$ | $\in^2$ | $\in^3$ | $\in^4$ | 1 |
| $\Gamma_2$ | $\in^2$ | $\in^4$ | $\in$ | $\in^3$ | 1 |
| $\Gamma_3$ | $\in^3$ | $\in$ | $\in^4$ | $\in^2$ | 1 |
| $\Gamma_4$ | $\in^4$ | $\in^3$ | $\in^2$ | $\in$ | 1 |
| $\Gamma_5$ | 1 | 1 | 1 | 1 | 1 |

$$\sum_R[\Gamma_p(R)][\Gamma_q(R)]^* = [(\in^p)^* \in^p . \in^r + (\in^{2p})^* \in^{2p}\in^{2r} +$$

$$[(\in^{3p})^*(\in^{3p} . \in^{3r})] + [(\in^{4p})^* \in^{4p}\in^{4r}] + [(\in^{5p})^* \in^{5p}\in^{5r}] \qquad ...(4.21)$$

Since $\in^{ix} \cdot \in^{-ix} = \in^{ix} \cdot (\in^{ix})^* = \in^{ix-ix} = \in^0 = 1$, then (eq. 4.21) becomes,

$$\sum_R[\Gamma_p(R)]\ [\Gamma_q(R)]^* = \in^r + \in^{2r} + \in^{3r} + \in^{4r} + \in^{5r} \qquad ...(4.22)$$

Further, if $p = q$, we have eq. (4.22) (as $q - p = r = 0$) [*i.e.*, $\Gamma_p$, $\Gamma_p^*$)]

$\in^0 + \in^0 + \in^0 + \in^0 + \in^0 = 1 + 1 + 1 + 1 + 1 = 5$. This means that all the representations are normalised.

If $p$ and $q$ are different then $r$ is 1, 2, 3, or 4.

We also have,

$$\in^5 = \cos(2\pi i5/5) + \sin(2\pi i5/5) = 1 \qquad ...(4.23)$$

and ,

$$\in^7 = \in^5 \cdot \in^2 = \in^2, \textit{ etc}$$

Therefore, eq. (4.22) becomes,

$$\in^r + \in^{2r} + \in^{3r} + \in^{4r} + \in^{5r} = \in + \in^2 + \in^3 + \in^4 + \in^5$$

$$= \sum_{n=1}^{5} \exp(2\pi in/5) \qquad ...(4.24)$$

[*e.g.*, for $r = 1$, we have $(\in^1 + \in^2 + \in^3 + \in^4 + \in^5)$

for $r = 2$, we have $(\in^2 + \in^4 + \in^6 + \in^8 + \in^{10}) = (\in^3 + \in^1 + \in^4 + \in^2 + \in^5)$

for $r = 3$, we have $(\in^3 + \in^6 + \in^9 + \in^{12} + \in^{15}) = (\in^2 + \in^4 + \in^1 + \in^3 + \in^5)$

and so on.]

Now, we have two trigonometric relations

$$\left.\begin{array}{ll} 1. & \sum_{n=1}^{n=5} \cos(2\pi n/5) = 0 \\ \text{and} & \\ 2. & \sum_{n=1}^{n=8} \sin(2\pi n/5) = 0 \end{array}\right\} \qquad ...(4.25)$$

The above identity shows that the equation (4.25) equals zero.

Using above results, one can write the character table of $C_5$ as given [*viz*. Tables 4.3 and 4.4].

It is obvious that $(\in^x)^* = \in^{5-x}$ [*e.g.*, $(\in^4)^* = \in^1$, *etc.*]

Using this relationship, we can rewrite Tables 4.3 and 4.4 after rearranging the terms.

**Table 4.5**

| $I.R/\Gamma_p$ | $E$ | $C_5$ | $C_5^2$ | $C_5^3$ | $C_5^4$ | |
|---|---|---|---|---|---|---|
| $\Gamma_5$ | 1 | 1 | 1 | 1 | 1 | |
| $\Gamma_1$ | 1 | $\in$ | $\in^2$ | $\in^{*2}$ | $\in^*$ | $E_1$ |
| $\Gamma_4$ | 1 | $\in^*$ | $\in^{2*}$ | $\in^2$ | $\in$ | |
| $\Gamma_2$ | 1 | $\in^2$ | $\in^*$ | $\in$ | $\in^{2*}$ | $E_2$ |
| $\Gamma_3$ | 1 | $\in^{2*}$ | $\in$ | $\in^*$ | $\in^2$ | |

In this manner, one can form the character tables of all the cyclic groups (*viz.* $C_n$).

## 4.4 RELATIONSHIP OF THE REDUCIBLE AND IRREDUCIBLE REPRESENTATIONS

In the last chapter, we have posed a couple of problems regarding reducible and I.Rs of any group. One of the problems is that if one is given a reducible representation of any point group, how one can find the numbers and the types of I.Rs that the reducible representation contains. It implies that we should know a relationship between a reducible representation of a group and its I.Rs. This will be of utmost importance in the practical applications of group theory to molecular problems. We will realise it when we will discuss the applications of group theory to some of the molecular problem in the book. We know that on choosing arbitrarily a basis for the representation of a group, we usually get a reducible representation [cf. Chapter Three]. We have also seen that for any reducible representation, one can find some similarity transformation which will reduce each matrix of the reducible representation to the one consisting of blocked-matrices of smaller dimensions along the diagonal. The process of reduction of reducible representation matrices can be reached to a stage when there is no further reduction of the blocked diagonal matrices. At this stage, each blocked diagonal matrix (one matrix representing one symmetry operation), belongs to an I.R of the group. We have also learnt that any similarity transformation does not change the trace (or the character) of a matrix. Thus, we can write

$$\chi^{(\text{red})}(R) = \sum_i a_i \chi_i^{\Gamma_i}(R) \quad \text{...(4.26)}$$

In eq. (4.26), $\chi^{(\text{red})}(R)$ is the character of the matrix associated with the symmetry operation $R$ in the *Reducible representation* and $a_i$ represents the number of times the blocked diagonal matrix belonging to the $i^{\text{th}}$ I.R ($\Gamma_i$) appears along the diagonal when the reducible representation is completely reduced by the similarity transformation. [One should realise that we do not bother at all about the process of similarity transformation to get the value of $a_i$. We will be only concerned about the ultimate result of the matrices]. The value of $a_i$ can be easily determined by using the orthogonal properties of the I.Rs of the group. We will be concerned with the character table of the groups. The value of $a_i$ can be found in the following way by using the equation (4.26):

(*a*) Multiply by $\chi_i^{\Gamma_i}$ on both sides of the eq. (4.26) and then sum each side over all the symmetry operations. That is, (*b*) $$\sum_R \chi_j^{\Gamma_i}(R)\chi^{\text{red}}(R) = \sum_R\sum_i a_i\chi_i^{\Gamma_i}(R)\chi_j^{\Gamma_i}(R) \quad ...(4.27)$$

$$= \sum_i\sum_R a_i\chi_j^{\Gamma_j}(R)\chi_i^{\Gamma_i}(R) \quad ...(4.28)$$

(*c*) In the sum on RHS, we will be having the sum of *i* number of terms in which each term will be equal to

$$\sum_R a_i\chi_i^{\Gamma i}(R)^*\chi_j^{\Gamma j}(R) = a_i\sum_R \chi_i^{\Gamma i}(R)[\chi_j^{\Gamma j}(R)]^* \quad ...(4.29)$$

Because of the orthogonality of I.Rs, the RHS of eq. 4.29

$= [a_i h]\,\delta_{ij}$ (*h* is the order of the group) ...(4.30)

That is,

RHS of eq. (4.28) = 0 for $i \neq j$

RHS of eq. (4.28) = $a_i \cdot h$ for $i = j$ ...(4.31)

Thus, when we sum up over all *i* terms in eq. (4.28), only those terms will be non-zero in which $i = j$ and in that case the sum in eq. (4.28) will be equal to

$$\sum_R \chi^{\text{red}}(R)\cdot[\chi_j^{\Gamma j}(R)]^* = a_i h$$

or $$a_i = \frac{1}{h}\sum_R \chi^{\text{Red}}(R)\cdot[\chi_j^{\Gamma j}(R)]^* = \frac{1}{h}\left[\sum_K g_k\chi_k^{\text{red}}(R)\cdot\chi_i^{\Gamma i}(R)\right]^* \quad ...(4.32)$$

where $g_k$ is the number of operation in class *k*.

Knowing the values of $\chi^{\text{red}}(R)$ and the values of $[\chi_i^{\Gamma i}(R)]^*$ from the character table of the point group, we can find the value of $a_i$, (*i.e.*, the number of time the $i^{\text{th}}$ I.R occurs in a reducible representation. (Here $\chi_k^{\text{red}}(R)$ and $[\chi_k^{\Gamma i}(R)]^*$ are the characters in class *k* and $R_k$ is one of the symmetry operator in the class *k*.

The above concept can be exemplified by the following examples of point group $C_{3v}$. In this example, we take the values of $[\chi^{\text{red}}(R)]$ from the literature. We write the Character Table 4.6 of $C_{3v}$:

From eq. 4.32, we find in $\Gamma_x^{\text{red}}$ (first example in the Table 4.6)

$a_1^{\Gamma_1}$ = 1/6 [1.5.1 + 2.2.1 + 3.(–1)(1)] = 1

$a_2^{\Gamma_2}$ = 1/6 [1.5.1 + 2.2.1 + 3.(–1)( –1)] = 2

$a_3^{\Gamma_3}$ = 1/6 [2.5.1 + 2.2.(–1) + 3.(–1)(0)] = 1

**Table 4.6:** Character Table of $C_{3v}$

| $C_{3v}$ | $E$ | $2C_3$ | $3\sigma_v$ | |
|---|---|---|---|---|
| $\Gamma_1$ | 1 | 1 | 1 | |
| $\Gamma_2$ | 1 | 1 | –1 | |
| $\Gamma_3$ | 2 | –1 | 0 | |
| $\Gamma_x^{red}$ | 5 | 2 | –1 | Example I |
| $\Gamma_y^{red}$ | 7 | 1 | –3 | Example II |

We find the components in $\Gamma_Y^{red}$ in example I

$$\Gamma_x^{red} = 1\Gamma_1 + 2\Gamma_2 + 1\Gamma_3$$

In the example II

$$a_1^{\Gamma_1} = 1/6\ [1.1.7 + 2.1.1 + 3.1(-3)] = 0$$

$$a_2^{\Gamma_2} = 1/6\ [1.1.7 + 2.1.1 + 3.(-1).(-3)] = 3$$

$$a_3^{\Gamma_3} = 1/6\ [1.2.7 + 2.(-1).1 + 3.(0)(-3)] = 2$$

That is, $\Gamma_y^{red} = 0\Gamma_1 + 3\Gamma_2 + 2\Gamma_3$

Thus, we have found that in the reducible representation $\Gamma_x^{red}$ and $\Gamma_y^{red}$, the number of times the I.R ($\Gamma_1$, $\Gamma_2$ and $\Gamma_3$) of $C_{3v}$ appears in the $\Gamma_i^{red}$. Or we may also write them as,

$$\Gamma_x = \Gamma_1 + 2\Gamma_2 + \Gamma_3$$
$$\Gamma_y = 0\Gamma_1 + 3\Gamma_2 + 2\Gamma_3 = 3\Gamma_2 + 2\Gamma_3$$

## 4.5 CHARACTER TABLE

We will be using character tables of the point groups throughout in the book particularly in Applications of Symmetry Properties.

**Table 4.7:** Character Table of $C_{3v}$ point group

| | $C_{3v}$ | $E$ | $2C_3$ | $3\sigma_v$ | ***Functions*** | |
|---|---|---|---|---|---|---|
| $\Gamma_1$ | $A_1$ | 1 | 1 | 1 | $z$ | $x^2 + y^2$ |
| $\Gamma_2$ | $A_2$ | 1 | 1 | –1 | $R_z$ | |
| $\Gamma_3$ | $E$ | 2 | –1 | 0 | $(x, y)(R_x R_y)$ | $[(x^2, y^2)], [xz, yz]$ |
| | $X$ | | $Y$ | | $Z_1$ | $Z_2$ |

These tables are of ulmost importance because without them, it is very difficult to proceed and to utilise the symmetry properties of molecules in our mathematical manipulations. It will therefore be better to understand, in detail, the character table of a point group in order to explain the meaning of numbers given in the table and the useful information one can obtain from it. We take the character table of $C_{3v}$ point group as an example. It is reproduced in Table 4.7. To explain the table, we have divided it into four parts ($X$, $Y$, $Z_1$ and $Z_2$). In the $X$-part of the table, we have written the nomenclature of the I.Rs. Till now we have represented them as $\Gamma_1, \Gamma_2, \Gamma_3, \ldots, \Gamma_n$. But in the general usage, we represent them as $A$, $B$, $E$, $T$, $H$, which have a certain meaning attached to them. Thus,

(*a*) All one-dimensional representations are designated by the letter $A$ or $B$, two-dimensional representations by $E$, three-dimensional representations by $T$ (or sometimes by $F$), four-dimensional by $H$, *etc.* (For our purpose $A$, $B$, $E$ and $T$ are sufficient.

(*b*) The systems which are symmetric with respect to rotation operation about the principal axis $C_n$ are represented as $A$ and the character of $C_n$ operation by (+1). The systems which are antisymmetric with respect to rotation operations about the principal axis ($C_n$), are designated by $B$ and the character of $C_n$ operations is represented by (–1).

(*c*) We also write 1 or 2 subscripts on $A$ and $B$ (*i.e.*, $A_1$ or $A_2$ $B_1$ or $B_2$). The number 1 or 2 represents the symmetry properties of the systems with respect to rotation operation about the $C_2$ axis, perpendicular to the principal axis $C_n$. If the systems are symmetric with respect to $C_2$ operation, then the number 1 is written ($A_1$ or $B_1$), if it is antisymmetric, then designated by number 2 (*viz.* $A_2$ or $B_2$). The characters are shown by (+1) for symmetric and (–1) for antisymmetric with respect to the $C_2$ operation.

(*d*) In case $C_2$ axis is missing from the system, then we examine the system with respect to the mirror plane $\sigma_v$ containing the principal axis $C_n$. If the systems are symmetric with respect to $\sigma_v$ plane then subscript 1 is written on $A$ or $B$ (*viz.* $A_1$ or $B_1$). If the system is antisymmetric then the subscript 2 is written (*viz.* $A_2$ or $B_2$). Their respective characters $\chi(\sigma_v)$ are (+1) or (–1) respectively.

(*e*) The symmetric or antisymmetric properties of the system with respect to $\sigma_h$ operation, are represented by (′) and (″) on $A$ and $B$. Thus, if the systems are symmetric, it is represented by (′) (*i.e.*, by $A'$ or $B'$), if they are antisymmetric, they are represented by double dash (″) and are represented as ($A''$ or $B''$). The characters under $\sigma_h$ are given as (+1) or (–1) respectively.

(*f*) If the system contains an inversion centre ($I$ or $i$), then, the systems which are symmetric with respect to inversion operation are indicated by writing $g$ as subscript on $A$ or $B$ (*i.e.*, $A_g$ or $B_g$). In case these are antisymmetric with respect to ($I$ or $i$), the letter $u$ is written as subscript on $A$ or $B$ (*i.e.*, $A_u$ or $B_u$). The characters under ($I$ or $i$) are written as +1 and –1 respectively.

(*g*) The subscripts 1 and 2 are also written on $E$ and $T$, but these are not properly defined.

**Columns *Y*:** In these columns, the characters of the I.Rs of the group for various symmetry operations have been given.

**Columns $Z_1$ and $Z_2$:** In this area, we write bases or the functions ($x, y, z, x^2, y^2, z^2$ and also $R_x$, $R_y$ and $R_z$). $x, y, z$ are the coordinates of any vector and $R_x$, $R_y$ and $R_z$ stand for rotations about $x$, $y$ and $z$-axes respectively. We know that any set of algebraic functions or vectors may serve as a basis for representation. These can be components which can be transformed by the symmetry operations. As a result of these transformations, one gets matrices representing a group. Thus using $x, y, z$ as bases for $C_{3v}$ point group, one gets the following matrices for one operation in each class (Do it as an exercise, *see* also chapter three). Matrices for the operations $E$, $C_3$, $\sigma_v$ are,

$$E = \left|\begin{array}{cc|c} 1 & 0 & 0 \\ 0 & 1 & 0 \\ \hline 0 & 0 & 1 \end{array}\right|, \quad C_3^1 = \left|\begin{array}{cc|c} \cos 120 & \sin 120 & 0 \\ -\sin 120 & \cos 120 & 0 \\ \hline 0 & 0 & 1 \end{array}\right| \quad \text{and} \quad \sigma_v \left|\begin{array}{cc|c} 1 & 0 & 0 \\ 0 & -1 & 0 \\ \hline 0 & 0 & 1 \end{array}\right|$$

These matrices are in the reduced form. One can see that $z$ forms an I.R on its own while in the second representation, for $C_3^1$ ($x$, $y$) coordinates are mixed and are two-dimensional with respect to ($x$, $y$) as basis for representation and second of one-dimension with $z$ coordinate as a basis. It means that the coordinate $z$ form a basis for $A_1$ (identity) representation, while ($x$, $y$) coordinates as basis for $E$ representation. This is shown in the $Z_1$ part of the character table for $C_{3v}$ point group. Similarly, when $R_x$, $R_y$ and $R_z$ (basis) are mixed like ($x$, $y$). These give an $E$-type of I.R.

In $Z_2$-column, we have shown the similar types of assignments for transformation properties of the product functions like ($xy$, $yz$, $xz$, $x^2$, $y^2$, $z^2$). This will be easily worked out.

# 5

# Bases, Projection Operators and SALCs

In earlier chapters, we were concerned with abstract symmetry operations. It has also been shown there that these can be replaced by the matrices that represent their effects on the basis vectors $(\overline{e}_1, \overline{e}_2, \ldots, \overline{e}_n)$ or on the coordinates $(x_1, x_2, \ldots, x_n)$ of a vector in the physical space of the molecule. The set of matrices shows one to one correspondence with the symmetry operations of the molecular point group and therefore forms a representation of the group. The set of basis vectors $(\overline{e}_1, \overline{e}_2, \ldots., \overline{e}_n)$ or the components $(x_1, x_2, \ldots, x_n)$ of a vector with respect to which we have derived the matrices form the *BASIS* of the representation. This concept arose from the idea that we have regarded a symmetry operation as an operator which operates on a position vector $\overline{r}$ producing a new vector. The idea is not limited to the physical space only but can be extrapolated to any vector $\overline{P}$ of any type of vector space, *e.g.*, function space, configuration space, *etc. For example,* in the function space of the wave functions belonging to a particular energy level of a molecule, if one associates with symmetry operator $\hat{R}$ of the molecule a corresponding linear operator $\hat{O}_R$ which can act on any vector of space (or function) to produce new vector (or function) belonging to the same space according to some defined rules, one should be able to obtain a representation of the molecular point group. This correspondence between the symmetry operations of the molecular point group and the operator $\hat{O}_R$ implies that the set of operators will also form a group having the same structure as that of the point group. Just as we introduced a coordinate system in physical space with bases $(\overline{e}_1, \overline{e}_2, \ldots., \overline{e}_n)$ which enabled us to form a matrix representation. We can also introduce a coordinate system into the function space with base vectors

$(\bar{f}_1, \bar{f}_2, \bar{f}_3, ..., \bar{f}_n)$. This will allow us to form a matrix representation of the group of operator $\hat{O}_R$ and so ultimately of the molecular point group by a set of matrices representing its symmetry operations. The set of base vectors $(\bar{f}_1, \bar{f}_2, \bar{f}_3, ..., \bar{f}_n)$ will form the basis for the matrix representation in the function space. Usually the basis vectors are orthonormal but not essential. We will discuss the formation of the matrix representation of the point groups in function space and configuration space in this chapter.

Another concept which is of ulmost importance in the application of group theory in mathematical manipulation of quantum mechanics is "how and why many molecular properties can be discussed in terms of "I.Rs." This will be discussed in the later chapters. Towards the end of chapter three, we have posed two big problems. One we have already solved in chapter four. The second problem concerning the basis of representation is that we usually arbitrarily choose basis vectors on which we find the effects of symmetry operations of the point group that give us matrices. The set of matrices, one for each operation forms a "*Reducible Representation*" of the molecular point group. By carrying out a similarity transformation on these matrices by an operator $\hat{A}$, all the matrices of the reducible representation appear in reduced form in which all the matrices are in the blocked diagonal form. The reducible matrix representation can thus be expressed as direct sum of the blocked matrices belonging to the I.Rs of the point group with respect to the newly formed basis of the vector space. This is carried out by the similarity transformation by applying the operator $\hat{A}$ on the matrices of the reducible representation. The matrix A transforms the bases of the reducible representation to the bases of the I.Rs. This change of basis of the reducible representation to the basis of I.Rs. is carried out by projection operators of the point group. The bases of the I.Rs thus obtained are the *symmetry adapted linear combinations* (SALCs) of the basis vectors used to form the reducible representation. The formation of such SALCs by projection operators will be discussed in this chapter.

## 5.1 OPERATORS IN FUNCTION SPACE

We have discussed the salient features of function and configuration spaces in chapter three. In this section, we will describe the results of application of symmetry operations on the vectors of function space and the extrapolation of the results to the configuration space.

Suppose that there are functions of positions in the physical space. A typical function is a function $[f(x_1, x_2, ..., x_n)]$ of coordinates $(x_1, x_2, ..., x_n)$ of a point $P$ in a physical space. That is, the function assigns numerical value $[f(x_1, x_2, ..., x_n)]$ to each point $P$ of physical space with coordinates $(x_1, x_2, ..., x_n)$. We apply a symmetry operation $R$ on the point vector $\bar{P}$, that shifts the point $P$

to the point $P'$ with coordinates $f(x_1', x_2', ..., x_n')$. This shift of coordinates of $P$ to those of $P'$ will change the basis function from $f$ to a new function $[\hat{O}_R(f)]$. We define $\hat{O}_R$ as an operator that is associated with symmetry operator $\hat{R}$. The basic premise on which we will be working in the function space, is that the numerical value of the function $f(x_1, x_2, ..., x_n)$ assigned to point $P$ is the same as the value assigned to the point $P'$ with coordinates $(x_1', x_2', ..., x_n')$ of the new transformed function $[\hat{O}_R(f)]$. The relationship between these two functions $f(x_1, x_2, x_3)$ and $\hat{O}_R(f)\ (x_1', x_2', x_3', ..., x_n')$ in 3-dimensional space is expressed as

$$[\hat{O}_R(f)]\ (x_1', x_2', x_3') = f(x_1, x_2, x_3) \qquad ...(5.1)$$

In this relationship, the coordinates $(x_1, x_2, x_3)$ and $(x_1', x_2', x_3')$ are related as given in chapter three, *viz.*

$$\hat{R}\begin{vmatrix} x_1 \\ x_2 \\ x_3 \end{vmatrix} = \begin{vmatrix} x_1' \\ x_2' \\ x_3' \end{vmatrix} = \begin{vmatrix} r_{11} & r_{12} & r_{13} \\ r_{21} & r_{22} & r_{23} \\ r_{31} & r_{32} & r_{33} \end{vmatrix} \begin{vmatrix} x_1 \\ x_2 \\ x_3 \end{vmatrix} \qquad ...(5.2)$$

In eq. (5.2), the values of the coefficients $(r_{ij})$ are determined by the specific symmetry operation $R$. [in $r_{ij}$, $i$ = indicates the serial number of row while $j$ determines the same for column.]

*The actual form* of $[\hat{O}_R(f)]$ can therefore be obtained by the use of eq. (5.2) for a particular symmetry operation. This is done by substituting $(x_1, x_2, x_3)$ of the function $f(x_1, x_2, x_3)$ in terms of $(x_1', x_2', x_3')$ respectively whereby on both sides of equation (5.1), the same variables $(x_1, x_2, x_3)$ appear. Thus, we have defined through equation (5.1) the effect of $\hat{O}_R$ on a function $f(x_1', x_2', x_3')$. These operators $\hat{O}_R$ appear in the form of matrices. We can further prove that the operators $[\hat{O}_R]$ defined by eq. (5.1) are linear and multiply together according to the multiplication rules and follow the group multiplication table. We will therefore adopt this definition of $[\hat{O}_R(f)]$ without any reservation. One can further extrapolate and generalise the results obtained for function of three variables to functions of any number of variables provided we are able to define their transformation properties under the action of symmetry operations. We can therefore, apply this definition on the vectors of configuration space. In this case, we should know as to how the coordinates $x$ of a point $P$ in configuration space are related to the coordinates of a another point $P'$, $x'$ (Here $x$ and $x'$ stand for the complete specification of all the coordinates of points $P$ and $P'$

respectively). Thus, the definition of operator $[\hat{O}_R]$ acting on a function $f$ of coordinate $x$ is given by,

$$[\hat{O}_R(f)](x') = f(x) \qquad ...(5.3)$$

We have obtained a *rule* of defining $\hat{O}_R$ acting on a function $f(x)$ to produce a new function $f(x')$ which can be used to apply the general theory of obtaining representation of the point group of the molecule in function space. A few examples are given below as illustrations.

**Example 1:** Consider the five-dimensional spaces whose basis functions are five $d$-orbitals

$$\left.\begin{aligned}
&1.\ d_{xy} = [cF_{(r)}]\left[\frac{xy}{r^2}\right] = d_2 \\
&2.\ d_{yz} = [cF_{(r)}]\left[\frac{yz}{r^2}\right] = d_4 \\
&3.\ d_{xz} = [cF_{(r)}]\left[\frac{xz}{r^2}\right] = d_3 \\
&4.\ d_{x^2-y^2} = [cF_{(r)}]\left[\frac{x^2-y^2}{r^2}\right] = d_1 \\
&5.\ d_{z^2} = [cF_{(r)}]\left[\frac{3z^2-r^2}{r^2}\right]\left(\frac{\sqrt{3}}{2}\right) = d_5
\end{aligned}\right\}$$

We omit the factor $c[F(r)/r^2]$, which is common to all the $d$-orbital wave functions and remains invariant by all the symmetry operations of the point group. By using $x_1$, $x_2$, $x_3$ in place of $x$, $y$, $z$ respectively, we have the $d$-orbitals in the following form.

$$d_1 = \frac{1}{2}(x_1^2 - x_2^2);\ d_2 = x_1x_2;\ d_3 = x_1x_3,\ d_4 = x_2x_3 \text{ and } d_5 = \frac{3x_3^2 - r^2}{r^2}\frac{\sqrt{3}}{2}$$

*Let us use these* five $d$-orbitals as basis vectors and find the effects of $\hat{C}_\alpha$ symmetry operators on them. We have (*see* Chapter 3)

$$\hat{C}_\alpha = \begin{vmatrix} \cos\alpha & -\sin\alpha & 0 \\ \sin\alpha & \cos\alpha & 0 \\ 0 & 0 & 1 \end{vmatrix}$$ is the matrix of transformation with respect to the basis coordinates $(\bar{e}_1, \bar{e}_2, \bar{e}_3)$. That is,

$$\begin{vmatrix} x_1' \\ x_2' \\ x_3' \end{vmatrix} = \begin{vmatrix} \cos\alpha & \sin\alpha & 0 \\ -\sin\alpha & \cos\alpha & 0 \\ 0 & 0 & 1 \end{vmatrix} \begin{vmatrix} x_1 \\ x_2 \\ x_3 \end{vmatrix} \text{ with respect to } (x_1, x_2, x_3) \qquad ...(5.4)$$

or

$$\left.\begin{aligned} x_1' &= x_1 \cos\alpha + x_2 \sin\alpha \\ x_2' &= -x_1 \sin\alpha + x_2 \cos\alpha \\ x_3' &= x_3 \end{aligned}\right\} \qquad ...(5.5)$$

[eq. (5.5) can be solved for $x_1$, $x_2$ and $x_3$ in terms of $x_1'$, $x_2'$ and $x_3'$. We get,

$$\left.\begin{aligned} x_1 &= x_1' \cos\alpha - x_2' \sin\alpha \\ x_2 &= x_1' \sin\alpha + x_2' \cos\alpha \\ x_3 &= x_3' \end{aligned}\right\} \qquad ...(5.6)$$

Consider the operation $\hat{C}_\alpha$ and let us assume that its corresponding operator is $\hat{O}_{C_\alpha}$ (in short $\hat{O}_c$). In these five-dimensional spaces of $d$-orbitals, we use eq. (5.1) to see the effect of $\hat{O}_c$ on the $d$-orbitals. We have,

$$(a)\ [\hat{O}_c(d_1)](x_1', x_2', x_3') = d_1(x_1, x_2, x_3) = \frac{1}{2}(x_1^2 - x_2^2) \qquad ...(5.7)$$

Substitute $x_1$ and $x_2$ in terms of $x_1'$, $x_2'$ from eq. (5.6), we get,

$$[\hat{O}_c d_1][x_1, x_2, x_3] = \frac{1}{2}[(x_1' \cos\alpha + x_2' \sin\alpha)^2 - (-x_1' \sin\alpha + x_2' \cos\alpha)^2]$$

$$= \frac{1}{2}[x_1'^2 \cos^2\alpha + x_2'^2 \sin^2\alpha + 2x_1'x_2' \sin\alpha\cos\alpha - x_1'^2 \sin^2\alpha$$

$$-x_2^2 \cos^2\alpha + 2x_1'x_2' \sin\alpha\cos\alpha]$$

$$= \frac{1}{2}[x_1'^2 \cos^2\alpha - \sin^2\alpha] - x_2'^2[\cos^2\alpha - \sin^2\alpha] + 4x_1'x_2' \sin\alpha\cos\alpha]$$

$$= (\cos 2\alpha)\frac{(x_1'^2 - x_2'^2)}{2} + 2x_1'x_2' \sin 2\alpha$$

$$= (\cos 2\alpha)d_1 + (\sin 2\alpha)\, d_2$$

Thus, we have

$$\hat{O}_c d_1 = (\cos 2\alpha)d_1 + (\sin 2\alpha)d_2 \qquad ...(5.8)$$

In the same way, one gets the following relationships

$$\hat{O}_c d_2 = (\sin 2\alpha)d_1 + (\cos 2\alpha)d_2 \quad ...(5.9)$$

$$\hat{O}_c d_3 = (\cos \alpha)d_3 + (\sin \alpha)d_4 \quad ...(5.10)$$

$$\hat{O}_c d_4 = (-\sin \alpha)d_3 + \cos \alpha\, d_4 \quad ...(5.11)$$

and $\hat{O}_c d_5 = d_5$

Using the eqs. (5.8)–(5.11), one gets the matrix representation of $\hat{C}_\alpha$ in five-dimensional function space with respect to $d$-orbital functions as basis.

$$\hat{C}_\alpha = \begin{vmatrix} \cos 2\alpha & -\sin 2\alpha & 0 & 0 & 0 \\ \sin 2\alpha & \cos 2\alpha & 0 & 0 & 0 \\ 0 & 0 & \cos\alpha & -\sin\alpha & 0 \\ 0 & 0 & \sin\alpha & \cos\alpha & 0 \\ 0 & 0 & 0 & 0 & 1 \end{vmatrix} \quad ...(5.12)$$

Next we take the $\sigma_v$ reflection plane and find its matrix representation (*i.e.*, the plane containing the $C_3$-axis and which makes an angle $\beta$ with the $e_1$ coordinate axis). As a result of reflection operation, the point $P(x_1, x_2, x_3)$ shifts to a new point $p'(x_1', x_2', x_3')$ and the effect of shift gives a matrix representation for $\sigma_v$ reflection. The matrix, thus obtained, is given below:

$$\sigma_v = \begin{vmatrix} \cos 4\beta & \sin 4\beta & 0 & 0 & 0 \\ \sin 4\beta & -\cos 4\beta & 0 & 0 & 0 \\ 0 & 0 & \cos 2\beta & \sin 2\beta & 0 \\ 0 & 0 & \sin 2\beta & -\cos 2\beta & 0 \\ 0 & 0 & 0 & 0 & 1 \end{vmatrix} \quad ...(5.13)$$

Similar matrices can also be calculated for all the operations of other point groups.

As an illustration, we take a specific point group $C_{3v}$. It has six symmetry operations $[E, C_3^1, C_3^2, \sigma_v^{(1)}, \sigma_v^{(2)}, \sigma_v^{(3)}]$. With the help of above equations, one can calculate the matrix representation of $C_{3v}$ point group for a five-dimensional function space with respect to five $d$-orbitals as basis vectors. These matrices are:

$$1.\ E = \begin{vmatrix} 1 & 0 & 0 & 0 & 0 \\ 0 & 1 & 0 & 0 & 0 \\ 0 & 0 & 1 & 0 & 0 \\ 0 & 0 & 0 & 1 & 0 \\ 0 & 0 & 0 & 0 & 1 \end{vmatrix};$$

$$2.\ C_3^1\ (\alpha = 120°) = \begin{vmatrix} -c & s & 0 & 0 & 0 \\ -s & -c & 0 & 0 & 0 \\ 0 & 0 & -c & -s & 0 \\ 0 & 0 & s & -c & 0 \\ 0 & 0 & 0 & 0 & 1 \end{vmatrix}$$

$$3.\ c_3^2 = (c_3^2)^{-1} = (c_3^1)^{-1} = \begin{vmatrix} -c & -s & 0 & 0 & 0 & 0 \\ s & -c & 0 & 0 & 0 & 0 \\ 0 & 0 & -c & s & 0 & 0 \\ 0 & 0 & -s & -c & 0 & 0 \\ 0 & 0 & 0 & 0 & 0 & 1 \end{vmatrix};$$

$$4.\ \sigma_v^{(1)} = \begin{vmatrix} 1 & 0 & 0 & 0 & 0 \\ 0 & -1 & 0 & 0 & 0 \\ 0 & 0 & 1 & 0 & 0 \\ 0 & 0 & 0 & -1 & 0 \\ 0 & 0 & 0 & 0 & 1 \end{vmatrix}$$

$$5.\ \sigma_v^{(2)} = \begin{vmatrix} c & s & 0 & 0 & 0 \\ s & c & 0 & 0 & 0 \\ 0 & 0 & -c & -s & 0 \\ 0 & 0 & -s & c & 0 \\ 0 & 0 & 0 & 0 & 1 \end{vmatrix};\ \text{and}\ \sigma_v^{(3)} = \begin{vmatrix} -c & -s & 0 & 0 & 0 \\ -s & c & 0 & 0 & 0 \\ 0 & 0 & -c & s & 0 \\ 0 & 0 & s & c & 0 \\ 0 & 0 & 0 & 0 & 1 \end{vmatrix}$$

(In the above matrices, we have put $c = \cos 60° = 1/2$ and $s = \sin 60°$ $= \sqrt{3/2}$ )

These matrices form a five-dimensional representation of $C_{3v}$. The salient features of this representation are:

1. These multiply together according to the rules of the formation of group.

2. The matrices (all of them) have the same block diagonal form. That is, all their non-vanishing elements occur along the block diagonal.
3. The $d_1$ and $d_2$ functions by themselves form the basis of two-dimensional representation. Similarly $d_3$ and $d_4$ functions form another two-dimensional representation and $d_5$ function forms one-dimensional representation for the $C_{3v}$ point group. Thus, they form three representations *viz.* $\{d_1, d_2\}$, $\{d_3, d_4\}$, $\{d_5\}$.
4. Each of the above three representations cited in the case (3) above forms a $C_{3v}$ point group because the matrices in each representation multiply together the same way as the operations of the group.

## 5.2 SET OF ORTHONORMAL FUNCTIONS

This is a very important concept while dealing with the basis functions. We have already discussed this aspect while discussing the basis in the physical vector space. In chapter four, in the case of function space, we stated that two functions are orthogonal if the following relationship holds good.

$$\int f_i^* f_j \partial\tau = \delta_{ij} \qquad ...(5.14)$$

Where $\partial\tau$ is an infinitesimal volume element corresponding to the variables of the functions $f_1$ and $f_2$. If in addition, either set of functions possesses unit norm.

A set of function $f_1, f_2 \ldots f_n$ is said to be orthonormal.

$$\int f_i^* f_j \partial\tau = \delta_{ij} \qquad ...(5.15)$$

As an example of the orthonormal sets, we have the following normalised functions defined over the

$$(i)\ f_n(x) = \frac{1}{\sqrt{L}} \sin \frac{\pi n x}{L}, \quad n = 1, 2, 3 \ldots \qquad ...(5.16)$$

and

$$(ii)\ g_n(x) = \frac{1}{\sqrt{L}} \cos \frac{\pi n x}{L}, \quad n = 1, 2, 3 \ldots \qquad ...(5.17)$$

## 5.3 SOME PROPERTIES OF BASIS FUNCTIONS

1. Any set of $n$ vectors or functions can be chosen to define an $n$-dimensional vector space.
2. The effects of the symmetry operations of a point group on these functions in the form of matrices form usually a reducible representation of the group.

3. By making proper linear combinations of these basis vectors, one can form an orthonormal set.
4. One can have any number of sets of orthonormal vectors or functions in a vector space.
5. An orthonormal set of vectors or of functions, remains orthonormal after a linear transformation, if and only if, the transformation is unitary.
6. *One can construct* orthonormal bases from any set of orthonormal basis vectors or of functions, such that the bases transform by the symmetry operations of a group according to the I.Rs of the group. These are known as SALCs. The formation of such bases is carried out by using projection operators which will be discussed in the next section.
7. If we have two sets of functions $(f_1, f_2, \ldots, f_n)$ and $(g_1, g_2, \ldots, g_n)$ and if these form two representations (whether reducible or irreducible ($\Gamma_m$ and $\Gamma_n$) of a group, then the direct product of the functions, *viz.* $(f_1g_1, f_1g_2, f_2g_1, \ldots, f_mg_n)$ form a basis for the direct product representation ($\Gamma_m \otimes \Gamma_n$) where matrices are $|f_i| \times |g_j|$ $i = 1, 2, \ldots, m, j = 1, 2, \ldots, n$, for the formation of the group.
8. The characters of the direct product representation ($\Gamma_m \otimes \Gamma_n$) are the products of the characters of $\Gamma_m$ and $\Gamma_n$.
9. *The set* $(f_1, f_2, \ldots, f_n)$ and the set $\{g_1, g_2, \ldots, g_n\}$ of above property no. (7) are orthogonal to each other unless in the reduction of $\Gamma_m \otimes \Gamma_n$ representation into its constituents I.Rs, there is present an Identity I.R. If it does not contain an identity I.R component, then the two sets are orthogonal [*i.e.*, $f_i g_j \partial\tau = 0$]. If $\Gamma_m$ and $\Gamma_n$ both I.Rs are present, the $\Gamma_m \otimes \Gamma_n$ will contain an identity representation only if both $\Gamma_m$ and $\Gamma_n$ are equivalent. If $\Gamma_m$ and $\Gamma_n$ are the same I.Rs then $f(x)$ and $g(x)$ are orthogonal if they belong to the different rows in the matrices of the representations. Thus, if $f_i$ and $g_i$ are the same functions then $f_i$ and $f_j$ are orthogonal when $i \neq j$.
10. We have three sets of basis functions *viz.* $(f_1, f_2, \ldots, f_m)$ and $(g_1, g_2, \ldots, g_n)$ and $(q_1, q_2, \ldots, q_n)$. The basis may be a set of functions or a set of operators. These three sets form three representations [*viz.* $\Gamma_n, \Gamma_m, \Gamma_p$] respectively. Then, the integral $\int f_i q_j g_k \partial\tau$ is zero. It is non-zero only when the direct product $\Gamma_n \Gamma_m \Gamma_p$ contains an identity component.
11. Suppose we have an operator $\hat{Q}$ (*e.g.*, $\hat{H}$ = Hamiltonian) which is invariant under all the operations of a group and it forms the basis for

identity representation. Further, suppose we have two sets of basis functions $(f_1, f_2, f_3, ..., f_n)$ and $(g_1, g_2, g_3, ..., g_n)$ which are the bases for two I.Rs of the same group, then the integral $\int f_i^* Q_j g_k \partial\tau$ is zero unless $f_i$ and $g_k$ belong to the same row of the same I.R.

12. If the set of function $(f_1, f_2, ..., f_n)$ form a basis for a unitary representation, then the functions,

$$f = f_1^* f_1 + f_2^* f_2, ..., f_n^* f_n \qquad ...(5.18)$$

is invariant under the operations of a group. That is, the basis for the identity representation *Unitary Representation* is a representation formed by unitary matrices. That is, when the matrix $|u|^{-1} = (u^t)^*$. Thus, an orthogonal matrix is a unitary matrix with all elements real.

## 5.4 PROJECTION OPERATOR

We have discussed the "Great Orthogonality" theorem which shows a relationship between matrix elements of the I.Rs of a group. This relation implies certain properties of the basis vectors which will help us to solve our second problem "to find a linear combination of our arbitrarily chosen basis vectors". These linearly combined basis vectors will form a new basis for the representation in *which the matrices of the representation are in the completely* reduced form. Each completely reduced representation belongs to one of the I.Rs of the group. Thus, these linearly combined base vectors are the basis vectors for the I.Rs of the point group. These are known as SALCs (Symmetry Adapted Linear Combinations). Thus, we take one or more sets of orthonormal functions forming a reducible representation and make their orthogonal linear combinations in such a way that the linearly combined bases form the basis for the I.Rs of the molecular point group. In forthcoming chapters, we will be using these SALCs very often while dealing with chemical bonding, molecular vibrations, *etc.*

Suppose, we have a group $G$ of order $h$, consisting of operations $(R_1, R_2, R_3, ..., R_h)$. It has I.Rs $\Gamma_1, \Gamma_2, ..., \Gamma_k$ where $k$ is the number of classes of the group. We wish to derive projection operators which as a result of operations on the chosen basis convert them into SALCs that belong to I.Rs. This can be done in the following steps.

**Step 1:** Let us suppose that we have the base vectors for each I.R given in the following tabular form.

**Table 5.1:** I.Rs, Base Vectors *etc.*, for a Group

| *I.R* | *Base Vectors* | *Components or elements of transformation matrix* | *No. of matrices* | *IR* | *$C_{3v}$ as example* | *Components* |
|---|---|---|---|---|---|---|
| $\Gamma_1$ | $f_1^1, f_2^1, f_3^1, ..., f_r^1$ | $[D^1(R)]_{ij}$ | $h$ | $A_1$ | $A_1$ one $f_1$ dimensional | 6 components 1-dimensional |
| $\Gamma_2$ <br> $\vdots$ | $f_1^2, f_2^2, f_3^2, ..., f_s^2$ <br> $\vdots$ | $[D^2(R)]_{ab}$ <br> $\vdots$ | $h$ | $A_2$ | $A_2$ one $f_2$ dimensional | 6 components 1-dimensional one vector |
| $\Gamma_\mu$ <br> $\vdots$ | $f_1^\mu, f_2^\mu, f_3^\mu, ..., f_t^\mu$ <br> $\vdots$ | $[D^\mu(R)]_{mn}$ <br> $\vdots$ | $h$ <br> $\vdots$ | $E$ <br> $\vdots$ | $f_1^E, f_2^E$ two dimensional | 4 sets of components in each set. |
| $\Gamma_\nu$ | $f_1^\nu, f_2^\nu, f_3^\nu, ..., f_u^\nu$ | $[D^\nu(R)]_{cd}$ | $h$ | $f_1^\nu, f_2^\nu ... f_n^\nu$ $n$-dimensional | | |

The effects of symmetry operations $R_i$ on any base vector [*e.g.*, $[f_q^\nu]$ of I.R. $\nu$] are expressed as,

$$\hat{R}[f_q^\nu] = \sum_{p=1}^{n_\nu}[D^\nu(R_i)_{pq} f_p^\nu \quad ...(5.19)$$

which can be written in the expanded form as shown in eq. (5.20)

$$R_1[f_q^\nu] = [D^\nu(R_1)]_{1q} f_1^\nu + [D^\nu(R)]_{2q} f_2^\nu + ... + [D^\nu(R_i)]_{pq} f_p^\nu \quad ...(5.20)$$

In eq. (5.19) $[D^\nu(R_i)]_{pq}$ is the matrix-element at the cross point of $p^{th}$ row and $q^{th}$ column of the matrix for the $R_i$ of I.R $\Gamma_\nu$ of the representation.

We further clarify this point by taking one particular case of one-dimensional representation. Suppose for the point group $C_{3v}$ we have, one-dimensional $A_2$ and $E$ two-dimensional representation with the following matrices given below.

**Table 5.2:** Representation for $A_2$ and E I.Rs. of $C_{3v}$ point group

| *I.Rs.* | *E* | $C_3^1$ | $C_3^2$ | $\sigma_v^{(1)}$ | $\sigma_v^{(2)}$ | $\sigma_v^{(3)}$ |
|---|---|---|---|---|---|---|
| $A_2$ | 1 | $[B_1]$ | $[B_2]$ | $[C_1]$ | $[C_2]$ | $[C_3]$ |
| $E$ | $\begin{vmatrix} 1 & 0 \\ 0 & 1 \end{vmatrix}$ | $\begin{vmatrix} b_{11}^1 & b_{12}^1 \\ b_{21}^1 & b_{22}^1 \end{vmatrix}$ | $\begin{vmatrix} b_{11}^2 & b_{12}^2 \\ b_{21}^2 & b_{22}^2 \end{vmatrix}$ | $\begin{vmatrix} b_{11}^3 & b_{12}^3 \\ b_{21}^3 & b_{22}^3 \end{vmatrix}$ | $\begin{vmatrix} b_{11}^4 & b_{12}^4 \\ b_{21}^4 & b_{22}^4 \end{vmatrix}$ | $\begin{vmatrix} b_{11}^5 & b_{12}^5 \\ b_{21}^5 & b_{22}^5 \end{vmatrix}$ |

The $A_2$ I.R is one-dimensional and therefore we will have one base vector, say $g_1^{A_2}$ I.R, $E$ is two-dimensional so we will have two base vectors $f_1^E, f_2^E$.

The effects of $R_i$ (*e.g.*, $\sigma_v^{(1)}$) on these base vectors (according to eq. (5.20) will be

$$\left.\begin{aligned}\sigma_v^{(1)}[f_1^E] = b_{11}^3 f_1^E + b_{12}^3 f_2^E \\ \sigma_v^{(1)}[f_2^E] = b_{21}^3 f_1^E + b_{22}^3 f_2^E\end{aligned}\right\} \quad ...(5.21)$$

**Step 2:** As a second step, we multiply on both sides by another matrix element say $[D^{\mu}(R_x)]_{ij}$. As a result, we get

$$[D^{\mu}(R_x)]_{ij}[R_y[f_q^{\nu}] = [D^{\mu}(R_x)]_{ij}\sum_{p=1}^{n_\nu}[D^{\nu}(R_y)]_{pq}]f_p^{\nu} \quad ...(5.22)$$

Adding all the equations for all the operations ($R_y$), we get

$$\sum_{R_y=1}^{R}[D^{\mu}(R_x)]_{ij}[R_y(f_q^{\nu})] = \sum_{R_y=1}^{R}\sum_{p=1}^{n_\nu}[D^{\mu}(R_x)]_{ij}[D^{\nu}(R_y)]_{pq}f_p^{\nu} \quad ...(5.23)$$

Eq. (5.23) can also be written as

$$\sum_{p=1}^{n_\nu}\sum_{R_y=1}^{R}[D^{\mu}(R_x)]_{ij}[R_y(f_q^{\nu})] = \sum_{p=1}^{n_\nu}\sum_{R_y=1}^{R}[D^{\mu}(R_x)]_{ij}[D^{\nu}(R_y)]_{pq}f_q^{\nu}$$

...(5.24)

Now according to orthogonality theorem, RHS of eq. (5.24) becomes

$$\sum_{R_y=1}^{R}[D^{i}(R_x)]_{ij}[R_y(f_q^{\nu})] = \sum_{p=1}^{n_\nu}\left[\frac{g}{\sqrt{n_\nu n_\mu}}\right][\delta_{\mu\nu}\delta_{ip}\delta_{jq}][f_q^{\nu}] \quad ...(5.25)$$

Establishing the equivalence of two sides

The coefficients of $(f_q^{\nu})$ vanishes for $\mu \neq \nu$. In case $\mu = \nu$, the RHS is equal to

$$\text{RHS} = \sum_{p=1}^{n_\nu}\frac{g}{n_\nu}[\delta_{ip}\cdot\delta_{jq}](f_p^{\nu}) \quad ...(5.26)$$

the only non-zero term on the RHS of eq. (5.25) is when $p = q$. In that case, all the terms in the summation vanish except

$$\sum_{R}[D^{\mu}(R)]_{ij}R[f_q^{\nu}] = \frac{g}{n_\nu}[\delta_{jq}]f_q^{\nu} \quad ...(5.27)$$

$$\left.\begin{array}{ll} \text{or} & p_{ij}^{\mu} f_q^{\mu} = \dfrac{g}{n_\nu}[\delta_{jq}] f_q^{\nu} \\ \text{where} & p_{ij}^{\mu} = \sum\limits_R [D^{\mu}(R)]_t^* \hat{R}. \end{array}\right\} \quad ...(5.28)$$

Here $p_{ij}^{\mu}$ is an operator which is a definite linear combination of the operator $R$ with coefficients that depend upon the matrices of $\Gamma^{\mu}$. The summation is over all the symmetry operations $\hat{R}$ of the group.

The projection operator in the form $\left[\sum\limits_R [D^{\mu}(R)]_{ij} \hat{R}\right]$ is not very convenient mathematically. We therefore consider it in another form, as below:

$$p_{ii} = \sum_R [D^{\mu}(R)]_{ii}^* \hat{R} \quad ...(5.29)$$

where $i = 1, 2, 3\ldots n_\nu$. If we sum all the $P_{ii}$ terms from $i = 1$ to $n_\nu$, we get,

$$\sum_{i=1}^{n_\nu} p_{ii} = \sum_R \sum_{n_\nu = 1}^{\nu} [D^{\mu}(R)]_{ii}^* \hat{R} \quad ...(5.30)$$

or

$$p^{\mu} = \sum_R \sum_{i=1}^{n_\nu} [D^{\mu}(R)]_{ii}^* \hat{R} = \sum_R (\chi_R^{\mu})^* \hat{R} \quad ...(5.31)$$

In this definition, the coefficients of $\hat{R}$ is the complex conjugate of the characters of $\hat{R}$ in $\Gamma^{\mu}$. One gets coefficients of all the I.Rs from the character table of the group for a particular $R$ in its I.R. [In case one wishes to use the operators in the form $p_{ij}^{\mu}$ one has to have the matrices for all the operations $R$ belonging to all the I.Rs of the group]. Thus, one obtains.

$$p_{ij}{}^{\mu} f_i^{\mu} = \frac{g}{n_\mu} f_q^{\mu}, \text{ where } i = j \text{ (otherwise zero)} \quad ...(5.32)$$

We, thus have found $\hat{P}$, an operator which annihilates any vector that does not belong to $\Gamma^{\mu}$ space, but which when applied to a vector $f^{\mu}$ belonging to $\Gamma^{\mu}$ simply produces a numerical multiple of $f^{\mu}$. This is known as projection operator.

**CASE (*a*): Basis vectors belonging to I.Rs**

**Example 1:** In this example, we have chosen basis vectors which belong to the I.Rs of the point group. The examples that we start with are base vectors which give reducible representations.

Let us choose $\phi_1$, $\phi_2$ and $\phi_3$, $\phi_4$ base vectors of the system having $C_{3v}$ point group. Let us assume that $\phi_1$ belongs to $A_1$ type I.R., $\phi_2$ to $A_2$ type I.R., $\phi_3$ and $\phi_4$ to $E$ type of I.R of $C_{3v}$ point group. Let us assume the transformation properties of these base vectors as shown in Table 5.3. The projection operators of $C_{3v}$ point group are

$$P^{A_1} = E + C_3^1 + C_3^2 + \sigma_v^{(1)} + \sigma_v^{(2)} + \sigma_v^{(3)}$$

$$P^{A_2} = E + C_3^1 + C_3^2 - \sigma_v^{(1)} - \sigma_v^{(2)} - \sigma_v^{(3)}$$

$$P^E = 2E - C_3^1 - C_3^2$$

**Table 5.3:** Transformation Properties of base vectors

| $C_{3v}$ | $E$ | $C_3^1$ | $C_3^2$ | $\sigma_v^{(1)}$ | $\sigma_v^{(2)}$ | $\sigma_v^{(3)}$ |
|---|---|---|---|---|---|---|
| $\phi_1$ | $\phi_1$ | $\phi_1$ | $\phi_1$ | $\phi_1$ | $\phi_1$ | $\phi_1$ |
| $\phi_2$ | $\phi_2$ | $\phi_2$ | $\phi_2$ | $-\phi_2$ | $-\phi_2$ | $-\phi_2$ |
| $\phi_3$ | $\phi_3$ | $\in\phi_3$ | $\in^*\phi_3$ | $\phi_4$ | $\in\phi_4$ | $\in^*\phi_4$ |
| $\phi_4$ | $\phi_4$ | $\in^*\phi_3$ | $\in\phi_4$ | $\phi_3$ | $\in^*\phi_3$ | $\in\phi_3$ |

It is clear that $\phi_1$, $\phi_2$, $\phi_3$ and $\phi_4$ do not mix among each other, because these base vectors belong to one of the three I.Rs of $C_{3v}$ point group. It is the property of such base vectors (I.Rs. base vectors) that under the symmetry operation, they transform among themselves with ± coefficients. The two base vector belonging to $E$-type of I.Rs. ($\phi_3$ and $\phi_4$) get mixed up under symmetry operation, but these do not mix with $\phi_1$ and $\phi_2$.

The application of these projection operators to the base vectors $\phi_1$, $\phi_2$, $\phi_3$ and $\phi_4$ gives the following equations, all in accord with the equation

$\left[P^{\mu} f_i^{\nu} = \dfrac{g}{n_{\mu}} f^{\nu}\right]$. When $\mu = \nu$, then $P^{\mu} f_i^{\nu} = \dfrac{g}{n_{\mu}} f^{\nu}$, when $\mu \neq \nu$, then $P^{\mu} f_i^{\nu} = 0$.

This is shown below.

$$P^{A_1}\phi_1 = \phi_1 + \phi_1 + \phi_1 + \phi_1 + \phi_1 + \phi_1 = 6\phi_1 = \frac{6}{1}\phi_1$$

$$P^{A_2}\phi_1 = \phi_1 + \phi_1 + \phi_1 - \phi_1 - \phi_1 - \phi_1 = 0, \text{ because } \mu \neq \nu \ (A_1 \neq A_2),$$

$$P^E\phi_1 = 2\phi_1 - \phi_1 - \phi_1 = 0, \text{ because } \mu \neq \nu \ (A_1 \neq E)$$

$$P^{A_1}\phi_2 = \phi_2+\phi_2+\phi_2-\phi_2-\phi_2-\phi_2=0, \text{ because } (A_1 \neq A_2),$$

$$P^{A_2}\phi_2 = \phi_2+\phi_2+\phi_2-(-\phi_2)-(-\phi_2)-(-\phi_2)$$

$$= 6\phi_2\frac{g}{1}\,(A_2 = A_2 \; i.e., \mu=\nu)$$

Similarly, we can now take $\phi_3$ and $\phi_4$;

$$P^{A_1}(\phi_3) = P^{A_1}(\phi_4)=\phi_3+\in\phi_3+\in^*\phi_3+\phi_4$$

$$+\in\phi_4+\in^*\phi_4+=0 \quad (\in+\in^*=-1)$$

$$P^{A_2}(\phi_3) = P^{A_2}(\phi_4)=0 \text{ (calculate)}$$

$$P^E(\phi_3) = 2\phi_3-\in\phi_3-\in^*\phi_3=3\phi_3=\frac{6}{2} \text{ [because } E=E \text{ or } \mu=\nu]$$

$$P^E(\phi_4) = 2\phi_4-\in\phi_4-\in^*\phi_4=3\phi_4=\frac{6}{2} \text{ [because } E=E \text{ or } \mu=\nu]$$

Thus, we have shown that projection operator when applied to basis vectors give the vectors which belongs to I.R.

**Example 2:** Let us now take $g_1$, $g_2$, $g_3$ and $g_4$ as basis vectors belonging to one of the irreducible representations of $D_4$ point group. We wish to find linear combinations of $g_1$, $g_2$, $g_3$ and $g_4$ which belong to one of the I.Rs of the point group $D_{4.}$ We do it in the following steps.

**Step 1:** We first calculate the projection operators for the various I.Rs of $D_4$ group (order 8) whose character table is given below $\left[P^{\mu}=\sum_R \chi_R^{*\mu}(R)\right]$

$$p^{A_1} = E+C_4^1+C_4^{-1}+C_2+C_2'^a+C_2'^b+C_2''^{(a)}+C_2''^{(b)}$$

$$p^{A_2} = E+C_4^1+C_4^{-1}+C_2-C_2'^a-C_2'^b-C_2''^{(a)}-C_2''^{(b)}$$

$$p^{B_1} = E-C_4^1-C_4^{-1}+C_2-C_2'^a-C_2'^b+C_2''^{(a)}+C_2''^{(b)}$$

$$p^{B_2} = E-C_4^1-C_4^{-1}+C_2-C_2'^a-C_2'^b-C_2''^{(a)}-C_2''^{(b)}$$

$$p^E = 2E+0+0-2C_2+0+0+0+0$$

**Table 5.4:** Character Table of $D_4$ group and effects of symmetry operations of $g_1$, $g_2$, $g_3$ and $g_4$

| $D_4$ | $E$ | $C_4$ | $C_4^{-1}$ | $C_2$ | $C_2'^a$, | $C_2'^b$ | $C_2''^a$ | $C_2''^b$ | |
|---|---|---|---|---|---|---|---|---|---|
| $A_1$ | 1 | 1 | 1 | 1 | 1 | 1 | 1 | 1 | |
| $A_2$ | 1 | 1 | 1 | 1 | –1 | –1 | –1 | –1 | |
| $B_1$ | 1 | –1 | –1 | 1 | –1 | –1 | 1 | 1 | |
| $B_2$ | 1 | –1 | –1 | 1 | –1 | –1 | –1 | –1 | |
| $E$ | 2 | 0 | 0 | –2 | 0 | 0 | 0 | 0 | |
| $g_1$ | $g_1$ | $g_2$ | $g_4$ | $g_3$ | $-g_4$ | $-g_2$ | $-g_1$ | $-g_3$ | $\Gamma_{\text{red}} = \sum_{i=1}^{k} a_i \Gamma^i_{\text{I.R}}$ |
| $g_2$ | $g_2$ | $g_3$ | $g_1$ | $g_4$ | $-g_3$ | $-g_1$ | $-g_4$ | $-g_2$ | |
| $g_3$ | $g_3$ | $g_4$ | $g_2$ | $g_1$ | $-g_2$ | $-g_4$ | $-g_3$ | $-g_1$ | |
| $g_4$ | $g_4$ | $g_1$ | $g_3$ | $g_2$ | $-g_1$ | $-g_3$ | $-g_2$ | $-g_4$ | |
| $\Gamma_{\text{red}}$ | 4 | 0 | 0 | 0 | 0 | 0 | –2 | –2 | $\Gamma_{\text{red}} = A_2 + B_2 + E$ |

**Step 2:** We take orthogonal basis vectors $(g_1, g_2, g_3, g_4)$ which give reducible representation $[\Gamma_{\text{red}} = 4, 0, 0, 0, 0, 0, -2, -2]$ of the group. In the character table, the effects of the symmetry operations of the $D_4$ group on these base vectors are given, forming a four-dimensional reducible representation. The application of the projection operators on the base vectors gives the following equations.

1. $$p^{A_1}(g_1) = E(g_1) + C_4^1(g_1) + C_4^{-1}(g_1) + C_2(g_1) + C_2'^a(g_1) + C_2'^b(g_1) + C_2''^a(g_1) + C_2''^b(g_1)$$
$$= g_1 + g_2 + g_4 + g_3 - g_4 - g_2 - g_1 - g_3 = 0,$$
(It implies that the $g_1$ does not belong to $A_1$ I.R. Similarly, the following calculations show that $g_2$, $g_3$ and $g_4$ also do not belong to $A_1$ I.R. of the group. As shown below:

2. $$p^{A_1}(g_2) = E(g_2) + C_4^1(g_2) + C_4^{-1}(g_2) + C_2(g_2) + C_2'^a(g_2) + C_2'^b(g_2) + C_2''^a(g_2) + C_2''^b(g_2)$$
$$= g_2 + g_3 + g_1 + g_4 - g_3 - g_1 - g_4 - g_2 = 0,$$
Implying $g_2$ does not belong to $A_1$.

3. $$p^{A_1}(g_3) = E(g_3) + C_4^1(g_3) + C_4^{-1}(g_3) + C_2(g_3) + C_2'^a(g_3) + C_2'^b(g_3) + C_2''^a(g_3) + C_2''^b(g_3)$$
$$= g_3 + g_4 + g_2 + g_1 - g_2 - g_4 - g_3 - g_1 = 0$$
(It implies that $g_3$ does not belong to $A_1$)

4. $p^{A_1}(g_4) = E(g_4)+C_4^1(g_4)+C_4^{-1}(g_4)+C_2(g_4)+C_2'^a(g_4) + C_2'^b(g_4)+C_2''^a(g_4)+C_2''^b(g_4)$

$= g_4 + g_1 + g_3 + g_2 - g_1 - g_3 - g_2 - g_4 = 0$

(It implies that $g_4$ does not belong to $A_1$)

We next take the projection operators of $A_2$ I.R. and find its effects on $g_1$, $g_2$, $g_3$ and $g_4$

$$p^{A_2}(g_1) = E(g_1)+C_4^1 g_1 + C_4^{-1} g_1 + C_2 g_1 - C_2'^a(g_1) - C_2'^b(g_1) - C_2''^a(g_1) - C_2''^b(g_1)$$

$$= g_1 + g_2 + g_4 + g_3 + g_4 + g_2 + g_1 + g_3$$

$$= 2(g_1 + g_2 + g_3 + g_4)$$

(It implies that $g_1$ belongs to $A_2$ type of I.R.)

Similarly,

$$p^{A_2}(g_2) = g_2 + g_3 + g_1 + g_4 + g_3 + g_1 + g_4 + g_2$$

$$= (2g_1 + 2g_2 + 2g_3 + 2g_4)$$

It implies that $g_2$ belongs to $A_2$ type of I.R.

Similarly, we start with $g_3$ and $g_4$ belonging to $A_2$ I.R

Now we take operator $B_1$

$$p^{B_1}(g_1) = E(g_1)-C_4^1(g_1)-C_4^{-1}(g_1)+C_2(g_1)-C_2'^a(g_1) - C_2'^b(g_1)+C_2''^a(g_1)+C_2''^b(g_1)$$

$$= g_1 - g_2 - g_4 + g_3 + g_4 + g_2 - g_1 - g_3 = 0,$$

$$p^{B_1}(g_2) = E(g_2)-C_4^1(g_2)-C_4^1(g_2)-C_2(g_2)-C_g'^a(g_2) - C_2'^b(g_2)+C_2''^a(g_2)+C_2''^b(g_2)$$

$$= g_2 - g_3 - g_1 + g_4 + g_3 + g_1 - g_4 - g_2 = 0$$

$$p^{B_1}(g_3) = E(g_3)-C_4^1(g_3)-C_4^1(g_3)+C_2(g_3)-C_g'^a(g_3) - C_2'^b(g_3)+C_2''^a(g_3)+C_2''^b(g_3)$$

$$= g_3 - g_4 - g_2 + g_1 + g_2 + g_4 - g_3 - g_1 = 0$$

$$p^{B_1}(g_4) = E(g_4)-C_4^1(g_4)-C_4^1(g_4)+C_2(g_4)-C_g'^a(g_4) - C_2'^b(g_4)+C_2''^a(g_4)+C_2''^b(g_4)$$

$$= g_4 - g_1 - g_3 + g_2 + g_1 + g_3 - g_2 - g_4 = 0.$$

Let us calculate for one more base vector *viz.* $g_1$ and find base vector for $B_2$.

$$p^{B_2}(g_1) = E(g_1) - C_4^1(g_1) - C_4^1(g_1) + C_2(g_1) + C_g'^a(g_1) - C_2'^b(g_1) - C_2''^a(g_1) - C_2''^b(g_1)$$

$$= g_1 - g_2 - g_4 + g_3 + g_4 + g_2 + g_1 + g_3 = 2(g_1 + g_3)$$

$$p^{B_2}(g_2) = 2(g_2 + g_4),$$

$$p^{B_2}(g_3) = 2(g_1 + g_3),$$

$$p^{B_2}(g_4) = 2(g_2 + g_4),$$

Let us see if $g_1$, $g_2$, $g_3$, $g_4$ belong to $E$ type I.R.

$$p^E(g_1) = 2E(g_1) - 2C_2(g_1)$$
$$= 2g_1 - 2g_3,$$

and
$$p^E(g_2) = 2E(g_2) - 2C_2(g_2)$$
$$= 2g_2 - 2g_4$$

Thus, $g_1$, $g_2$, $g_3$ and $g_4$ belong to $E$ type I.R.

Thus, the orthogonal base vectors ($g_1$, $g_2$, $g_3$, $g_4$ belong to I.Rs. $A_2$, $B_1$ and $E$ I.Rs. and do not belong to $A_1$ and $B_2$ I.Rs. We have therefore seen that the projection operators annihilate all those basis vectors which do not belong to $\Gamma_\mu$ space (I.Rs. space) and give all those vectors belonging to $\sqrt{\mu}$ space, a numerical multiple of the basis vectors *viz.* $g_1$, $g_2$, $g_3$, $g_4$. We have calculated LCAO of $g_i$'s belonging to I.Rs of $D_2$.

**CASE (*b*):** The above calculations have been carried out for those basis vectors which belong to $\Gamma_{i.r}$ I.Rs. Let us now consider a reducible representation $\Gamma_{red}$. In this case, we have chosen an arbitrary set of vectors *e.g.*, ($\phi_1$, $\phi_2$, ..., $\phi_n$) where $n$ defines the dimensionality of the vector space. Since the representation is reducible, it implies that the vector space, $V$ contains a number of invariant subspaces belonging to I.Rs. of the group $G$. Let us assume that it contains. $\nu^{(1)}$, $\nu^2$ ... $\nu^{(v)}$ sub spaces that is,

$$\nu = a_1\nu^{(1)} + b_1\nu^{(2)} + \ldots p\nu^{(p)}$$

Let us further suppose that $\nu^{(1)}$, $\nu^{(2)}$ ... $\nu^p$ subspaces are $n^1$, $n^2$ ... $n^p$ dimensional space respectively and belong $\Gamma^1$, $\Gamma^2$, ... $\Gamma^n$ I.Rs. respectively of the point group G. We assume that these I.Rs. do not occur more than once. $n\Gamma = \Gamma^1$, $\Gamma^2$ ..., $\Gamma^n$. So in order to define $\Gamma_{red}$, representation, we require $(\phi_1^{(1)}, \phi_2^{(1)} \ldots \phi_n^{(1)}) + (\phi_1^{(2)}, \phi_2^{(2)} \ldots \phi_n^{(2)}) \ldots (\phi_1^{\nu}, \phi_2^{\nu} \ldots \phi_{n\nu}^{\nu})$ basis vectors which serve as a new basis for the reducible representation $\Gamma_{red}$. So, if we have taken ($\phi_1$, $\phi_2$ ... $\phi_n$) as an arbitrary original basis for the formation of the reducible representation, then we have to transform the original basis $\{\phi_i\}$ to a new basis, *viz.* a set of ($\phi_n{}^i$) where $i = 1, 2, \ldots v$ and $n = 1, 2, 3 \ldots n$) using some particular similarity transformation. Thus, we can express the vectors $\phi_i$ as linear combinations of the orthonormal basis vectors, *viz.* $(\phi_1^1, \phi_2^1 \ldots\ldots \phi_1^{\nu}, \phi_2^{\nu}, \phi_3^{\nu} \ldots \phi_n^{\nu})$.

We can therefore write,

$$\phi_i = \sum_{\nu} \sum_{i=1}^{n_\nu} C_{\nu o}^{(s)} (\phi_i^{\nu}) \qquad ...(5.33)$$

In eq. (5.33) $(C_{\nu i}^{S_i})$ are numerical coefficients. This equation can be written as

$$\phi_i = \sum_{v} [F_s^{v}] \qquad ...(5.34)$$

Where $F_s^v$ is a vector of the subspace $\nu_i$ which is one of the invariant subspaces *viz.* ($v$ = 1, 2, 3 … ν) and belong to a particular I.R. of the group. So our original vector $\phi_i$ of eq. 5.33 become,

$$\phi_i = F_s^1 + F_s^2 + F_s^3 + ... F_s^v \qquad ...(5.35)$$

We further calculate the I.Rs which are present in the original $\Gamma_{\text{red}}$.

$$\Gamma_{\text{red}} = \Gamma_1 + \Gamma_2 + ... \Gamma_s \qquad ...(5.36)$$

Now, we choose one of the I.Rs (say $\Gamma^\mu$) and apply its projection operator ($p^\mu$) to $\phi_I$ [eq. (5.34)]. From eq. (5.30) we will see that

$$p^\mu(\phi_i) = \left(\frac{g}{n_\mu}\right)(F_s^\mu) \qquad ...(5.37)$$

Since, the only term that survives on the RHS of eq. (5.35) will be the one in which ν = μ with a constant ($g/n$).

As we know that $P^\mu$ is a linear combination of the operators $\hat{R}$ of the group G. Since, the original reducible representation $\Gamma_{\text{red}}$ is defined with respect to $\phi_1$, $\phi_2$, ..., $\phi_n$ as basis then $R(g_i)$ is a known linear combination of ($\phi_1$, $\phi_2$, ..., $\phi_n$). The eq. (5.37) tells us that this linear combination of $\phi_1, \phi_2, \ldots, \phi_n$ is proportional to a vector $F_S^\mu$ which belongs to the irreducible $\Gamma^\mu$ representation. If we apply $P^\mu$ to each of the basis vectors {$\phi_1$, $\phi_2$, ..., $\phi_n$}, we will obtain $n$-linear combinations ($\phi_1$, $\phi_2$ , ..., $\phi_n$) which belong to the invariant subspace and which can be taken as the basis for this subspace. In this fashion, we will be able to obtain the bases for the I.Rs. by the application of projection operators. We will explain the whole process by taking the following examples. In these examples, we have chosen base vectors which give $\Gamma_{\text{red}}$.

**Note:** One may skip over the matter given in the last two paragraphs and read the examples given below.

**Example 3:** Let us take a vector space with orthogonal functions $\phi_1$, $\phi_2$, $\phi_3$, $\phi_4$, $\phi_5$ and $\phi_6$ which provide us six-dimensional reducible representation $\Gamma_{\text{red}}$ of $C_{3v}$.

The effects of the symmetry operations of $C_{3v}$ *viz.* ($E, C_3^1, C_3^2, \sigma_v^{(1)}, \sigma_v^{(2)}, \sigma_v^{(3)}$) are given in the following Table.

**Table 5.5:** Projection operators and effects on base vectors by symmetry operations of $C_{3v}$

| $\boldsymbol{P^{A_1} = E + C'_3 + C_3^2 + \sigma_v^{(1)} + \sigma_v^{(2)} + \sigma_v^{(3)}}$ | ***Base vector*** | $\boldsymbol{E}$ | $\mathbf{C'_3}$ | $\boldsymbol{C_3^2}$ | $\sigma'_v$ | $\sigma_v^{(2)}$ | $\sigma_v^{(3)}$ |
|---|---|---|---|---|---|---|---|
| $P^{A_2} = E + C'_3 + C_3^2 - \sigma_v^{(1)} - \sigma_v^{(2)} - \sigma_v^{(3)}$ | $\phi_1$ | $\phi_1$ | $-\frac{1}{2}(\phi_3 - \sqrt{3}\phi_4)$ | $-\frac{1}{2}(\phi_5 - \sqrt{3}\phi_6)$ | $-\phi_1$ | $\frac{1}{2}(\phi_5 - \sqrt{3}\phi_6)$ | $\frac{1}{2}(\phi_3 - \sqrt{3}\phi_4)$ |
| $P^{B} = 2E - C'_3 - C_3^2$ | $\phi_2$ | $\phi_2$ | $-\frac{1}{2}(\sqrt{3}\phi_3 - \phi_4)$ | $+\frac{1}{2}(\sqrt{3}\phi_5 - \phi_6)$ | $\phi_2$ | $\frac{1}{2}(\sqrt{3}\phi_5 - \phi_6)$ | $-\frac{1}{2}(\sqrt{3}\phi_3 - \phi_4)$ |
| | $\phi_3$ | $\phi_3$ | $-\frac{1}{2}(\phi_5 - \sqrt{3}\phi_6)$ | $-\frac{1}{2}(\phi_1 - \sqrt{3}\phi_2)$ | $-\phi_5$ | $\frac{1}{2}(\phi_3 - \sqrt{3}\phi_4)$ | $\frac{1}{2}(\phi_1 - \sqrt{3}\phi_2)$ |
| | $\phi_4$ | $\phi_4$ | $-\frac{1}{2}(\sqrt{3}\phi_5 - \phi_6)$ | $+\frac{1}{2}(\sqrt{3}\phi_1 - \phi_2)$ | $\phi_6$ | $\frac{1}{2}(\sqrt{3}\phi_3 - \phi_4)$ | $-\frac{1}{2}(\sqrt{3}\phi_1 + \phi_2)$ |
| | $\phi_5$ | $\phi_5$ | $-\frac{1}{2}(\phi_1 - \sqrt{3}\phi_2)$ | $-\frac{1}{2}(\phi_3 - \sqrt{3}\phi_4)$ | $-\phi_3$ | $\frac{1}{2}(\phi_1 + \sqrt{3}\phi_2)$ | $\frac{1}{2}(\phi_5 - \sqrt{3}\phi_6)$ |
| | $\phi_6$ | $\phi_6$ | $-\frac{1}{2}(\sqrt{3}\phi_1 - \phi_2)$ | $+\frac{1}{2}(\sqrt{3}\phi_3 - \phi_4)$ | $\phi_4$ | $\frac{1}{2}(\sqrt{3}\phi_1 + \phi_2)$ | $\frac{1}{2}(\sqrt{3}\phi_5 - \phi_6)$ |

Character Table + $C_{3v}$

| | $E$ | $2C_2$ | $2C_3\ 3\sigma v$ |
|---|---|---|---|
| $A_1$ | 1 | 1 | 1 |
| $A_2$ | 1 | 1 | –1 |
| $E$ | 2 | –1 | 0 |

Let us calculate $\{ p_{\phi_1}^{A_1}, p_{\phi_2}^{A_1}, \ldots\} \{ p_{\phi_1}^{A_2}, p_{\phi_2}^{A_2}, \ldots\}$ *etc.*

From the Table 5.5, we obtain

$$P^{A_1}\phi_1 = \phi_1 - \frac{\phi_3}{2} + \frac{\sqrt{3}}{2}\phi_4 - \frac{\phi_5}{2} - \frac{\sqrt{3}}{2}\phi_6 - \phi_1 + \frac{\phi_5}{2} + \frac{\sqrt{3}}{2}\phi_6 + \frac{\phi_3}{2} - \frac{\sqrt{3}}{2}\phi_4 = 0$$

(Since, we do not get any linear combinations of $P^{A_1}\phi_1 = 0$, it implies that $\phi_1$ does not belong to the invariant subspace of $A_1$ type in $\Gamma^{A_1}$ That $\Gamma^{A_1} = 0$.

$$P^{A_1}\phi_2 = \phi_2 - \frac{\sqrt{3}}{2}\phi_3 - \frac{\phi_4}{2} + \frac{\sqrt{3}}{2}\phi_5 - \frac{\phi_6}{2} + \phi_2 + \frac{\sqrt{3}}{2}\phi_5 - \frac{1}{2}\phi_6 - \frac{\sqrt{3}}{2}\phi_3 - \frac{\phi_4}{2}$$

$$= 2\phi_2 - \sqrt{3}\phi_3 - \phi_4 + \sqrt{3}\phi_5 - \phi_6$$

We have got a linear combination of $\phi_i$ which is equal to

$$\frac{1}{2\sqrt{3}}[2\phi_2 - \sqrt{3}\phi_3 - \phi_4 + \sqrt{3}\phi_5 - \phi_6]$$

$$P^{A_1}\phi_3 = \phi_3 - \frac{\phi_5}{2} + \frac{\sqrt{3}}{2}\phi_6 - \frac{\phi_1}{2} - \frac{\sqrt{3}}{2}\phi_2 - \phi_5 + \frac{\phi_3}{2} + \frac{\sqrt{3}}{2}\phi_4 + \frac{\phi_1}{2} - \frac{\sqrt{3}}{2}\phi_2$$

$$= \sqrt{3}\phi_2 + \frac{3}{2}\phi_3 + \frac{\sqrt{3}}{2}\phi_4 - \frac{3}{2}\phi_5 + \frac{\sqrt{3}}{2}\phi_6$$

We have another linear combination

$$\frac{1}{\sqrt{3}}\left[-\phi_2 + \frac{\sqrt{3}}{2}\phi_3 + \frac{\phi_4}{2} - \frac{\sqrt{3}}{2}\phi_5 + \frac{\phi_6}{2}\right]$$

$$= -\frac{1}{2\sqrt{3}}[+2\phi_2 - \sqrt{3}\phi_3 - \phi_4 + \sqrt{3}\phi_5 - \phi_6] = -P^{A_1}\phi_2$$

$$P^{A_1}\phi_4 = \phi_4 - \frac{\sqrt{3}}{2}\phi_5 - \frac{\phi_6}{2} + \frac{\sqrt{3}}{2}\phi_1 - \frac{\phi_2}{2} + \phi_6 + \frac{\sqrt{3}}{2}\phi_3 - \frac{\phi_4}{2} - \frac{\sqrt{3}}{2}\phi_1 - \frac{\phi_2}{2}$$

$$= \phi_2 + \frac{\sqrt{3}}{2}\phi_3 + \frac{\phi_4}{2} - \frac{\sqrt{3}}{2}\phi_5 + \frac{\phi_6}{2}$$

$$= \frac{1}{2\sqrt{3}}[-2\phi_2 + \sqrt{3}\phi_3 + \phi_4 - \sqrt{3}\phi_5 + \phi_6] = P$$

$$P^{A_1}\phi_5 = \phi_5 - \frac{\phi_1}{2} + \frac{\sqrt{3}}{2}\phi_2 - \frac{\phi_3}{2} - \frac{\sqrt{3}}{2}\phi_4 - \frac{\phi_3}{2} + \frac{\phi_1}{2}$$

$$+ \frac{\sqrt{3}}{2}\phi_2 + \frac{\phi_5}{2} - \frac{\sqrt{3}}{2}\phi_6$$

$$= \sqrt{3}\phi_2 - \frac{3}{2}\phi_3 - \frac{\sqrt{3}}{2}\phi_4 + \frac{3}{2}\phi_5 - \frac{\sqrt{3}}{2}\phi_6$$

$$= \frac{1}{2\sqrt{3}}[2\phi_2 - \sqrt{3}\phi_3 - \phi_4 + \sqrt{3}\phi_5 - \phi_6] = P^{A_1}\phi_5$$

$$P^{A_1}\phi_6 = \phi_6 - \frac{\sqrt{3}}{2}\phi_1 - \frac{\phi_2}{2} + \frac{\sqrt{3}}{2}\phi_3 - \frac{\phi_4}{2} + \phi_4$$

$$+ \frac{\sqrt{3}}{2}\phi_1 - \frac{\phi_2}{2} - \frac{\sqrt{3}}{2}\phi_5 - \frac{\phi_6}{2}$$

$$= -\phi_2 + \frac{\sqrt{3}}{2}\phi_3 + \frac{\phi_4}{2} - \frac{\sqrt{3}}{2}\phi_5 + \frac{\phi_6}{2}$$

$$= \frac{1}{2\sqrt{3}}[-2\phi_2 + \sqrt{3}\phi_3 + \phi_4 - \sqrt{3}\phi_5] = P^{A_1}\phi_6$$

Thus, the result of applying $P_1^A$ to $\phi_1$, $\phi_2$, $\phi_3$, $\phi_4$, $\phi_5$, $\phi_6$ gives a linear combinations of $\phi_2$, $\phi_3$, $\phi_4$, $\phi_5$ and $\phi_6$ $\pm$ $[2\phi_2 - \sqrt{3\phi_3} - \phi_4 + \sqrt{3\phi_5} + \phi_6]$ and belongs to I.R, $A_1$ which may be taken as a basis for $A_1$ subspace.

Similarly one can calculate $P^{A_2}\phi_i's$ and get another linear combinations of $\phi_i s$. This linear combination is taken as a basis for the $A_2$ subspace. This linear combination will be orthogonal to the one obtained from the calculation of $[P^{A_1}\phi_1]$. For $A_2$ subspace, one gets the basis $[2\phi_1 - \phi_3 + \sqrt{3\phi_4} - \phi_5 - \sqrt{3\phi_4}]/2\sqrt{3}$.

By applying $P^E$ to the $\phi_i's$ one gets two choices of two orthogonal basis vectors for the $E$ representations. One has to choose one out of the two choices.

Thus, we have got a method of finding the basis for the I.R′s, of any point group in terms of linear combinations of the orthogonal basis forming reducible representation of the point group. We will be using this method very often while applying symmetry properties of the molecules to solve various problems related to vibration and bonding properties of molecule.

## 5.5 SALCs (SYMMETRY ADAPTED LINEAR COMBINATION OF BASE VECTORS)

In the foregoing paragraphs, we have already constructed symmetry adapted linear combinations of the orthogonal basis coordinates ($\phi_1$, $\phi_2$, ..., $\phi_6$) of six-dimensional vector space. Since SALCs play very important role in our study [discussed later], we will give a few more examples of constructing SALCs using projection operators.

We first take the example of cyclopropenyl ion $[C_3H_3]^+$. It has 3-orbitals [$\phi_1$, $\phi_2$ and $\phi_3$] which are used as orthonormal basis vectors and which form a reducible representation of $D_{3h}$ point group to which the molecule belongs. The operations associated with this point group are $E, C_2, C_2', C_2'', \sigma_h, S_3^1, S_3^5$, $\sigma_v^1, \sigma_v^2, \sigma_v^3$. Using three $\phi_i's$ orbitals as basis vectors one can easily obtain the reducible representation as given below.

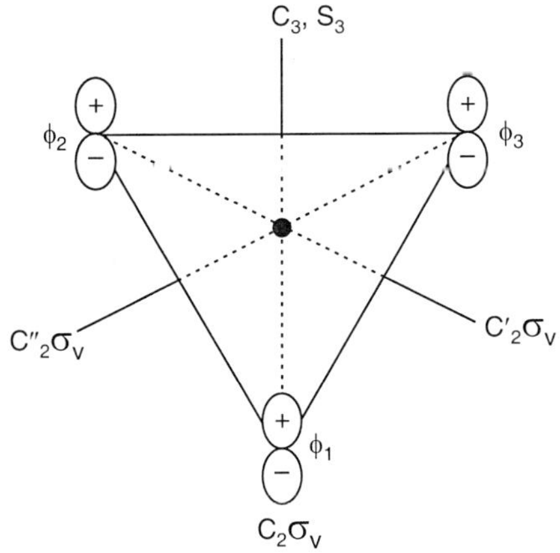


**Fig. 5.1** π-orbitals of cyclopropenyl ion $[C_3H_3]^+$.

**Table 5.6:** Character Table of $D_{3h}$

| $D_{3h}$ | $E$ | $2C_3$ | $3C_2$ | $\sigma_h$ | $2S_3$ | $3\sigma_v$ | *I.R.* |
|---|---|---|---|---|---|---|---|
| $A_1'$ | 1 | 1 | 1 | 1 | 1 | 1 | |
| $A_2'$ | 1 | 1 | 1 | 1 | 1 | –1 | |
| $E'$ | 2 | –1 | 0 | 2 | –1 | 0 | |
| $A_1''$ | 1 | 1 | 1 | –1 | –1 | –1 | |
| $A_2''$ | 1 | 1 | –1 | –1 | –1 | 1 | |
| $E''$ | 2 | –1 | 0 | –2 | 1 | 0 | |
| $\Gamma_{red}$ | 3 | 0 | –1 | –3 | 0 | 1 | $A_2'' + E''$ |

Using this reducible representation and the character table of $D_{3h}$, $\Gamma_{red}$ is reduced to I.Rs. components ($A_2'' + E''$) by using the $a_v = \frac{1}{g}\sum_{k=1} g_k \chi_R^{\Gamma_{red}} . \chi_R^{*I.R}$.

The components of I.Rs. are shown in the table given below.

Since, the $\Gamma_{red}$ is reduced to I.R. components $A_2'' + E''$ so we will only be using the projection operators of $A_2'' + E''$. These are given below.

**Table 5.7:** Character for $\Gamma_{red}$ of $[C_3H_3]^+$ ion

| *Operations* | $E$ | $C_3^1$ | $C_3^2$ | $C_2$ | $C_2'$ | $C_2''$ | $\sigma_h$ | $S_3^1$ | $S_3^5$ | $\sigma_v^1$ | $\sigma_v^2$ | $\sigma_v^3$ |
|---|---|---|---|---|---|---|---|---|---|---|---|---|
| $\chi_R$ for $\Gamma_{red}$ | 3 | 0 | 0 | –1 | –1 | –1 | –3 | 0 | 0 | 1 | 1 | 1 |

$$P^{A_2''} = \frac{1}{12}\left[\begin{matrix} E + C_3^1 + C_3^2 - C_2 - C_2' - C_2'' - \sigma_h - S_3^1 - S_3^5 \\ +\sigma_v^{(1)} + \sigma_v^{(2)} + \sigma_v^{(3)} \end{matrix}\right] \quad ...(5.38)$$

and $$P^{E''} = \frac{1}{12}\left[\begin{matrix} 2E - C_3^1 - C_3^2 - 0C_2 + 0C_2' - 0C_2'' - 2\sigma_h + S_3^1 \\ +S_3^5 + 0\sigma_v^{(1)} + 0\sigma_v^{(2)} + \sigma_v^{(3)} \end{matrix}\right] \quad ...(5.39)$$

Since in the table of $\Gamma_{red}$ no complex quantity appears so we ignore the complex conjugate of $(\chi_{(R)}^{*(IR)})$ and use the real quantity, *viz.* $(\chi_{(R)}^{(IR)})$. We shall ignore the numerical constant 1/12 in (eqs. 5.38 and 5.39) because it can be taken care of when we normalise the SALCs.

We apply $P^{A_2''}$ and $P^{E''}$ operators on the basis functions, *viz.* $\phi_1$, $\phi_2$, $\phi_3$. The results are given below.

We thus obtain,

$$P^{A_2''} = \left[\begin{matrix} E\phi_1 + C_3^1\phi_1 + C_3^2\phi_1 - C_2\phi_1 - C_2'\phi_1 - C_2''\phi_1 - \sigma_h\phi_1 \\ - S_3^1\phi_1 - S_3^5\phi_1 + \sigma_v^{(1)}\phi_1 + \sigma_v^{(2)}\phi_1 + \sigma_v^{(3)}\phi_1 \end{matrix}\right]$$

$$= \underbrace{\phi_1}_{E} + \underbrace{\phi_2 + \phi_3}_{C_2} + \underbrace{\phi_1 + \phi_2 + \phi_3}_{3C_2} + \underbrace{\phi}_{\sigma_h} + \underbrace{\phi_2 + \phi_3}_{2S_3} + \underbrace{\phi_1 + \phi_2 + \phi_3}_{\sigma_v}$$

$$(f^{A_2}) = 4(\phi_1 + \phi_2 + \phi_3).$$

$$\text{Normalised SALC} = \frac{1}{\sqrt{3}}(\phi_1 + \phi_2 + \phi_3) \qquad ...(5.40)$$

[Normalisation of $f^{A_2}$ is carried out by the condition $\int_{-\infty}^{+\infty}(f^{A_2} f^{A_2})\partial\tau$ $= N^2$. Since we have taken $\phi_i's$ as orthonormal set, we have $\phi_i\phi_j = \delta_{ij}$. So

$$\int f^{A_2}.\, f^{A_2}\partial\tau = \int(\phi_1 + \phi_2 + \phi_3)\,(\phi_1 + \phi_2 + \phi_3)\partial\tau$$

Here $\int \phi_i\phi_j\partial\tau = 0$, and $\int \phi_i\phi_i\partial\tau = 1$, So

$\int f^{A_2} f^{A_2}\partial\tau = \int \phi_1^2\partial\tau + \int \phi_2^2\partial\tau + \int \phi_3^2\partial\tau = 3$ and hence eq. (5.40)]

One can also apply $P^{A_2''}$ to the basis functions $\phi_2$ and $\phi_3$ in place of $\phi_1$. One will find that $P^{A_2''}\phi_1 = P^{A_2''}\phi_2 = P^{A_2}\phi_3 = \frac{1}{\sqrt{3}}(\phi_1 + \phi_2 + \phi_3)$. So that $A_2''$ symmetry adapted linear combination of $\phi_1$, $\phi_2$, $\phi_3$ is $\frac{1}{\sqrt{3}}(\phi_1 + \phi_2 + \phi_3)$.

As a next step we wish to form SALCs belonging to $E''$ type of I.R. We again take $\phi_1$ as a starting function and apply the operator $P^E\,\phi_1$ on this using eq. (5.39). We get,

$$P^E\phi_1 = [2E\phi_1 - C_3^1\phi_1 - C_3^2\phi_1 - 2\sigma_h\phi_1 - S_3^1\phi_1 - S_3^2\phi_1] \qquad ...(5.41)$$

$$- 2\phi_1 - \phi_2 - \phi_3 + 2\phi_1 - \phi_2 - \phi_3$$

$$f_2^{E''} = 2\phi_1 - \phi_2 - \phi_3 \qquad ...(5.42)$$

$$\text{Normalised } f_2^{E''} = \frac{1}{\sqrt{6}}[2\phi_1 - \phi_2 - \phi_3] \qquad ...(5.43)$$

Eq. (5.43) gives only one of the two $E$-type functions (we need two SALCs to form the basis for $E$-representations). There are several general methods to find its partner functions. The following two procedures are easy to find the second function.

**Method 1:** It is based on the principle that if we have two function $f_1$ and $f_2$ belonging to $E$ type of I.R and if we apply the symmetry operation of the point group to any one of them as a result of operation, it will either go as +1 or –1 times itself or into its partner function as (a linear combination of $f_1$ and $f_2$ functions). What we do now is that we choose the symmetry operation which takes $f_1$ or $f_2$ into a different function. Since, one of the symmetry operation is usually a rotation operation in all the point groups, we choose $C_3^1$ operation. Applying $C_3^1$ operation on the function that we have obtained (*i.e.*, given by eq. 5.43), we get,

$$C_3^1\left(\frac{1}{\sqrt{6}}[2\phi_1 - \phi_2 - \phi_3]\right) = \frac{1}{\sqrt{6}}[2\phi_2 - \phi_3 - \phi_1] \quad ...(5.44)$$

This function is neither ±1 times of the first function nor it is orthogonal to it. It should therefore be a linear combination of the first function *viz.* $\left(\frac{1}{\sqrt{2}}[2\phi_1 - \phi_2 - \phi_3]\right)$ and the unknown second function. Thus, ignoring normalisation factor, we have a relationship,

$$(2\phi_1 - \phi_2 - \phi_3) \pm C(f_2^E) = (2\phi_2 - \phi_3 - \phi_1) \quad ...(5.45)$$

or

$$(2\phi_2 - \phi_3 - \phi_1) \pm C'(2\phi_1 - \phi_2 - \phi_3) = f_2^E \quad ...(5.46)$$

By choosing $C'$ as $\left(-\frac{1}{2}\right)$, we get

$$2\phi_2 - \phi_3 - \phi_1 - \frac{1}{2}(2\phi_1 - \phi_2 - \phi_3) = \frac{3}{2}(\phi_2 - \phi_3) \quad ...(5.47)$$

We normalise the second function $f_2^E$ and get,

$$f_2^E = \frac{1}{\sqrt{2}}(\phi_2 - \phi_3) \quad ...(5.48)$$

It can be shown that $f_2^E = \left[\frac{1}{\sqrt{2}}(\phi_2 - \phi_3)\right]$ is orthonormal to $f_1^E$ = [*viz* $(2\phi_1 - \phi_2 - \phi_3)$]. [Choosing the constant $C'$ in eq. 5.46 is however a little tricky, but can be easily guessed after a little experience].

**Method 2:** Another method which is easy to apply is as follows. The general principle is that instead of using the whole group (*viz.* $D_{3h}$) we use only its rotational subgroup (*viz.* $C_3$). That is, we restrict our attention only to the pure rotational symmetry about the principal axis and use it to obtain SALCs. In our example, we use $C_3$ point group. This group being Abelian, will have only one-dimensional representation. In this subgroup ($C_3$), the set of three $\phi_i$'s orbitals [$\phi_1$, $\phi_2$, $\phi_3$] spans $A$ and $E$ representations but the $E$-[I.R] appears in $C_3$ as two associated one-dimensional representations and the projection operator $p^\mu$ can be written in terms of each of the two one-dimensional I.Rs indirectly. We will get two separated SALCs for $E$-representation using two projection operators. We show below its working methodology.

The projection operators for $A$ and $E$ I.R. for $C_3$ point group are,

$$P^{A_1} = (E + C_3^1 + C_3^2) \quad ...(5.49)$$

$$P^{E_1} = (E + \in C_3^1 + \in^* C_3^2) \quad ...(5.50)$$

$$P^{E_2} = (E + \in^* C_3^1 + \in C_3^2) \quad ...(5.51)$$

Applying $P^{A_1}$ operator on $\phi_1$, we get,

$$P^{A_1}\phi_1 = (E\phi_1 + C_3^1\phi_1 + C_3^2\phi_1)$$

$$f_1^{A_1} = \phi_1 + \phi_2 + \phi_3$$

After normalisation we get,

$$f_1^{A_1} = \frac{1}{\sqrt{3}}(\phi_1 + \phi_2 + \phi_3) \quad ...(5.52)$$

$$(P^{E_1}\phi_1) = (E + \in C_3^1\phi_1 + \in^* C_3^2\phi_1)$$

$$f_1^{E_1} = \phi_1 + \in\phi_2 + \in^*\phi_3 \quad ...(5.53)$$

Similarly,

$$P^{E_2}\phi_1 = (E + \in^* C_3^1\phi_1 + \in C_3^2\phi_1)$$

$$f_1^{E_2} = \phi_1 + \in^*\phi_2 + \in\phi_3$$

Normalisation gives $f_1^{E_2} = \dfrac{1}{\sqrt{1 + \in^2 + \in^{*2}}}[\phi_1 + \in^*\phi_2 + \in\phi_3]$ ...(5.54)

**Table 5.8:** Character Table of $C_3$

| $C_3$ | $E$ | $C_3^1$ | $C_3^2$ | *I.Rs* |
|---|---|---|---|---|
| $A$ | 1 | 1 | 1 | |
| $E$ $E_1$ | 1 | $\in$ | $\in^*$ | |
| $E$ $E_2$ | 1 | $\in^*$ | $\in$ | |
| $\Gamma_{red}$ | 3 | 0 | 0 | $A + E$ |

So the SALCs belonging to $A$ is the same as that obtained previously when the whole group has been used to form $f^{A_2}$ function.

The $E$-type SALCs, though can be used as they are, but we wish to have the coefficients as real numbers instead of imaginary ones ($\in_1$ and $\in^*$). This can be done by adding and subtracting function $f^{E_1}, f^{E_2}$ term by term whereby the imaginary part is eliminated leaving back the real parts. The addition and subtraction process are like taking another linear combination. Thus, we have by addition

$$\phi_1 + \in\phi_2 + \in^*\phi_3 + \phi_1 + \in^*\phi_2 + \in\phi_3 = 2\phi_1 + (\in + \in^*)\phi_2 + (\in + \in^*)\phi_3$$

(Here, $\in + \in^* = 2\cos\dfrac{2\pi}{3} = -1$, So we have,

$$f^{E_1} = 2\phi_1 - \phi_2 - \phi_3 \qquad \text{...(5.55)}$$

By subtraction, we get,

$$\phi_1 + \in\phi_2 + \in^*\phi_3 - \phi_1 + \in^*\phi_2 + \in\phi_3 = (\in - \in^*)\phi_2 - (\in - \in^*)\phi_3$$

Now $\qquad \in - \in^* = 2i\sin\dfrac{2\pi}{3} = \sqrt{3i}$

So $\qquad f^{E_2} = \sqrt{3}i(\phi_2 - \phi_3)$

Neglecting $\sqrt{3}i$, and normalising $f^{E_2} = \dfrac{1}{\sqrt{2}}(\phi_2 - \phi_3)$ ...(5.56)

Thus, we get another SALC.

So the SALCs for cyclopropenyl ion are

1. $f^{A_2''} = \dfrac{1}{\sqrt{3}}(\phi_1 + \phi_2 + \phi_3)$

2. $f^{E_1''} = \dfrac{1}{\sqrt{6}}(2\phi_1 - \phi_2 - \phi_3)$ ...(5.57)

and 3. $f^{E_2''} = \dfrac{1}{V}(\phi_2 - \phi_3)$

By the same procedure one can get SALCs for other delocalised cycloalkene systems. We give below some of them.

1. *Cyclobutadiene. Point group* $D_{4h}(\phi_1, \phi_2, \phi_3, \phi_4)$ *basis vectors*

$$\Gamma_{red} = A + E + B.$$

SALCs are

$$f^A = \frac{1}{2}(\phi_1 + \phi_2 + \phi_3 + \phi_4)$$

$$f^B = \frac{1}{2}(\phi_1 - \phi_2 + \phi_3 - \phi_4)$$

$$f^{E_1} = \frac{1}{\sqrt{2}}(\phi_1 - \phi_3)$$

$$f^{E_2} = \frac{1}{\sqrt{2}}(\phi_2 - \phi_4)$$

2. *Cyclopentadiene* ($C_5H_5$) ($D_{5h}$) ($\phi_1$, $\phi_2$, $\phi_3$, $\phi_4$, $\phi_5$) *basis vectors.*

$$\Gamma_{red} = A + E_1 + E_2$$

SALCs are

$$f^A = \frac{1}{\sqrt{5}}(\phi_1 + \phi_2 + \phi_3 + \phi_4 + \phi_5)$$

$$f_1^{E_1} = \sqrt{\frac{2}{5}}(\phi_1 + \phi_2 \cos\omega + \phi_3 \cos\omega + \phi_4 \cos\omega + \phi_5 \cos\omega)$$

$$f_2^{E_1} = \sqrt{\frac{2}{5}}(\phi_1 + \phi_2 \sin\omega + \phi_3 \sin\omega - \phi_4 \sin\omega + \phi_5 \sin\omega)$$

$$f_2^{E_1} = \sqrt{\frac{2}{5}}(\phi_1 + \phi_2 \sin\omega + \phi_3 \sin\omega + \phi_4 \sin\omega + \phi_5 \sin\omega)$$

$$f_2^{E_2} = \sqrt{\frac{2}{5}}(\phi_2 \sin 2\omega - \phi_3 \sin 2\omega + \phi_4 \sin 2\omega - \phi_5 \sin 2\omega)$$

$$\left(\omega = \frac{2\pi}{5}\right)$$

3. *Benzene* ($C_6H_6$) $D_{6h}$($\phi_1$, $\phi_2$, $\phi_3$, $\phi_4$, $\phi_5$, $\phi_6$) *basis vectors,*

$$\Gamma_{red} = A + B + E_1 + E_2$$

SALCs are

$$f^A = \frac{1}{\sqrt{6}}(\phi_1 + \phi_2 + \phi_3 + \phi_4 + \phi_5 + \phi_6)$$

$$f^B = \frac{1}{\sqrt{6}}(\phi_1 - \phi_2 + \phi_3 - \phi_4 + \phi_5 - \phi_6)$$

$$f_1^{E_1} = (\phi_1 + \in\phi_2 - \in^*\phi_3 - \phi_4 - \in\phi_5 + \in^*\phi_6)$$

$$f_2^{E_1} = (\phi_1 + \in\phi_2 - \in\phi_3 - \phi_4 - \in^*\phi_5 + \in\phi_6)$$

$$f_1^{E_2} = (\phi_1 - \in^*\phi_2 - \in\phi_3 + \phi_4 - \in^*\phi_5 - \in\phi_6)$$

$$f_2^{E_2} = (\phi_1 - \in\phi_2 - \in^*\phi_3 + \phi_4 - \in\phi_5 - \in^*\phi_6)$$

Let us take another example from the vibrational spectroscopy and see how the SALCs are formed. Let us take the example of $PF_5$ molecule.

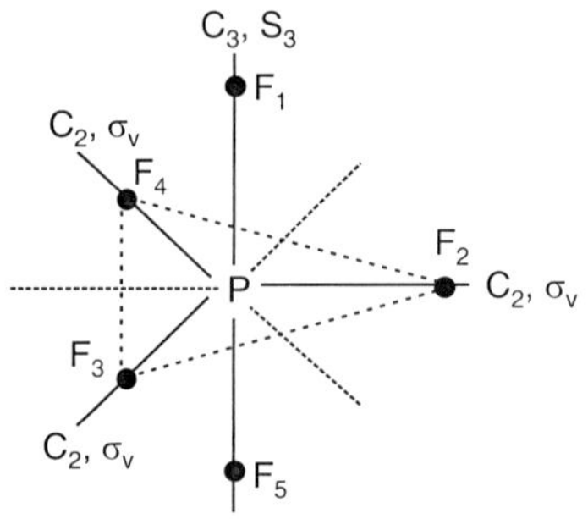


**Fig. 5.2** Symmetry operations.

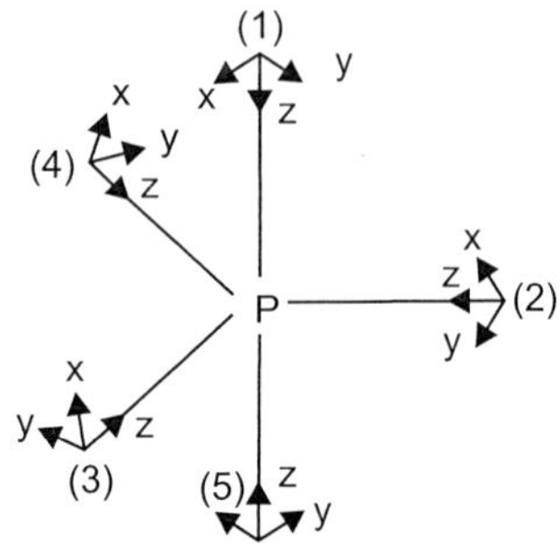


**Fig. 5.3** Coordinates of $PF_5$ used as basis coordinates.

In this case, the basis vectors are the displacement vectors $\Delta r_i$ indicating the $i^{th}$ atom's displacements from their equilibrium position. These displacement vectors are resolved along the *x*, *y*, *z*-axes of the coordinate systems fixed at the $i^{th}$ atoms with axes in certain directions. Therefore, basis vector are displacements $\Delta x^a$, $\Delta y^a$, $\Delta z^a$. We thus have $3n$ basis vectors in the form of $\Delta x^a$, $\Delta y^a$, $\Delta z^a$ where '*a*' varies from 1 to 6 *i.e.*, the number of atoms in the molecule. These cartesian displacement components form a reducible representation ($\Gamma_{red}$) of point group of the molecule. One can in principle, find the I.Rs present in $\Gamma_{red.}$ With the application of the projection operators [$P^{IR}$] one is able to calculate the linear combinations of $\Delta x^a$, $\Delta y^a$, $\Delta z^a$ belonging to a particular I.R and obtain SALCs of the molecule. It is however mathematically cumbersome and instead of cartesian basis vectors ($\Delta x$, $\Delta y$, $\Delta z$) one prefers to use internal coordinates and then to the normal coordinates. This topic will be discussed in detail in chapter twelve on molecular vibrations. In this paragraph, we would like to emphasize that the basis vectors can be of any nature. These vectors should be suitable for particular system. These basic coordinates are then transformed in SALCs which are then used in the calculation.

# 6

# Direct Products and Vanishing Integrals

So far we have discussed the representations of the symmetry operations of molecular point groups through a set of matrices which multiply in the same way as the symmetry operations. These matrices are obtained as a result of finding the effects of the symmetry operations on the vectors of vector space. It led us to classify and to characterize once for all the point groups through irreducible representations which help us in the application of group theory in quantum mechanical calculations. However, there are cases where the products $n_\mu n_\nu$, $\left(\phi_i^\mu \psi_j^\nu\right)$, ($i = 1, 2, ..., n_\mu j = 1, 2, ..., n_\nu$ and ν, μ are the two representations of the group) form a basis for the direct product representations ($\Gamma^\mu \times \Gamma^\nu$). (It is immaterial if the representations $\Gamma^\mu$ and $\Gamma^\nu$ are reducible or irreducible). It is also notable to mention that one can also extend the direct product concept to more than two representations. This chapter deals with the discussion of such direct product representations and of their applications (though brief and introductory) in quantum mechanics.

## 6.1 DIRECT PRODUCT OF TWO MATRICES

There are two types of matrix multiplication (*i*) simple multiplication represented as $|A| \cdot |B|$ and (*ii*) Direct product of two matrices denoted by $|A| \otimes |B|$.

(*i*) Simple product of two matrices is carried out as follows:

$$\begin{vmatrix} a_{11} & a_{12} & a_{13} \\ a_{21} & a_{22} & a_{23} \\ a_{31} & a_{32} & a_{33} \end{vmatrix} \circ \begin{vmatrix} b_{11} & b_{12} \\ b_{21} & b_{22} \\ b_{31} & b_{32} \end{vmatrix} = (\text{Row of } |A| \cdot \text{column of } |B|) \quad ...(6.1)$$

$$= \begin{vmatrix} a_{11}b_{11} + a_{12}b_{21} + a_{13}b_{31} & a_{11}b_{12} + a_{12}b_{22} + a_{13}b_{32} \\ a_{21}b_{11} + a_{22}b_{21} + a_{23}b_{31} & a_{21}b_{12} + a_{22}b_{22} + a_{23}b_{32} \\ a_{31}b_{11} + a_{32}b_{21} + a_{33}b_{31} & a_{31}b_{12} + a_{32}b_{22} + a_{33}b_{32} \end{vmatrix} \quad ...(6.2)$$

$$= \begin{vmatrix} (a_{11}a_{12}a_{13})(b_{11}b_{21}b_{31}) & (a_{11}a_{12}a_{13})(b_{12}b_{22}b_{32}) \\ (a_{21}a_{22}a_{23})(b_{11}b_{21}b_{31}) & (a_{21}a_{22}a_{23})(b_{12}b_{22}b_{32}) \\ (a_{31}a_{32}a_{33})(b_{11}b_{21}b_{31}) & (a_{31}a_{32}a_{33})(b_{12}b_{22}b_{32}) \end{vmatrix} \begin{matrix} \rightarrow \text{element of I row of } A \circ I \text{ column of } B\,(I \text{ row of A})\,(\text{2nd column of } B) \\ \rightarrow \text{2nd row of } A \circ I \text{ column of } B\,(\text{2nd row of } A)\,(\text{2nd column of } B) \\ \rightarrow \text{3rd row of } A \circ I \text{ column of } B\,(\text{3rd row of } A)\,(\text{2nd column of } B) \end{matrix}$$

...(6.3)

In general,

$$|A| \cdot |B| = |a_{ij}b_{jk}| = |\sum_{i=1}^{3} (i^{th} \text{ row of } |A|) \sum_{j=1}^{n} (j^{th} \text{ column of } |B|) \quad ...(6.4)$$

*The simple product of matrices* have the following properties.

1. Trace $|A| \cdot |B|$ = Trace of $|B| \cdot |A|$ = (Trace of $A$) $\cdot$ (Trace of $B$) ...(6.5)

This is possible only for square matrices of the same dimensions of $A$ and $B$. That is,

Dimension of the row of $|A|$ = Dimension of the column of $|B|$. ...(6.6)

2. Order of the matrix $|A| \cdot |B|$ = (Number of elements in the row of $|A|$) $\times$ (Number of elements in column of $|B|$) *e.g.*, in above case order of the product is $|AB| = |3 \times 2|$ *i.e.*, it has 3 rows and 2 columns.

(*ii*) *Direct product of matrices* $|A|$ *and* $|B|$. It is carried out as given below. Suppose we have two matrices $|A|$ and $|B|$ where $|A|$ and $|B|$ are

$$|A| = \begin{vmatrix} a_{11} & a_{12} & a_{13} \\ a_{21} & a_{22} & a_{23} \\ a_{31} & a_{32} & a_{33} \end{vmatrix} \text{ and } |B| = \begin{vmatrix} b_{11} & b_{12} \\ b_{21} & b_{22} \end{vmatrix} \quad ...(6.7)$$

The direct product of $|A|$ and $|B|$ [*i.e.*, $|A| \otimes |B|$] is equal to

$$|A| \otimes |B| = \begin{vmatrix} a_{11} & a_{12} & a_{13} \\ a_{21} & a_{22} & a_{23} \\ a_{31} & a_{32} & a_{33} \end{vmatrix} \otimes \begin{vmatrix} b_{11} & b_{21} \\ b_{12} & b_{22} \end{vmatrix}$$

$$\neq \begin{vmatrix} b_{11} & b_{21} \\ b_{12} & b_{22} \end{vmatrix} \otimes \begin{vmatrix} a_{11} & a_{12} & a_{13} \\ a_{21} & a_{22} & a_{23} \\ a_{31} & a_{32} & a_{33} \end{vmatrix} \quad ...(6.8)$$

[The above relationship is usually not commutative].

$|B| \otimes |A|$ is equal to

$$= \begin{vmatrix} b_{11}|A| & b_{12}|A| \\ b_{21}|A| & b_{22}|A| \end{vmatrix} \quad \ldots(6.9)$$

$$= \begin{vmatrix} b_{11}\begin{vmatrix} a_{11} & a_{12} & a_{13} \\ a_{21} & a_{22} & a_{23} \\ a_{31} & a_{32} & a_{33} \end{vmatrix} & b_{12}\begin{vmatrix} a_{11} & a_{12} & a_{13} \\ a_{21} & a_{22} & a_{23} \\ a_{31} & a_{32} & a_{33} \end{vmatrix} \\ b_{21}\begin{vmatrix} a_{11} & a_{12} & a_{13} \\ a_{21} & a_{22} & a_{23} \\ a_{31} & a_{32} & a_{33} \end{vmatrix} & b_{22}\begin{vmatrix} a_{11} & a_{12} & a_{13} \\ a_{21} & a_{22} & a_{23} \\ a_{31} & a_{32} & a_{33} \end{vmatrix} \end{vmatrix} \quad \ldots(6.10)$$

$$= \begin{vmatrix} b_{11}a_{11} & b_{11}a_{12} & b_{11}a_{13} & b_{12}a_{11} & b_{12}a_{12} & b_{12}a_{13} \\ b_{11}a_{21} & b_{11}a_{22} & b_{11}a_{23} & b_{12}a_{21} & b_{12}a_{22} & b_{12}a_{23} \\ b_{11}a_{31} & b_{11}a_{32} & b_{11}a_{33} & b_{12}a_{31} & b_{12}a_{32} & b_{12}a_{33} \\ b_{21}a_{11} & b_{21}a_{12} & b_{21}a_{13} & b_{22}a_{11} & b_{22}a_{12} & b_{22}a_{13} \\ b_{21}a_{21} & b_{21}a_{22} & b_{21}a_{23} & b_{22}a_{21} & b_{22}a_{22} & b_{22}a_{23} \\ b_{21}a_{31} & b_{21}a_{32} & b_{21}a_{33} & b_{22}a_{31} & b_{22}a_{32} & b_{22}a_{33} \end{vmatrix} \quad \ldots(6.11)$$

It has order $|2 \times 3| = |6 \times 6|$

One can obtain the direct product of two matrices of any order. We can write in general, the direct product of two matrices of $n$ and $n'$ dimensions as follows:

$$|A| \otimes |B| = |(n \times n)| \otimes |n' \times n'| \quad \ldots(6.12)$$

$$= \begin{vmatrix} a_{11}|B| & a_{12}|B| \ldots\ldots a_{1n}|B| \\ a_{21}|B| & a_{22}|B| \ldots\ldots a_{2n}|B| \\ \vdots & \\ a_{n1}|B| & a_{n2}|B| \ldots\ldots a_{nn}|B| \end{vmatrix} = |C| \quad \ldots(6.13)$$

The general element of this matrix $c_{ikjl}$ is given by

$$c_{ikjl} = a_{ij} \cdot b_{kl} \quad \ldots(6.14)$$

Thus, the first second, third, ..., $n^{\text{th}}$ column (or rows) of the direct product matrix are number (1, 1), (1, 2), ...., (1, $n'$), (2, 1), (2, 2), ... (2, $n'$) ... $n1$, $n2$, ...., ($n$ $n'$), respectively.

Thus, according to eq. (6.13), the element in (4 2) row and (5 6) column, *i.e.*, $c_{4\,2,\,5\,6}$ will be equal to

$$c_{4526} = a_{42} \cdot b_{56}$$

(This is in accordance with eq. (6.14).

Some of the properties of direct product of two matrices are

(*a*) The direct product of two diagonal matrices is a diagonal matrix.

(*b*) The direct product of two unit matrices is a unit matrix.

(*c*) The trace of the direct product of two matrices is equal to the product of the traces of the matrices *i.e.*,

$$\text{Trace of } |A \otimes B| = \text{Trace of } |A| \otimes \text{Trace of } |B| \qquad \ldots(6.15)$$

(*d*) The direct product of two matrices, in general, is not commutative. That is,

$$|A| \otimes |B| \neq |B| \otimes |A| \qquad \ldots(6.16)$$

(*e*) The following relationship holds good for direct products of matrices

1. $|A| \otimes (|B| \otimes |C|) = (|A| \otimes |B|) \otimes |C| = (|A| \otimes |B| \otimes |C|)$ ...(6.17)

   (Associative law holds good.)

2. $|A| \otimes (|B| + |C|) = |A| \otimes |B| + |A| \otimes |C|$ ...(6.18)
3. $(|A| + |B|) \otimes |C| = |A| \otimes |C| + |B| \otimes |C|$ ...(6.19)

   (Distributive law holds good)

4. $(|A| \otimes |B|) \cdot (|C| \otimes |D|) = (|A| \otimes |C|) \cdot (|B| \otimes |D|)$ ...(6.20)
5. $(|A| \otimes |B|)^{-1} = (|A|^{-1} \otimes |B|^{-1})$ ...(6.21)
6. $(|A| \otimes |B|)' = |A|^t \otimes |B|'$ ...(6.22)

(Here $A^t$ = transpose of the matrix $A$ obtained by interchanging elements of rows and columns).

## 6.2 DIRECT PRODUCT OF SUBGROUPS

Prior to discussing the role of direct product in the multiplication of various subgroups, let us first obtain the direct product of *two row matrices*,

$$|a_{11}\, a_{12}\, a_{13} \ldots a_{1n}| \otimes |b_{11}\, b_{12}\, b_{13} \ldots b_{1m}| \qquad \ldots(6.23)$$
$$= |A| \otimes |B|$$

According to eq. (6.7), the direct product in eq. (6.23) is equal to

$$= |a_{11}|B|a_{12}|B|a_{13}|B| \ldots a_{1n}|B|$$
$$= |a_{11}(b_{11}b_{12} \ldots b_{1m})a_{12}\,(b_{11}b_{12} \ldots b_{1m}) \ldots a_{1n}\,(b_{11}b_{12} \ldots b_{1m})$$
$$= |a_{11}b_{11}a_{11}b_{12} \ldots a_{11}b_{1m},\ a_{12}b_{11}a_{12}b_{12} \ldots a_{12}b_{1m},$$
$$\ldots a_{1n}\, b_{11}a_{1n}\, b_{12} \ldots a_{1n}\, b_{1m}| \qquad \ldots(6.24)$$

That gives the row matrix with $(n \times m)$ elements.

We consider the direct product of two or more than two point groups. We emphasize at this point that the role of the sets of symmetry operations is like a set of basis coordinates of a vector that define the symmetry-space of the system. Thus, some properties of the system with definite symmetry operations can be expressed in terms of the basis vectors (*e.g.*, symmetry operations) that span the symmetry space (point group symmetry operations of the system) and the group symmetry operations themselves form a space having dimensions of the order of group. If we have two groups, it implies that we have two spaces. The direct product of these two spaces is equivalent to the direct product of the basis coordinates (symmetry operations) of these combined spaces. This concept lies at the base of direct product of two point groups. But there are two conditions that have to be satisfied in obtaining the direct product of two or more groups. In the following paragraph, we will apply it in understanding the product of two or more subgroups. If there is a group G having a number of subgroups $H_1$, $H_2$, ..., $H_p$ then we have,

$$G = H_1 \otimes H_2 \otimes H_3 \otimes \ldots \otimes H_p \qquad \ldots(6.25)$$

The Eq. (6.25) is satisfied only when the following two conditions are satisfied.

1. Every symmetry operation of each subgroup (*viz.* $H_1$, $H_2$, ..., or $H_p$) commutes with every symmetry operation of all the other subgroups.
2. Each element $g_i$ of $G$ can be expressed in a unique way in the form

$$g_i = (h_1, h_2, h_3, \ldots, h_m) \qquad \ldots(6.26)$$

where, $h_i$ is an element of $H$.

**Example 1:** Let us take the $C_{3h}$ point group (Group G). It has the symmetry operations $\{E, C_3^1, C_3^2, \sigma_h, S_3^1, S_3^5\}$, $[\{S_3^1 = \sigma_h.C_3^1 = C_3^1.\ \sigma_h)$ and $S_3^5 = \sigma_h \cdot C_3^2 = C_3^2 \cdot \sigma_h)\}]$ and it has two subgroups, *viz.* $C_3 = (E, C_3^1, C_3^2)$ and $C_s = (E, \sigma_h)$ besides $E$ as a third subgroup. In the latter two subgroups, $[C_3$ and $C_s]$, $E$ and $\sigma_h$ of $C_s$ subgroup commute with all the elements of $C_3$. Thus, the condition (1) is satisfied for the formation of cross product. We thus have,

$$\left.\begin{aligned} C_s \otimes C_3 &= (E, \sigma_h) \otimes (E, C_3^1, C_3^2) = \{E(C_3), \sigma_h(C_3)\} \\ &= \{E(E, C_3^1, C_3^2), \sigma_h(E, C_3^1, C_3^2\} \\ &= \{EE, EC_3^1, EC_3^2, \sigma_h E, \sigma_h C_3^1, \sigma_h C_3^2\} \\ &= \{E, C_3^1, C_3^2, \sigma_h, S_3^1, S_3^5\} \\ &= C_{3h} \end{aligned}\right\} \qquad \ldots(6.27)$$

We as per hypothesis will get the same result and in the same order if the cross product $(E, C_3^1, C_3^2) \otimes (E, \sigma_h)$ is obtained.

**Example 2:** Now let us take $C_{3v} = (E, C_3^1, C_3^2, \sigma_v^{(1)}, \sigma_v^{(2)}, \sigma_v^{(3)})$ where $[\sigma_v^{(1)}.\ C_3^2 = \sigma_v^{(2)}, \sigma_v^{(1)}.\ C_3^2 = \sigma_v^{(3)}]$. It has four subgroups, *viz.* $C_3 = (E, C_3^1, C_3^2)$, $C_s^{(1)} = [E, \sigma_v^{(1)}]$ $C_s^{(2)} = [E, \sigma_v^{(2)}]$ and $C_s^{(3)} = [E, \sigma_v^{(3)}]$ besides the subgroup $[E]$. Let us take the cross products of $(C_3 \otimes C_s^{(1)})$, $(C_3 \otimes C_s^{(2)})$ and $(C_3 \otimes C_s^{(3)})$. We first obtain the cross product of $[C_3 \otimes C_s^{(1)}]$. In this case, the elements of $C_s$ (. $E, \sigma_v$) do not commute with elements of $C_3$ $(C_3^{(1)}\ C_3^{(2)})$ but all the elements of $C_{3v}$ are obtained in the cross product. That is,

$$\left.\begin{aligned} |E, \sigma_v^{(1)}| \otimes |E C_3^1 C_3^2| &= |E||E, C_3^1, C_3^2)|\sigma_v^{(1)}||E, C_3^1, C_3^2| \\ &= |E \cdot E(E \cdot C_3^1), (E \cdot C_3^2)(\sigma_v^{(1)} \cdot E), \\ &\qquad (\sigma_v^{(1)} \cdot C_3^1), (\sigma_v^{(1)} C_3^2| \\ &= [E, C_3^1, C_3^2, \sigma_v^{(1)}, \sigma_v^{(2)}, \sigma_v^{(3)}] \\ &= C_{3v} \end{aligned}\right\} \quad \text{...(6.28)}$$

Similarly, $|C_s^{(2)}| \otimes |C_3|$ and $|C_s^{(3)}| \otimes |C_3|$ give the same symmetry operations, but in different order. Further $\sigma_v^{(2)}, \sigma_v^{(3)}$ also do not commute with $C_3^1, C_3^2$ operations of $C_3$ point group. Thus, the cross products are different from those of $(E \,.\, \sigma_h) \times (E\, C_3^1, C_3^2)$, whenever all the elements of one or other subgroups do not commute among each other, but in the cross product(s), all elements of group $G$ are obtained, then such types of cross products are known as *Semi Direct Products* and denoted by $H_1 \wedge H_2$ (*e.g.*, $C_3 \wedge C_s^{(1)}$).

A number of point groups are the direct or semi direct products of the subgroups of the main group $G$. Some examples are,

1. $S_{2n} = C_{n/2} \otimes C_i$ (when $n$ is divisible by 4, (*e.g.*, $S_4$, $S_8$, *etc.*)
2. $C_{nh} = C_n \otimes C_i$ ( for any natural numbers $n$),
3. $D_{nh} = D_n \otimes C_i$ ($n$ = even), $D_{nh} = D_n \otimes C_S$ (any $n$),
4. $D_{nd} = D_n \otimes C_i$ ($n$ = odd),
5. $T_h = T \otimes C_s$
6. $O_h = O \otimes C_s$; $I_h = I \otimes C_s$; $K_h = K \otimes C_s$ ... (6.29)

In above discussion we have shown that certain point groups are cross products of their subgroups (*e.g.*, $C_n \otimes \sigma_h = C_{nh}$). It can also be shown that the character table of the main group (*e.g.*, $C_{nh}$) is also the direct product of the character tables of their subgroups (*e.g.*, $C_n$ and $C_s$). Thus,

$$C_S \otimes C_n = C_{nh} \quad \text{...(6.30)}$$

(Character Table of $C_s$) ⊗ (Character Table of $C_n$) = Character Table of $C_{nh}$. For example,

$$\underbrace{\begin{vmatrix} 1 & 1 \\ 1 & -1 \end{vmatrix}}_{C_s} \otimes \underbrace{\begin{vmatrix} 1 & 1 & 1 \\ 1 & \in & \in^* \\ 1 & \in^* & \in \end{vmatrix}}_{C_3} = \begin{Vmatrix} 1\begin{vmatrix} 1 & 1 & 1 \\ 1 & \in & \in^* \\ 1 & \in^* & \in \end{vmatrix} & 1\begin{vmatrix} 1 & 1 & 1 \\ 1 & \in & \in^* \\ 1 & \in^* & \in \end{vmatrix} \\ 1\begin{vmatrix} 1 & 1 & 1 \\ 1 & \in & \in^* \\ 1 & \in^* & \in \end{vmatrix} & -1\begin{vmatrix} 1 & 1 & 1 \\ 1 & \in & \in^* \\ 1 & \in^* & \in \end{vmatrix} \end{Vmatrix} \quad \text{...(6.31)}$$

$$
= \begin{vmatrix} 1 & 1 & 1 & 1 & 1 & 1 \\ 1 & \in & \in^* & 1 & \in & \in^* \\ 1 & \in^* & \in & 1 & \in^* & \in \\ 1 & 1 & 1 & -1 & -1 & -1 \\ 1 & \in & \in^* & -1 & \in & -\in^* \\ 1 & \in^* & \in & -1 & -\in^* & -\in \end{vmatrix} \begin{matrix} A' \\ \left.\begin{matrix} \\ \\ \end{matrix}\right\} E' \\ A'' \\ \left.\begin{matrix} \\ \\ \end{matrix}\right] E'' \end{matrix} \quad \ldots(6.32)
$$

$$E \quad C_3^1 \quad C_3^2 \quad \sigma_h \quad S_3^1 \quad S_3^5$$

$= C_{3h}$ Character Table

## EXERCISE

**1.** There are two spaces. One space is two-dimensional with base vector $(x_1, x_2)$ and the other space is three-dimensional with base vector $(y_1, y_2, y_3)$. We now define a six-dimensional space obtained by the cross product of these two spaces. Their base vectors are given as
$(\phi_1 = x_1 y_1, \phi_2 = x_1 y_2; \phi_3 = x_1 y_3, \phi_4 = x_2 y_1; \phi_5 = x_2 y_2, \phi_6 = x_2 y_3)$ Now, if there is a transformation $M_1$ by which $(x_1, x_2)$ base vectors are transformed as

$$\begin{vmatrix} x_1' \\ x_2' \end{vmatrix} = |\hat{M}_1| \begin{vmatrix} x_1 \\ x_2 \end{vmatrix} \quad \ldots(6.33)$$

and there is another transformation $\hat{M}_2$ by which $(y_1, y_2, y_3)$ base vectors are transformed as,

$$\begin{vmatrix} y_1' \\ y_2' \\ y_3' \end{vmatrix} = |\hat{M}_2| \begin{vmatrix} y_1 \\ y_2 \\ y_3 \end{vmatrix} \quad \ldots(6.34)$$

Prove that the base vector $(\phi_1, \phi_2, ..., \phi_6)$ are transformed as

$$|\phi_i'| = |\hat{T}|\,|\phi_i| \quad \ldots(6.35)$$

where $$|\hat{T}| = |\hat{M}_1| \otimes |\hat{M}_2| = |\hat{M}_2 \otimes \hat{M}_1| \quad \ldots(6.36)$$

$|\hat{M}_1 \otimes \hat{M}_2|$ is the cross product of $|\hat{M}_1|$ and $|\hat{M}_2|$

## 6.3 DIRECT PRODUCT REPRESENTATIONS

In the above exercise, we have shown how the cross products of two representations behave. We elaborate the concept a little further. Suppose, we have two spaces of dimensions $n$ and $m$, respectively with basis vectors $(f_1, f_2, ..., f_n)$ and $(g_1, g_2, ..., g_m)$, respectively. We now define a space of $(n \otimes m)$ dimensions with the base vectors as

$[f_1(g_1, g_2, ..., g_m), f_2(g_1, g_2, ..., g_m), ... f_n(g_1, g_2, ..., g_m)]$.

The basis functions are $(f_1, g_1), (f_1, g_2), (f_1\ g_3), ..., (f_1, g_m), (f_2, g_1)(f_2, g_2) ... (f_2, g_m)\ (f_n, g_1)\ (f_n, g_2), ..., (f_n, g_m)$ a total of $(m \times n)$ basis vectors.

Using the $(n \otimes m)$ basis vectors, we form a representation of a group G $(E, R_1, R_2, ..., R_{h-1})$ of order $h$. Applications of these symmetry operations separately will give $h$ matrices, each of $(mn \times mn)$ dimensions. We have the following facts.

| *Group G* | *E* | $R_1$ | $R_2$ | $R_3$ | ... $R_{h-1}$ | *Dimension of matrices* | |
|---|---|---|---|---|---|---|---|
| $\Gamma^{\nu}$[base vectors $(f_1, f_2, ..., f_n)$] | $M_E^{\nu}$ | $M_{R_1}^{\nu}$ | $M_{R_2}^{\nu}$ | $M_{R_3}^{\nu}$ | ... $M_{R_h}^{\nu}$ | $\|n \times n\|$... | ...(6.37) |
| $\Gamma^{\mu}$[base vectors $(g_1, g_2, ..., g_m)$] | $M_E^{\mu}$ | $M_{R_1}^{\mu}$ | $M_{R_2}^{\mu}$ | $M_{R_3}^{\mu}$ | ... $M_{R_h}^{\mu}$ | $\|m \times m\|$... | ...(6.38) |
| $\Gamma^{\mu\otimes\nu}$[base vectors $(f_1g_1, f_2g_2, ..., f_ng_n)$] | $M_E$ | $M_{R_1}$ | $M_{R_2}$ | $M_{R_3}$ | ... $M_{R_{h-1}}$ | $\|mn \times mn\|$... | ...(6.39) |

The matrices in eq. (6.39) form a representation for the group $G$ with bases $(f_i\ g_j)$ and it will be a direct product representation of $(\Gamma^{\nu} \otimes \Gamma^{\mu})$. The effects of $R_i$ on these product basis vectors can be calculated from matrices in eq. (6.37)–(6.38).

Since a relationship between $[M(R_i)]_{st}^{(\mu\times\nu)}$ and $[M^{\nu}(R_i)]_{\mu i}$ and $[M^{\mu}(R_i)]_{\nu i}$ matrices is a little complicated to derive, we would like to derive the relationship in terms of their characters, which we generally use and can easily obtain from the character table of the group. Also we have already shown, a relationship in the previous section between the character of $R_i$ in $[\Gamma^{\mu} \otimes \Gamma^{\nu}]$ (*i.e.*, the direct product of $[\Gamma^{\mu} \otimes \Gamma^{\nu}]$ as given below.

$$\begin{aligned}\chi^{(\mu\otimes\nu)}(R_i) &= \chi^{\nu}(R_i)\,\chi^{\mu}(R_i)\\ &= \chi^{\mu}(R_i)\chi^{\nu}(R_i)\end{aligned} \qquad ...(6.40)$$

Eq. (6.40) shows that the products of $\chi^{\mu}$ and $\chi^{\nu}$ is commutative and the order of $\nu$ and $\mu$ is immaterial.

Thus, the character of

$$\Gamma^{\mu} \otimes \Gamma^{\nu} = \Gamma^{\nu} \otimes \Gamma^{\mu} \qquad \dots (6.41)$$

The discussion so far, indicates the representations $\Gamma^{\mu}$ and $\Gamma^{\nu}$ as I.Rs but both $\Gamma^{\text{red}}$ and $\Gamma^{\text{I.R}}$ can be used to form the direct product. In quantum mechanical problems, it is usually I.Rs which are used. The characters of $\Gamma^{\nu}$ and $\Gamma^{\mu}$ are present in the character table of the group and thus to calculate them one takes data from the character table. Further, even though $\Gamma^{\nu}$ and $\Gamma^{\mu}$ may be taken as I.Rs; but their direct product representation is generally reducible. In this reducible representation, one can find the numbers and types of I.Rs, by the use of equation (6.39).

That is,

$$a_{\text{I.R}} = \frac{1}{g}\sum_{i=1}^{k} g_i \chi_i^{*\text{I.R}} \cdot \chi_i^{\text{Red}}$$

(Here $g$ = order of group; $g_i$ = number of elements in class $i$; $\chi_i^{*\text{IR}}$ is the conjugate character in class $i$ of the I.R; $\chi_i^{\text{Red}}$ is the character in reducible representation of the class $i$. In this function,

$$\Gamma^{\nu} \times \Gamma^{\mu} = a_1\Gamma_1 + a_2\Gamma_2 + \dots \qquad \dots (6.42)$$

**Example 3:** Let us take the point group '$O$' and determine the various direct product representations.

**Direct products of $O$ point group:** In the character table of on point group, the direct product of $T_2 \otimes T_1$ are given in the last row which is obtained by multiplying the characters of $T_1$ and $T_2$. In this reducible representation of $T_2 \otimes T_1$, one can get the following I.Rs. using eq. (6.39).

$$a_{A_1} = (1/24)\,[1 \times 9 + 8 \times 0 \times 1 + 3 \times 1 \times 1 - 6 \times 1 \times 1 - 6 \times 1 \times 1$$
$$= (1/24)\,[9 + 0 + 3 - 6 - 6] = 0 \qquad \dots (6.43)$$

$$a_{A_1} = (1/24)\,[1 \times 1 \times 9 + 8 \times 0 \times 1 + 3 \times 1 \times 1 + 6 + 6] = 1 \qquad \dots (6.44)$$

**Table 6.1:** Character Table of point group $O$

| $O$ | $E$ | $8C_3$ | $3C_2$ | $6C_4$ | $6C_2'$ |
|---|---|---|---|---|---|
| $A_1$ | 1 | 1 | 1 | 1 | 1 |
| $A_2$ | 1 | 1 | 1 | –1 | –1 |
| $E$ | 2 | –1 | 2 | 0 | 0 |
| $T_1$ | 3 | 0 | –1 | 1 | –1 |
| $T_2$ | 3 | 0 | –1 | –1 | 1 |
| $T_1 \otimes T_2$ | 9 | 0 | 1 | –1 | –1 |
| $E \otimes T_1$ | 6 | 0 | –2 | 0 | 0 |

The reduction of $T_1 \otimes T_2$ is $T_1 \otimes T_2 = E + T_1 + T_2 + A_2$

$$a_E = (1/24)[2 \times 9 \times 1 + 8 \times 0 \times -1 + 1 \times 2 \times 3 + 0 + 0\,] = 1 \quad \ldots(6.45)$$

$$a_{T_1} = (1/24)[3 \times 9 \times 1 + 0 \times 8 \times -3 \times -6 + 6\,] = 1 \quad \ldots(6.46)$$

$$a_{T_2} = (1/24)[3 \times 9 \times 1 + 0 - 3 + 6 - 6] = 1 \quad \ldots(6.47)$$

$$T_1 \otimes T_2 = \mathrm{A}_2 + E + T_1 + T_2 \quad \ldots(6.48)$$

Similarly, one also determines

$$T_2 \otimes T_1 = E + T_1 + T_2 + A_2$$
$$E \otimes T_2 = T_1 \times T_2$$
$$E \otimes E = A_1 + A_2 + E \quad \ldots(6.49)$$
$$A_1 \otimes T_1 = T_2 \quad \text{and} \quad A_1 \otimes A_2 = A_2$$
$$A_2 \otimes E = E \quad \text{and} \quad A_1 \otimes E = E$$

We can also verify that

$$\begin{aligned} E \otimes (T_1 \otimes T_2) &= E \otimes (A_2 + E + T_1 + T_2) \\ &= E \otimes A_2 + E \otimes E + E \otimes T_1 + E \otimes T_2 \\ &= E + A_1 + A_2 + E + T_1 + T_2 + T_1 + T_2 \\ &= A_1 + A_2 + 2E + 2T_1 + 2T_2 \end{aligned} \quad \ldots(6.50)$$

## 6.4 APPLICATION OF DIRECT PRODUCT IN QUANTUM MECHANICS (VANISHING INTEGRALS)

In many applications of quantum mechanics, we come across integrals of the type

$$\int \psi^*_\sigma(x) F_\lambda(x) \psi_\rho \, \partial\tau \quad \ldots(6.51)$$

where $\psi^*_\sigma$ and $\psi_\rho$ are wave functions belonging to energy values $E_\sigma$ and $E_\rho$ respectively of a molecule and $\mathrm{F}_\lambda(x)$ is any function like $H(x_1, x_2, ..., x_n)$ of the configuration space of variables $(x_i)$. The integral runs over all the variables of the space. Though in some case we may like to have the actual value of integral, but in most cases; the interest lies in the fact "what factors the integral must satisfy if it is not to vanish". For example, the intensity of a band in the spectrum of a molecule appeared owing to transition between two states. It depends upon the integral given by eq. (6.49) in which $F_\lambda$ is electric dipole moment of the molecule. Thus, if this integral is zero, the transition is forbidden. The value of the integral (6.51) depends upon the symmetry of $\psi_\rho$, $\psi^*_\sigma$ and $F_\lambda$. It can be shown that the product of the form (6.51) may be expressed as a linear combination of the base functions of the direct product representation $\Gamma_\sigma \times \Gamma_\lambda \times \Gamma_\mu$ which is reduced into the I.Rs of the molecular points group. It can be shown that if $\Gamma_\sigma \times \Gamma_\mu \times \Gamma_\lambda$ does not contain the identity ($A_1$-type) irreduceble representation then the integral is zero. If it has $A_1$ representation then the integral has some value and the transition is allowed. The conclusion of this argument is that "if $\psi_\rho$, $\psi^*_\sigma$ and $F_\lambda$ are functions of the space variable whose basis generate the I.Rs. $\Gamma_\sigma \times \Gamma_\lambda \times \Gamma_\mu$ and if $\Gamma_\lambda \times \Gamma_\mu$ does not contain the $\Gamma_\sigma$ then the integral $\int \psi^*_\sigma F_\lambda \psi_\rho \partial\tau = 0$. Some other applications of the direct product representation will be given later in the book.

# 7

# Symmetry in Quantum Chemistry and Hybridisation

Schrodinger wave equation for any dynamical system is given as

$$H\Psi = E\Psi \qquad \text{...(7.1)}$$

where the operator $\hat{H}$, termed as Hamiltonian operator, implies a certain operation to be carried out on a function $\Psi$, (an eigenfunction or a vector in a function space), $E$ is the energy of the system (eigenvalue)[1].

The Hamiltonian operator is written as the sum of the operators for the potential energy ($V$) and the kinetic energy (K.E.) of the system (an expression for the classical energy). These potential and kinetic energy expressions are transformed into an operator form by replacing the momentum term in which the K.E. is expressed by a differential operator according to quantum mechanics postulates. We are not concerned about the formation of the Hamiltonian; the only property which concerns us the most is its symmetry properties and its relationship with molecular symmetry. The particles of the system are electrons moving in the field of the nuclear charge binding them to form atoms. Besides, there are other dynamic processes like rotation and translation of the molecule as a whole plus other dynamic properties of the nuclei which will not be discussed in the chapter.

While studying the symmetry properties of the potential and the kinetic energy, we determine the effects of symmetry operations on these energy expressions. The symmetry operations carry the system from its original

1. We do not expect the students to have an intensive training in quantum mechanics, but a little familiarity of the field is necessary to understand the concept given in the chapter.

configuration to another equivalent configuration which is indistinguishable from the original one. Therefore, it is expected that the potential and the kinetic energy after applying the symmetry operations on the system should remain invariant. It implies that the Hamiltonian operator and the symmetry-operator should commute.

## 7.1 THE GROUP OF HAMILTONIAN

**(*a*) Potential and Kinetic Energy:** As told in the previous paragraph, the Hamiltonian is composed of K.E. and potential energy terms. Each term, in general, is some function of the coordinates ($x$, $y$, $z$). The K.E. term is given as $\left(-\frac{h^2}{2m_i}\right)\left[\frac{\partial^2}{\partial x_i^2}+\frac{\partial^2}{\partial y_i^2}+\frac{\partial^2}{\partial z_i^2}\right]$ and the potential energy term as $V(xyz)$. These possess the symmetry of the physical system under consideration. Let us find what basis vectors are used in two cases *viz.* vibrational and electronic studies.

**(*b*) Symmetry Properties of Potential Energy:** According to the Born-Oppenheimer approximation, we can deal with the electronic and vibrational problems separately. In both cases, we start with two different configuration spaces (Chapter 3). While discussing electronic structure, we make an assumption that the electrons in the molecule move with respect to a fixed framework formed by the nuclei of the atoms located at their respective equilibrium positions. The potential energy of the electrons is the sum of the electrostatic attraction and repulsion energies of the all the electrons by their nuclei and among the electrons. Thus, at a particular moment, one sets up a configuration space with coordinates $\{(x_1^1x_2^1x_3^1), (x_1^2x_2^2x_3^2), \ldots (x_1^nx_2^nx_3^n)\}$. Here $(x_1^px_2^px_3^p)$ are ($x$, $y$, $z$) coordinates of $p^{\text{th}}$ electron with respect to the chosen set of coordinate axes ($XYZ$) (Chapter 3). In this coordinate system, the origin is chosen on the nucleus (or on the point) that remains unmoved during the application of symmetry operations. The same axes are used to find the coordinates of all the electrons. Thus, the set of all the values of $(x_1^1x_2^1x_3^1)\ (x_1^2x_2^2x_3^2)\ \ldots (x_1^nx_2^nx_3^n)$ describes a specific configuration of electrons in the molecule. It implies that all the electrons are moving in a three-dimensional physical space. Instead of all the electrons moving in their three-dimensional space, we now assume that these numbers or coordinates represent $3n$ components of one single vector in a $3n$-dimensional space. In this fashion, the points in that $3n$-dimensional configuration space represent all possible configurations of a molecule. One must realise here that in this space, each particle's coordinates are referred to its own separate individual coordinate system in physical space with bases $e_1^k, e_2^k, e_3^k$ as basis. Thus, any particular state of all the electrons in a molecule will be described by a function of the coordinates of all the electrons. Therefore, we are concerned with a $3n$-dimensional vector space. The electronic structure of the molecule can be discussed in that $3n$-dimensional configuration space. The electronic wave functions are then considered as vectors in that space with variables $(x_1^1 x_2^1 x_3^1,\ x_1^2 x_2^2 x_3^2, \ldots x_1^n x_2^n x_3^n)$ of $n$-electrons keeping the coordinates of the nuclei as fixed constant parameters.

Similarly, when we consider the vibrational motion of the atoms in a molecule, we consider the displacements $(\Delta x_1^p \Delta x_2^p \Delta x_3^p)$ of the nuclei from their respective equilibrium positions. These displacements form the components of a vector in the configuration space of $3N$ dimension ($N$ = number of nuclei). We then consider the function space formed by the vibrational wave functions of variables $\{(\Delta x_1^p \Delta x_2^p \Delta x_3^p)\ldots\}$ of $N$-atoms and the coordinates of the electrons as fixed parameters. These displacements of atoms are obtained with respect to a set of three axes with their respective origins at the equilibrium positions of each nucleus. Thus, we set up '$N$' number of coordinate axes, forming a $N$-coordinate systems. In this type of coordinate system, all the $x$-axes, $y$-axes and $z$-axes, at the $N$-atoms are respectively parallel to each other. These are also parallel to the $x$-, $y$-, $z$-axis that we take at the centre of mass point.

In this fashion, we set up different configuration spaces while dealing with the problems of the electronic and vibrational structures. Each set of coordinates [*viz.* $(x_1^p, x_2^p, x_3^p)$ and $(\Delta x_1^p, \Delta x_2^p, \Delta x_3^p)$] describes a particular configuration of the electrons or the nuclei. In both approximations ($x_i^p$ or $\Delta x_i^p$), the assumption is that the potential energy is a functions of the coordinates which associates a certain value [$V(x)$] with each configuration $X$. It implies that a point in both the configuration spaces define a value of potential energy at one moment of time. That is the configuration space of electrons, describes the potential energy of electrons and the configuration space of vibration, the potential energy of the vibrational motion.

The second important point concerning the potential energy is intimate relationship of the framework formed by the nuclei of the atoms in equilibrium positions with the point group of the molecule. We show this relationship in the following paragraphs.

## 7.2 THE RELATIONSHIP OF $V_{(x)}$ WITH THE POINT GROUP OF THE MOLECULE

Suppose, we have symmetry operations $R_i$ belonging to the point group of the molecule. The effect of application of these symmetry operations on the vectors of the configuration space is that the configuration ($X$) of the molecule is changed to a new configuration $X'$. The new configuration $X'$ is related to the old configuration $X$. The relationship is given as follows.

(*a*) **Electronic Structure:** In this case, nuclear framework of the molecule remains fixed. We simultaneously apply the symmetry operation on the coordinates of all the electrons so that each electron is rotated or reflected into new position and all the electrons are moved in the same way.

(*b*) **Vibrational Problem:** We assume that in the equilibrium state, the nuclei are at some fixed positions and do not consider the effects of symmetry operation on the electron coordinates. During vibration, at a certain moment, each nuclei is displaced to a new position by a distance ($\Delta r_i$) from its equilibrium position. It implies a definite displacement at that moment which defines a configuration $X$ of the molecule. Let this configuration be represented as shown in Fig. (7.1)

In the configuration $X$, each nucleus has a definite displacement (*viz.* $\Delta x_1, \Delta x_2, \Delta x_3, \Delta x_4$). This is shown geometrically by the arrows from the equilibrium positions of each nuclei to the position, it occupies at that moment. We choose the coordinate system at each nucleus.We take the $x$-axes, $y$-axes and $z$-axes of the coordinate system at each nucleus to be parallel to each other. The directions of these axes remain the same even after applying symmetry operations. (Why do we have this assumption? Explain). (Fig. 7.1)

Now, we apply the symmetry operation ($R_i$) to these arrows representing displacements, keeping the nuclear framework and the axes fixed. The effects of the symmetry operations on these displacement arrows attached to each nucleus are to get the new arrows. (*See* Fig. 7.1 $X'$ configuration) which produces a new configuration $X'$. We assume that, the effect of symmetry operation on these arrows will be to tranfer the arrow ($\Delta x_i$) from the atom at position $i$ to the postion $j$. One *must* note here, that we are applying the symmetry operations only to the arrows (*i.e.*, displacements of the nuclei, [($\Delta x_i$'s)]. Thus, the displacement ($\Delta x_i$) of the nucleus at position $i$, is transferred to position $j$ which is the position of another atom that was present at positon $j$ before operation. In this manner, all the displacements of ($n$) number of nuclei are transferred to the positions of other atoms, present before the operation. Thus, the changed position of arrows ($\Delta x_i$'s) give the new configuration $X'$.

If Fig. 7.1 $\Delta x_1$ which represent displacement of *nucleus* at *position A* is transferred to position $B$, at which another nucleus was present prior to symmetry operation ($C_3^1$). In this manner, only displacements are shifted keeping the *position* $A$, $B$, $C$ fixed. ($\Delta x_1$ is shifted from position at $A$ to $B$, that of $B(\Delta x_2)$ from position $B$ to $C$ and that of $C$ to position $A$. The position of $O$ remains fixed. Thus, gives a new configuration $X'$ of the system.

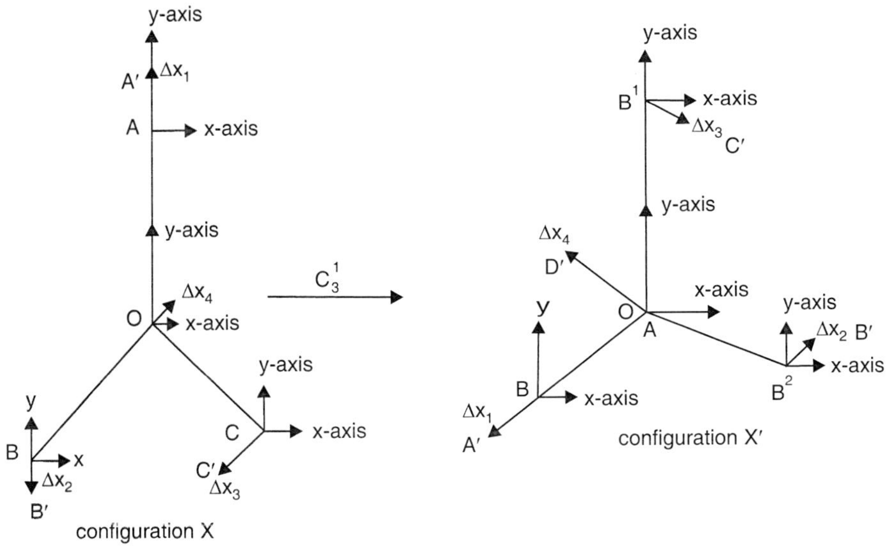


**Fig. 7.1** Effect of rotation on displacements.

In this process of symmetry operation we assume that

1. The position of coordinate system at equlibrium remain fixed (*i.e.*, the positions of $A$, $B$, $C$ and $O$ remain fixed) during the symmetry operation.
2. On the application of $\hat{R}$ symmetry operation, only the displacements $\Delta x_1$, $\Delta x_2$, $\Delta x_3$ shift to $B$, $C$, $A$ respectively. For example, the displacement $\Delta x_1$, present at $A$ now shifts to the position $B$ in the new configuration $X'$, the displacement $\Delta x_1$ shifts to the $B$. Its direction will be the one obtained by rotating the displacement by 120° from its old positons at $B$, similarly $\Delta x_2$ and $\Delta x_3$ at $B$ and $C$ are shifted to the positions of $A$ and $B$ of the configuration $X$.

Since position of $A$ will not be shifted, the displacements at $B(\Delta x_3)$ will be shifted to position $A$, but in rotated position (120° rotation).

We can draw these displacements as given in (Fig. 7.2). It defines the application of symmetry operation $\hat{R}$ to a configuration $X$. It is possible to show that the potential energies of the configurations $X$ and $X'$ will be the same. (We will not prove it here). It follows that the potential energy function has the property, that when any symmetry operation is applied on a molecule, the configuration changes $X$ to $X'$ that is,

$$V(X)' = V(X) \qquad \ldots(7.2)$$

The relationship [eq. (7.2)] is obtained on geometrical ground. We now go from configuration space of the molecule to a function space $F(x)$ which is a function of coordinates specifying the configuration. We also specify the symmetry operator $\hat{R}$, by an operator $[P_R]$ which we define as follows.

$$[P_R]\, F(X') = F(X) \qquad \ldots(7.3)$$

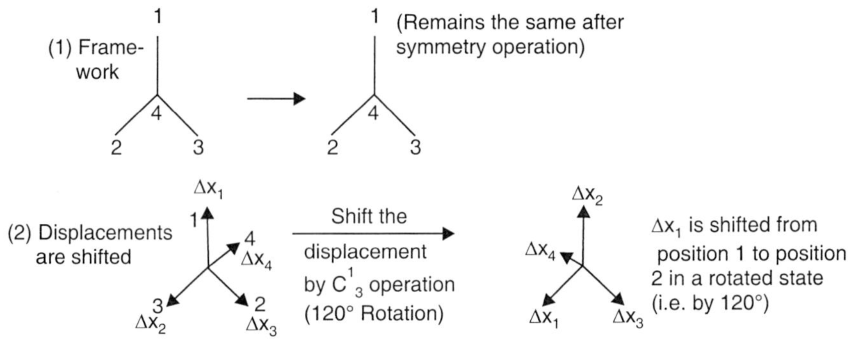


**Fig. 7.2** $C_3$ Operation on the displacement (in two steps).

After applying the definition [eq. (7.3)] to our potential energy function. We get,

$$[P_R(V)](x_1' x_2' x_3' ...) = V(x_1 \, x_2 \, x_3 \, ... \, x_n) \qquad ...(7.4)$$

This definition is applicable to any case and for any configuration ($X$) or ($X'$). Hence, in any configuration, we expect that the transformed functions $[P_R(V)]$ has the same value as $V$. Or,

$$[P_R(V) = V \qquad ...(7.5)$$

This equation is valid for all the symmetry operations ($R_i$) of the molecular point group. Thus, we arrive at the conclusion. **"The potential energy V possesses all the symmetries of the molecular framework. That is, it is invariant under all the operations of the point group".**

The above conclusion holds for both vibrational as well as electronic functions. Further we can also apply it (eq. 7.5) to any function of the function space. Thus, if $F$ is any function of coordinates and take the product function ($VF$), then the application of operator $P_{(R)}$ on this function, will give,

$$P_R(VF) = (P_R V)(P_R F) \qquad ... (7.6)$$

Eq. (7.6) implies that by applying eq. (7.5), we can also write

$$(P_R V)(P_R F) = V(P_R F) \qquad ...(7.7)$$

One can also prove similar relationship for the K.E. expression of the molecule. That is,

$$P_R(TF) = T(P_R F) \qquad ...(7.8)$$

where $T$ is K.E. operator of the molecule. Or,

$$P_R\left[\frac{1}{m^s}\nabla_s^2 F\right] = \frac{1}{m^q}\nabla_q^2 (P_R F) \qquad ...(7.9)$$

(Here $s$ and $q$ are symmetrically equivalent atoms or nuclei in the molecule and hence, they have the same mass). Hence, eq. (7.8) can be written as

$$\sum_{s} P_R \left[ \frac{1}{m^s} \nabla_s^2 F \right] = \sum_{q} \frac{1}{m^q} \left[ \nabla_q^2 (P_R F) \right] \qquad ...(7.10)$$

## 7.3 REPRESENTATION AND SCHRODINGER EQUATION

We have stated in the beginning of the chapter that the Hamiltonian operator of the Schrodinger equation for electrons or the nuclei of the molecule or the atom represents the total energy [(K.E.+ Potential energy]. That is,

$$H = \text{K.E.} + \text{Potential energy}$$
$$= T + V \qquad ...(7.11)$$

T is the K.E. operator [eq. (7.10)] and V = Potential Energy [eq. (7.2) or (7.3)] We have also indicated in eqs. (7.5–7.10) that

$$\hat{P}_R(HF) = H(\hat{P}_R F) = E(P_R F) \qquad ...(7.12)$$

[where $F$ is any eigen function of coordinates of electrons or of the nuclei, $R$ is one of the symmetry operators of the molecular point group.] The equation (7.12) also implies that $H$ is invarient under all the operations of the point group. This invariance property and a few other properties of $H$ allow us to apply symmetry properties or group theoretical principles to the various problems of the molecules.

Schrodinger equation is a partial differential equation ($H\Psi = E\Psi$) where $\psi$ is the state of the molecule having the energy $E$, $\Psi$ is a function of the coordinates of all the particles. (*viz.* electrons or/and nuclei of the molecule). From the Schrodinger equation, one can obtain the physical possible BOUND states of the molecule by solving the equation under certain boundary conditions. These conditions are (1) $\Psi$ goes to zero at infinity (2) There should be no singularity in the wave function '$\Psi$' and (3) total probability of finding the particle (electron or nuclei) in the whole space is one. Because of these conditions, the number of solutions of the equation is limited. It can be solved only for a certain values of $E$, (say $E$, $E_2$, $E_3$ ... $E_n$). These energy values represent the various possible energy states in which the molecule can stay.

Another important point about the solution of the equation is that there is a possibility of finding many $\Psi$-functions, say ($\Psi_1$, $\Psi_2$, ..., $\Psi_k$) for one value of $E$. Thus, if there are $k$ such solutions then, we have

$$H\Psi_1 = E_k\Psi_1; H\Psi_2 = E_k\Psi_2, H\Psi_3 = E_k\Psi_3 \ldots H\Psi_k = E_k\Psi_k \qquad ...(7.13)$$

Since $H$ is a linear operator, one can also have,

$$H(a_1\Psi_1 + a_2\Psi_2 + a_3\Psi_3 + \ldots a_k\Psi_k) = E_k(a_1\Psi_1 + a_2\Psi_2 + \ldots a_k\Psi_k) \qquad ...(7.14)$$

Here ($a_1$, $a_2$ ... $a_k$) can take any value (from 0 to 1). This linearly combined wave function (*viz.* $a_1\Psi_1 + a_2\Psi_2 \dots a_k\ \Psi_k$) is also a solution of the Schrodinger equation and also satisfies the same three boundary conditions. It means that the set $\Psi_1$, $\Psi_2$ ... $\Psi_k$) forms a function space of $k$-dimensions associated with $E_k$ energy value. One can, then choose ($\Psi_1$, $\Psi_2$, $\Psi_3$, ..., $\Psi_k$ orthonormal functions as a basis for the $k$-dimensional space and each function $\Psi_i$ satisfies the boundary conditions. The dimension $k$ of the function space associated with energy $E_k$ is termed as "$k$-fold degeneracy" of the orbitals.

"What is the relationship between the wave functions of a molecule and its symmetry? The link between these two depends upon the invariance property of the Hamiltonian". "The operator $\hat{P}_R$ which satisfies eq. (7.12) are those symmetry operators which correspond to the symmetry operations of the point group of the molecule. If we could show that all these operated functions [*viz.* $(P_R\Psi_i)$] with ($i$ = 1, 2, ... $k$), belong to the functions space of energy $E_k$, then one can express $(\hat{P}_R\Psi_i^{\nu})$ as a linear combination of $(\Psi_1^{\nu}, \Psi_2^{\nu}, \Psi_3^{\nu}, \dots, \Psi_k^{\nu})$, That is,

$$\hat{P}_R(\Psi_g^{\nu}) = \sum_{i=1}^{\infty} [D^{\nu}(R)]_{ij}\,\Psi_i^{\nu} \qquad \dots(7.15)$$

If eq. 7.15 is satisfied, then the $\Psi_j^{\nu}$ forms a basis for the representation of the molecular point group and $[D^{\nu}(R)]_{ij}$ is a matrix formed from the coefficients of eq. (7.15). The preceding statement can be easily proved by the following arguments. We have already proved that

$$\hat{H}(P_R\Psi_j^{\nu}) = P_R(H\Psi_j^{\nu}) = P_R(E_{\nu}\Psi_j^{\nu}) = E_{\nu}(P_R\Psi_j^{\nu}) \qquad \dots(7.16)$$

Thus $[P_R\Psi_j^{\nu}]$ is a solution of the Schrodinger equation ($H\Psi = E\Psi$), with $E = E_v$ and satisfies the boundary conditions. It further proves that the function $[P_R\Psi_j^{\nu}]$ must belong to the function space associated with energy level $E_v$. Accordingly $[P_R\Psi_j^{\nu}]$ is a linear combination of $\Psi_i^{\nu}{}'s$ belonging to the energy $E_v$.We have thus established a fundamental relationship between the solutions of the Schrodinger equation of any molecule and the molecular symmetries. It says that "The wave functions $(\Psi_i^{\nu})$ which belong to the energy level $E_v$ of a molecule form a function space". This function space generates a "Representation" of the point group of the molecule.

The above statement, two important questions that we have not answered yet. These are,

1. Whether the representation, formed by applying the symmetries of the point group is reducible or irreducible.

2. In all the above derivations, we have shown that the Hamiltonian operator $\hat{H}$ is invariant under the symmetries of the molecular framework; but we have not shown:
   (*a*) whether the Hamiltonian is invariant under the product of two or more than two symmetry operations [*i.e.*, $(P_{R_1}) \cdot (P_{R_2})$],
   (*b*) does the set of operators under which *H* is invariant have an inverse of the operator [$P_{R_1}$ *i.e.*, $(P_{R_1})^{-1}$] which is also a member of the set of symmetry operations of the group,
   (*c*) does the set have an identity operation,
   (*d*) if the members of the set obey the associative law.

These four criteria are essential for forming a group out of a set of symmetry operations of the molecular framework.

Although the above two questions are not answered, but while applying the group theory, the following assumptions will be made.

1. $\Gamma_v$ (*i.e.*, representation formed by the matrices of eq. 7.15) is irreducible.
2. The result of application of two or more than two symmetry operations applied successively to the Hamiltonian will also keep it invariant.
3. The result of application of two or more than two symmetry operations in succession will be the same as that obtained by the application of the third symmetry operation that also belongs to the set of symmetry operations forming the point group and which also keeps the Hamiltonian invariant.

Thus, the set of all such symmetry operations will form a group. This is designated as "The group of symmetry operations of the Hamiltonian" and will be the same as the "Point Group of the molecular framework".

These assumptions are applicable to the "Hamiltonian *H* of the Schrodinger equation".

We have also shown that the set of eigenfunctions associated with the same eigenvalue (energy $E_v$) form a function space, whose dimension is the same as the degeneracy of the energy state. This function space also remains invariant under all the symmetry operations of Hamiltonian group and all the eigenfunctions, $\Psi_k$ form the basis for the representation of this group. It implies that to each eigenvalue $E_k$ of the Hamiltonian *H*, there is associated a vector space of dimension *k* where *k* stands for the number of linearly independent wave fucntions associated with the eigenvalue $E_k$, *i.e.*,

$$H\Psi_k^m = E_m\Psi_k^m \quad \text{...(7.17)}$$

($\Psi_k^m = k$ number of wave functions associated with the eigenvalue $E_m$). The eq. (7.17) implies that there will be $m$ number of such orthogonal vector spaces where ($m = 1, 2, 3, \ldots \infty$). The number ($k$) varies for each eigenvalue $E_m$. We thus associate one function space, (say $k$ of eigenvalue $E_k$) with one I.R. of group.

We can therefore label and describe an energy level of the system and its eigenfunctions associated with it by the I.Rs of the group of Hamiltonian. Thus if $\Psi_k^m{}'s$ are the $k$ number of eigenfunctions (with eigenvalue $E_m$), then each set of eigenfunctions $\Psi_k$ with eigenvalue $E_m$, can be labeled by one of the I.Rs of the Hamiltonian group $G$.

As a corollary of the above theorem, we can also state that if a Hamiltonian is invariant under the symmetry transformations of a Hamiltonian group $G$, then the eigenfunctions of the Hamiltonian ($\Psi_k^m$) which transform according to one I.R of the Hamiltonian group $G$, has the same eigenvalue (*i.e.*, $E_m$). As an example, let us take the three $p$ wave funtions of hydrogen atom. The point group of hydrogen atom is $R(3)$ and its Hamiltonian $H = \left[\frac{1}{m_H}\nabla^2 + \frac{e^2}{r}\right]$. It is invariant under all the "rotations (*viz.* $C_\infty$) of $R(3)$ group. Among the "number of such eigenfunctions, the three $p$-orbitals eigenfunctions (*viz.* $p_x, p_y, p_z$) belong to the same three-dimensional function space and to one of the I.Rs of the $C_\infty$-point group (*i.e.*, $\pi$-I.R) and the same energy value is associated with all three of them.

One can question here "Do the eigenfunctions belonging to different I.Rs of the Hamiltonian group (*i.e.*, irreducible vector spaces) always belong to different energy levels. As a matter of fact, the answer must be in affermative. But there are cases where two functions belonging to different I.Rs can have the same energy value. This irregularity arises under the two cases. These are,

1. **Accidental degeneracy:** In this situation, the energy of two or more than two wave functions belonging to different I.Rs can have the same energy. It implies that there has to be some symmetry transformation of the Hamiltonian which produces the two different vector spaces (I.R.) to be associated with the same energy value $E_n$ (*i.e.*, degeneracy of two spaces).
2. **Dynamic symmetry:** If we have included all possible symmetries in the Hamiltonian group $G$, then we must have different irreducible vector spaces of eigenfunctions to be associated with different energy levels. The reason is that no symmetry transformation is left that is not included in the group $G$ which make them to have the same energy value. But there are cases where there are hidden symmetries. These symmetries

are not included in the group of Hamiltonian (See Dynamic Symmetry in Chapter 2). In such cases, we have a certain limitations of the above corollary and the modified generalisation in that case is "If in the group G, we include all possible symmetry operations (including hidden) of the Hamiltonian $H$, then the eigenfunction of each energy level transforms irreducibly under the transformation of the group $G$, apart from the **Accidental degeneracy.**"

## 7.4 MATRIX REPRESENTATION OF HAMILTONIAN

In the matrix representation of quantum mechanics, we write the Hamiltonian in the matrix form similar to that of operator $[P_{(R)}]$.

In the eigenvalue equation of Schrodinger,

$$H\Psi^m = E_m\Psi_m \qquad \text{...(7.18)}$$

the Hamiltonian of any system can be written exactly but the wave function $\Psi_n$ is not known. Therefore, after we set up the experssion for $H$, in terms of energy operators, we *arbitrarily* select a complete set of orthonormal functions $\phi_i$ which obey the three conditions for being a good function [*viz.* (1) absence of singularity, (2) goes to zero at infinity (3) $\int_{-\infty}^{+\infty} \phi\,\phi^* \partial T = 1$]. These chosen functions are not the real solutions of the Schrodinger equation (eq. 7.18). We assumed that the arbitrarily chosen functions $\Psi_m$ which are the solution of the equation (7.18) can be expressed approximately as a linear combination of $\phi_i$'s. That is,

$$\Psi_m = \sum_{i=1}^{p} a_{mi}\phi_i \qquad \text{...(7.19)}$$

(Here $m = 1, 2, 3, \ldots p$ and $a_{mi}$ are constants). We then represent eq. 7.18 as,

$$\hat{H}\sum_{i=1}^{p}(a_{mi})\phi_i \cong E_m \sum_{i=1}^{p} a_{m_i}\phi_i \qquad \text{...(7.20)}$$

Multiply every term of eq. (7.20) by $\phi_j$ and integrate over the whole space. We get,

$$\int \phi_j^* \hat{H} \sum_{i=1}^{p}(a_{mi}\phi_i)\partial\tau \cong E_m \int \phi_j^* \sum_{m=1}^{p} a_{mi}\phi_i \partial\tau \qquad \text{...(7.21)}$$

or,

$$\sum_{i=1}^{p} a_{mi} \int \phi_j^* \hat{H} \phi_i \partial\tau \cong E_m \sum_{i=1}^{p} \int \phi_j^* \phi_i \partial\tau \qquad \text{...(7.22)}$$

or,
$$\sum_{i=1}^{p} a_{mi}\hat{H}_{ji} \cong E_m \sum_{i=1}^{P} a_{mi} S_{ji} \qquad \text{...(7.23)}$$

(Where $H_{ji} = \int \phi_j^* H \phi_i \partial\tau$ and $S_{ji} = \int \phi_j^* \phi_i \partial\tau$ )

Eq. (7.23) can also be written as,

$$\sum_{i=1}^{p} (H_{ji} - E_m S_{ji}) a_{mi} \cong 0 \qquad \text{...(7.24)}$$

This is the resulting equation when we multiplied by $\phi_j^*$ and integrated the eq. (7.20). We will get a set of $p$-equations similar to (7.24) if we multiply (eq. 7.20) successively by other functions, (*viz.* $\phi_1^*, \phi_2^* ... \phi_j^* ... \phi_p^*$ ) and integrate them seperately one by one. In this fashion, we get the following $p$ number of equations

$$\left.\begin{aligned} \sum_{i=1}^{p} (H_{1i} - E_m S_{1i}) a_{mi} &= 0 \\ \sum_{i=1}^{p} (H_{2i} - E_m S_{2i}) a_{mi} &= 0 \\ \sum_{i=1}^{p} (H_{3i} - E_m S_{2i}) a_{mi} &= 0 \end{aligned}\right\} \qquad \text{...(7.25)}$$

and so on. The final equation will be

$$\sum_{i=1}^{p} (H_{pi} - E_m S_{pi}) a_{mi} \cong 0 \qquad \text{...(7.26)}$$

For non-trivial solutions $a_{mi} \neq 0$, the determinant of the coefficients of $a_{mi}$ has to be zero and therefore, we get a determinant of order ($p \times p$) which is given below. This determinant has to be zero. That is,

$$\begin{vmatrix} H_{11} & H_{12} & H_{13} & \cdots & H_{1p} \\ H_{21} & H_{22} & H_{23} & \cdots & H_{2p} \\ \vdots & & & & \\ H_{p1} & H_{p2} & H_{p3} & \cdots & H_{pp} \end{vmatrix} \begin{vmatrix} a_{m1} \\ a_{m2} \\ \vdots \\ a_{mp} \end{vmatrix}$$

$$= E_m \begin{vmatrix} S_{11} & S_{12} & S_{13} & \cdots & S_{1p} \\ S_{21} & S_{22} & S_{23} & \cdots & S_{2p} \\ \vdots & & & & \\ S_{p1} & S_{p2} & S_{p3} & \cdots & S_{pp} \end{vmatrix} \begin{vmatrix} a_{m1} \\ a_{m2} \\ \vdots \\ a_{mp} \end{vmatrix} \qquad \text{...(7.27)}$$

This matrix is of order ($p \times p$) *i.e.*, $p$ is equal to number of basis functions which we have taken. However, if the number of basis functions increases to infinite, then the matrix will be of an infinite order. In practice, however, we often use a finite set of functions, (see Chapter 8 and 9) because it is not possible to solve any matrix of infinite order.

Eq. (7.25) is therefore an approximate equation. It does not give the exact eigenvalue, [*viz* $E_m$]. We therefore substitute $\lambda_m$ in place of $E_m$, ($\lambda_m \cong E_m$). Our final equation then becomes,

$$\sum_{i=1}^{p}(H_{pi} - \lambda_m S_{pi})a_{mi} = 0 \qquad \ldots(7.28)$$

(eqs. 7.27) are a set of $p$-linear homogeneous equations with coefficients $a_{m_1}, a_{m_2}, \ldots a_{mp}$ as unknowns. Thus for a set of nonzero $a_{mi}$, we must have the determinant of this matrix to be zero.

$$| H_{ji} - \lambda_m S_{ji} | = 0 \qquad \ldots(7.29)$$

Equation (7.29) is a polynomial equation of $p^{\text{th}}$ order in $\lambda_m$ and we will obtain $p$ number of roots of $\lambda_m$. We substitute these roots of $\lambda_m$ in eq. (7.27) one at a time and solve the equation for $a_{pi}$. In this manner, we will get one set of $a_{pi}$ for one root of $\lambda_i$ and for $p$ number of roots of $\lambda_m$ we will get $p$-sets. These $p$-sets of $\lambda_{mi}$ will generate $p$-vectors (*viz.* $a_{p1}\phi_1 + a_{p2}\phi_2 + a_{p3}\phi_3 \ldots a_{pi}\phi_p$ ). Thus for each vector, we will get one state associated with the energy value of $\lambda_m (\cong E_m)$. These energy values (*viz.* roots of $\lambda_m$) will be higher than the actual experimental value $E_p$. (This is in accordance with the variation principle). These are generalised values of $\lambda_p$ when we have taken $\phi_i$ to be non-orthogonal. If the set of assumed $\phi_i$ functions are orthonormal then, the problem is more simplified. Under these conditions we have,

$$S_{ij} = \delta_{ij}, \qquad \ldots(7.30)$$

$$H_{ij} = H_{ji} \text{ and } \langle \phi_j | H | \phi_i \rangle$$

$$= \sum_i \langle \phi_j / \phi_i \rangle H_{ij} = \sum_l \delta_{ij} H_{j'i} = \langle \phi_i H \phi_j \rangle$$

This gives a simplified equation, *viz.*,

$$| H_{ij} - \lambda_m | = 0 \qquad \ldots(7.31)$$

The solution of eq. (7.26) is further simplified by the application of the symmetry operations of the molecule. The procedure of simplification will be given later.

## 7.5 USE OF SYMMETRY PROPERTIES OF HAMILTONIAN

In Section [7.4], we have shown that the eigenvalues $\lambda_n(\approx E_n)$ and the eigen-

functions $\left(\sum_i a_{ni}\phi_i\right)$ can be obtained by solving the $p^{th}$ degree polynomial

equation in $\lambda_n$. The solution of the polynomial in $\lambda_n$ is very difficult if $n$ is greater than four. This means that we need to use such a basis set of $\phi_i$ that factors the secular equation (or determinant) so that the polynomial equation can be factored, and can be solved separately. The principle that allows to do so is to use the SALCs of the basic functions of the symmetry group of Hamiltonian. The method of formation of SALCs is given below.

We have already seen that the Hamiltonian of a molecule remains invariant under all the symmetry operations of the point group of the molecule (Sec. 7.1 and 7.2). This implies that the corresponding Hamiltonian matrix (see. Sec. 7.4) might take a very useful form when the basis functions are taken as SALCs of the symmetry group of the molecule. Since, the eigenfunctions of Hamiltonian $H$ form the bases for the representation (normally irreducible) of the symmetry group of $H$ and further partners of a given degenerate I.R. will have the same eigenvalue (*e.g.*, two functions of $E$ type I.R, will have the same eigenvalue), we will have degeneracies not higher than the dimensions of the I.Rs of the point group of the molecule. For example, a molecule with $C_{2v}$ point group will not have any degenerate I.Rs and while that of $C_{3v}$ will have one doubly degenerate I.Rs. However, the degeneracies and therefore, the degree of the polynomial $\lambda_n$ cannot be greater than the dimensions of the I.R. of the point group owing to the following reasons.

Suppose we employ a basis of $\phi_i's$, which are symmetry adopted linearly combined set. That is, these are basis functions for the I.Rs of the symmetry group of Hamiltonian $H$. Then the matrix elements $H_{ij}$ will have the following properties.

$$H_{ij} = \int \phi_i * H\phi_j \partial\tau \qquad \ldots(7.32)$$

This integral will be zero unless $\phi_i = \phi_j$ (Chapter 6) or these are from two different basis sets (SALCs) which belong to the same I.Rs and are the corresponding (important) partners in these two sets belonging to the same I.R. Let us clarify it further by an example,

Assume that the point group of the Hamiltonian is $C_{3v}$ and we start with a nine dimensional basis *viz.* ($\phi_1$, $\phi_2$ ... $\phi_9$). Using this basis, we form a reducible representation $\Gamma_{\text{red}}$ of $C_{3v}$. We determine the number and type of I.Rs that $\Gamma_{\text{red}}$ contains. Suppose it is reduced in the following manner.

$$\Gamma_{\text{red}} = 3E + 2A_2 + A_1 \qquad ...(7.33)$$

We now apply projection operators (Chapter 5) to the $\phi_i$ functions of this set and form the following SALCs for $E$-type of I.Rs. There will be three sets because according to equation (7.33) one set will belong to one $E$-representation. Since, there are three functions of $E_S$ so there will be three sets. Each set will contain two functions (a total of six SALCs). Let these three sets be

Set 1: $\phi_1^E(1), \phi_2^E(1)$

Set 2: $\phi_1^E(2), \phi_2^E(2)$

Set 3: $\phi_1^E(3), \phi_2^E(3)$

Similarly we will have two sets of SALCs of $A_2$ type of I.Rs and one set of $A_1$ type of I.Rs. Suppose these sets are

Set 4: $[\phi^{A_2}(1)]$

Set 5: $[\phi^{A_2}(2)]$

Set 6: $[\phi^{A_2}(1)]$

The total number of SALCs are nine. We use these SALCs as basis.

Using these sets, we will have nonzero integrals as follows:

We will be having $(a \times a)$ matrix elements of the type $\int \psi_i^* H \psi_j \partial\tau$ where $\psi_s^1$ are SALCs.

We will have nonzero matrix elements which will be as under:

$$\int \phi^{*E}_1(1) H \phi_1^E(2) \partial\tau, = \int \phi^{*E}_1(2) H \phi^E(1) \partial\tau = H_1^E(1,2) = H_1^E(2,1) \quad ...(7.34)$$

Thus, if we take these SALC's in order

$$\underbrace{\phi_1^E(1), \phi_1^E(2), \phi_1^E(3), \phi_2^E(1), \phi_2^E(2), \phi_2^E(3)}_{\text{Six functions of } E\text{-type}};\ \underbrace{\phi^{A_2}(1), \phi^{A_2}(2)}_{\text{Two functions of } A_2\text{-type}} \text{ and } \underbrace{\phi^{A_1}(1)}_{\text{One functions of } A_1\text{-type}}$$

The Hamiltonian matrix will be as given below.

| | $\Phi_1^E(1)\ \Phi_1^E(2)\ \Phi_1^E(3)$ | $\Phi_2^E(1)\ \Phi_2^E(2)\ \Phi_2^E(3)$ | $\Phi_1^{A_2}(1)\ \Phi_1^{A_2}(2)$ | $\Phi_1^{A_1}(1)$ |
|---|---|---|---|---|
| $\Phi_1^E(1)$ | $H_1^E(1,1)\ H_1^E(1,2)\ H_1^E(1,3)$ | 0 0 0 | 0 0 0 | 0 0 |
| $\Phi_1^E(2)$ | $H_1^E(2,1)\ H_1^E(2,2)\ H_1^E(2,3)$ | 0 0 0 | 0 0 0 | |
| $\Phi_1^E(3)$ | $H_1^E(3,1)\ H_1^E(2,3)\ H_1^E(3,3)$ | 0 0 0 | 0 0 0 | 0 0 |
| $\Phi_2^E(1)$ | 0 0 0 | $H_2^E(1,1)\ H_2^E(1,2)\ H_2^E(1,3)$ | 0 0 0 | 0 0 |
| $\Phi_2^E(2)$ | 0 0 0 | $H_2^E(2,1)\ H_2^E(2,2)\ H_2^E(2,3)$ | 0 0 0 | |
| $\Phi_2^E(3)$ | 0 0 0 | $H_2^E(3,1)\ H_2^E(3,2)\ H_2^E(3,3)$ | 0 0 0 | |
| $\Phi_1^{A_2}(1)$ | 0 0 0 | 0 0 0 | $H^{A_2}(1,1)\ H^{A_2}(1,2)$ | 0 |
| $\Phi_1^{A_2}(2)$ | 0 0 0 | 0 0 0 | $H^{A_2}(2,1)\ H^{A_2}(2,2)$ | 0 |
| $\Phi_1^{A_1}(1)$ | 0 0 0 | 0 0 0 | 0 0 0 | $H^{A_1}(1,1)$ |

= 0 ...(7.34*a*)

By using SALCs of $C_{3v}$, we have converted a nine degree polynomial to two cubic equations (third degree polynomial) one quadratic equation (two order polynomial) and one one degree equation. These equations can be solved easily.

We state another principle for simplification related to these equations. The number of coefficients in any $p^{\text{th}}$ degree polynomial, correspond to the number of coefficients for a particular root of $\lambda_m$. Accordingly in eq. (7.34*a*) for the first (3 × 3) matrix, we will have functions $\Psi_{11}^E, \Psi_{12}^E$ and $\Psi_{13}^E$. These eigenfunctions are the three linear combinations of $\Psi_1^E(1), \Psi_1^E(2)$ and $\Psi_1^E(3)$, having the eigenvalues $E_1$, $E_2$ and $E_3$. Let the three eigenfunctions $\Psi_{21}^E, \Psi_{22}^E$ and $\Psi_{23}^E$ arise from the second (3 × 3) matrix. These are three different linear combinations of $\Psi_2^E(1), \Psi_2^E(2)$ and $\Psi_2^E(3)$. Since $E$ irreducible representation is doubly degenerate, the eigenvalue of the second set must be identical to that of first, *i.e.*, $E_1$, $E_2$ and $E_3$. So the roots of second cubic equation (2[nd] matrix) must be either the same or these differ from the first set of roots by a constant factor. Therefore, we have to solve only one cubic equation instead of two.

Similarly, the quadratic equation (3[rd] matrix) will give two roots. Since $A_2$ is singly degenerate, its two roots will give two eigenvalues corresponding to the two I.Rs in (eq. 7.34*a*). The fourth matrix corresponding to $A_1$ will give one root or one eigenvalue.

All functions will be linear combinations of $\phi_1^E(1)\phi_2^E(1)...\phi_1^{A_2}(1), \phi_1^{A_2}(2)$ and $\phi^{A_1}$. There will be six pairs of basis functions of symmetry $E$ giving three eigenvalues ($E_1$, $E_2$, $E_3$) and six eigenfunctions which are degenerate in pairs. Each of these will be the six linear combinations of three basis functions *e.g.*, $4E$ in place of $3E$. If we have used four pairs of basis functions, we have to solve fourth degree equation and each $\Psi_n$ function (*i.e.*, eigenfunction) will be a linear combinations of four basis functions.

A number of examples related to the above method of finding eigenfunctions of proper I.Rs and eigenvalues will be given in the application of symmetry in vibrational spectroscopy and in chemical bonding chapters of the book. We postpone providing examples of the above concepts to those chapters.

## 7.6 ATOMIC ORBITALS

In this section, we will exploit the symmetry of a one electron atom by considering the electronic charge interaction with the field of nuclear charge. We will develop a theoretical model which will enable us to understand the origin of the labels (*s*, *p*, *d*, *etc.*) and of the quantum numbers ($n$, $l$, $m$) of the atom by using the symmetry properties of the atom.

The preceeding sections describe the method to calculate the eigenvalues and eigenfunctions of molecules. The method is greatly simplified with the aid of symmetry properties of the system. In the process, we have chosen a set of wave functions or orbitals as basis (not necessarily eigenfunctions of the Hamiltonian) to construct a reducible or irreducible representation. It is then followed by the use of projection operators of the point group of the molecule, which let us form SALCs. The latter set of functions (SALCs) brought the matrix representation into a diagonal blocked form along the diagonal. As a result, we were able to calculate the energy levels of the M.Os of the molecules.

The above summarised process can be applied only to obtain various eigen-states of those molecules that belong to finite groups. But the atomic system is a spherical system where one has to use continuous full-rotation group $R(3)$. In such system, the above procedure cannot be applied and the actual process is a little complicated.

We assume in an atom that the electron is moving in a spherical symmetric field of nucleus. Thus, the Hamiltonian is considered to be invariant under the group of symmetry operations of a sphere, which is a direct product of

$(R_3 \rightarrow S_2)$, where $R_3$ is the continuous full rotation group and $S_2$ is the finite group containing $E$ and $i$ as operations. We will consider only the sketchy description of the process used to derive eigenvalues and the eigenfunctions of the system having $R_3$ group in the next section.

## 7.7 FORMATION OF HAMILTONIAN, BASIS FUNCTIONS, AND I.Rs OF R(3)

The Hamiltonian of an atom in spherical polar coordinates $(r, \theta, \phi)$ in terms of $\nabla$ and $V$ is given as,

$$H = \left(\nabla^2 + \frac{e^2}{r}\right), \qquad \ldots(7.35)$$

where

$$\nabla^2 = \frac{1}{r^2}\left\{\frac{\partial}{\partial r}\left(r^2\frac{\partial}{\partial r}\right) + \frac{1}{\sin\theta}\frac{\partial}{\partial\theta}\left(\sin\theta\frac{\partial}{\partial\theta}\right) + \frac{1}{\sin^2\theta}\frac{\partial^2}{\partial\phi^2}\right\} \qquad \ldots(7.36)$$

In the central field approximation, the potential energy is a function of $(r)$ only and it is not affected by the changes in $\theta$ and $\phi$ part of Hamiltonian. The angular part of the Hamiltonian in the operator form is given as,

$$-\left\{\frac{1}{\sin\theta}\frac{\partial}{\partial\theta}\left(\sin\theta\frac{\partial}{\partial\theta}\right) + \frac{1}{\sin^2\theta}\frac{\partial}{\partial\phi^2}\right\} \qquad \ldots(7.37)$$

Let us express it by $L^2$. That is,

$$L^2 = -\left\{\frac{1}{\sin\theta}\frac{\partial}{\partial\theta}\left(\sin\theta\frac{\partial}{\partial\theta}\right) + \frac{1}{\sin^2\theta}\frac{\partial^2}{\partial\theta}\right\} \qquad \ldots(7.38)$$

The eigenfunctions of $L^2$ is the spherical harmonics $Y_l^m(\theta, \phi)$ with the eigenvalue $[l(l + 1)]$. It is expressed as,

$$Y_l^m(\theta, \phi) = (-1)^{1/2}(m+|m|)\left[\frac{(2l+1)(1-|m|)!}{4\pi(l+|m|)!}\right]^{1/2} p_l^{|m|}(\cos\theta)e^{im\Phi} \qquad \ldots(7.39)$$

[$P_l^{|m|}(\cos\theta)$ are the associated Legendre functions (or polynomial)]. We can now write,

$$L^2Y_l^m(\theta, \phi) = l(l+1)Y_l^m(\theta, \phi) \qquad \ldots(7.40)$$

In all the above functions, the label $(l)$ stands for positive integers including zero and the label $(m)$ is another integer which can take all the values from $[-l \mid \ldots 0 \ldots \mid +l]$, *i.e.*, $(2l + 1)$ values (integers). [Equation (7.40) is satisfied by the $(2l + 1)$ spherical harmonics.] It implies that there is $(2l + 1)$ fold degeneracy associated with one eigenvalue of $(l)$.

Let us now use the symmetry properties of the Hamiltonian (or that of the atom $[R(3)]$, having spherical symmetry). Let $\hat{P}_R^{\alpha}$ be a symmetry operator of point group $[R(3)]$. This symmetry operation is associated with a rotation by an angle $\alpha$ about any axis of sphere passing through its centre. The symmetry operator ($\hat{P}_R^{\alpha}$) commutes with the Hamiltonian . That is,

$$\hat{P}_R^{\alpha}(H) = H(\hat{P}_R^{\alpha}) \qquad \ldots(7.41)$$

(The relationship (7.41) has been proved in the earlier sections of the chapter.) The Hamiltonian operator ($H$) also involves those functions which depend upon $r$ besides $\theta$ and $\phi$. So these variables *viz.* ($r$, $\theta$, $\phi$) should be involved in the determination of eigenvalues and eigenfunctions. However, the $r$-dependent function is invariant under the operations of the group $[R(3)]$. ($R$-does not vary because of rotation or reflection operation). It then implies that we should take only ($\theta$, $\phi$) part of the function as basis functions to determine the effects of symmetry properties of $[R(3)]$ operators. Therefore, we have

$$(\hat{P}_R)L^2Y_l^m(\theta, \phi) = \hat{P}_R[l(l+1)]Y_l^m(\theta, \phi) \qquad \ldots(7.42)$$

or

$$L^2(\hat{P}_R)Y_l^m(\theta, \phi) = l(l+1)\hat{P}_RY_l^m(\theta, \phi) \qquad \ldots(7.43)$$

Eqs. 7.42 and 7.43 suggest that $\hat{P}_RY_l^m(\theta, \phi)$ are also eigenfunctions of $L^2$. These eigenfunctions are either the same functions $Y_l^m(\theta, \phi)$ or are other spherical harmonics $Y_l^m(\theta, \phi)$ or are linear combinations of spherical harmonic which form the degenerate set of $(2l + 1)$ functions (for one value of $l$, we have $(2l + 1)$ functions which give the same eigenvalue $l(l + 1)$, resulting in a $(2l + 1)$ degeneracy). We therefore, conclude that the spherical harmonics $Y_l^m(\theta, \phi)$ act as basis functions for a $(2l + 1)$ dimensional representation of $[\hat{P}_R(\Gamma_R)]$ which is represented by the matrices $D^l(R)$ because of the following relationship,

$$\hat{P}_R[(Y_l^m(\theta, \phi))] = \underbrace{|D^l(R)|}_{\substack{\text{matrix of } (l\times l)\\ \text{dimension}}} \; \underbrace{|Y_l^m(\theta, \phi)|}_{\text{Column matrix}} \qquad \ldots(7.44)$$

In equation (7.43), we have $l$ to be a whole number integer. It means that $(2l + 1)$ will be odd. So the dimensions of $D^l(R)$ will be odd number matrix. That is why all the I.Rs ($s, p, d, f$ ...) are odd ordered representations (1, 3, 5, 7 ... *etc*). These are the representations of $[R(3)]$ point group when spherical harmonics are taken as basis for the representation. (Even order representations are obtained when spin functions are also included in the eigenfunctions.)

It is a very difficult exercise to obtain the matrices $[D^l(R)]$ in our case. This difficulty can be taken care of by considering the property of a class of the group. The property is that the characters of all symmetry operations belonging to one class are the same. We must also realise that all rotations about *ANY* axis by the same angle (*e.g.*, $\alpha$) belong to the same class and hence have the same character (why ?). This property allows us to choose $z$-coordinate axis as our rotation axis. With the assurance that the character of the matrices $[D^l(\alpha)]$ which represents the rotation by an angle $\alpha$ about the $z$-coordinate axis, is equal to the character which we would obtain from the matrix $[D^l(\alpha)]$.

The above said statement allows us to assume the Abelian point group $C_\infty$ to the system.

The effect of $\alpha$-angle rotation about the $z$-axis on $Y_l^m(\theta, \phi)$ gives the following relationship

$$Y_l^m(\theta, \phi \pm \alpha) = \left(e^{\pm im\alpha}\right) Y_l^m(\theta, \phi) \quad \ldots(7.45)$$

which will produce $|[D^l(\alpha)]|$ matrix representing $(P_R)$.

We thus have the following representation for $[P_R]$. (rotation by an angle $\alpha$ about $z$-axis). In other words, we say that the system is quantised about $z$-axis.

$$[P(R)_\alpha][Y_l^l(\theta, \phi), Y^{(l-1)}_{(\theta, \phi)} \ldots Y^{-l}_{(\theta, \phi)}]$$

$$\begin{vmatrix} e^{-il\alpha} & 0 & 0 & 0 & 0 & \\ 0 & e^{-i(l\text{-}1)\alpha} & & & & \\ 0 & & e^{-i(l\text{-}2)\alpha} & & & \\ & & & \ddots & & \\ & & & & \ddots & \\ \vdots & & 0 & & & e^{+il\alpha} \\ 0 & & & & & \end{vmatrix} \quad \ldots(7.46)$$

and the character for $\hat{R}(C_\alpha)$ rotation is given by

$$\chi^l(R_\alpha) = \left[e^{-il\alpha} + e^{-i(l-1)\alpha} + e^{-i(l-2)\alpha} + \ldots e^{+il\alpha}\right]$$

$$= e^{-il\alpha}[1 + e^{l\alpha} + e^{2i\alpha} + \ldots e^{2il\alpha}] \quad \ldots(7.47)$$

$$= e^{-il\alpha}\frac{[1 - e^{i(2l+l)\alpha}]}{1 - e^{i\alpha}} \quad \ldots(7.48)$$

$$= \frac{e^{2\alpha/2}[e^{i\left(l+\frac{1}{2}\right)\alpha} - e^{-1\left(l+\frac{1}{2}\right)\alpha}]}{(e^\alpha - 1)} = \frac{\sin\left(l + \dfrac{1}{2}\right)\alpha}{\sin\alpha/2} \quad \ldots(7.49)$$

This is the character for the representation of $R_\alpha$ but the question is whether this character is for the formation of $\Gamma_{\text{red}}$ or $\Gamma_{\text{I.R.}}$. It can be shown that these matrices are for I.Rs, on the ground that it is impossible to produce a single matrix $|A|$ which will bring simultaneously all the representation matrices in the blocked form by a similarity transformation [*viz.* $A^{-1}D^1(R)A$.] We therefore, conclude that the spherical harmnoics used as basis functions give $(2l + 1)$ dimensional I.Rs of $[R(3)]$ group. Therefore, one can label the basis functions of Hamiltonian $H$ with quantum number $n$, $l$, $m_l$ where $l$ denotes the I.R of the

Hamiltonian group and $m$ denotes the column to which the function belongs. Since $l$ has the eigenvalue $[l(l+1)]$ of the spherical harmonics, so we can also associate to each I.R. $[\hat{O}^l(R)]$ a distinct eigenvalue ($l$).

## 7.8 TOTAL ROTATION-INVERSION GROUP OF $\hat{H}[R(3) \times S]$ GROUP

In the above discussion, we have assumed $[R(3)]$ point group for the Hamiltonian. Since the potential energy function in a hydrogen-like atom is centrosymmetric, Hamiltonian should also be invariant under the symmetry operations of $S_2$ point group (*viz.* $E$ and $i$). It is well established that $E$ and $I$ are commutative operators therefore, have two classes. Thereby $S_2$ group should have two I.Rs of one-dimension ($A_1$ and $A_2$). Consequently, the eigenfunctions of the Hamiltonian of an atom should also be the basis functions for the $S_2$ point group representation. That is,

$$[P_R]S_2\Psi_{\text{eig}} = \pm\Psi]$$

or

$$\left.\begin{aligned}\hat{E}\Psi_{\text{eig}} &= +\Psi_{\text{eig}}\\ \hat{i}\Psi_{\text{eig}} &= \pm\Psi_{\text{eig}}\end{aligned}\right\} \qquad \text{...(7.50)}$$

Thus, $\Psi_1$ [with $\hat{E}\Psi_{\text{eig}} = +\Psi_{\text{eig}}$ and $\hat{i}\Psi_{\text{eig}} = +\Psi_{\text{eig}}$] will be an even function and $\Psi_2$ [with $\hat{E}\Psi_{\text{eig}} = +\Psi_{\text{eig}}$ and $\hat{i}\Psi_{\text{eig}} = -\Psi_{\text{eig}}$] will be an odd function.

In $C_\infty$ point group symmetry transformations, we have used spherical harmonics as basis which gave $(2l+1)$ dimensional I.Rs. The same eigenfunctions (spherical harmonics) should also be the basis functions for $S_2$ group representation. Employing them as basis, for $S_2$ point group representation we have,

$$\hat{E}Y_l^m(\theta,\phi) = Y_l^m(\theta,\phi) \qquad \text{...(7.51)}$$

and

$$\hat{i}Y_l^m(\theta,\phi) = [Y_l^m(\pi-\theta)(\phi+\pi)$$

$$= [Y_l^m(\theta,\phi)\cdot(-1)^l \qquad \text{...(7.52)}$$

(Eq. (7.52) can be derived by writing the spherical harmonics expressions and substituting $(\pi - \theta)$ and $(\phi + \pi)$ for $\theta$ and $\phi$ respectively).

According to eqs. (7.51) and (7.52), we will be having for even value of $l[i\cdot Y_l^m(\theta,\phi)] = +Y_l^m(\theta,\phi)$ and for odd value of $l$ $[iY_l^m(\theta,\phi) = -Y_l^m(\theta,\phi)]$. This will give us two following expressions,

$$\left.\begin{aligned}
&\text{for even values of } l,\\
&\hat{E}Y_l(\theta,\phi) = Y_l(\theta,\phi)\\
&\hat{i}Y_l(\theta,\phi) = +Y_l(\theta,\phi)\\
&\text{And for odd values of } l\\
&\hat{E}Y_l(\theta,\phi) = Y_l(\theta,\phi)\\
&\hat{i}Y_l(\theta,\phi) = (-1)Y_l(\theta,\phi)
\end{aligned}\right\} \quad \ldots(7.53)$$

The point group of Hamiltonian is the direct product group of [$R(3)$] and $S_2$ (*i.e.*, $R(3) \times S_2$). Therefore, the matrices for the representation of Hamiltonian point group will be the direct product of the matrices of [$R(3)$] and $S_2$ point groups representations. This will give us,

$D^{l \to g}(A \cdot R) = D^A(A) \times D^l(R)$, where (A) is the matrix for the $S_2$ group representation. The conclusion is,

$$\left.\begin{aligned}
& D^{l \otimes g}(E \cdot R) = D^l(R)\\
\text{or,}\quad & X^{l \otimes g}(E \cdot R) = X^l(R)\\
\text{and when } l = \text{odd}\quad & D^{l \otimes u}(E \cdot R) = D^l(R) \quad \text{and}\\
& D^{l \otimes u}(i \cdot R) = -D^l(R)\\
\text{or,}\quad & X^{l \otimes u}(i \otimes R) = X^l(R)
\end{aligned}\right\} \quad \ldots(7.54)$$

Eq. (7.54) determines the parity of the functions. For even value of $l$, the parity is even ($g$) and for odd values, it is odd ($u$), consequently $s$, $d$, $g$ eigenfunctions are of even parity ($l = 0, 2, 4...$) and $p$, $f$, $h$ eigenfunctions are of odd parity ($l = 1, 3, 5...$).

For many electron systems, the value of $l$ does not determine the parity of the functions. In those cases, parity is determined by the values of capital $L$.

One should realise that we have classically defined the angular momentum of electrons with the quantum number $l$ and $m_l$ and shown a relationship of $L^2$ and $L_z$. But [$R(3)$] group is a continuous group, known as Lie group. In this group, the angular momentum is related to the infinitesimal operators. Since we are chiefly concerned with finite descrete groups, we will not discuss the Lie groups at this stage and satisfy ourselves with the classical picture only.

## 7.9 HYBRIDISATION

In the preceeding description, we have taken the electron/s to be present in spherical field of nucleus ([$R(3)$] point group) and their behaviour described in terms of the I.Rs of [$R(3)$] group (*viz.* $s$, $p$, $d$...). These orbitals have certain definite dierctionalities. For example, an electron present in the $p_x$-orbital will be concentrated or directed along the $x$-axis. However, when we consider molecules, the atomic orbitals of two or more atoms overlap, and form $\sigma$ or $\pi$-type of bonds. These primarily decide the geometry of the molecule. If we assume that these $\sigma$ or $\pi$-types of bonds are formed by the overlap of pure, $s$, $p$

or *d*-orbitals of the bonded atoms, then the angles between such bonds do not match with the angles that are found experimentally. Therefore, a new bonding model has to be proposed which is based on hybrid orbitals, rather than of pure *s, p, d*-orbitals. In most cases, the directional characteristics of the localised bonds (mainly σ-type of bonds) are described conveniently on the basis of hybrid orbital model. For example, in the $AX_n$ type of molecules with *A* as the central atom, *A* forms *n* number of σ bonds by the overlap of one orbital of atom *X* of proper symmetry, with *n* number of orbitals on *A* of proper symmetries. These *n* localised σ-bonds decide the geometry of the molecule (ignoring π-bonds). One can understand how the geometry of the molecule with certain angle between the bonds is a consequence of the presence of the probability maximum of bonded electron density. The latter is based on the criterion of maximum overlap for the largest contribution of bond energy. For example, in $H_2O$ molecule, the bond angle between two O – H bond is 105°. If one assumes that O – H bond is formed by the overlap of 1*s* orbital of hydrogen with the *p*-orbitals of oxygen (*e.g.*, $p_x$ and $p_y$), the angle between two O – H bonds should have been 90°. The presence of 105° angle in the actual $H_2O$ molecule is consistant with the assumption of involvement of *hybrid atomic orbitals*, formed by LCAO of 2*s*- and 2*p*-orbitals on oxygen atom. We therefore, define "Hybridisation" as a phenomenon of forming new orbitals by LCAO of *s, p, d* orbitals belonging to different angular momentum quantum numbers on the same atomic centre in a molecule. These hybrid orbitals give maximum overlap to grant the maximum energy to the bond. The direction of these hybrid orbitals are different depending on the type, amount and kind of orbitals involved in the hybridisation process. Mathematically, hybridisation involves an orthogonal transformation (or making a linear combination) of a given basis set (*e.g.*, $s, p_x, p_y, p_z, d_z$ *etc.*) to another basis set (*e.g.*, $sp$, $sp^2$, *sd...etc.*). In a simple language, it involves making a linear combination of two orbitals (*e.g.*, *s, p, s, d*), or three orbitals (*e.g.*, $s, p_x, p_y$), or four orbitals (*e.g.*, $s, d_{xy}, d_{yz}, d_{xz}$) and so on. These orbitals are defined as $sp$, $sp^2$, $d^3s$, $d^2sp^3$ *etc.*, hybrid atomic orbitals directed towards different angles.

## We Wish to know the Answer of the Following Two Questions:

**Q.1:** It is a qualitative question. Which ones of the atomic orbitals on the central atom *A* of the $AB_n$ type molecules, are required to construct hybrid orbitals for σ and/or for π-bonding to the set of pendent atomic *B* orbitals?

**Q.2:** How to write the explicit expressions for each of the hybrid orbitals which will state as to how much each A.O. contributes in each hybrid orbital?

**Ans.:** A partial answer of these questions will be given in the following paragraphs.

To answer the question one, let us start with an example. Let us take the case of a tetrahedral molecule $AB_4$ belonging to $T_d$ point group. We wish to find which orbitals on atom $A$ are suitable for forming four σ-bonds. (σ type orbitals of atom $A$ which have their lobes directed towards $B$-atom to fulfil the criterion of maximum overlap). The set of these four orbitals on $A$, will form the basis for the formation of representation of $T_d$ point group. Let us represent these orbitals by four vectors, (*viz.* $p_1, p_2, p_3$ and $p_4$) (Fig. 7.3) and determine the characters of the $\Gamma_{red}$ representations of the point group.

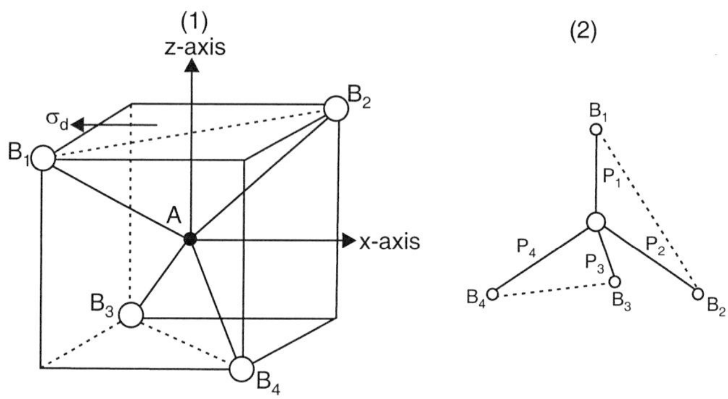


**Fig. 7.3** Tetrahedral $AB_4$ molecule.

There is a very simple rule for finding the characters of matrices which are associated with the symmetry operations and which form the reducible representation of point group. We therefore, do not have to go through any mathematical manipulations for the formation of the matrices. The rule is as follows:

We assume the set of hybrid orbitals on $A$ pointing towards the pendent atoms $B_i$. These orbitals form the basis for the formation of representation (Reducible or I.R). We have designated them as $p_1$, $p_2$, $p_3$ and $p_4$ vectors (Fig. 7.3). We now carry out symmetry operations of the point group [$T_d$($E$, $8C_3$, $3C_2$, $6S_4$, $6S_4$, $6\sigma_d$)]. If on carrying out the symmetry operations on these hybrid orbitals or on the set of the vectors, (*e.g.*, on $p_1, p_2, p_3$ and $p_4$), a certain vector remains invariant or unshifted, then in the matrix associated with the symmetry operation, there appears 1 as the element on its diagonal. The character for the symmetry element is then 1. In case no vector remains unshifted, then the character of the symmetry element in the reducible representation is zero.

We apply the above rule to determine the characters of symmetry operations of $T_d$, [*e.g.*, ($AB_4$) with $p_1, p_2, p_3$ and $p_4$ as basis vectors]. The set of characters for the representation thus generated are,

| $T_d$ | $E$ | $8C_3$ | $3C_2$ | $6S_4$ | $6\sigma_d$ |
|---|---|---|---|---|---|
| $\Gamma_{red}$ for $T_d$ | 4 | 1 | 0 | 0 | 2 |

These numbers are obtained as follows.

(*a*) *E* (*identity operation*): By carrying *E* operation, on all the four vectors $(p_1, p_2, p_3, p_4)$, these remain unshifted. Hence character for *E* is 4 (*i.e.*, four vectors $p_1, p_2, p_3, p_4$ remain unshifted by the operation *E*).

(*b*) $C_3$ *operations*: There are four $C_3$ axis (*viz.* $AB^1, AB^2, AB^3, AB^4$), so the number of operations are eight $[(C_3^1, C_3^2) \times 4]$. By carrying out either $C_3^1$ or $C_3^2$, it is one vector in each operation that remains invariant (Fig. 7.4). Hence, the character for $C_3$ is one.

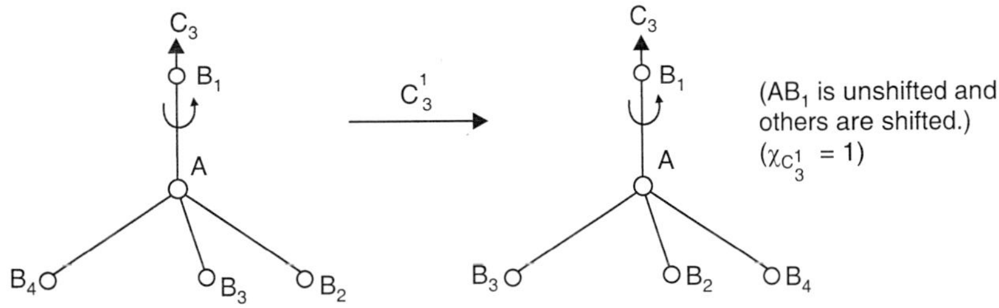


**Fig. 7.4** $C_3^1$ operations

(*c*) $C_2$ *and* $S_4$ *operations*: There are three $C_2$-axes. These are *x*-, *y*-and *z*-coordinate axes (Fig. 7.3). All the vectors are shifted. For example, rotation about *z*-axis by 180°, will shift $B_1$ to $B_2$, and $B_3$ to $B_4$. Similar results will be obtained as a result of rotation by 180° about *x*- and *y*-axes. Since by these rotation operations, all the vectors are shifted, so $\chi_{C_2}$ is zero. Identical results will be obtained by $S_4$ rotation operations because $C_2$ and $S_4$ axes are the same. Hence, $\chi(S_4)$ is also zero.

(*d*) $\sigma_d$ *operations*: There are $6\sigma_d$ planes. All the planes passing through the diagonal of all the faces (6 faces → 6 diagonals → $6\sigma_d$ planes), By reflecting the molecule $AB_4$ in any $\sigma_d$ planes, two vectors remain unshifted. Hence, $\chi(\sigma_d)$ is equal to 2.

By applying the above rule, we are thus able to find the characters $[\chi_{\text{oper}}]$ for all the symmetry operations of the $T_d$ point group associated with the $\Gamma_{\text{red}}$ representation.

Refering to the character table of $T_d$ point group and applying the formula $\left[a_{\Gamma i} = \frac{1}{g}\sum_{i=1}^{k} g_i \chi^*_{\text{I.R.}}(R) \bullet \chi_{\text{red}}(\dot{R})\right]$, one finds that this representation is reduced into the following I.R components:

$$\Gamma_{\text{red}}(T_d \text{ point group}) = A_1 + T_2 \qquad \ldots(7.55)$$

Eq. (7.55) suggests that the four atomic orbitals must be combined to form a set of four hybrid orbitals. These atomic orbitals must be chosen so as to include one orbital of $A_1$ symmetry and a set of three orbitals which belong to

$T_2$-I.R of the group. To know these orbitals (*i.e.*, of $A_1$ and $T_2$ symmetry or I.Rs), we again look at the character table of $T_d$ group. This will tell us as to which orbitals on the atom $A$ fall into the categories of $A_1$ and $T_2$ I.R (Chapter 3). A look at the last column of the Table, gives the needed orbitals.

| Orbitals | $A_1$ | $T_2$ |
|---|---|---|
| | $s$ | $p_x, p_y, p_z$ or $d_{xy}, d_{yz}, d_{xz}$ |

...(7.56)

We now take one orbital (*viz. s*) of $A_1$(IR) and three orbitals of $T_2$ because $A_1$ is non-degenerate and $T_2$ is triply degenerate. Our set of orbitals is thus either $sp^3$ or $sd^3$. (Here $d = d_{xy}, d_{yz}, d_{xz}$ and not $d_{x^2-y^2}, d_{z^2}$ ). [The symmetry properties of the molecule will not be able to decide one among these two combination.]

One of the ways to choose one out these two alternative combinations (*viz.* $sp^3$ or $sd^3$) is based on energy consideration of these orbitals on the central atom $A$. Thus, if *d*-orbitals are of very high energy relative to *p*-orbitals, the hybridisation of *s* and *p*-orbitals is favoured over $sd^3$. (for example in carbon compounds). But $sd^3$ hybridisation is favoured over $sp^3$, if *d*-orbital energy of atom $A$ is lower than that of *p*-orbitals (*e.g.*, $[MnO_4]^-$ ion). Besides such extreme cases, ( if the energies of *p*- and *d*-orbitals are comparable then the hybridisation involves a mixture of both.)

Let us take a couple of more examples to illustrate the above concept, We take the example of $PF_5$ ($D_{3h}$ point group with $2C_3$, $3C_2$, $3C_2$, $\sigma_h$, $3S_3$, and $3\sigma_v$} as symmetry operations) (Fig. 7.5). In the figure, we have shown the rotation axes, and reflection planes.

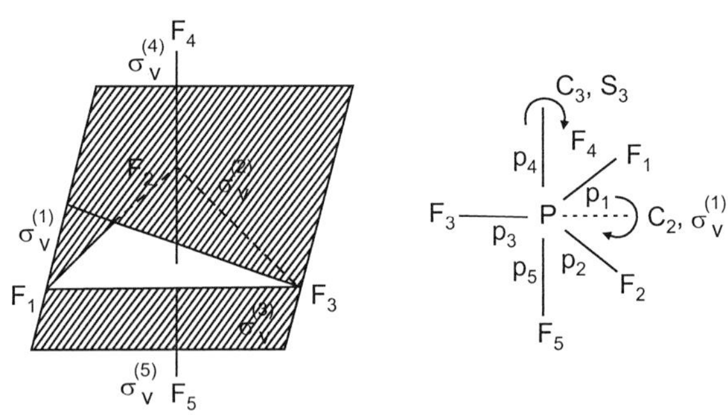


**Fig. 7.5** Triangular bipyramid $AB_5$ molecule.

The set of five hybrid σ-orbitals on phosphorus atom form the basis of reducible representation of $D_{3h}$ group.We designate the hybrids by $p^1$, $p^2$, $p^3$, $p^4$, $p^5$ (Fig. 7.5). We again apply the rule cited above to determine the characters $\chi_{(R)}$ for each operation forming the reducible representation $\Gamma_{(red)}$ of the group. It is given ahead.

| $D_{3h}$ | $E$ | $2C_3$ | $3C_2$ | $\sigma_h$ | $2S_3$ | $3\sigma_v$ |
|---|---|---|---|---|---|---|
| $\chi_R$ for $\Gamma_{red}$ | 5 | 2 | 1 | 3 | 0 | 3 |

...(7.57)

Thus, all bonds remain unshifted by the operation $E$, ($\chi_E = 5$). By $C_3$-operations $p^{(4)}$ and $p^{(5)}$ remain unshifted and hence $\chi_{C_3} = 2$. For $C_2$ operation only one ($p^{(1)}p^{(2)}$ or $p^{(3)}$) remains unshifted. Hence $\chi_{C_2} = 1$, for $\sigma_h$, three bonds ($p^{(1)}p^{(2)}$ and $p^{(3)}$) remain unshifted. $\sigma_v$ leaves three bonds [*e.g.*, $f^{(4)}, f^{(5)}, f^{(1)}$ *etc.*] unshifted, Hence $X_{\sigma_v} = 3$ and so on.

The reducible representation (*e.g.*, $\Gamma_R$ = 5, 2, 1, 3, 0, 3) (eq. 7.57) is then reduced into the following I.R, components.

$$\Gamma_{red} = 2A_1' + A_2'' + E' \qquad ...(7.58)$$

Reference to the character Table of $D_{3h}$ shows the orbitals on phosphorus which should be combined to make the set of five σ-type hybrid orbitals. These are given below. These fall into the categories of $A_1'$, $A_2'$ and $E'$

| $A_1'$ | $A_2'$ | $E'$ |
|---|---|---|
| $s$ | $p_z$ | $(P_x, P_y)$ |
| $d_{z^2}$ | – | $(d_{xy}, d_{xz}, d_{yz})$ |

Thus to make the five hybrids any one of the following combinations could be used.

$$\left.\begin{array}{lll} 1. & ns, (n+1)s, & p_z \begin{cases} (p_x p_y) & (a) \\ (d_{xy}, d_{x^2-y^2}) & (b) \end{cases} \\ 2. & nd_{z^2}, (n+1)d_{z^2}, & p_z \begin{cases} (p_x p_y) & (a) \\ (d_{xy}, d_{x^2-y^2}) & \end{cases} \\ 3. & ns, d_{z^2}, & p_z \begin{cases} (p_x p_y) & (a) \\ (d_{xy}, d_{x^2-y^2}) & \end{cases} \end{array}\right\} \qquad ...(7.59)$$

The choice will depend upon the energy consideration of the various orbitals of the $P$ atom. For example, the choice for $PF_5$ is 3*a*, *i.e.*, the combination of [*viz.* $(s, d_{z^2}, p_x, p_y, p_z)$ *i.e.*, $dsp^3$]. On the other hand for $MoCl_5$ in gaseous state, possibly a mixture of 3*a* and 3*b* might be involved.

We give below hybridisation of orbitals of atom $A$ in $AX_n$ type of some important molecules.

1. $AB_4$ [Tetragonal pyramid, $C_{4v}(E, 2C_4, C_2, 2\sigma_v, 2\sigma_d)$]

| $C_{4v}$ | $E$ | $2C_4$ | $C_2$ | $2\sigma_v$ | $2\sigma_d$ |
|---|---|---|---|---|---|
| $\Gamma_{red} = \chi_R$ | 5 | 1 | 1 | 3 | 1 |
| $\Gamma_{red}$ is reduced | $\Gamma_{red} = 2A_1 + B_1 + E$ | | | | |

Orbitals required for hybridisation on atom $A$ in $AB_4$

| $A_1$ | $B_1$ | $E$ |
|---|---|---|
| $s$ | $d_{x^2-y^2}$ | $(p_x, p_y)$ |
| $p_z$ | | $(d_{xz}, d_{yz})$ |
| $d_{z^2}$ | | |

Choices are (1) $ns, (n+1)s, d_{x^2-y^2}, p_x, p_y$

(i) $s, p_z, d_{x^2-y^2} \begin{cases} (p_x, p_y) \\ (d_{xz}, d_{yz}) \end{cases}$

(ii) $s, d_{z^2}, d_{x^2-y^2} \begin{cases} (p_x, p_y) \\ (d_{xz}, d_{yz}) \end{cases}$

2. $AB_3$ (planar) [$D_{3h}$ ($E$, $2C_3$, $3C_2$, $\sigma_h$, $2S_3$, $3\sigma_v$)

| $D_{3h}$ | $E$ | $2C_3$ | $2C_2$ | $\sigma_h$ | $2S_3$ | $3\sigma_v$ |
|---|---|---|---|---|---|---|
| $\chi_{Red}$ for $\Gamma_{red}$ | 3 | 0 | 1 | 3 | 0 | 1 |
| IR (Comp.) | $A'' + E'$ | | | | | |

Probable combinations are $(s, p_x, p_y)$, $(s, d_{xy}, d_{x^2-y^2})$, $(d_{z^2}, p_x, p_y)$, $(d_{z^2}, d_{xy}, d_{x^2-y^2})$ (*i.e.*, $sp^2$, $sd^2$, $dp^2$, $d^3$)

3. $AB_4$ (planar, $D_{4h}(E, 2C_4, C_2, 2C_2', 2C_2'', i, 2S_4, \sigma_h, 2\sigma_d\, 2\sigma_v)$

| $D_{4h}$ | $E$ | $C_3$ | $C_2$ | $2C_2'$ | $2C_2''$ | $i$ | $2S_4$ | $\sigma_n$ | $2\sigma_v$ | $2\sigma_d$ |
|---|---|---|---|---|---|---|---|---|---|---|
| $\chi_{(R)}$ in $\Gamma_{red}$ | 4 | 0 | 0 | 2 | 0 | 0 | 0 | 4 | 2 | 0 |
| IR Components in $\Gamma_{red}$ | $A_{1g} + E_g + T_{1g}$ | | | | | | | | | |

Possible combinations are: $(s, p_x, p_y, d_{x^2-y^2})$, $(d_{z^2}, d_{x^2-y^2}, p_x, p_y)$ *i.e.*, $dsp^2$ or $d^2p^2$

4. $AB_6$ [Octahedral, $O_h$ (E, $8C_3$, $6C_2$, $6C_4$, $3C_2$, $i$, $6\sigma_4$, $8S_6$, $3\sigma_h$, $6\sigma_d$)]

| $O_h$ | $E$ | $8C_3$ | $6C_2$ | $6C_4$ | $3C_2$ | $i$ | $6S_4$ | $8S_6$ | $3\sigma_h$ | $3\sigma_d$ |
|---|---|---|---|---|---|---|---|---|---|---|
| $\chi_{(R)}$ in $\Gamma_{red}$ | 6 | 0 | 0 | 2 | 2 | 0 | 0 | 0 | 4 | 2 |
| IR Components in $\Gamma_{red}$ | $A_{1g} + E_g + T_{2g}$ | | | | | | | | | |

Possible combinations are $s, p_x, p_y, p_z, d_{z^2}, d_{x^2-y^2}$ and $d^2sp^3$.

After knowing the atomic orbitals of the central atom $A$ for constructing the hybrid orbitals for $\sigma$ bond formation with the pendent atoms $B_i$ we now answer the second question *viz.* How to write the explicit expression for each hybrid orbital?

Let us take the case of $AB_3$ type molecule (*e.g.*, $BF_3$) belonging to $D_{3h}$ point group for which *AOs* on *A* of *A′* and *E′* symmetry (I.R) are required, (*See* foregoing paragraph). For this $s$, $p_x$ and $p_y$ orbitals meet the requirement for forming $sp^2$-hybrid orbitals on *A* assuming *xy* plane as molecular plane. We make the following three LCAOs of $s$, $p_x$ and $p_y$ orbitals (*i.e.*, $\Psi_1$, $\Psi_2$, $\Psi_3$).

$$\left.\begin{aligned} \Psi_1 &= C_{11}s + C_{12}p_x + C_{13}p_y \\ \Psi_2 &= C_{21}s + C_{22}p_x + C_{23}p_y \\ \Psi_3 &= C_{31}s + C_{32}p_x + C_{33}p_y \end{aligned}\right\} \qquad \ldots(7.60)$$

In this equation, we will determine the coefficients $C_{ij}$ to get the hybrid orbitals ($\Psi_1$, $\Psi_2$, $\Psi_3$). We can write eq. (7.60) as

$$\begin{vmatrix} \Psi_1 \\ \Psi_2 \\ \Psi_3 \end{vmatrix} = \begin{vmatrix} C_{11} & C_{12} & C_{13} \\ C_{21} & C_{22} & C_{23} \\ C_{31} & C_{32} & C_{33} \end{vmatrix} \begin{vmatrix} s \\ p_x \\ p_y \end{vmatrix} \qquad \ldots(7.61)$$

$$= |C||\text{orbitals}|$$

Taking the inverse of the matrix $|C|$, we get

$$\begin{vmatrix} s \\ p_x \\ p_y \end{vmatrix} = \begin{vmatrix} C_{11} & C_{12} & C_{13} \\ C_{21} & C_{22} & C_{23} \\ C_{31} & C_{32} & C_{33} \end{vmatrix}^{-1} \begin{vmatrix} \Psi_1 \\ \Psi_2 \\ \Psi_3 \end{vmatrix} \qquad \ldots(7.62)$$

or

$$\begin{vmatrix} s \\ p_x \\ p_y \end{vmatrix} = \begin{vmatrix} d_{11} & d_{12} & d_{13} \\ d_{21} & d_{22} & d_{23} \\ d_{31} & d_{32} & d_{33} \end{vmatrix} \begin{vmatrix} \Psi_1 \\ \Psi_2 \\ \Psi_3 \end{vmatrix} \qquad \ldots(7.63)$$

$|D|$ matrix is the inverse of $|C|$ matrix since $|C|$ and $|D|$ matrices are orthogonal matrices, $|C|^{-1} = |D| = |C|^{\text{Transpose}} = |C|^T$.

We have learnt a method in chapter six for writing $|D|$ according to which a set of three equivalent basis function [*viz.* $\sigma_1$, $\sigma_2$, $\sigma_3$ or $\Psi_1$, $\Psi_2$, $\Psi_3$] can be transformed into a set of SALCs using projection operators of the I.Rs of the group. Such SALCs are linear combinations of $\Psi_i$'s whose coefficients are the elements of $|D|$ matrix. We use three equivalent $\sigma$-orbitals present either on the atom *A* or on the pendent atoms *B* and apply projection operators of *A′* and *E′* I.Rs of $D_{3h}$ point group. We get three SALCs, the coefficients of which give the matrix $|D|$, the inverse of which will give matrix $|C|$.

We calculate the elements of matrix | $D$ | in the following three steps:

**Step 1:** Form SALCs from the set of three equivalent orbitals on the pendent atoms ($B^1$, $B^2$, $B^3$). [Fig. 7.6($a$, $b$, $c$)]. (cf. Chapter 5). Or, we could also use orbitals on the central atom $A$. ($e.g.$, boron). In $BF_3$ these are hybrids of $sp^2$ type as discussed earlier.

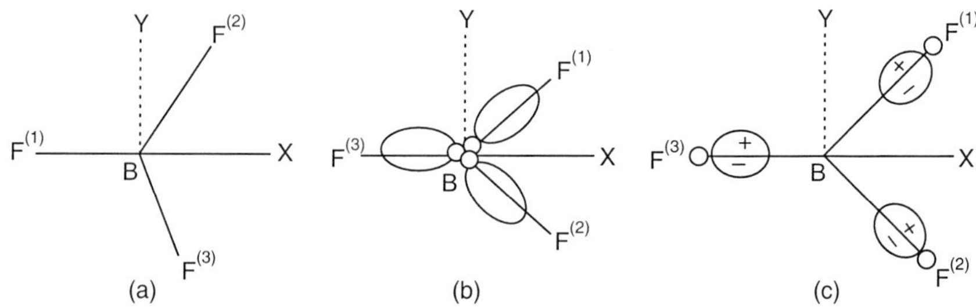


**Fig. 7.6** Trigonal $AB_3$ molecule.

Application of projection operators of $A_1'$ and $E'$ I.Rs of $D_{3h}$ will give the following equations. ($\Gamma_{red} = A_1' + E'$ for $D_{3h}$ point group $AB_3$ type molecule. The results obtained are

$$\left.\phi_1(A_1') = \frac{1}{\sqrt{3}}(\sigma_1 + \sigma_2 + \sigma_3)\right\} \quad \text{...(7.64)}$$

($\sigma_1$, $\sigma_2$ and $\sigma_3$ are three equivalent hybrid orbitals on atom $A$ or on pendent atoms $B$)

$$\text{Partners}\left\{\begin{matrix} \phi_2(E') = \dfrac{1}{\sqrt{6}}(2\sigma_1 - \sigma_2 - \sigma_3) \\ \phi_3(E') = \dfrac{1}{\sqrt{2}}(\sigma_2 - \sigma_3) \end{matrix}\right\} \quad \text{...(7.65)}$$

**Step 2:** The matrix | $D$ | in terms of coefficients of eqs. (7.63 and 7.64) are,

$$|D| = \begin{vmatrix} 1/\sqrt{3} & 1/\sqrt{3} & 1/\sqrt{3} \\ 2/\sqrt{6} & -1/\sqrt{6} & -1/\sqrt{6} \\ 0 & 1/\sqrt{2} & -1/\sqrt{2} \end{vmatrix} \quad \text{...(7.66)}$$

Inverse of | $D$ | $i.e.$, $|D|^{-1} = |C|$ which is transpose of matrix | $D$ |, is

$$|C| = \begin{vmatrix} 1/\sqrt{3} & 2/\sqrt{6} & 0 \\ 1/\sqrt{3} & -1/\sqrt{6} & 1/\sqrt{2} \\ 1/\sqrt{3} & -1/\sqrt{6} & -1/\sqrt{2} \end{vmatrix} \quad \text{...(7.67)}$$

**Step 3:** The $\psi_1$, $\psi_2$, $\psi_3$ according to eq. (7.60) is

$$\begin{vmatrix} 1/\sqrt{3} & 2/\sqrt{6} & 0 \\ 1/\sqrt{3} & -1/\sqrt{6} & 1/\sqrt{2} \\ 1/\sqrt{3} & -1/\sqrt{6} & -1/\sqrt{2} \end{vmatrix} \begin{vmatrix} s \\ p_x \\ p_y \end{vmatrix} = \begin{vmatrix} \Psi_1 \\ \Psi_2 \\ \Psi_3 \end{vmatrix}$$

or

$$\left.\begin{aligned} \Psi_1 &= 1/\sqrt{3}s + 2/\sqrt{6}p_x \\ \Psi_2 &= 1/\sqrt{3}s - 1/\sqrt{6}p_x + 1/\sqrt{2}p_y \\ \text{and} \quad \Psi_3 &= 1/\sqrt{3}s - 1/\sqrt{6}p_x - 1/\sqrt{2}p_y \end{aligned}\right\} \quad \ldots(7.68)$$

Thus, we have got three hybrid orbitals as linear combinations of A.Os on $A$, (*viz.* $s$, $p_x$ and $p_y$).

## Some Comments on Hybridisation

**[A]** The model based on hybridisation of orbitals to explain the structures and shapes of molecules, although, is quite successful, but there are some problems associated with it. In this model, the concept is to increase the extent of overlapping of the lobes of the hybrid orbital in one direction at a particular angle (Fig. 7.7). Therefore, when one uses such orbitals for bond formation, the overlap of the orbitals (involved on two atoms in bonding) is increased in one direction (bond direction) because of the increased surface of the hybrid in that direction. This is because that the (+ve) value of one orbital (*e.g.*, *p*-orbital) in one direction superimposes on the +ve value of the other combining orbital at the cost of cancellation of +ve value with the (–ve) value of the combining orbitals (Fig. 7.7). Therefore the MOs, formed by the overlap of these orbitals will be lower in energy or stronger in comparison to those formed by the overlap of pure ($s$, $p_x$, $p_y$, $p_z$ *etc.*) orbitals because of increased probability of finding the bonded pair of electrons in the overlap region.

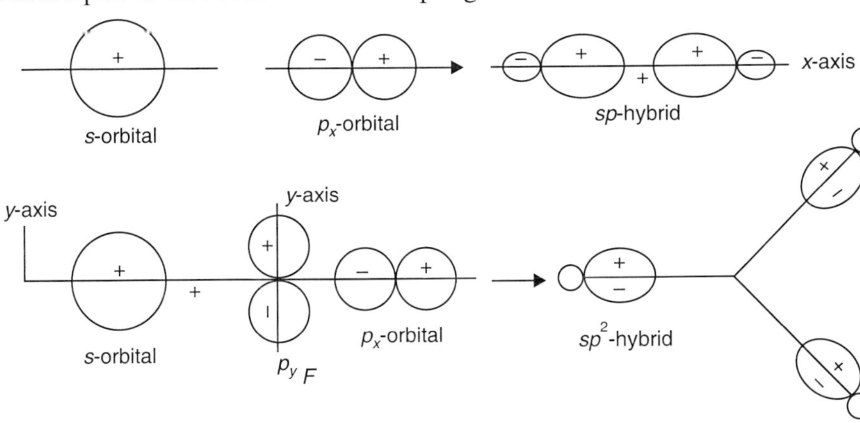


**Fig. 7.7** Formation of *sp*- and *sp*²- hybrids.

**[B]** It is obvious from the values of coefficients $C_{ij}$ in the eq. (7.66) that one can form an infinite number of hybrids, depending on the choice of the values of these coefficients *viz.* $C_{ij}'s$. One should, however, realise that the number of hybrids formed will be the same as the number of A.Os, one uses to construct them. For example, we form three hybrids of $sp^2$ types by combining one $s$ and two $p$-orbitals. Further in a hybrid $sp^n$, the superscript $n$, refers to the ratio of $p$-orbital-character and the $s$-character in terms of the coefficients $C_{ij}$.

**[C]** No hybridisation is possible to increase the sideway overlaps so that pure $p$- or $d$-orbitals of proper symmetry are usually used to form $\pi$-bonds. The remaining orbitals are hybridised to improve the $\sigma$ overlap and the directionality. The hybridization of orbitals of any (octet) atom (I, II, III period elements) is given by the equation (7.69) after the number of $\pi$-bonds used in multiple bonds that the atom forms.

$$\text{Hybridisation} = sp^{3-N_{p\pi}} = sp^{3-N_{\text{double}}-2N_{\text{triple}}} \qquad \ldots(7.69)$$

**[D]** Hybridisation affects some of properties of the atoms and molecules, (*e.g.*, the bond length, electronegativity, acid-character, *etc.*) The reason is that the $s$-orbital (used in hybridisation) has a definite probability to be present at the nucleus while $p$-orbitals do not have such a probability. Therefore, the electron present in the hybrid orbital containing $s$-orbital is relatively more attracted towards the nucleus than that present in pure $p$- or $d$-orbital. Thus, (C – H) single bond length decreases from ethane [($sp^3$)(1.1. A°)] to $C_2H_4$($sp^2$ (1.1A°)] and in $C_2H_2$[($sp$), (1.08A°)].

The electronegativity of the above hybrid orbitals in a molecule also changes with the degree of $s$-character. The order is as follows $p < sp^3 < sp^2 < sp$. Higher the $p$-character the lower is the electronegativity.

Another important aspect of hybridisation is that when the atoms in a molecule vibrate due to thermal energy, the hybridisation of the orbitals of the vibrating atoms must also change because of the changes in the bond lengths and bond angles as a result of vibration. We calculate bond force constants at the equilibrium position of the atoms in a molecule, so when the atoms vibrate, the hybridisation on the central atom (*e.g.*, of $A$ in $AX_n$ molecule) continuously change during the vibration time. It is therefore, not possible to find the exact molecular force constant. We should, therefore, expect that the degree of hybridisation must follow the change of bond length during vibration. It implies that when valence angle and bond length with the change during vibration, there has to be some forces which come into play with the change of bond angles and bond length. Therefore, we must have some time-dependent cross terms in their potential energy expressions used in molecular interaction calculations.

There are other effects of hybridisation. For example, the dipole moment should change because of hybridisations. One may refer other books like

Coulsons valence (Oxford Press) which has an excellent description on Hybridisation.

In conclusion, one must realise that hybridisation is not a real phenomenon. It is something like resonance phenomenon between two structures, where each structure is not real, but it is only is a mathematical model. One must, however, realise the following points in the hybridisation context beside the aforesaid points.

(*i*) It is the most effective way of preserving the concept of localised bond with perfect pairing of orbitals between two bonded atoms.

(*ii*) It is a restricted form of resonance because there are molecules in which just one structure with localised bond only is not sufficient to explain all the properties of the molecule. In such molecules, there has to be some interactions between various bonds which has to be explained in terms of delocalised M.Os. Thus in cobalt ammines, where the polarity is about 50%, there must be considerable interactions between electron distribution in various bonds. There has to be some resonance between pure covalent and pure ionic bond. In such cases, the concept of complete delocalised M.O. becomes conceptually the simplest model to use and the concept of localised electron pair bonds starts breaking with the interaction among bonding electrons and therefore hybridisation is no more a good model. It is a good model for molecules like $CH_4$, $H_2O$ *etc.* That is those molecules where we emphasise only localised bonds and all their properties are explainable on that model. In the next chapter we start with a simple model of M.O., where the delocalisation of the bonding electrons takes over the localised bonds (*i.e.*, Lewis theory).

# 8

# Symmetry Properties of Molecular Orbitals of $AX_n$ Type Molecules ($\sigma$-Type Bonds)

The statement that bonds are localised between atoms in a molecule involves two assumptions:

1. The bonded pair of electrons is concentrated between the bonded atoms.
2. There is no interaction between one bonded pair and the other pairs of electrons.

The molecular shapes are not guaranteed by localised bonded atoms. We decide them by our chemical intuition. It is only the properties of the molecules that make us believe the electron interaction between various pairs of electrons involved in bonding. The molecular orbital theory starts with a premise that all the interacting orbitals extend over the whole molecule. Although it is, however, possible that the electrons may be concentrated between two atoms in a particular case. But this will be a special case and does not support the localised bond theory. The concept of localisation is not the "build in concept" of MO theory. In this chapter, we will introduce the formation of MOs, and the symmetry properties of $AX_n$ type molecules with no $\pi$-bonding. The previous chapters provide the background concepts which will be referred to as and when, needed.

## 8.1 ENERGY CALCULATIONS OF THE MOLECULAR ORBITALS

In earlier chapters, we have indicated that the starting point for energy calculation of any system is the solution of time-independent Schrodinger equation, *viz.*,

$$H\Psi = E\Psi \qquad \ldots(8.1)$$

In equation (8.1), $H$ is the Hamiltonian operator, $E_n$ is constant energy (eigenvalue) for the $n^{\text{th}}$ quantum state represented by $\psi_n$ (eigenfunctions). Since for any system, it is not possible to find the exact eigenfunctions, we have to employ the approximate wave functions ($\phi_n$) to obtain eigenvalues $E_n$. These calculated eigenvalues are always higher, but never less than the experimental values (variation theorm). The closer are the arbitrarily chosen approximate functions to the exact functions, the nearer are the eigenvalues to the experimental ones.

We can also employ the Schrodinger equation to explain the bonding in molecules. But the problem is again the same, *viz.* the choice of eigenfunctions so that the eigenenergy can be calculated. For this two models have been proposed to choose the approximate molecular eigenfunctions. These are (1) valance bond model and (2) molecular orbital model (MO). In the following sections, we employ these methods to explain the bonding in di- and polyatomic molecules.

**(A) Diatomic Molecules:** Let us take the case of diatomic molecules ($X_2$ or $XY$) as first example. In *MO* approach, the chosen wave functions are approximated as a linear combination of the valence orbitals on two interacting atoms. For example, $\Psi_{MO}$ for the $H_2$ molecule may be taken as

$$\Psi_{MO}(H_2) = C_1\phi_{1s}(H^{(1)}) + C_2\phi_{1s}(H^{(2)})$$

or

$$\Psi_{MO}(H_2) = C_1\phi_{1s}(1) + C_2\phi_{1s}(2) \qquad \ldots(8.2)$$

In eq. (8.2) the $\phi_{1s}(1)$ and $\phi_{1s}$ are the 1s atomic orbital present on $H^{(1)}$ and $H^{(2)}$ atoms respectively. $C_1$ and $C_2$ are the weighing coefficients to be adjusted in order to get the best $\Psi_{MO}$ function for the $H_2$ molecule.

Prior to energy calculation (eigenvalue) for $H_2$ molecule we generally normalise the $\Psi_{MO}$ wave function in order to simplify the mathematical manipulations. That is,

$$\int \Psi^*_{MO} \cdot \Psi_{MO} \partial\tau = N^2\left[\int [C_1\phi_{1s}(1) + C_2\phi_{1s}(2)][C_1\phi_{1s}(1) + C_2\phi_{1s}(2)\right]\partial\tau = 1$$

or $$\int (C_1\phi_{1s}(1))^2\partial\tau + \int (C_2\phi_{1s}(2))^2\partial\tau + 2C_1C_2\int (\phi_{1s}(1)\phi_{1s}(2)\partial\tau) = 1/N^2$$

or $$[C_1^2 + C_2^2 + 2C_1C_2S_{12}] = 1/N^2 \qquad \ldots(8.3)$$

[Here $S_{12}$ = overlap integral = $\int \phi_{1s}(1)\phi_{1s}(2)\partial\tau$. Since $\phi_{1s}(1)$ and $\phi_{1s}(2)$ are normalised, we get the normalised $\Psi^N_{MO}$ as,

$$\text{Normalised } \Psi_{MO}^{N} = \left[\frac{1}{\sqrt{C_1^2 + C_2^2 + 2C_1C_2S_{12}}}\right][C_1\phi_{1s}(1) + C_2\phi_{1s}(2)] \qquad \ldots(8.4)$$

Using the normalised wave function $\Psi_{MO}^{N}$, the energy of $H_2$ molecule is calculated by substituting it in the following expression (cf. chapter seven).

$$E = \frac{\int \Psi_{MO}^{*N} H \Psi_{MO}^{N} \partial\tau}{\int \Psi_{MO}^{*N} \cdot \Psi_{MO}^{N} \partial\tau} \qquad \ldots(8.5)$$

The integration is over the whole space, since the domain of the electron in the $H_2$ molecule is the whole space which is also the domain of variables of the wave functions $\Psi_{MO}^{N}$. Substituting the expression of $\Psi_{MO}^{N}$ (eq. 8.4) into (eq. 8.5), we obtain,

$$E = \frac{1}{(C_1^2 + C_2^2 + 2C_1C_2S_{12})}\left[\frac{\int [C_1\phi_{1s}(1) + C_2\phi_{1s}(2)]H[C_1\phi_{1s}(1) + C_2\phi_{1s}]\partial\tau}{\frac{\int [C_1\phi_{1s}(1) + C_2\phi_{1s}(2)][C_1\phi_{1s}(1) + C_2\phi_{1s}(2)]\partial\tau}{(C_1^2 + C_2^2 + 2C_1C_2S_{12})}}\right] \qquad \ldots(8.6)$$

Since, we have taken $\Psi_{MO}^{N}$ as normalised function *i.e.*, $\left[\int \Psi_{MO}^{N}\Psi_{MO}^{N}\partial\tau = 1\right]$, the denominator of eq. 8.6 is equal to one, $E$ is therefore given as

$$E = \frac{\int [C_1\phi_{1s}(1) + C_2\phi_{1s}(2)]H[C_1\phi_{1s}(1) + C_2\phi_{1s}(2)]\partial\tau}{(C_1^2 + C_2^2 + 2C_1C_2S_{12})} \qquad \ldots(8.7)$$

$$= \frac{C_1^2H_{11} + C_2^2H_{22} + 2C_1C_2H_{12}}{(C_1^2 + C_2^2 + 2C_1C_2S_{12})} \qquad \ldots(8.8)$$

Here $H_{ij}$ is matrix element of Hamiltonian in matrix form (cf. chapter seven) and is given as $\int \Psi_i^* H \Psi_j \partial\tau$. Thus knowing $\phi_{1s}(1)$ and $\phi_{1s}(2)$, one can obtain the energy of $H_2$ molecule. We will carry out this manipulation a little later. However, we would like to generalise the concept at this point. Regarding eq. (8.8) the following points are worth noting.

1. The integrals $H_{ij}$ and $S_{ij}$ are definite integrals, so these are numbers and the energy of the molecule is a function of $C_1$, $C_2$ and $S_{12}$. $C_1$ and $C_2$ are indirectly function of $Z_1$ and $Z_2$ (nuclear charges of the atoms and internuclear distance $R_{A-B}$).

2. The equation (8.8) is applicable to any diatomic molecule ($X_2$) or ($XY$) or even to polyatomic molecule with polynuclear centres (to be discussed later in the same chapter).
3. Energy obtained by equation (8.8) will be a function of $C_1$, $C_2$, $S_{AB}$ and $R_{A-B}$ for the wave functions of the atoms which are bonded. Since we are interested in the minimum value of energy (ground state of the molecule), we should know these values of $C_1$, $C_2$ when $R_{A-B}$ is the equilibrium distance [($A - A$ or $A - B$) at $R_{eq}$].
4. To know $E_{min}$ at $R_{eq}$ we carry out the following mathematical manipulation. We can write eq. 8.8 as

$$E[(C_1)^2 + (C_2)^2 + 2C_1C_2S_{12}] = [C_1^2H_{11} + C_2^2H_{22} + 2C_1C_2H_{12}] \quad \ldots(8.9)$$

Differentiating equation (8.9) with respect to $C_1$ and $C_2$ separately, putting them equal to zero for the minimum energy value, we get,

$$\frac{\partial}{\partial C_1}\left[E(C_1^2 + C_2^2 + 2C_1C_2S_{12}) - (C_1^2H_{11} + C_2^2H_{22} + 2C_1C_2H_{12})\right] = 0$$

or $$\frac{\partial E}{\partial C_1}(C_1^2 + C_2^2 + 2C_1C_2S_{12}) + E(2C_1 + 2C_2S_{12}) - 2C_1H_{11} - 2C_2H_{12}) = 0$$

$$\ldots(8.10)$$

In the same way, one gets, (when we differentiate with respect to $C_2$)

$$\frac{\partial E}{\partial C_2}(C_1^2 + C_2^2 + 2C_1C_2S_{12}) + E(2C_2 + 2C_1S_{12}) - 2C_2H_{22} - 2C_1H_{12} = 0$$

$$\ldots(8.11)$$

We know that, $\frac{\partial E}{\partial C_1}$ and $\frac{\partial E}{\partial C_2}$ must be equal to zero at the minimum (or extreme) value of $E$ so the eqs. (8.10) and (8.11) become

$$C_1(H_{11} - E) + C_2(H_{12} - ES_{12}) = 0 \quad \ldots(8.12)$$

$$C_1(H_{12} - ES_{12}) + C_2(H_{22} - E) = 0 \quad \ldots(8.13)$$

Eqs. (8.12) and (8.13) are homogeneous linear equations in $C_1$ and $C_2$, so for non-trivial solutions of $C_1$ and $C_2$ (or $C_1 = C_2 \neq 0$), we must have the following determinant to be zero,

or, $$\begin{vmatrix} H_{11} - E & H_{12} - ES_{12} \\ H_{21} - ES_{21} & H_{22} - E \end{vmatrix} = 0 \quad \ldots(8.14)$$

One can easily show that $H_{12} = H_{21}$.

Eq. 8.14 can be written as

$$(H_{11} - E)(H_{22} - E) - (H_{12} - ES_{12})(H_{21} - ES_{21}) = 0 \quad \ldots(8.15)$$

Eq. (8.15) is a quadratic equation in $E$, which will give following two roots,

$$E = \frac{(H_{11} \pm H_{12})}{(1 \pm S_{12})} \qquad ...(8.16)$$

Eq (8.16) is true for any homonuclear diatomic molecule, so we can write

$$E \pm = \frac{(H_{ii} \pm H_{ij})}{(1 \pm S_{ij})} \qquad ...(8.17)$$

Here $S_{ij}$ or $S_{12} < 1$ for partial overlap, $S_{12} = 1$ for 100% overlap and $S_{12} = 0$, when the atoms are far apart (no interaction). Under partial overlap ($S_{12} < 1$), we have two values of energy one is lower than $H_{11}$ or $H_{ii}$) and another higher than $H_{ii}$ or $H_{11} = H_{22}$ (for $H_2$ or $X_2$ molecule). As a result when $E < H_{ii}$ (or $H_{11}$), it is energy for the bonding orbital and when $E > H_{ii}$ (or $H_{22}$) it is for antibonding orbital. Thus for two hydrogen atoms (or two $A$ atoms) we have the situation shown in Fig. 8.1.

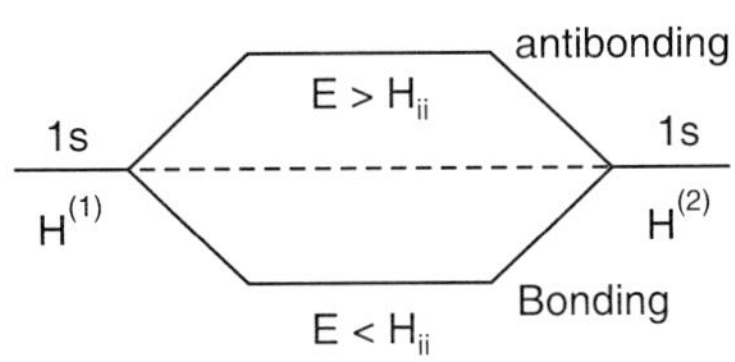


**Fig. 8.1** Interaction of two *A*-atoms.

The sign of $H_{11}$ or $H_{22}$ ($H_{ii}$) is always negative as it is the ionisation energy of atom ($A$), $H_{12}$ is also negative in the case when $S_{12}$ has +ve value (overlap is +ve).

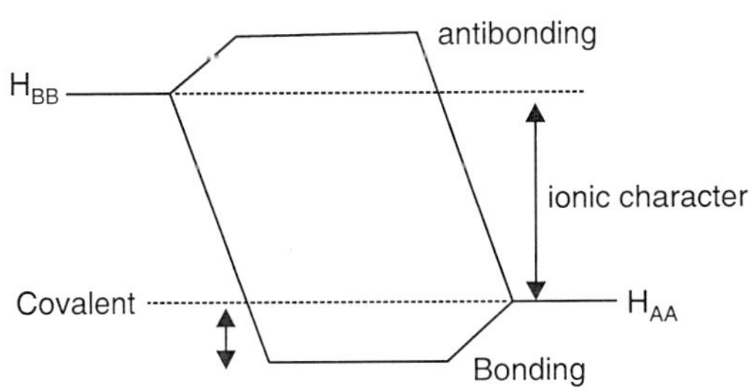


**Fig. 8.2** Interaction of two different atoms (*AB*).

Employing these two roots of energy values, one can calculate the coefficients, $C_1$ and $C_2$ by substituting the roots in eqs. (8.12) and (8.13). This will give for $H_2$ molecule,

$$C_1 = -C_2 \frac{[H_{12} - ES_{12}]}{[H_{11} - E]} \qquad ...(8.18)$$

and $$C_1 = -C_2 \frac{[H_{11} - E]}{[H_{12} - ES_{12}]} \qquad \ldots(8.19)$$

These eqs. (8.18) and (8.19) will give

$$C_1 = \pm C_2 \qquad \ldots(8.20)$$

From the relationship of eq. (8.20), one gets for $H_2$ molecule.

$$\left.\begin{aligned} \Psi_{MO} &= \frac{1}{\sqrt{2+2S}}[\phi_{1s}^{(1)} + \phi_{1s}^{(2)}] \ldots \text{ Bonding orbital} \\ \text{and} \quad \Psi_{MO} &= \frac{1}{\sqrt{2-2S}}[\phi_{1s}^{(1)} - \phi_{1s}^{(2)}] \ldots \text{ Antibonding orbital} \end{aligned}\right\} \qquad \ldots(8.21)$$

In the general case of *AB* type molecules, when the orbital on *A* overlaps with the proper orbital of another atom *B*, we will get the following expressions

$$\left.\begin{aligned} E_1 &= H_{AA} - \frac{H_{AB}^2}{H_{BB} - H_{AA}} \\ E_2 &= H_{BB} - \frac{H_{AB}^2}{H_{BB} - H_{AA}} \end{aligned}\right\} \qquad \ldots(8.22)$$

From the equation (8.22) one gets ($H_{AA} << H_{BB}$)

(*a*) one orbital slightly more stable than $\phi_A$ and

(*b*) another orbital slightly less stable than $\phi_B$ (it is shown in Fig. 8.2).

It is also important to realise that the value of overlap integral $\left(\int \phi_A^* \cdot \phi_B \partial \tau\right)$ is very much dependent on the symmetries of $\phi_A$ and $\phi_B$. If the symmetries of the two orbitals are different, the overlap is zero (see Chapter 6). If the symmetry is same, overlap is positive.

**(B) Polyatomic Molecules:** The above-described method is limited to two atom molecules only. It can however be applied to *n*-atomic molecules. We again assume a **trial** wave function which is represented by $\Psi_{MO}$. Let this functions be

$$\Psi_{MO} = C_1\phi_1 + C_2\phi_2 + C_3\phi_3 + \ldots\ldots C_n\phi_n \qquad \ldots(8.23)$$

where $\phi_1$, $\phi_2$ ... $\phi_n$ are the hydrogenic wave functions on the atoms 1, 2, ... *n* respectively and the coefficients $C_1$, $C_2$ ... $C_n$ are variable parameters which have to be optimised for obtaining the minimum energy of the molecule.

By substituting the trial wave function in the eq. (8.5), the energy of the molecule can be calculated. Similar to the case of diatomic molecules, we minimise the the energy expression by differentiating with respect to $C_1$, $C_2$, $C_3$

... $C_n$, consequently we will get the following $n$ number of linear homogeneous equations in $E$.

$$\left.\begin{array}{l} C_1(H_{11}-ES_{11})+C_2(H_{12}-ES_{12})+C_3(H_{13}-ES_{13})...C_n(H_{1n}-ES_{1n})=0 \\ C_1(H_{21}-ES_{21})+C_2(H_{22}-ES_{22})+C_3(H_{23}-ES_{23})...C_n(H_{2n}-ES_{2n})=0 \\ \vdots \\ C_1(H_{n1}-ES_{n1})+C_2(H_{n2}-ES_{n2})+C_3(H_{n3}-ES_{n3})...C_n(H_{nn}-ES_{nn})=0 \end{array}\right\} \quad ...(8.24)$$

For non-trivial solution (*i.e.*, $C_1$, $C_2$, ... $C_n \neq 0$), the following secular determinant has to be zero. That is,

$$\begin{vmatrix} (H_{11}-ES_{11}) & (H_{12}-ES_{12}) & ... & (H_{1n}-ES_{1n}) \\ (H_{21}-ES_{21}) & (H_{22}-ES_{22}) & ... & (H_{2n}-ES_{2n}) \\ \vdots & & & \\ (H_{n1}-ES_{n1}) & (H_{n2}-ES_{n2}) & ... & (H_{nn}-ES_{nn}) \end{vmatrix} = 0 \quad ...(8.25)$$

The solution of the secular determinant (8.25) gives $n$ number of roots of energy $E$. Each root of $E$, when substituted in eq. 8.24, gives one set of $C_1$, $C_2$, $C_3$, ... $C_n$ and thereby for $n$, number of roots of $E$ we will get $n$ sets of $C_1$, $C_2$, ... $C_n$).

These $n$-sets of $C_i$'s will thus generate $n$-different molecular orbitals $\Psi^1_{MO}$'s which will be $n$ different energy states of the molecule.

According to variational principle, if the energy values thus obtained do not correspond to the experimental values, then we have to further modify our basis functions $\phi_i$'s. This process of modification of $\phi_i$'s is continued until we get energy values, very close to the experimental ones.

Another big problem in the solution of the secular determinant (8.25), is that it is impossible to solve any polynomial of degree higher than four. This gives an unsolvable difficulty. The situation is however, not so bad, because symmetry of the molecule gives a way to solve the problem. In such situation, we make the symmetry adapted linear combinations (SALCs) of the starting $\phi_i$ functions. The method of carrying it out is given in chapter six using projection operators. Employing these SALCs of $\phi_i$ functions, one obtains the secular determinant in the diagonally block-form Eq. (8.26) whereby each block of the matrix of small dimension gives separate solution of the secular equation.

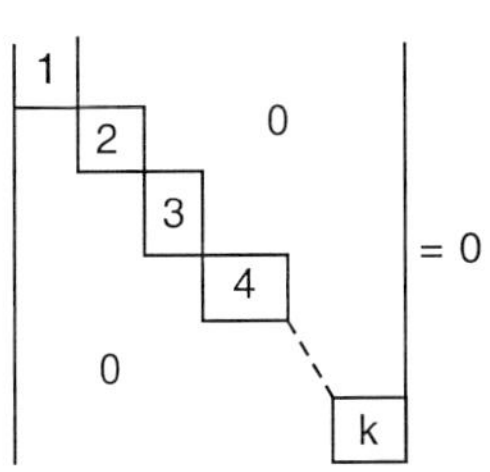


...(8.26)

In this equation, each of the matrix of ($l$, $m$, $n$, ... $k$) of smaller dimensions than $n$, can be separately solved for $k$-energy values and thereby one can calculate the $n$-number of $\Psi'^{i}_{MO}$ functions (orbitals).

After generalisation of the whole process of calculating $\Psi'^{i}_{MO}$ and their energy values $E_i$, we employ it to calculate the $\Psi_{MO}$ functions for the various $AX_n$ molecules.

## 8.2 SYMMETRY PROPERTIES OF MOLECULES AND THEIR APPLICATIONS IN EIGENVALUES AND EIGENFUNCTIONS CALCULATIONS

### (A) Diatomic Molecules (Homonuclear)

There are two types of diatomic molecules. These are

1. Homonuclear $A_2$-type with point group $D_{\infty h}$, like $H_2$, $B_2$, $C_2$, $N_2$, *etc.*
2. Heteronuclear $AB$-type with point group $D_{\infty v}$, like HBr, HCl, HF *etc.*

If we assume $z$-axis to be the molecular axis, then it is an infinite fold rotation axis and there will be $\infty$ number of vertical reflection planes containing molecular axis. In homonuclear molecules, the mid point of the $A$–$A$ bond is the centre of inversion and has a horizontal plane perpendicular to the molecular axis. These symmetry elements generate the $D_{\infty h}$ point group. Thus, the set of symmetry operations in $D_{\infty h}$ are ($E$, $C_\infty$, $\sigma_v$, $i$, ($iC_\phi$), ($i\sigma_v$) and those in $C_{\infty v}$ $A-B$ type are $\{E, C_\infty, \sigma_v\}$.

In general, all the symmetry operations of the molecular point group change only the number of equivalent atoms and do not alter the values of any physical quantities. Therefore, the molecular integral that are invariant under the molecular symmetry operations, are different from zero. For example, the wave functions $\Psi_{MO's}$, obtained in section [8.1], are $\left[\dfrac{1}{\sqrt{2\pm 2s}}\right][\phi_{1s}(1)\pm\phi_{1s}(2)]$. Since the term

$\left[\dfrac{1}{\sqrt{2\pm 2s}}\right]$ is a constant number, we can write it as $N[\phi_{1s}(1)\pm\phi_{1s}(2)]$ for $H_2$ molecule. According to the above said statement about the symmetry operation, if $R_i$ is the symmetry operation of the $H_2$ (or $A_2$) molecule, we can write the statement as $R[\Psi^{(i)}_{M.O}]=\lambda[\Psi'^{i}_{M.O}]$, where $\lambda = \pm 1$. Thus, we can write the characters of the symmetry operations in the tabular form as given in the right hand margin. On comparing the set of these characters with those given in $D_{\infty h}$ character table, one realises that $\Psi_{MO}^{(1)}$ behaves like $\sum_g^+$ (or $\sigma_g^+$) I.R, and $\Psi_{MO}^{(2)}$ like $\sum_u^+$ (or $\sigma^+$) I.R. These two functions (*viz.* $\Psi_{MO}^{(1)}$ and $\Psi_{MO}^{(2)}$) form the basis for the I.R $\sum_g^+$ and $\sum_u^+$ I.Rs of $D_{\infty h}$ point group.

Similarly, if we take the case of heteronuclear $A - B$ type molecules, and find out the symmetry properties of $\Psi_{MO}^{(1)}$ and $\Psi_{MO}^{(2)}$, and using the character table of $C_{\infty v}$, we will find that $\Psi_{MO}^{(1)}$ $(A - B)$ and $\Psi_{MO}^{(2)}$ $(A - B)$ functions belong to $\sum_g(\sigma_g)$ and $\sum_u(\sigma_u)$ types of I.Rs respectively.

Therefore, $\psi^1 MO$ and $\psi^2 MO$ orbitals are σ-type $\left(\sum \text{ type}\right)$ orbitals and are represented as $\sigma_{1s}$ and $\sigma_{1s}^*$ orbitals $\left[\sum_u^+(\sigma_u)\right]$ respectively.

| $D_{\infty h}$ | $E$ | $C_\phi$ | $\sigma_v$ | $i$ | $iC_\phi$ | $i\sigma_v$ |
|---|---|---|---|---|---|---|
| $\psi_{MO}^1$ | 1 | 1 | 1 | 1 | 1 | 1 |
| $\psi_{MO}^2$ | 1 | 1 | 1 | –1 | –1 | –1 |

Analogous to 1[st] orbitals used to form linear combinations in order to form $\Psi_{MO}^{(1s)}$, we now examine *p*-orbitals. The various types of orbital overlaps are given in Fig. 8.3. In principle, there are two basic criteria for two orbitals to have positive overlap. (1) The energies of the overlapping orbitals must be similar (if not exact). (2) The symmetries of the combining orbitals (*a* and *b*) must be the same with respect to the bond axis. These two criteria are obvious from the [Fig. 8.3]. The symmetries of $(\overset{(a)}{P_x}, \overset{(b)}{P_x}), (\overset{(a)}{P_y}, \overset{(b)}{P_y}), (\overset{(a)}{P_z}, \overset{(b)}{P_z})$, $(\overset{(a)}{S}, \overset{(b)}{P_x}), (\overset{(a)}{S}, \overset{(b)}{d_{z^2}}), (\overset{(a)}{d_{x^2-y^2}}, \overset{(b)}{d_{x^2-y^2}})$ are the same with respect to bond axis (*i.e.*, *z*-axis) and hence the overlaps between *a* and *b* is positive ($S_{ab} > 0$). The symmetry of $(\overset{(a)}{p_x}, \overset{(b)}{p_y}), (\overset{(a)}{s}, \overset{(b)}{z}), (\overset{(a)}{s}, \overset{(b)}{p_y}), (\overset{(a)}{s}, \overset{(b)}{p})(\overset{(a)}{s}, \overset{(b)}{d_{x^2-y^2}})$, *etc.*, are not the same with respect to the bond axis (*viz.* *z*-axis) and hence, the overlap of these orbitals are zero (*i.e.*, no bond formation).

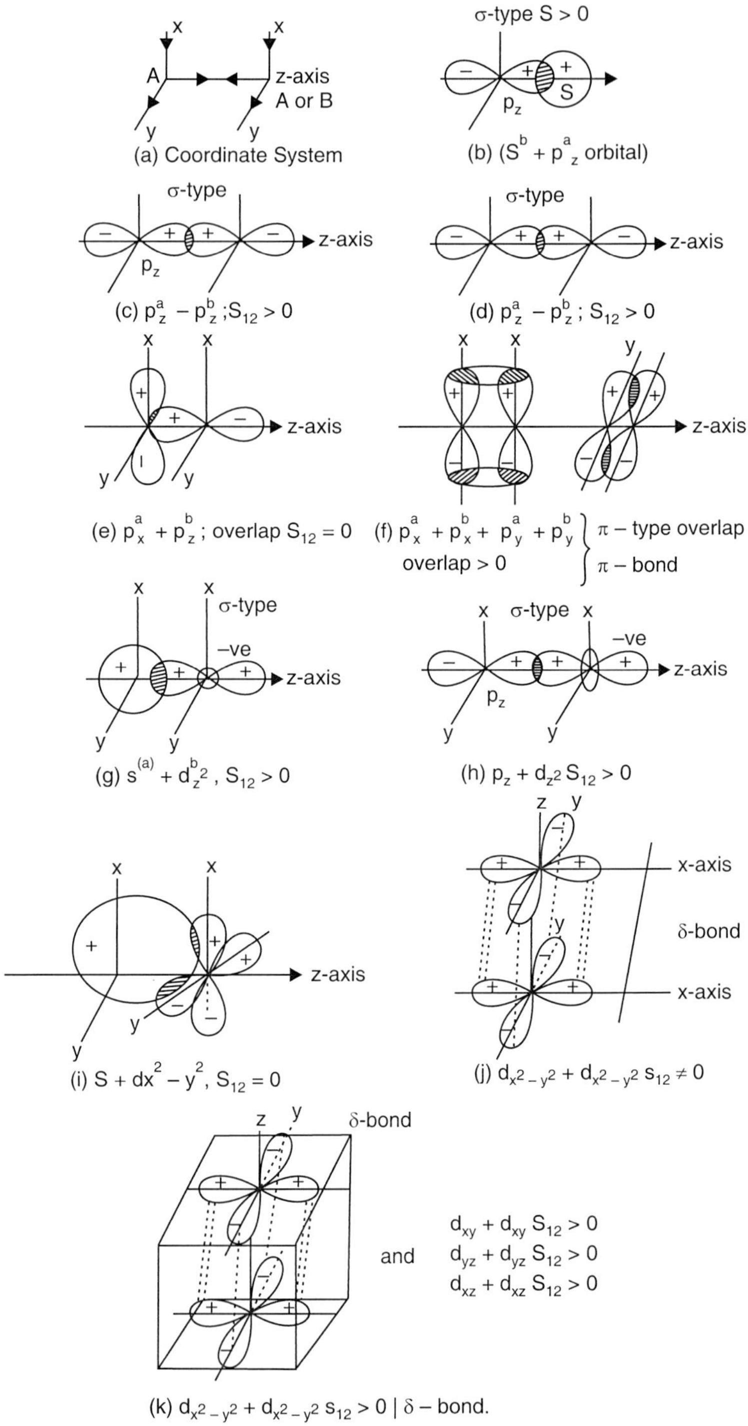


**Fig. 8.3** Mutual overlap of $S$, $p_x$, $p_y$, $p_z$, $d_{x^2-y^2}$, $d_{xy}$, $d_{yz}$, $d_{xz}$ and $d_{z^2}$ orbitals.

Another point to be realised is that the symmetries of the overlaps with respect to bond axis is not the same in every case. Consider the overlap between $(\overset{(a)}{s}, \overset{(b)}{p_z}), (\overset{(a)}{s}, \overset{(b)}{s}), (s, \overset{(b)}{d_{z^2}}), (\overset{(a)}{p_z}, \overset{(b)}{p_z})$. The symmetry of the overlap in these cases has a cylindrical symmetry, that is, the overlap is symmetric with respect to rotation around the bond axis (*i.e.*, *z*-axis) and there is no nodal plane. Therefore in the molecule $(A - A)$ or $(A - B)$, if the overlap is cylindrical symmetric, then, the bond will be σ-type. It means that the bond (or the overlap) will transform according to $\sum^{+}_{(g)}$ or $\sum_{u}$ type of I.Rs of the $D_{\infty v}$ (or $A_2$) point group of the molecule. Or, it will form σ-type of bonds. Another type of overlap, that is present in the $A_2$ or $AB$ molecules is that shown by $(\overset{(a)}{p_y}, \overset{(b)}{p_y}), (\overset{(a)}{p_x}, \overset{(b)}{p_x})$, *etc.*, overlaps, which does not have the cylindrical symmetry. In this case, the overlap is sideways and has a nodal plane containing the bond axis (*z*-axes) and lies perpendicular to two side overlaps (one overlap lies above the plane and one, below the plane). The transformation properties of such type of overlap is that of π-type or I.R of $D_{\infty h}$ and $C_{\infty v}$ point group of $A_2$ or $AB$ molecule. The bonds formed by such overlaps are termed as π-bonds. We have also shown the third type of overlaps between $(\overset{(a)}{d_{x^2-y^2}}, \overset{(b)}{d_{x^2-y^2}}), (\overset{(a)}{d_{xy}}, \overset{(a)}{d_{xy}}), (\overset{(a)}{d_{yz}}, \overset{(b)}{d_{yz}})$, $(\overset{(a)}{d_{xz}}, \overset{(b)}{d_{xz}})$, orbitals, such overlaps have two nodal planes perpendicular to the bond axis. These transform as Δ type of I.Rs of the $D_{\infty h}$ and $C_{\infty v}$ point groups of $A_2$ or $AB$ molecules. Such bonds are known as δ-bonds.

The third important point regarding (σ, π or δ) types of bonds is that the order of amount of overlap is σ > π > δ. Since the amount of overlap between two orbitals is a measure of bond strength, it implies that the bond strength of σ bond is more than that of π-bond, which inturn is more strong than δ-bonds and less strong than σ-bond.

The fourth important point regarding different overlaps is that whenever two orbitals overlap, we form two molecular orbitals. We have seen this when we are considering the overlap of two 1s orbitals in $H_2$ molecule we form a stable bond, which is termed as bonding orbital (σ, π or δ-type) and another *MO* of higher energy, which is termed as antibonding orbital ($\sigma^*$, $\pi^*$ or $\delta^*$). The amount of stabilisation of bonding orbital is nearly equal to the amount destabilisation of antibonding *MO* orbital with respect to the energy of the combining orbitals ($\Delta E$ (destabilisation $\geq \Delta$E (stabilisation). When we consider the order $\Delta E(\sigma) > \Delta E(\pi) > \Delta E(\delta)$ it also implies that $\Delta E(\sigma^*) > \Delta E(\pi^* > \Delta E(\delta^*)$ orbitals (Fig. 8.4).

We have described the formation of $\sigma$ and $\sigma^*$ bonds in the case of $H_2$ molecule. The same procedure along with the above description of various orbital-overlap can be extended to polyelectronic diatomic and polyatomic molecules. Let us take the case of $C_2$ molecule as an illustrative example. We have the following information about this system.

1. Electronic configuration of each carbon atom is $1s^2\ 2s^2\ 2p^2$. If we do not take $1s^2$ electrons on each carbon atom into consideration, we have eight electrons to be occupied into the MOs formed by $s$ and $p$ orbitals of carbon atoms.
2. Choose $z$-axis as $MO$ axis with origin as the mid point of $C - C$ bond
3. We have $2s$, $2p_x$, $2p_y$, $2p_z$ orbitals on each carbon atom. Let us represent them $s^1, s^2, p_x^2, p_x^1, p_y^2, p_y^1$ (1 and 2 are the carbon atoms. The coordinate system is shown in [Fig. 8.5].
4. Point group of $C_2$ is $D_{\infty h}$.
5. In [Fig. 8.4], we have defined $z$-axis in such a manner that $z^{(1)}$-axis on $C^{(1)}$ points toward $C^{(2)}$ and $z^{(2)}$-axis points towards $C^{(1)}$, so that positive lobe of $p_z^{(1)}$ points towards $C^{(2)}$ and the positive axis of $p_z^{(2)}$ points towards $C^{(1)}$. Thus when $C^1$ approaches towards $C^{(2)}$, there is a (+ve) overlap between $p_z^{(1)}$ and $p_z^{(2)}$ orbitals (Fig. 8.5).

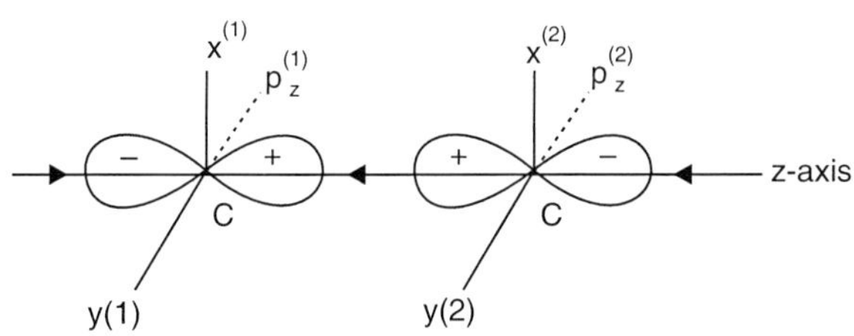


**Fig. 8.4** Coordinate System for $C_2$.

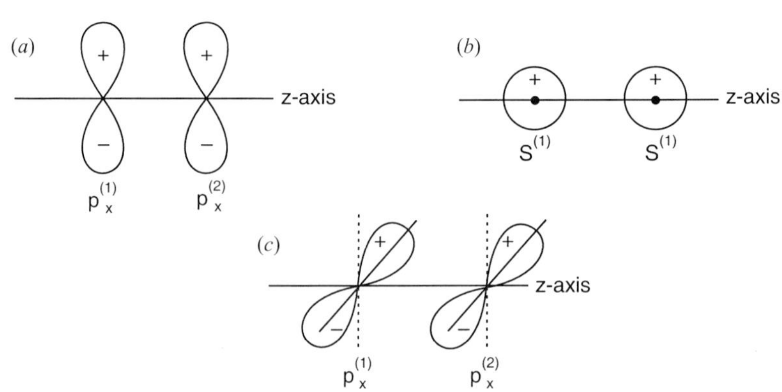


**Fig. 8.5** $p_x^{(1)}$, $p_x^{(2)}$ and other orbitals of $C_2$.

6. Now we take three sets of orbitals which form the basis for the representations of $D_{\infty h}$ point group. These sets are (1) $[p_z^{(1)} p_z^{(2)}]$ (2) $[p_x^{(1)}, p_x^{(2)}], [p_y^{(1)}, p_y^{(2)}]$ and (3) $[2s^1, 2s^2]$.

7. Using $[p_z^{(1)} p_z^{(2)}]$ we form $\Gamma_{red}$ for the point group $D_{\infty h}$ whose symmetry operations are $\{E, C_\phi, \sigma_v, i, iC_\phi, \delta\sigma_v\}$

$$E\begin{vmatrix} p_z^{(1)} \\ p_z^{(2)} \end{vmatrix} = \begin{vmatrix} 1 & 0 \\ 0 & 1 \end{vmatrix}\begin{vmatrix} p_z^{(1)} \\ p_z^{(2)} \end{vmatrix}, (\chi_E = 2); \quad C_\phi\begin{vmatrix} p_z^{(1)} \\ p_z^{(2)} \end{vmatrix} = \begin{vmatrix} 1 & 0 \\ 0 & 1 \end{vmatrix}, (\chi_{C\phi} = 2);$$

$$i\begin{vmatrix} p_z^{(1)} \\ p_z^{(2)} \end{vmatrix} = \begin{vmatrix} 0 & 1 \\ 1 & 0 \end{vmatrix}\begin{vmatrix} p_z^{1} \\ p_z^{(2)} \end{vmatrix}, (\chi_i = 0) \quad iC_\phi\begin{vmatrix} p_z^{(1)} \\ p_z^{(2)} \end{vmatrix} = \begin{vmatrix} 1 & 0 \\ 0 & 1 \end{vmatrix}, (\chi_{iC\phi} = 0)$$

$$iC_\phi\begin{vmatrix} p_z^{(1)} \\ p_z^{(2)} \end{vmatrix} = i\begin{vmatrix} 1 & 0 \\ 0 & 1 \end{vmatrix}\begin{vmatrix} p_z^{(1)} \\ p_z^{(2)} \end{vmatrix} = \begin{vmatrix} 0 & 1 \\ 1 & 0 \end{vmatrix}\begin{vmatrix} 1 & 0 \\ 0 & 1 \end{vmatrix}\begin{vmatrix} p_z^{(1)} \\ p_z^{(2)} \end{vmatrix}$$

$$= \begin{vmatrix} 0 & 1 \\ 1 & 0 \end{vmatrix}\begin{vmatrix} p_z^{(1)} \\ p_z^{(2)} \end{vmatrix}, (\chi_{iC\phi});$$

$$\sigma_v\begin{vmatrix} p_z^{(1)} \\ p_z^{(2)} \end{vmatrix} = \begin{vmatrix} 1 & 0 \\ 0 & 1 \end{vmatrix}\begin{vmatrix} p_z^{(1)} \\ p_z^{(2)} \end{vmatrix}, (\chi_{\sigma_v} = 2);$$

and $$i\sigma_v\begin{vmatrix} p_z^{(1)} \\ p_z^{(2)} \end{vmatrix} = \begin{vmatrix} 0 & 1 \\ 1 & 0 \end{vmatrix}\begin{vmatrix} p_z^{(1)} \\ p_z^{(2)} \end{vmatrix}, (\chi_{i\sigma_v} = 0).$$

Thus we get $\Gamma_{red} = [2, 2, 2, 0\ 0\ 0]$ ...(8.27)

On reduction $\Gamma_{red}$, we get the I.R component as $\sum_g^+$ and $\sum_u^+$ and the corresponding SALCs are $\frac{1}{\sqrt{2}}\left(p_z^1 + p_z^2\right)$ and $\frac{1}{\sqrt{2}}(p_{z_2}^{(1)} - p_{z_2}^{(2)})$. These give $\sigma_p$ and $\sigma_p^*$ as molecular orbitals which are respectively bonding and antibonding orbitals.

By the same manipulation, as above, if we take $s^1$ and $s^2$ as basis orbitals we will get two molecular orbitals, (*viz.* $\Psi_{MO}^{(1)}$ and $\Psi_{MO}^{(2)}$). $\Psi_{MO}^{(1)}$ is equal to $\frac{1}{\sqrt{2}}(s^1 + s^2)$ and $\Psi_{MO}^{(2)} = \frac{1}{\sqrt{2}}(s^1 - s^2)$ which will be termed as $\sigma_{2s}$ and $\sigma_{2s}^*$ (bonding $\sigma$ and antibonding $\sigma^*$ MOs respectively.

Now we start with $(p_x^1, p_x^2, p_y^1, p_y^2)$ as basis orbitals to form the reducible representation for the $D_{\infty h}$ point group.

8. The effects of symmetry operations of $D_{\infty h}$ on these base vectors are given below

$$E\begin{vmatrix} p_x^{(1)} \\ p_x^{(2)} \\ p_y^{(1)} \\ p_y^{(2)} \end{vmatrix} = \underbrace{\begin{vmatrix} 1 & 0 & 0 & 0 \\ 0 & 1 & 0 & 0 \\ 0 & 0 & 1 & 0 \\ 0 & 0 & 0 & 1 \end{vmatrix}}_{R_E} \begin{vmatrix} p_x^{(1)} \\ p_x^{(2)} \\ p_y^{(1)} \\ p_y^{(2)} \end{vmatrix} (\chi_E = 4),$$

$$C_\phi\begin{vmatrix} p_x^{(1)} \\ p_x^{(2)} \\ p_y^{(1)} \\ p_y^{(2)} \end{vmatrix} = \underbrace{\begin{vmatrix} \cos\phi & -\sin\phi & 0 & 0 \\ \sin\phi & \cos\phi & 0 & 0 \\ 0 & 0 & \cos\phi & -\sin\phi \\ 0 & 0 & \sin\phi & \cos\phi \end{vmatrix}}_{R_{C\phi}} \begin{vmatrix} p_x^{(1)} \\ p_x^{(2)} \\ p_y^{(1)} \\ p_y^{(2)} \end{vmatrix} (\chi_{C\phi} = 4\cos\phi),$$

$$\sigma_v\begin{vmatrix} p_x^{(1)} \\ p_x^{(2)} \\ p_y^{(1)} \\ p_y^{(2)} \end{vmatrix} = \underbrace{\begin{vmatrix} \cos 2\beta & \sin 2\beta & 0 & 0 \\ \sin 2\beta & -\cos 2\beta & 0 & 0 \\ 0 & 0 & \cos 2\beta & \sin 2\beta \\ 0 & 0 & \sin 2\beta & -\cos 2\beta \end{vmatrix}}_{R\sigma_v} \begin{vmatrix} p_x^{(1)} \\ p_x^{(2)} \\ p_y^{(1)} \\ p_y^{(2)} \end{vmatrix} X_{\sigma_v} = 0;$$

$$i\begin{vmatrix} p_x^{(1)} \\ p_x^{(2)} \\ p_y^{(1)} \\ p_y^{(2)} \end{vmatrix} = \begin{vmatrix} 0 & 0 & -1 & 0 \\ 0 & 0 & 0 & -1 \\ -1 & 0 & 0 & 1 \\ 0 & -1 & 0 & 0 \end{vmatrix} \begin{vmatrix} p_x^{(1)} \\ p_x^{(2)} \\ p_y^{(1)} \\ p_y^{(2)} \end{vmatrix} X_{i\sigma_v} = 0.$$

Similarly $\chi_{i\sigma\phi} = 0$; and $\chi_{i\sigma_V} = 0$

Thus $\Gamma_{\text{red}} = [4, 4\cos\alpha, 0\ 0\ 0\ 0\ ]$.

On reducing $\Gamma_{\text{red}}$ into its I.R components, one will get $\Gamma_{\text{red}} = \pi_{ux} + \pi_{uy}$

Thus, $p_x^1$ and $p_x^2$ $p_y^1$ and $p_y^2$ will give two pairs of functions *viz.* $\frac{1}{\sqrt{2}}[p_x^1 \pm p_x^2]$ and $\frac{1}{\sqrt{2}}[p_y^1 \pm p_y^2]$ which are represented as $p_{\pi_u x}, p_{\pi_u y}, p_{\pi_g x}$ and $p_{\pi_g y}$. We will represent these functions as $\pi_{p_x}, \pi_{p_y}, \pi^*_{p_x}, \pi^*_{p_y}$.

Thus, the final results that we have obtained for $C_2$ molecule by making L.C A.O. of $p_x^1, p_y^1, p_z^1, p_x^2, p_y^2$ and $p_z^2$, $2s^1$, $2s^2$ are,

$$\Gamma_{\text{AO}} = 2\sum_g^+ + 2\sum_u^+ + \pi_g + \pi_u \qquad \ldots(8.28)$$

On substituting these MOs into the energy equation, [*viz.* $E = \dfrac{\int \Psi^*_{MO} H \Psi_{MO} \partial\tau}{\int \Psi^*_{MO} \Psi_{MO} \delta\tau}$, we get the energy of the molecular orbitals. Out of the eight MOs there will be four bonding MOs (*viz.* $\sigma_{2s}, \sigma_{2p_z}, \pi_{2p_x}$ and $\pi_{2p_y}$) and four antibonding orbitals of higher energy. (*viz.* $\overset{*}{\sigma}_{2s}\overset{*}{\sigma}_{2p_z}, \overset{*}{\pi}_{2p_x}$ and $\overset{*}{\pi}_{2p_y}$. This is as far as the group theoretical analysis based on the symmetry properties of the molecule can take us. For the relative energy calculations of these molecular orbitals ($\Psi_{MO}$'s) we have to fall back to the actual calculations using the equation $\left[\int \Psi^* H \Psi \partial\tau / \int \Psi^* \Psi \partial\tau\right]$.

In order to calculate the energies of the MOs, we assume the order of the energy levels of the atomic orbital as 1*s* < 2*s* < 2*p*. Therefore, we assume the $\Psi_1^{(1)}\left(=\sum_g^+\right)$ and $\Psi_{MO}^{(2)}\left(=\sum_u^+\right)$ are lower in energy than $\Psi_{MO}^3\left(\sum_g^+\right)$ and $\Psi_{MO}^4\left(=\sum_u^+\right)$ which, in turn, is lower in energy than $\Psi_{MO}^5\left(=\sum_g^+\right)$ and $\Psi_{MO}^6$. MOs $\Psi_{MO}^1\ \Psi_{MO}^2, \Psi_{MO}^3$ and $\Psi_{MO}^4$ have to be normalised. The energy level diagram of the 2nd period elements are given below (Fig. 8.6).

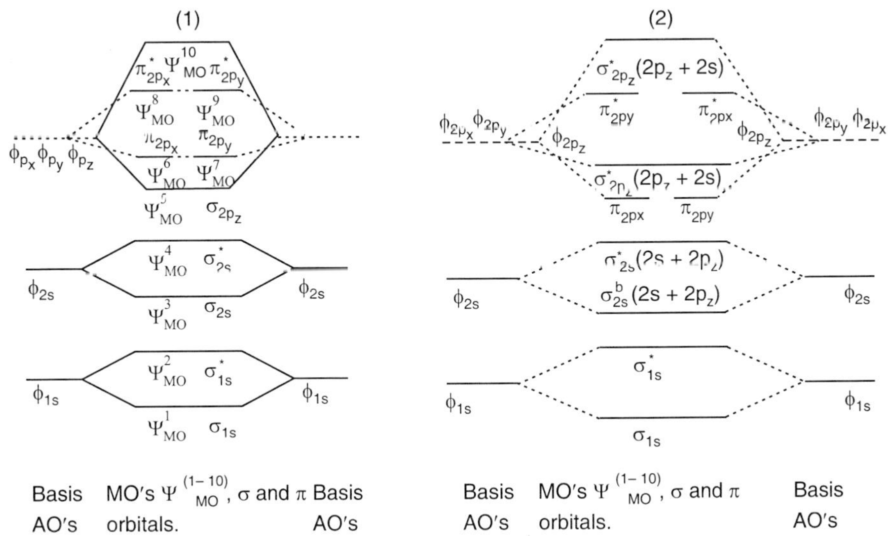


**Fig. 8.6** Energy level diagram of diatomic molecules (1) without configuration interaction and (2) with configuration interaction.

In Fig. (8.6), we have given two orders of energies of the MOs. In the first, we have assumed that the energy difference 2*s* and 2*p* orbitals is very large and therefore we take 1*s* and 2*s* orbitals as core orbitals. In such case, we can assume no interaction between 2*s* and 2*p* orbitals and we can safely take the *MO* energy given in [Fig. 8.6(1)]. This applies very well for the case $H_2$, $He_2$, $Li_2$, $Be_2$. But as we go to the next element *B* (boron), whose diatomic molecule is paramagnetic (2-unpaired electrons) the energy difference between 2*s* and 2*p* orbitals decreases. In that case, the $\sigma_{2s}$ and $\sigma_{2p}$ come close enough for interaction $\sigma_{2s}$ and $\sigma_{2p}$ MOs will interact because their symmetry is the same. These two identical symmetry orbitals ($\Psi_{MO}$) having similar energy, repel each other (configuration interaction) when they come close enough, then their mutual repulsive effect lets $\sigma_{2s}$ orbitals to become stable and $\sigma_{2p_z}$ orbitals become less stable. This makes $\pi_{2p_x}$ and $\pi_{2p_y}$ to fall below $\sigma_{2p_z}$ and thereby the order of $\sigma_{2p_z}$ and $(\pi_{2p_x} = 2\pi_{p_y})$ gets reversed as shown in [Fig. 8.6(2)] (Generally configuration interaction is proportional to the difference in energy of two interacting orbitals *e.g.*, $\sigma_{2s}$ and $2p_z$).

**Note:** We have labelled the MOs in Fig. 8.5 which appear more commonly in literature. However, there are two other systems of labelling the levels of diatomic molecules cited in the literature. We give below these systems for information.

$\sigma_{1S}$ $\overset{*}{\sigma}_{1S}$ $\sigma_{2S}$ $\overset{*}{\sigma}_{2S}$ $\sigma_{2p_z}$ $(\pi_{2p_x} = \pi_{2p_y})$ $(\overset{*}{\pi}_{2p_x} = \overset{*}{\pi}_{2p_y})$ $\overset{*}{\sigma}_{2p_z}$

– our space notation.

$1\sigma_g^+$ $1\sigma_u^+$ $2\sigma_g^+$ $2\sigma_u^+$ $3\sigma_g^+$ $\pi_u$ $\pi_u$ $\pi_g$ $\pi_g$ $3\sigma_u^+$

– notation according to I.R's of $D_{\infty h}$

– – $z\sigma$ $y\sigma$ $xw\pi$ $v\pi$ $v\pi$ $v\pi$ $4\sigma$ 1

– another arbitrary notation.

We can now easily assign the electrons of the molecule to different MOs taking Pauli's principle into consideration. We give below electronic configurations of a few molecules

$$Be_2(8) = \sigma_{1s}^2 < \sigma_{1s}^{*2} < \sigma_{2s}^2 < \sigma_{2s}^{*2} < \sigma_{2p_z}^0 \quad \text{(No unpaired electron)}$$

$$B_2(10) = \sigma_{1s}^2 < \sigma_{1s}^{*2} < \sigma_{2s}^2 < \sigma_{2s}^{*2} < \pi_{2p_x}^1 = \pi_{2p_y}^{(1)} \quad \text{(Two unpaired electrons)}$$

$$C_2(12) = \sigma_{1s}^2 < \sigma_{1s}^{*2} < \sigma_{2s}^2 < \sigma_{2s}^{*2} < \pi_{2p_x}^2 = \pi_{2p_y}^{(2)} \quad \text{(No unpaired electron)}$$

$$N_2(14) = \sigma_{1s}^2 < \sigma_{1s}^{*2} < \sigma_{2s}^2 < \sigma_{2s}^{*2} < \pi_{2p_x}^2 = \pi_{2p_y}^{(2)} < 2\sigma_{2p_z}^2$$

(No unpaired electrons)

$$O_2(16) = \sigma_{1s}^2 < \sigma_{1s}^{*2} < \sigma_{2s}^2 < \sigma_{2s}^{*2} < \pi_{2p_x}^2 = \pi_{2p_y}^2 < 2\sigma_{2p_z}^2 < \pi_{2p_x}^{*1} = \pi_{2p_y}^{*1}$$

(Two unpaired electrons)

$$F_2(18) = \sigma_{1s}^2 < \sigma_{s}^{*2} < \sigma_{2s}^2 < \sigma_{2s}^{*2} < \pi_{2p_x}^2 = \pi_{2p_y}^{(2)} < 2\sigma_{2p_z}^2 < \pi_{2p_x}^{*2} = \pi_{2p_y}^{*2}$$

(No unpaired electrons)

In the aforesaid paragraphs, we have given the interactions between $(s - s)$, $(p - s)$, $(p - p)$ orbitals. Let us now summarise what kind of interactions are possible in $(d - d)$ orbitals (diatomic molecules of transition metals). These have been shown in [Fig. 8.3]. These figures show three types of $(d - d)$ interactions (*viz.* σ, π and δ-type). Thus we have π-type of interactions among $(d_{xz} - d_{xz})$, $(d_{yz} - d_{yz})$, which give the following MOs.

$$\Psi^{11} MO(xz) = \pi_u^{(b)} \equiv \frac{1}{\sqrt{2}}[(d_{xz}^1) + (d_{xz}^2)] \quad \ldots(8.29)$$

$$\Psi^{12} MO(xz) = \pi_g^{(a)} = \frac{1}{\sqrt{2}}[(d_{xz}^1) - (d_{xz}^2)] \quad \ldots(8.30)$$

$$\Psi^{13} MO(yz) = \pi_u^{(b)} = \frac{1}{\sqrt{2}}[(d_{yz}^1) + (d_{yz}^2)] \quad \ldots(8.31)$$

$$\Psi^{14} MO(yz) = \pi_g^{(a)} = \frac{1}{\sqrt{2}}[(d_{yz}^1) - (d_{yz}^2)] \quad \ldots(8.32)$$

These π-type of overlaps of *xz* and *yz* orbitals on atoms *a* and *b* are shown in Fig. 8.7.

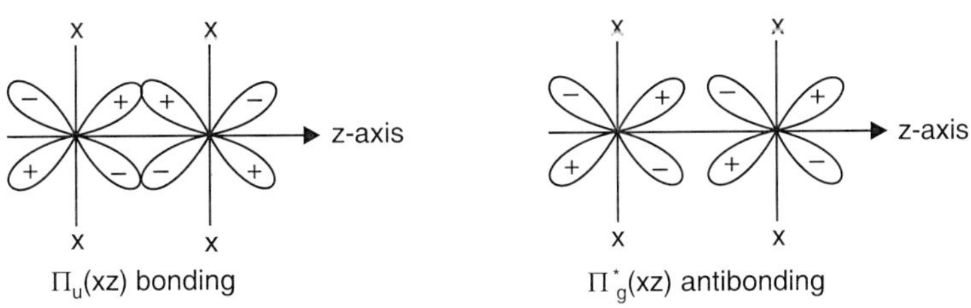


[Similarly $\Pi_u^{b^*}$ (yz) and $\Pi_g^{a}$ (yz) are formed in y-z plane.]

**Fig. 8.7** π-bonding in orbitals.

We now consider δ-bond formation by the overlap of *d*-orbitals by using $(d_{xy}, d_{x^2-y^2})$ orbitals. We will have,

$$\Psi^{15}MO(xy) = \delta_g^b = \frac{1}{\sqrt{2}}[(d_{xy}^1) + (d_{xy}^2)] \quad \text{...(8.33)}$$

$$\Psi^{16}MO(xy) = \delta_u^a = \frac{1}{\sqrt{2}}[(d_{xy}^1) - (d_{xy}^2)] \quad \text{...(8.34)}$$

$$\Psi^{17}MO((x^2 - y^2)) = \delta_g^b = \frac{1}{\sqrt{2}}[(d^1_{(x^2-y^2)}) + (d^2_{(x^2-y^2)})] \quad \text{...(8.35)}$$

$$\Psi^{18}MO((x^2 - y^2)) = \delta_u^a = \frac{1}{\sqrt{2}}[(d^1_{(x^2-y^2)} - (d^2_{(x^2-y^2)})] \quad \text{...(8.36)}$$

The δ-type of overlap is shown in [Fig. 8.4].

Similar molecular orbitals can be formed $\sigma(d_{z^2})$ types of bond. Here we will be having both bonding and antibonding types of orbitals.

Using these MOs and the usual energy equation one can calculate the energies of σ, π and δ-types of interactions among *d*-orbitals. We have five *d*-orbitals on each atom, with a total of ten *d*-orbitals on the two atoms of diatomic molecules. The order of these ten MOs energy is found to be

$$\sum\nolimits_g^{+(b)} < \pi_u^b(xz = yz) < \delta_g^b(xy = d_{x^2-y2})$$

$$< \delta_u^*(xy = d_{x^2-y2}) < \pi_g^*(xz = yz) < \sum\nolimits_u^{*+} \quad \text{...(8.37)}$$

This order is shown in Fig. 8.8.

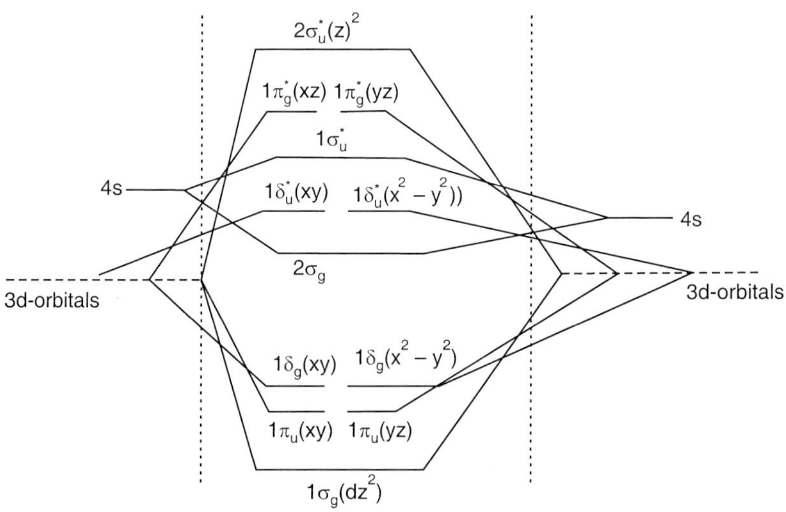


**Fig. 8.8** *d*-orbital interactions to form (σ, π and δ-types of bonds in *A–A* molecules).

**(B) Heteronuclear Diatomic Molecules (AB type):** The principles of combining orbitals on atom $A$ with those on atom $B$ for positive overlap are the same. These are

1. Similar energy of the combining orbitals.
2. Identical symmetry of the combining orbitals.
3. Large overlap of the combining orbitals. It implies similar charge clouds on the combining atoms.

Let us take the formation of a bond in $H-F$ molecule. We have $\phi_{1s}$ orbital on hydrogen atom. Fluorine, being more electronegative and polyelectron atom, the $1s$ and $2s$ orbitals are more stable than $1s$ orbital of hydrogen atom. It implies that $\Delta E$ between $1s$ and $2s$ orbital of $H$ and $1s$ or $2s$ orbitals of fluorine is too large for a good overlap (point one of the above conditions). It is only $2p$ orbital energy of fluorine that comes close to $1s$ orbital energy of hydrogen for a good overlap. Now we have to use one of the three $p$ orbitals of fluorine for bond formation. If we assume $z$ axis as the molecular axis then symmetry-wise it is $2p_z$ orbital which is suitable for interaction [Fig. 8.9].

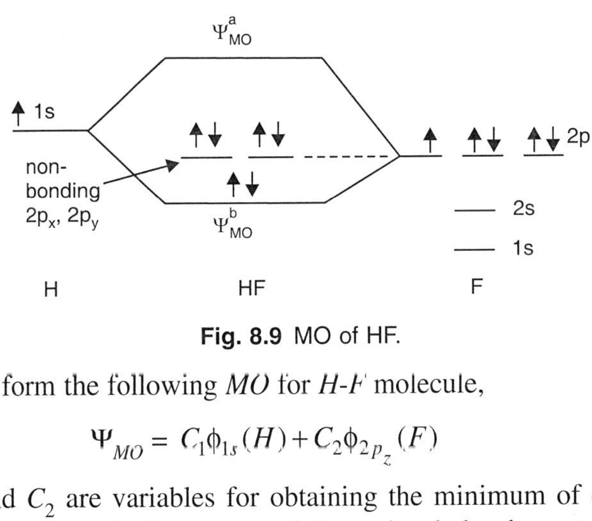


**Fig. 8.9** MO of HF.

We thus form the following $MO$ for $H$-$F$ molecule,

$$\Psi_{MO} = C_1\phi_{1s}(H) + C_2\phi_{2p_z}(F) \quad \ldots(8.38)$$

where $C_1$ and $C_2$ are variables for obtaining the minimum of energy of the molecule. One calculates the energy of the molecule by the energy expression

$\int \Psi^*_{MO} H \Psi_{MO} \partial\tau \Big/ \int \Psi^*_{MO} \Psi_{MO} \partial\tau$ with respect to the ground state energy of the combining orbitals. It will give two sets of $(C_1, C_2)$. One set will give bonding and another, antibonding orbital (sec. 8.2 ). We will thus get,

and
$$\left.\begin{aligned} \Psi^b_{MO} &= C_1'\phi_{1s}(H) + C_2'\phi_{2p_z}(F) \\ \Psi^a_{MO} &= C_1''\phi_{1s}(H) + C_2''\phi_{2p_z}(F) \end{aligned}\right\} \quad \ldots(8.39)$$

The molecular orbital diagram for *HF* is shown in Fig. 8.9. Since hydrogen atom does not have *p*-orbitals of proper symmetry that can interact with $2p_x$- and $2p_y$-orbitals of fluorine, there will be no π-bond formation in HF The *p*-electrons in fluorine ($2p_x$ and $2p_y$ orbitals thus remain as non-bonding. Similarly on the ground of energy alone the electrons in the 1*s*- and 2*s*-orbitals on fluorine are also non-bonding. Its electronic configuration is $\left[1s^2, 2s^2, 2\sigma_{HF}^2[2p_x^2 \equiv 2p_y^2]\right]$.

Let us take the case of those (*A* – *B*) type molecules, where the atoms *A* and *B* belong to the same period (*e.g.*, BN, CO, CN *etc.*), or those molecules in which atom *B* belongs to higher periods (*e.g.*, S, O, F, Cl, *etc.*). Such atoms have both *s* and *p*-type orbital having valence electrons.

Let us assume that in *A* – *B* molecule the electronegativity of *B* is higher than that of *A*. We, have therefore,

1. Electronegativity of *B* > *A*.
2. There is a difference in electronegativities of *A* and *B*; but the difference is not too large, In that case, the electrons in the more stable orbitals on *B* will not interact with the higher energy orbital of *A*. For example, in $BF_3$, the π-orbitals on $B(2p_x, 2p_y)$ will not interact with π-orbitals on fluorine $2p_x, 2p_y$)
3. In such cases, the MOs of *A* – *B* molecule will be *similar* to those of the homonuclear molecules of $A_2$ type. But the electron density in the *A* – *B* bond will be concentrated *more* on the atom *B* (more electronegative atom). The *MO* diagram of such molecules is shown in Fig. 8.11. (See for notation for labelling).

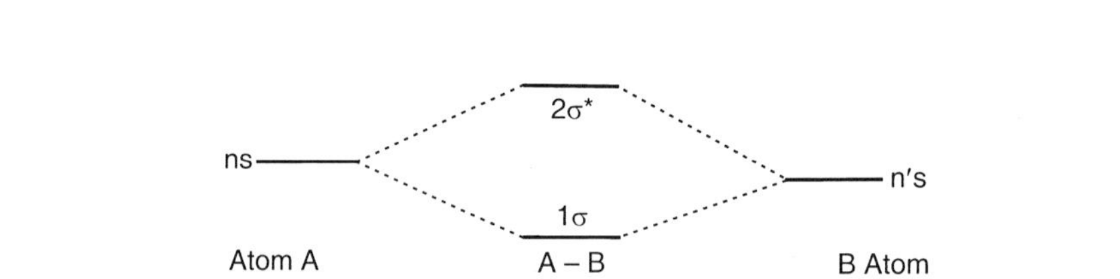


**Fig. 8.10** *MO* of *A* – *B* molecules (Electronegativity of *B* > *A*).

(*a*) Thus electronic configuration of CO (number of electrons = 14) is

$$[(C_{1s}, O_{1s})^4 1\sigma_{2s}^2 2\sigma_{2s}^{*2} (1\pi_{2p_x} = 1\pi_{2p_y})^4 3\sigma_{2p_z}^2].$$

(*b*) Electron configuration of CN (number of electrons = 13) is

$$[C_{1s}, N_{1s}]^4[1\sigma_{2s}^2 2\sigma_{2s}^{*2}(\pi_{2p_x} = \pi_{2p_y})^4 3\sigma_{2p_z}^1)]$$

In BO molecule, there is a configuration interaction. Therefore, $2\sigma^*$ orbital gets more destabilised and goes above $1\pi$-orbital. Hence its electronic configuration [number of electrons = 13)] is

$$\{[B_{1s}, O_{1s}]^4[1\sigma_{2s}^2 2\sigma_{2p_z}^2(\pi_{2p_x} = \pi_{2p_y})^4 3\sigma_{2p_z}^1)]\}$$ (See Fig. 8.11).

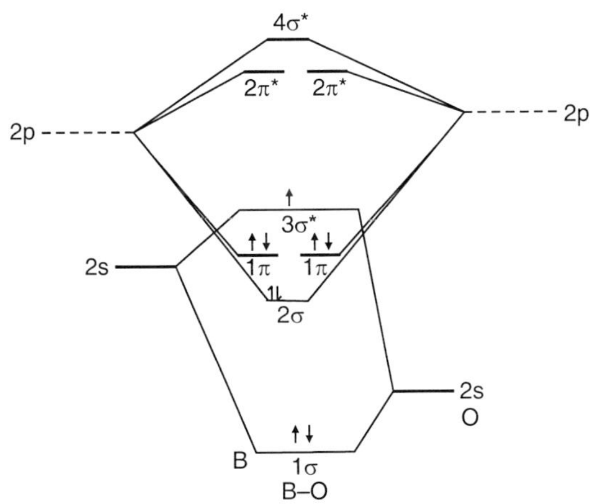


**Fig. 8.11** *MO* of (*B* – *O*).

## 8.3 POLYATOMIC MOLECULES ($AX_n$-TYPE)

The calculation of MOs of polyatomic molecules (number of atoms > 2) is mathematically prohibitive without the application of symmetry properties of the molecules. We give below a method to find the energy levels of $AX_n$ type molecules. There are two types of molecules (1) $AX_n$ (2) $A_xB_y$ type. In $AX_n$ molecules, one atom *A* lies at the centre of mass of the molecule which is also taken as the origin of the coordinate system. In $A_xB_y$ type molecules, no atom is present at the centre of mass which is usually taken as the origin of the coordinate system (we will discuss bonding in these molecules in the next chapter).

We proceed to determine the MOs of the polyatomic molecules of $AX_n$ type in the following steps.

| **Step 1** | **Determine the structure and point group of the molecule.** |
|---|---|
| (*a*) $BF_3$ | Triangular with *B* at the centre and F at the vertices of the triangle (Point group $D_{3h}$), (partial π-bonding ) |
| (*b*) $PF_5$ | Triangular bi-pyramid with *P* at the centre with no π-bonding (Point group $D_{3h}$) |
| (*c*) $PtCl_4$ | Square planar with Pt at the centre of square and Cl atoms at the corners of square (both σ and π-bonding) point group $D_{4h}$ |
| (*d*) $CCl_4$ | Tetrahedral with *C* at the centre of tetrahedron (no π-bonding) (Point group $T_d$) |
| (*e*) $MX_6$ | Octahedral with *M* at the centre and *X* at the corners of an octahedron. (both σ and π–bonding) Point group $O_h$. |

| **Step 2** | **If we do not use symmetry properties to simplify our mathematical computation, then the number of our chosen basis functions are very large. Thus, we have in,** |
|---|---|
| (*a*) $BF_3$ | 16 basis function (2*s*, and 2*p* orbitals on *B* and *F* (*i.e.*, 4 × 4) |
| (*b*) $PF_5$ | 24 basis functions (3*s* and 3*p* orbitals on *P* and 2*s* and 2*p* orbitals on fluorine atoms (total number of basis functions = 24) |
| (*c*) $PtCl_4$ | 21 basis functions 9 on Pt (4*d* + 5*s* + 5*p*) + 4 × 3 *p*-orbitals on Cl atoms) |
| (*d*) $CCl_4$ | 8 basis functions (2*s* + 2*p* on carbon + 4*p* orbital on Cl) |
| (*e*) $MX_6$ | 27 basis functions 9(*d* + *s* + *p*) orbitals on Pt + 18 orbitals on Cl (3 × 6)). |

Thus, if we use these large number of basis functions to form MOs by LCAO method, we have to solve $[n \times n]$ secular determinant which is mathematically impossible if $n > 4$. As an example, for the $PF_5$ molecule if we use 24 basis functions, we will make the following linear combination.

$$\Psi_{MO} = C_1\phi_1 + C_2\phi_2 + C_3\phi_3 \ldots C_{24}\phi_{24} \qquad \ldots(8.40)$$

Using this wave function Eq. (8.40) in our energy equation, *viz.*

$$E = \frac{\int \Psi^*_{MO} H \Psi_{MO} \partial\tau}{\int \Psi^*_{MO} \Psi_{MO} \partial\tau},$$ differentiating *E* with respect to $C_1, C_2 \ldots C_{24}$ followed by putting each one of *E* equations to zero we will obtain 24 linear equations in 24 unknowns *viz.* ($C_1, C_2, C_3 \ldots C_{24}$). The non trivial solution of these 24 equations is possibly only if we solve (24 × 24) secular equation (determinant) to get 24 roots (values) of energies. This will give a polynomial of 24th degree, which is impossible to solve. Hence, we can not get 24 MOs (*i.e.*, 24 sets of $C_1$, $C_2$, ..., $C_{24}$). Only procedure to simplify it is to bring the secular determinant in the

diagonal-block form. As a result, each block then is of lower dimensions and can be separately solved. This conversion of a bigger order determinant into blocked-diagonal form having determinant of lower dimensions is carried out by the use of symmetry properties of the molecule (*viz.* $PF_5$). This has been discussed and exemplified at a number of places in previous chapters. We will do it again here.

**Step 3:** Our task is to classify the MOs according to the I.Rs of the point group of the molecule and to produce an energy level diagram. We obtain it by the following method.

**Step 4:** We start with a set of $n$ basis functions which we have assumed and operate upon them by the symmetry operators of the point group. This will give us a reducible representation of the group ($\Gamma_{red}$) with respect to the chosen base vectors of the $n$-dimensional function space. Although characters for each operation of the representation can be calculated by the matrices of ($\Gamma_{red}$), but it can also be determined by a short-cut method defined in chapter six. Accordingly, we perform the operations of the point group of the molecule on each $\phi_i$ function. If $\phi_i$ is transformed into itself ($\phi_i$) then the character is +1, if it turns into $-\phi_i$ then the character is –1 and it is zero if it transforms into another orbital. We repeat this process of operation on all the $\phi_i$ functions. As a result of this operation, some of the $\phi_i s$ will give $\phi_i s$, some will give $-\phi_i s$ and the remaining will change to other $\phi_j s$. Each unchanged $\phi_i$ will give +1, those, changed to $-\phi_i s$ will give (–1) and the rest will give zero contribution to the value of the character for that operation. We, therefore take the algebraic sum of the +1 and –1. The sum total of these +1 and –1 will be the character of that operation. In this manner, we will get the characters for all the symmetry operations of the point group. The set of the characters for all the operations will form ($\Gamma_{red}$) of the group.

**Step 5:** If we do the operations on all the $\phi_i$ functions in step 4, one must keep in mind that a symmetry operation $R$ interchanges some of the ligand $X$ *atoms* either among themselves or it keeps them invariant. Similarly by the symmetry operation $R$, the orbitals of central atom $A$ of $AX_n$ molecule interchanges only among the orbitals of $A$. The orbitals of $A$ are never mixed with those of $X$ and vice-versa. We can, therefore, immediately divide our $\phi_i$ functions (functions space of AOs) into function subspaces of the orbitals of $A$ and the function subspace of AOs of $X$. Therefore, we can write

$$\Gamma_{red} \text{ (of } \phi_i \text{ functions)} = \Gamma_{red} \text{ (}A - \text{orbitals)} + \Gamma_{red} \text{ (}X \text{ orbitals)} \quad \ldots(8.41)$$

**Step 6:** We thus reduce the subspaces belonging to atom $A$ and that of atoms $X$ separately.

**Step 7:** We now choose a coordinate system ($x$-, $y$-, and $z$-axis) on the atom $A$ and on the atoms $X$. These are shown in Fig. 8.12 for our examples.

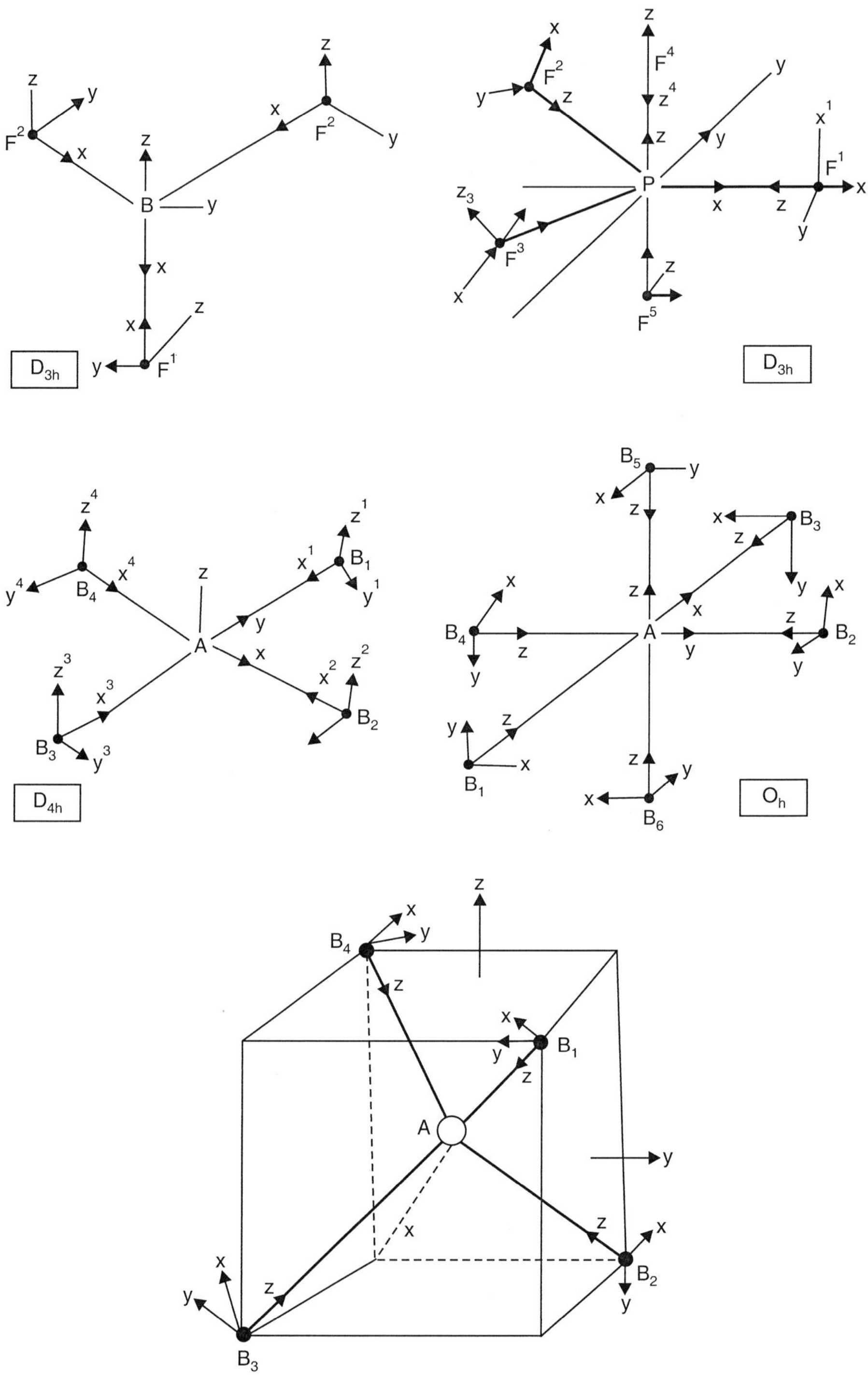


**Fig. 8.12** Coordinate systems of $BF_3$, $PF_5$, $AB_4$ and $AB_6$ systems.

**Step 8:** The proper valence shell orbitals on the central atom $A$ are then assumed as basis and a reducible representation $\Gamma^A_{\text{red}}$ of the point group of the molecule $AX_n$ is formed with respect to the chosen basis. $\Gamma^A_{\text{red}}$ is then reduced into its I.R components of the group. This process is however not necessary and by consulting the character table of the point group, one can find how the chosen valence shell orbitals of atom $A$ transform. Thus in our examples, we show the transformation properties of the chosen valence shell orbitals of $A$.

1. $BF_3$–($D_{3h}$ point group), chosen orbitals of $B$:2$s$, 2$p_z$, 2$p_x$, 2$p_y$. Their transformation properties are (From character table of $D_{3h}$)

$$\begin{aligned} A_1' &: 2s_B && \sigma\text{-type bond} \\ A_2'' &: 2p_z^B && \pi\text{-type bond} \\ \text{and} \quad E' &: 2p_x^B, 2p_y^B && \sigma\text{-type bond} \end{aligned} \qquad \ldots(8.42)$$

2. $PF_5$ ($D_{3h}$ point group), chosen orbitals of $P$:(3$s$, 3$p_z$, 3$p_x$, 3$p_y$, 3$d_{z^2}$, 3$d_{x^2-y^2}$, 3$d_{xy}$, 3$d_{xy}$, 3$d_{xz}$). Their transformation properties are (From character table of $D_{3h}$)

$$\begin{aligned} &A_1' : (3s, 3d_{z^2})\ \sigma\text{-type} \\ &A_2'' : 3p_z \rightarrow \pi\text{-type} \\ &E' : (3p_x, 3p_y); (3d_{xy}, d_{x^2-y^2}) \rightarrow \sigma\text{-type} \\ &E' : (3d_{xz}, 3d_{yz}) \rightarrow \pi\text{-type} \end{aligned} \qquad \ldots(8.43)$$

3. $XeF_4$ and[$PtCl_4$], ($D_{4h}$ Point group); chosen orbitals of Xe: [6$s$, 2$p_z$, 6$p_x$, 6$p_y$, 5$d_z$, 5$d_{xy}$, 5$d_{xz}$ 5$d_{yz}$ and 5$d_{x^2-y^2}$]. Their transformation properties are (From the character Table $D_{4h}$)

$$\left.\begin{aligned} &A_{1g} : d_{z^2}, s \quad \sigma - \text{type} \\ &B_{1g} : d_{x^2-y^2}, \sigma - \text{type} \\ &B_{2g} : d_{xy} - \pi - \text{type} \\ &E_g : d_{xz}, d_{yz} - \pi - \text{type} \\ &A_{2u} : p_z - \pi - \text{type} \\ &E_u : (p_x, p_y) - \sigma - \text{type} \end{aligned}\right\} \qquad \ldots(8.44)$$

4. $AB_6$ {*e.g.*, $[FeCl_6]^{3-}$, $[PtCl_6]^{2-}$} ($O_h$), *Chosen* orbitals of $A$ are [$s$, $d_{z^2}$, $d_{x^2-y^2}$, $p_x p_y p_z$, $d_{xy} d_{yz} d_{xz}$]. Their transformation properties are (From the character table of $O_h$ point group)

$$\left.\begin{array}{ll} A_{1g} : s & \to \sigma-\text{type} \\ E_g : d_{z^2}, d_{x^2-y^2} & \to \sigma-\text{type} \\ T_{1u} : p_x, p_y, p_z & \to \sigma-\text{type} \\ T_{2g} : d_{xy}, d_{yz}, d_{xz} & \to \pi-\text{type}) \end{array}\right\} \quad \ldots(8.45)$$

5. $AB_4$ [*e.g.*, $(CoCl_4)^{2-}$] [$(T_d)$, chosen orbitals of $A$ are ($s, p_x, p_y, p_z, d_{xy}, d_{yz}, d_{xz}, d_{x^2-y^2}, d_{z^2}$)]. Their transformation properties are (From the character table of $T_d$ point group)

$$\left.\begin{array}{ll} A_1 : s & \to \sigma-\text{type} \\ T_2 : (p_x, p_y, p_z)(d_{xy}, d_{yz}, d_{xz}) & \to \sigma-\text{type} \\ E : (d_{z^2}, d_{x^2-y^2}) & \to \pi-\text{type} \end{array}\right\} \quad \ldots(8.46)$$

The procedure of forming the $\Gamma_{red}$ from these orbitals of $A$ is given in the section on hybridisation in chapter seven.

**Step 9:** In the step 8, we have indicated the atomic orbitals on the atom $A$, which are used in the formation of $\sigma$ bonds ($A - B$) in the $AB_n$ molecules. (This has been done on the basis of discussion on hybridization in chapter seven). Besides these $\sigma$ type of bonds there is also a possibility of $\pi$-type of interaction between atom $A$ and atom $B$. The orbitals used for $\pi$ interactions on the atom $A$, will be derived in the following paragraphs.

We have indicated in chapter seven, that if a $\pi$-bond is formed between two atoms by the overlap of two orbitals (one on each atom), then the orbitals must be of $\pi$-types with respect to the internuclear axis. It implies that there will be a nodal plane containing the internuclear axis.

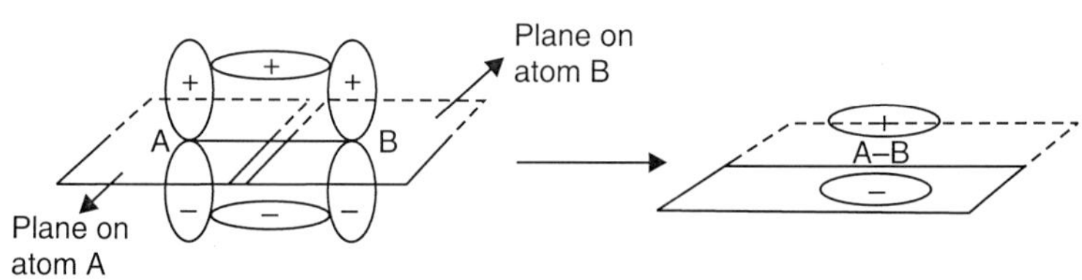


**Fig. 8.13** Formation of a $\pi$-bond.

There will be two nodal planes, one on each $\pi$-orbital of $A$ and $B$. These two nodal planes must coincide to form $\pi$-bond (Fig. 8.13). Thus, there will be only two orthogonal $\pi$-bonds between the two atoms $A$ and $B$ (Fig. 8.14). The nodal planes corresponding to these $\pi$-bonds will be perpendicular to each other (Fig. 8.14).

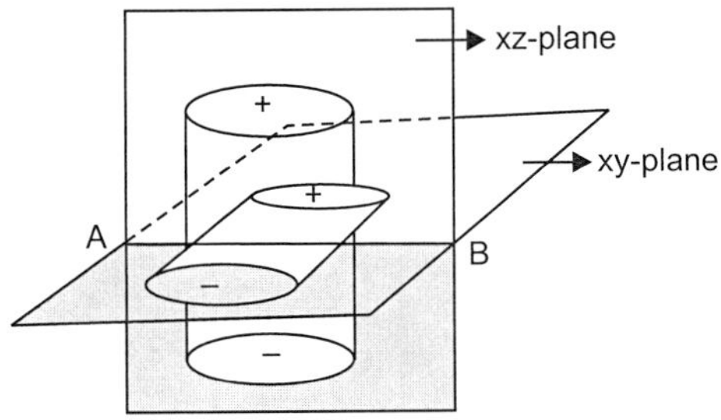


**Fig. 8.14** Two π-bonds of $A - B$ in two $\perp^{ar}$ planes.

According to the foregoing discussion, there can be two π-bonds on each $A - B$ σ-bond in two perpendicular planes. It implies that in $AB_n$ molecule, there should be $n - 6$ bonds and there will be $2n$ number of π-bonds associated with $n$ number of σ–bonds of $AB_n$. These $2n$ number of π-bonds require $2n$-orbitals (may be hybrid type) on atom $A$ and two π-forming orbitals on each $B$-atoms. These $2n$-hybrid orbitals on $A$, must match symmetrically with each of the π-orbitals on $B$, as shown in Fig. 8.14. These $2n$-hybrids on $A$ plus $2n$-orbitals on $B$, will both form bases for the same representations of the group of $AB_n$.

We can represent each orbital on atom $A$ and $B$ by an arrow. Each arrow head points towards the (+ve) lobe and arrow tail, towards the negative lobe of the orbital. These arrows are shown on each $A - B$ bond in $AB_n$ molecule in Fig. 8.12, where we have shown the coordinate systems of different $AB_n$ type molecules. Now, we take different molecules which we are discussing in this chapter, *viz.* $BF_3$, $AB_4$, *etc.*, and discuss the formation of π-bonds in them.

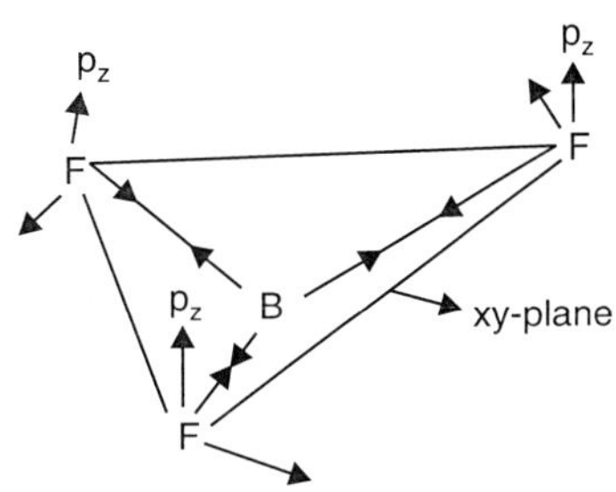


**Fig. 8.15** Orbitals of F in $BF_3$.

(*a*) $BF_3$-the point group is $D_{3h}$. We take three orbitals on each fluorine atom as $p_x$ $p_y$, $p_z$ and $s$ orbital. The six orbitals ($p_y$ and $p_z$ orbitals of the F-atom are shown as arrows on each fluorine atom. One ($p_x + s$) hybrid orbital is used as σ-bonding and the second, for non-bonding). Thus, we are left with two orbitals ($p_y$ and $p_z$) of fluorine for π-bond.

The set of six vector (or 6 π-orbitals of three fluorine atoms) is now used as basis for a representation [ $\Gamma^{\pi}_{red}$ of the point group $D_{3h}$. (In order to form $(\Gamma^{\pi}_{red})$ we use the same simple rule as stated earlier. Thus, any vector if shifted by the symmetry operation of the point group contributes zero to the character and if remained invariant contribute +1 and if left invariant and reverses in direction, contributes (–1). In this way one can easily form $(\Gamma^{\pi}_{red})$.

Another rule is that if we apply the above-said rule on the π-orbitals, one can easily *observe* that in no case is any vector lying perpendicular to the plane (*i.e.*, ($\perp^{ar}$) even interchanged with those in the plane. This implies that each set [*viz.* π(⊥) and π(||)] will give rise to a representation independent of the other.

Thus, we get $(\Gamma^{\pi}_{red})(\perp)$, $(\Gamma^{\pi}_{red})(||)$ and $(\Gamma^{\pi}_{red})$ as given in the following table:

$$\begin{aligned}&D_{3h} - [E, 2C_3, 3C_2, \sigma_h, 3S_3, 3\sigma_v]\\&\Gamma^{\pi}_{red}\,[6, 0, -2, 0, 0, 0],\ \Gamma^{\pi}_{red}(||)\,[3, 0, -1, 3, 0, -1]\\&\Gamma^{\pi}_{red}[3, 0, -1, -3, 0, 1]\end{aligned} \quad \ldots(8.47)$$

**Table 8.1**

| $D_{3h}$ | $E$ | $2C_3$ | $3C_2$ | $\sigma_h$ | $2S_3$ | $3\sigma_v$ | *I.Rs* |
|---|---|---|---|---|---|---|---|
| $\Gamma^{\pi}_{red}$ | 6 | 0 | −2 | 0 | 0 | 0 | $\Gamma^{\pi}_{red} = \Gamma^{\pi}_{red}(\perp) + \Gamma^{\pi}_{red}(\|\|)$ |
| $\Gamma^{\pi}_{red}(\|\|)$ | 3 | 0 | −1 | 3 | 0 | −1 | $\Gamma^{\pi}_{red}(\perp) = A_2'' + E''$ |
| $\Gamma_{red}(\perp)$ | 3 | 0 | −1 | −3 | 0 | 1 | $\Gamma^{\pi}_{red}(\|\|) = A_2' + E'$ |

The $(\Gamma^{\pi}_{red})$ and $[(\Gamma^{\pi}_{red}(||)]$ are reduced by the usual formula discussed earlier in chapter three. The I.Rs present in the three $\Gamma^{\pi}_{red}$ are given in the above table.

In this way, we have identified the AOs on atom *B*. If there is π-bond formation, then we must have (1) an orbital on the *A* showing the symmetry of either *A*′ or $A_2''$ and (2) a degenerate pair of AOs transforming either as *E*′ or *E*″ of point group $D_{3h}$.

Referring to (eq. 8.41) one finds that out of *s*-, *p*- or *d*-orbitals which meet this requirement, for the formation of *π*(⊥) bonds are,

$$\left.\begin{aligned}A_2'' &= p_z\\E'' &: (d_{xz}, d_{yz})\end{aligned}\right\} \quad \ldots(8.48)$$

Thus, either $A_2''$ or *E*″ or a hybrid of these can form π(⊥) bond with the π-orbitals of fluorine atoms. It will give π(⊥) bond.

Similarly for the π(||) bonds, we find from eq. (8.41) that AO of the required type are,

$$\left.\begin{aligned}A_2' &: \text{None}\\E' &: (p_x, p_y) \text{ and } (d_{x^2-y^2}, d_{xy})\end{aligned}\right\} \quad \ldots(8.49)$$

Eq. (8.48) shows, that there is no orbital of $A_2'$ type of I.R on the boron atom and hence, one can not have any $\pi(||)$ bond between $B$ and $F$ atoms. However, it does not mean that there can not be any $\pi(||)$ type bond between $B$ and $F$ atoms. It only means that there can be only two $\pi(||)$ bonds, shared equally between three fluorine atoms.

In response to the above discussion using equations (8.41–8.48), we have shown the orbitals on the atom $A$ in the $AB_3$ molecules capable of forming $\pi$-bonds with the orbitals on $B$ atoms.

(*b*) We take the second example ($AB_6$ type), illustrating the procedure to assign $\sigma$ and $\pi$-types orbitals on the central atom $A$ in $AB_n$ type molecules.

Similar to $BF_3$, we attach two vectors on each pendent atom ($B$) as shown in the Fig. 8.16. This shows the possibility of formation of twelve $\pi$ bonds. Unlike $BF_3$, we can not subdivide these vectors into two types *i.e.*, $\pi(\perp)$ and $\pi(||)$ because each vector in the octahedral case, is exchanged with each of the other eleven orbitals by one or another symmetry operation of the $O_h$ group. It implies mixing of all twelve orbitals under the symmetry operation of $O_h$. This means that all the twelve vectors (or orbitals) fall in the same set. Using these orbitals as basis, we form $\Gamma_{red}$ for the $O_h$ group. We get the character for the operations by the same method as we have got in the case of $BF_3$ by the use of simple rule (no shift by the operation). $\Gamma_{red}$ is as follows.

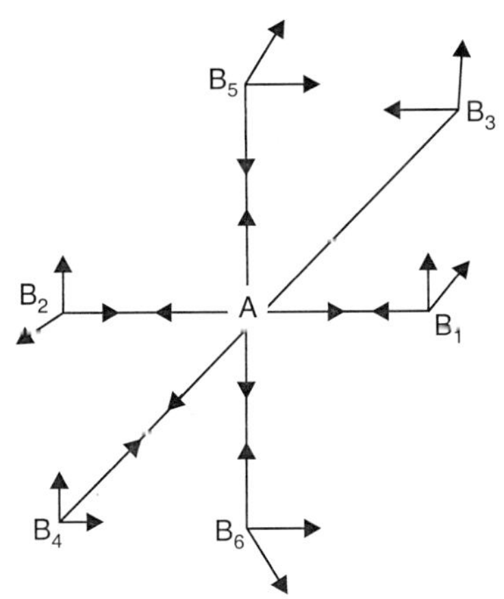


**Fig. 8.16** Twelve $\pi$-type orbitals on $B$ atom.

**Table 8.2:** $\Gamma_{red}$ and IRs contained in $\Gamma_{red}$ of $AB_6$ octahedral molecule

| $O_h$ | $E$ | $8C_3$ | $6C_2$ | $6C_4$ | $2C_2(=C_4^2)$ | $i$ | $6S_4$ | $3\sigma_h$ | $3\sigma_d$ | $8S_6$ | *I.Rs components* |
|---|---|---|---|---|---|---|---|---|---|---|---|
| $\Gamma_{red}$ | 12 | 0 | 0 | 0 | –4 | 0 | 0 | 0 | 0 | 0 | $\Gamma_{red}^{\pi} = T_{1g} + T_{2g} + T_{1u} + T_{2u}$ |

$\Gamma^{\pi}_{red}$ is then reduced by the usual formula, and the type and the number of I.Rs contained in $\Gamma_{red}$ is then calculated. These are given in Table 8.2.

Again we have a condition to be fulfilled. If π-bonds have to be formed between *A* and *B* then we must have orbitals on the atom *A* which transform according the I.Rs as given in Table 8.2.We have seen in equations (8.41–8.45), that the orbitals on atom *A* (*viz.* $p_x, p_y, p_z, d_{xy}, d_{yz}, d_{xz}, d_{x^2-y^2}, d_{z^2}$, ) transforms in $O_h$ point group as

$$\left.\begin{array}{ll} T_{1g}: & \text{None} \\ T_{2g}: & d_{xy}, d_{yz}, d_{xz} \\ T_{1u}: & (p_x, p_y, p_z) \\ T_{2u}: & \text{None} \end{array}\right\} \qquad \ldots(8.50)$$

Thus, we can draw the following conclusions from the above information.

1. One can not form all the twelve π-bonds, because on the central atom *A*, we do not have orbitals of $T_{1g}$ and $T_{2u}$ symmetry.
2. $p_x$- $p_y$- $p_z$-orbitals belong to $T_{1u}$ which can form π-bonds. These orbitals are also involved in σ-bond formation ($d^2sp^3$). So, if these orbitals are involved in σ-bond formation, we can not use them for π-bond formation. (This is a reasonable assumption). Thus, only three orbitals ($d_{xy}$, $d_{yz}$, $d_{xz}$) orbitals are available for π-bond formation. Hence, there are only three π-bonds shared between six atoms.

(*c*) We take a square planar $AB_4$ molecule as third example. The π-orbitals on the pendent *B* atoms of $AB_4$ molecule (Point group $D_{4h}$) are shown as vector in (Fig. 8.17.) Analogous to planar $AB_3$, we can divide the π-orbitals on atoms *B* into two sets. These are (1) A set of four π-orbitals which are perpendicular to molecular plane of $AB_4$ and (2) second set of four π-orbitals lying in the molecular plane. $\Gamma_{red}$ of these types of molecules can be subdivided into two $\Gamma_{red}$'s. One will be $[\Gamma^{\pi}_{red}(\perp)]$ and other $[\Gamma^{\pi}_{red}(||)]$. The representation of these two sets are then formed by employing the usual simple rule (no shift of vectors by the symmetry operation] by using the two set of vectors $[\Gamma^{\pi}_{red}(\perp)$ and parallel $[\Gamma^{\pi}_{red}(||)]$. The following results are obtained.

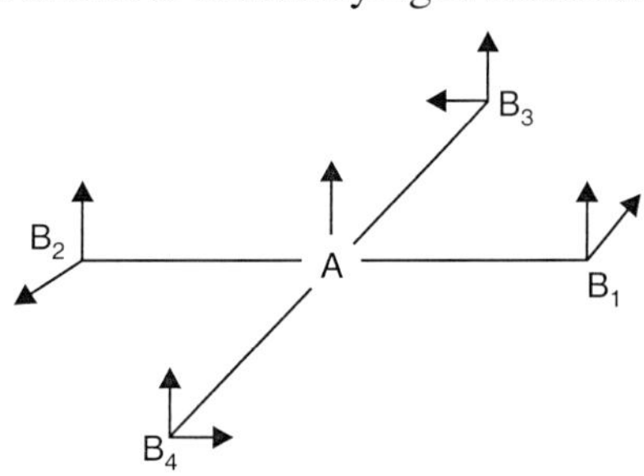

**Fig. 8.17** π-orbitals of $AB_4$.

**Table 8.3:** $\Gamma^{\pi}_{red}$, $\Gamma^{\pi}_{red}(\perp)$, $\Gamma^{\pi}_{red}(||)$ of $AB_4$ square planar molecule

| $D_{4h}$ | $E$ | $2C_4$ | $C_2$ | $2C_2'$ | $2C_2''$ | $i$ | $2S_4$ | $\sigma_h$ | $2\sigma_V$ | $2\sigma_d$ | *I.Rs components* |
|---|---|---|---|---|---|---|---|---|---|---|---|
| $\Gamma^{\pi}_{red}(\perp)$ | 4 | 0 | 0 | –2 | 0 | 0 | 0 | –4 | 2 | 0 | $A_{2u} + B_{2u} + E_g$ |
| $\Gamma^{\pi}_{red}(\|\|)$ | 4 | 0 | 0 | –2 | 0 | 0 | 0 | 4 | –2 | 0 | $A_{2u} + B_{2g} + E_u$ |

Referring to eqs. (8.41–8.45) and also the character Table of $D_{4h}$, we get the following information about the orbitals on the atom $A$

$$\begin{array}{l|l|l} A_{2u} : p_z & B_{2u} : \text{none} & E_g : (d_{xy}, d_{yz}) \\ A_{2g} : \text{none} & B_{2g} : d_{xy} & E_u : (p_x, p_y) \end{array} \quad \ldots(8.51)$$

From the above information, one can tell that (1) all the eight π-orbitals on atoms $B$ can not form π-bonds since no orbital of $A_{2g}$ and $A_{2u}$ symmetry are present on the atom $A$ and (2) only three π(⊥) and three π(||) bonds share among the four atoms of $B$, further $s$, $p_x$, $p_y$ and $d_{x^2-y^2}$ orbitals are used in σ-type bonds, only one π-bond π(⊥) [$d_{xy}$] will be shared among four atoms in molecular plane and three π(⊥) orbitals ($p_z$, $d_{xz}$, $d_{yz}$) will be shared among four atoms of $B$ in the perpendicular plane.

(*d*) Let us take the case of tetrahedral molecule $AB_4$ (Fig. 8.18). Since the eight vectors (orbitals) interchange among themselves by the symmetry operations of $T_d$ point group, all the *eight* vectors will be used to form π-bonds (not σ-type) or orbitals on the pendent atoms ($B$) used as basis to form the reducible representation ($\Gamma^{\pi}_{red}$). The representation is formed by employing the simple rule (No shift of vectors by symmetry operations for obtaining $\chi_{(R)}$). By this method, we get the following representation which is given below along with the components (irreducible) contained in $\Gamma^{\pi}_{red}$.

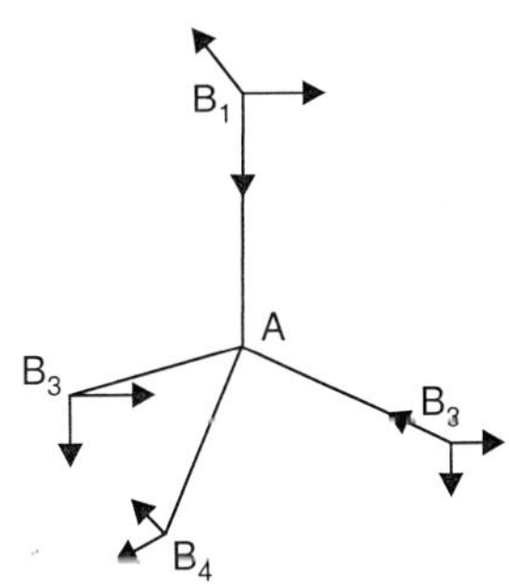

**Fig. 8.18**

**Table 8.4:** $\Gamma^{\pi}_{red}$ and $\Gamma^{\pi}_{I.R}$ of $AB_4$ tetrahedral molecule

| $T_d$ | $E$ | $8C_3$ | $3C_2$ | $6S_4$ | $6\sigma_d$ | *I.R Component* |
|---|---|---|---|---|---|---|
| $\Gamma^{\pi}_{red}$ | 8 | –1 | 0 | 0 | 0 | $E + T_1 + T_2 = \Gamma^{\pi}_{red}$ |

Referring to eqs. (8.41–45) and also the character Table of $T_d$ point group, one gets the following information

$$\left.\begin{array}{ll} E: & (d_{x^2-y^2}, d_{z^2}) \\ T_1: & \text{None} \\ T_2: & (p_x, p_y, p_z) \text{ and } (d_{xy}, d_{yz}, d_{xz}) \end{array}\right\} \qquad \ldots(8.52)$$

Thus, we conclude that out of the possible eight π-bonds which can be formed in tetrahedral $AB_4$ molecules, only five may be formed with $(d_{z^2}, d_{x^2-y^2})$ $(p_x, p_y, p_z)$ or $(d_{z^2}, d_{x^2-y^2})$ $(d_{xz}, d_{yz}, d_{xy})$ orbitals. The atom $A$ has $s$, $p$ and $d$-orbitals capable of forming π-bonds.

In case the set $(p_x, p_y, p_z)$ is used for σ-bond formation. ($sp^3$ hybridisation) $(d_{x^2-y^2}, d_{z^2})$ and $(d_{xy}, d_{yz}, d_{xz})$ will form π-bonds. If $(d_{xy}, d_{yz}, d_{xz})$ set is used to form σ-bond ($sd^3$-hybrid), then the set $(p_x, p_y, p_z)$ $(d_{x^2-y^2}, d_{z^2})$ can be used for π-bonding.

We are not taking the case of $PF_6$. Its π-bonding orbitals are given in Fig. 8.12. We leave it as an exercise for the students to practice.

**Step 10:** We now have two sets of orbitals (1) σ-bond forming (2) π(⊥) and π(||) bond-forming. Thus in $BF_3$ we have,

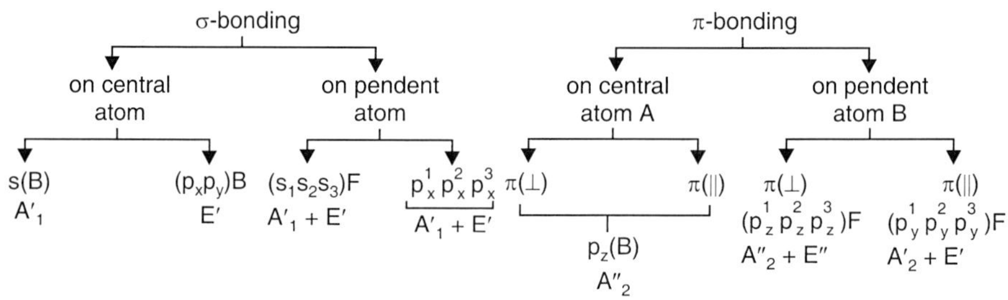


Thus for σ-and π-bond formations, orbitals of same symmetry on the central atom $A$ and pendent atoms ($F^1$, $F^2$, $F^3$) combined to form the σ-and π-bonds.

**Step 11:** The basis sets of the pendent atoms of the same symmetry are combined into SALCs corresponding to the I.Rs to which they belong. We have already given the procedure of forming SALCs in chapter six by employing projection operators of the I.Rs of the group. We have carried out the same procedure and form the SALCs corresponding to $A'$, $E'$, $A''$ and $E''$. These are given below.

$$\psi^{\sigma}_{A_2'} = \frac{1}{\sqrt{3}}(s_1 + s_2 + s_3); \frac{1}{\sqrt{3}}(p_x^1 + p_x^2 + p_x^3)$$

$$\psi_{E'}^{\sigma} = \frac{1}{\sqrt{6}}(2s_1 - s_2 - s_3) \text{ and } \frac{1}{\sqrt{2}}(s_2 - s_3);$$

$$\frac{1}{\sqrt{6}}(2p_x^1 - p_x^2 - p_x^3) \text{ and } \frac{1}{\sqrt{2}}(p_x^2 - p_x^3)$$

Similarly for π-bonds, we have,

$$\psi_{A_2'}^{\pi} = \frac{1}{\sqrt{3}}(p_y^1 + p_y^2 + p_y^3)$$

$$\psi_{A_2''}^{\pi} = \frac{1}{\sqrt{3}}(p_z^1 + p_z^2 + p_z^3)$$

$$\psi_{E'}^{\pi} = \frac{1}{\sqrt{6}}(2p_y^1 - p_y^2 - p_y^3), \text{ and } \frac{1}{\sqrt{2}}(p_y^2 - p_y^3)$$

$$\psi_{E''}^{\pi} = \frac{1}{\sqrt{6}}(2p_z^1 - p_z^2 - p_z^3), \text{ and } \frac{1}{\sqrt{2}}(p_z^2 - p_z^3)$$

Using these SALCs of the pendent fluorine atoms and the orbitals of boron atom into our secular determinant, we get the following blocks of matrices on the diagonal of the determinant.

1. $A_1'$ block of $A_1'$ symmetry (3 × 3): $[\Psi_{A_2'}^{\sigma}(s),\ \Psi_{A_2'}^{\sigma}(p_x),\ \Psi_{A'}^{\sigma}(B)p_x]$
2. $A_2'$ block of $A_2'$ symmetry (1 × 1): $[\Psi_{A''}\, p\pi(\text{II})]$
3. Two $E'$ (block) of $E'$ symmetry (4 × 4): $[\Psi_{E_1'}^{s\sigma}(s),\ \Psi_{E_1'}^{p\sigma}(p_x);$ $\Psi_{E_1'}^{p\pi(\text{II})}(p_y);\ \Psi_{E'}^{p\pi(1)}(p_z)]$ and $[\Psi_{E_2'}^{s\sigma}(s),\ \Psi_{E_2'}^{p\sigma}(p_x);\ \Psi_{E_2'}^{p\pi(\text{II})}(p_y);$ $\Psi_{E_2'}^{p\pi 1}(p_z)]$
4. $A_2''$ (block) of $A_2''$ symmetry (2 × 2): $[\Psi_{A_2'}^{p\pi(1)}(p_z),\ \Psi_{A_2''}^{p\pi(1)}(Bp_z)]$
5. Two $E''$ (blocks) (1 × 1): $\Psi_{E}^{p\pi 1}(p_z)$, and $\Psi_{E_b}^{p\pi(1)}(p_z)$.

   These blocks are then solved for the energy values. We give below a example of the secular determinant which is simplified. There are further simplification procedures which we will not discuss here. One such simplification is the Huckel method which we will describe in the next chapter. The second simplification technique is to assume the value of the off-diagonal matrix element $H_{ij}$ to be proportional to the overlap integrals.

The above diagonal matrices are given below. Each matrix will be equated to zero and the corresponding energy values will be the roots of $E$ obtained from the secular determinant.

$A_1'$ block

$$\begin{vmatrix} H_{sB,\,sB}-E & H_{sB,\,s\sigma} & H_{sB,\,p\sigma} \\ H_{sB,\,s\sigma} & H_{s\sigma,\,s\sigma} & H_{s\sigma,\,p\sigma} \\ H_{sB,\,p\sigma} & H_{s\sigma,\,p\sigma} & H_{p\sigma,\,p\sigma}-E \end{vmatrix} = 0 \qquad \ldots(8.53)$$

$A_2''$ block

$$E = H_{p\pi(\|),\,p\pi(\|)} = 0 \qquad \ldots(8.54)$$

$E'$ block

$$\begin{vmatrix} H_{p_x(B),\,p_x(B)}-E & H_{p_x(B),\,s\sigma} & H_{p_x(B),\,p(\sigma)} & H_{p_x(B),\,p\pi(\|)} \\ H_{p_x(B),\,s(\sigma)} & H_{s(\sigma)s(\sigma)}-E & H_{s(\sigma),\,p(\sigma)} & H_{s(\sigma),\,p\pi(\|)} \\ H_{p_x(B),\,p(\sigma)} & H_{s(\sigma)p(\sigma)} & H_{s(\sigma),\,p(\sigma)}-E & H_{p(\sigma),\,p\pi(\|)} \\ H_{p_x(B),\,p\pi(\|)} & H_{s(\sigma),\,p\pi(\|)} & H_{p(\sigma),\,p(\|)} & H_{p\pi,\,p\pi(\|)}-E \end{vmatrix} = 0$$

$$\ldots(8.55)$$

$A_2''$ block

$$\begin{vmatrix} H_{p_z(B),\,p_z(B)}-E & H_{p_z(B),\,p\pi(\perp)} \\ H_{p_z(B),\,p_z(\perp)} & H_{p_z(1),\,p_z(\perp)} \end{vmatrix} = 0 \qquad \ldots(8.56)$$

$E''$ block

$$E = H_{p\pi(\perp),\,p\pi(\perp)} = 0 \qquad \ldots(8.57)$$

Here the subscript of the matrices are the wave functions shown in eq. 8.53.

For example in $E'$ block, $H_{s(\sigma),\,p(\sigma)} = \int \Psi_{E'}^{s\sigma}(s) H \Psi_{E'}^{p(\sigma)}(p_x)\partial\tau$ *etc.* The roots of energy $E$ in these matrices are calculated and values of these roots are the energy values. We can draw MO energy level diagram from these values. It is given in Fig. 8.19($a$) and ($b$) for $BF_3$ molecule.

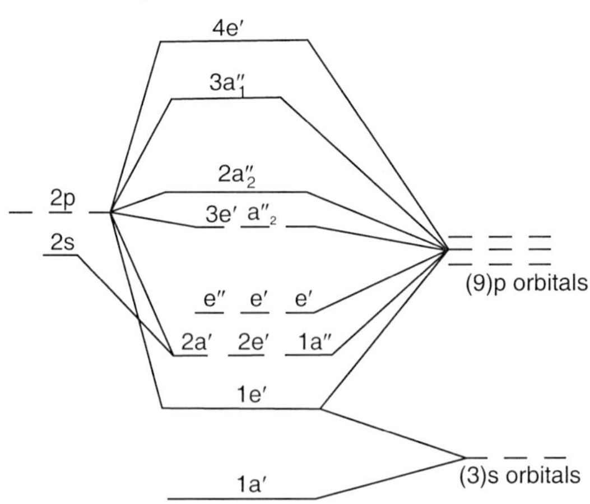


**Fig. 8.19** ($a$)

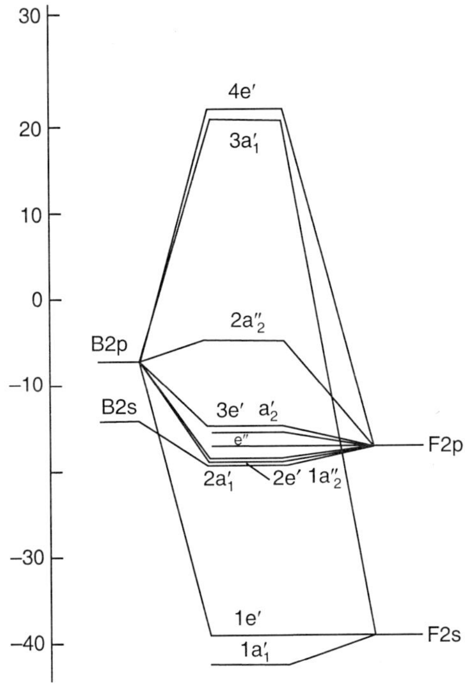


(*b*)

**Fig. 8.19** (*b*) Energy level diagram of $BF_3$.

Let us now take the case of octahedral molecule $AB_6$ type. Here we have the following symmetry orbitals.

| ***Central Atom (A)*** | ***Pendent Atoms (B) ($B_1$, $B_2$, $B_3$, $B_4$, $B_5$ and $B_6$)*** |
|---|---|
| $\left.\begin{array}{l} s\text{-orbitals} : A_{1g} \\ d_{x^2-y^2, dz^2} : E_g \\ (p_x, p_y, p_z) : T_{1u} \end{array}\right\} \sigma - \text{type}$<br>$(d_{xy}, d_{yz}, d_{xz}) : T_{2g} - \pi - \text{type}$ | Six $s$-orbitals – $A_{1g} + E_g + T_{1u}$<br>$- \sigma_s -$ type ($s$-orbital)<br>Six $p_z$-orbitals – $A_{1g} + E_g + T_{1u}$<br>$- \sigma_{pz} -$ type ($p_z$-orbital)<br>Twelve $(p_x + p_y)$ orbitals – $T_{1g} + T_{2g} + T_{1u}$<br>$+ T_{2u} \pi_p -$ type |

Examining the orbitals of the pendent atoms, one realises a large degree of degeneracy (both in σ and π-type of orbitals). The large degree of degeneracy causes a great difficulty in making SALCs of the orbitals. We will however, not plunge overselves into the procedures of making the SALCs of the orbitals of the pendent atoms of proper I.Rs, but will give their SALCs. These are given as follows:

$$\left.\begin{aligned}
\psi^1_{A_{1g}} &= \frac{1}{\sqrt{6}}(\phi_1+\phi_2+\phi_3+\phi_4+\phi_5+\phi_6)\\
\psi^a_{E_g} &= \frac{1}{\sqrt{6}}[2\sigma_5+2\sigma_6-\sigma_1-\sigma_2+\sigma_3+\sigma_4]\\
\psi^b_{E_g} &= \frac{1}{2}[\phi_1-\phi_2-\phi_3-\phi_4]\\
\psi^a_{T_{1u}} &= \frac{1}{\sqrt{2}}(\phi_5-\phi_6)\\
\psi^b_{T_{1u}} &= \frac{1}{\sqrt{2}}(\phi_3-\phi_4)\\
\psi^c_{T_{1u}} &= \frac{1}{\sqrt{2}}(\phi_1-\phi_2)
\end{aligned}\right\}\begin{array}{l}(\sigma\text{-type molecule})\\ \text{LCAO of orbitals}\\ \text{on pendent atoms } B\end{array}$$

The symmetries of $\psi^1_{A_{1g}}, (\psi^a_{E_g}, \psi^b_{E_g})$ and $T_{1u}$ matched with the symmetries of $s$, $(d_{x^2}, d_{x^2-y^2})$ and $(p_x, p_y, p_x)$ orbitals on the central atoms.

Finally, the LCAOs of the $\pi$-type orbitals on the pendent atoms, (*viz.* $p_x$ and $p_y$ orbitals on each $B$-atom) are given below.

$$\psi^a_{T_{2g}} = \frac{1}{2}(p^1_y + p^5_x + p^3_x + p^6_y)$$

$$\psi^b_{T_{2g}} = \frac{1}{2}(p^2_x + p^5_y + p^4_y + p^6_x) \text{ matching with}$$

$$\text{symmetry} \begin{Bmatrix} d_{xz} \\ d_{yz} \\ d_{xy} \end{Bmatrix} \text{ orbitals on atom } A$$

and
$$\psi^c_{T_{2g}} = \frac{1}{2}(p^1_x + p^2_y + p^3_y + p^4_x)$$

These SALCs are employed in the secular determinant whereby the latter is diagonally blocked. Each block secular determinant is equated to zero and energies of the MOs calculated. The MO diagram is shown in Fig. 8.20.

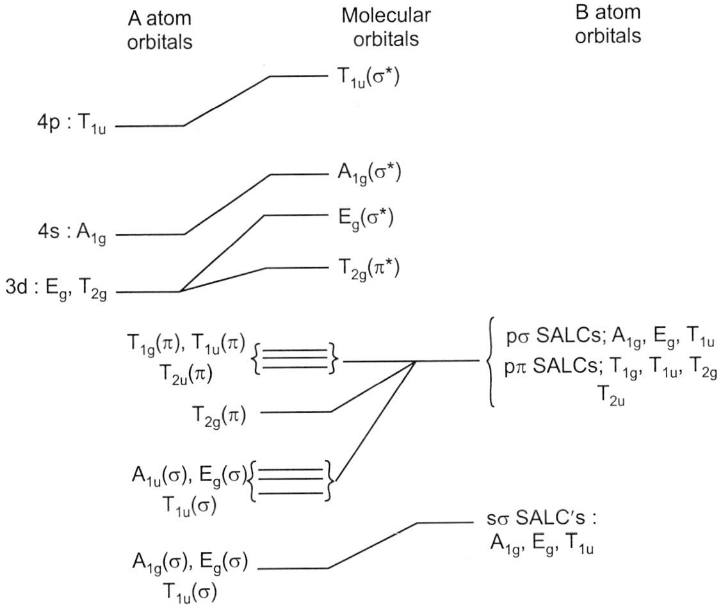


**Fig. 8.20** An Approximate M.O. diagram for an octahedral $AB_6$ molecule or ion (where $A$ is +2 or +3 ion from the first transition series and $B$'s are F or Cl atom).

**(C) $AB_4$ Tetrahedral Type Molecules:** The next chosen example for the formation of SALCs is the $AB_4$ (tetrahedral) molecules. We again assume that the orbitals of the central atom $A$ used for the formation of MOs are $s$, $p$ and $d$-orbitals. We also assume that the pendent atom $B$ uses $s$ and $p$ σ-type orbitals and two $2p\pi$-orbitals on each $B$-atom. Thus we have,

| *Central Atom (A)* | *Pendent Atom (B) ($B^1, B^2, B^3, B^4$)* |
|---|---|
| $s$ orbitals : $A_1$ | $4s$ σ type orbitals : $A_1 + T_2$ |
| $d_{z^2}, d_{x^2-y^2}$ : $E$ | $4p$ σ type orbitals : $A_1 + T_2$ |
| $p_x, p_y, p_z$: $T_2$ $d_{xy}, d_{yz}, d_{xz}$: $T_2$ | 8 π orbitals: $E + T_1 + T_2$. |

SALCs orbitals of pendent atoms are

$$A_1^{\sigma} : \Psi_1^{\sigma} = \frac{1}{2}(\phi_1 + \phi_2 + \phi_3 + \phi_4) \text{ matching } S^{A_1}, \text{ (Here } \phi_i = S_i \text{ or } p_z^i)$$

$$\left[ E: \begin{array}{l} \Psi_2 = \frac{1}{2}\left[p_x^1 - p_x^2 - p_x^3 + p_x^4\right] \\ \Psi_3 = \frac{1}{2}\left[p_y^1 - p_y^2 - p_y^3 + p_y^4\right] \end{array} \right] \text{ matching } d_{z^2} d_{x^2-y^2} \text{ respectively}$$

$$T_2^{\sigma}: \begin{bmatrix} \Psi_4 = \frac{1}{2}[\phi_1 - \phi_2 + \phi_3 - \phi_4] \\ \Psi_5 = \frac{1}{2}[\phi_1 + \phi_2 - \phi_3 - \phi_4] \\ \Psi_6 = \frac{1}{2}[\phi_1 - \phi_2 - \phi_3 + \phi_4] \end{bmatrix} \text{ matches respectively with } p_x, p_y \text{ and } p_z.$$

$$T_2^{(\pi)}: \begin{bmatrix} \Psi_7 = \frac{1}{4}\left[p_x^1 + p_x^2 - p_x^3 - p_x^4\right] + \frac{\sqrt{3}}{4}\left[-p_y^1 - p_y^2 + p_y^3 + p_y^4\right] \\ \Psi_8 = \frac{1}{4}\left[p_x^1 - p_x^2 + p_x^3 - p_x^4\right] + \frac{\sqrt{3}}{4}\left[p_y^1 - p_y^2 + p_y^3 - p_y^4\right] \\ \Psi_9 = -\frac{1}{2}\left[p_x^1 + p_x^2 + p_x^3 + p_x^4\right] \end{bmatrix}$$

$$\text{symmetry matches with the symmetry of } \begin{bmatrix} d_{yz} \\ d_{xz} \\ d_{xy} \\ \ldots\ldots \end{bmatrix}$$

Using these symmetry orbitals of the central atom $A$ and the pendent atoms $B$ in the secular equation, we form the MOs of the $AB_4$ type of tetrahedral molecules in Fig. 8.21.

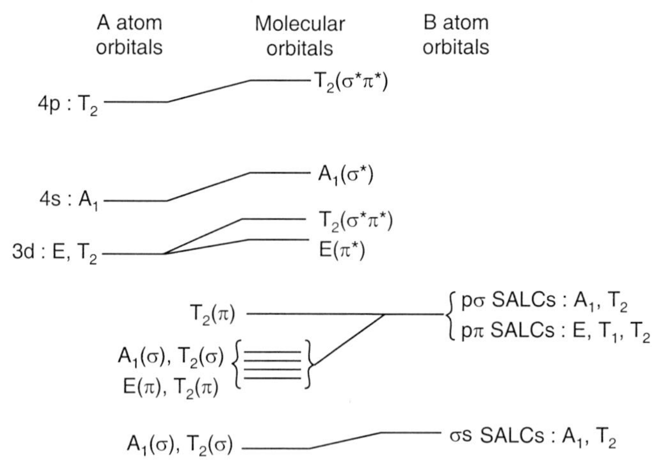


**Fig. 8.21** An Approximate MO diagram for a tetrahedral $AB_4$ molecule or complex (where $A$ is +2 ion from the first transition series and B's are F or Cl atom).

In the similar fashion, we can obtain the MOs of $AB_4$ square planar and $AB_5$-trigonal bipyramid molecules. Their MOs are given in Figs. 8.22 and 8.23 respectively.

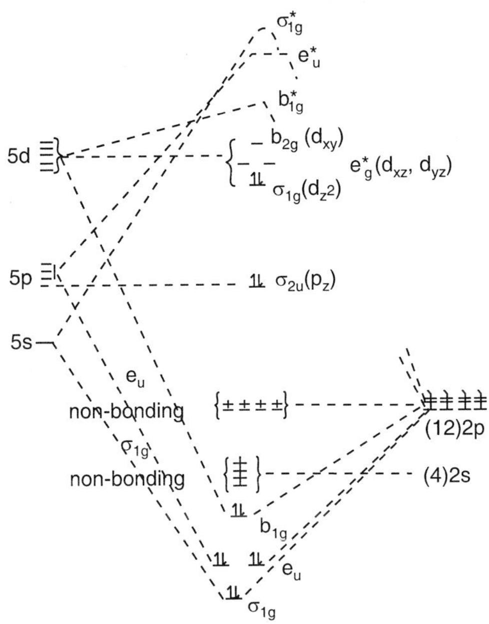


**Fig. 8.22** Qualitative MO energy level diagram for $XeF_4$.

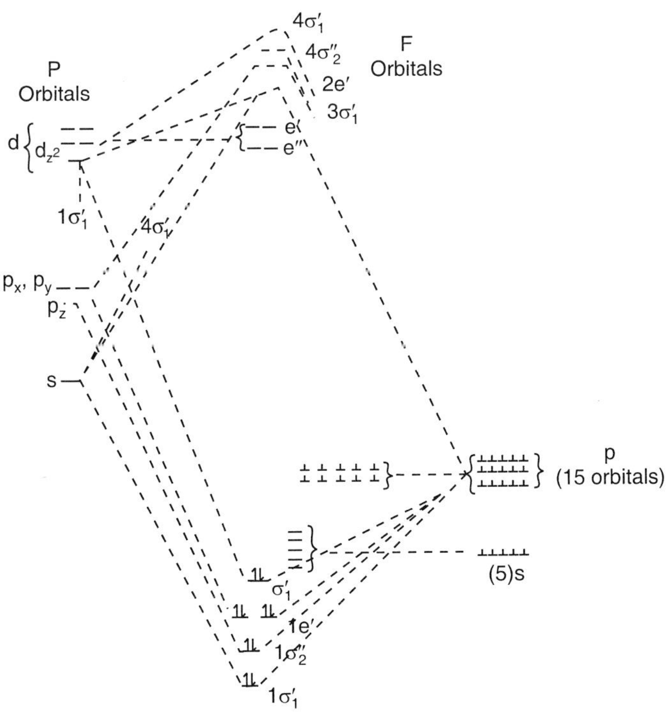


**Fig. 8.23** Qualitative MO energy level diagram for $PF_5$.

# 9

# Symmetry Aspects of Molecular Orbital Theory of Multicentre $A_MB_N$ Type Molecules

In chapter eight, we have discussed bonding in those molecules wherein, the bonds are described by the delocalised MOs formed in terms of SALCs. As a consequence, σ-bonds are formed by an electron pair between two σ-bonded atoms and π-bonds are delocalised over the entire molecule. The chosen molecules are of $AB_n$ type in which one atom (*A*) is present at the centre of the molecule linked separately, with *n*-pendant atoms (*B*) by *n*-different bonds to the central atom *A*. There are, however, many other molecules which do not have any central atom *A* and in which atoms are linked among each other and the π-orbitals are delocalised over the whole molecule (*e.g.*, $C_6H_6$). The bonding in some such molecules will be discussed in this chapter.

## 9.1 π-BONDS IN HYDROCARBON CHAIN MOLECULES ($C_nH_{n+2}$) CHAIN

Analogous to the methods already discussed in chapters seven and eight, the use of molecular symmetry *vis-a-vis* group theory in the MO determination of molecules can also be applied in the molecules of this chapter. Accordingly we:

1. Find the appropriate basis to generate $\Gamma_{red}$ of the molecular point group.
2. Reduce $\Gamma_{red}$ obtained in (1) into its constituent I.Rs.

3. Make linear combinations of the chosen basis vectors (using projection operators of the I.Rs of the point group, (cf. point 2), in such a way that each linear combination belongs to one of the I.Rs of the group (*i.e.*, forming SALCs).
4. Use these SALCs as basis in the secular determinant in order to obtain block diagonalised secular determinant in terms of smaller order determinants.
5. Obtain the energy of various MOs and finally,
6. Get the MO wave functions.

We now give the MO calculations of some $A_mB_n$ type molecules using the above method. In the chapter on SALCs, we have shown as to how one gets SALCs of some cyclic molecules. As a first example let us find the MOs and the energies of allyl $(CH_2CHCH_2)^-$ ion ($C_{2v}$ point group (Fig. 9.1). The two bonding electrons are delocalised over the three carbon atoms and one non-bonding electron pair is localised over the non-bonded *C*-atom (*i.e.*, $C_2$).

We assume that the molecule is planar in *YZ* plane, with one $p\pi$-orbital on each carbon atom, lying along the *y*-axis. All the three carbon atoms are linked with *H*-atoms by the overlap of 1*s*-orbital of *H*-atoms with $sp^2$-orbitals of carbon atoms. Three carbon atoms are linked by σ-bonds by the overlaps of one carbon atom ($sp^2$-orbital) to that of another carbon atom. This gives a bent framework of three *C*-atoms. Therefore, there is left one *p*-orbital on each carbon atom perpendicular to the *YZ* plane used for the π-bond formation. We thus take these three $p_y$-orbitals as basis. Let us represent them as $\phi_1$, $\phi_2$, $\phi_3$. These basis sets for π bonding are in two sets:

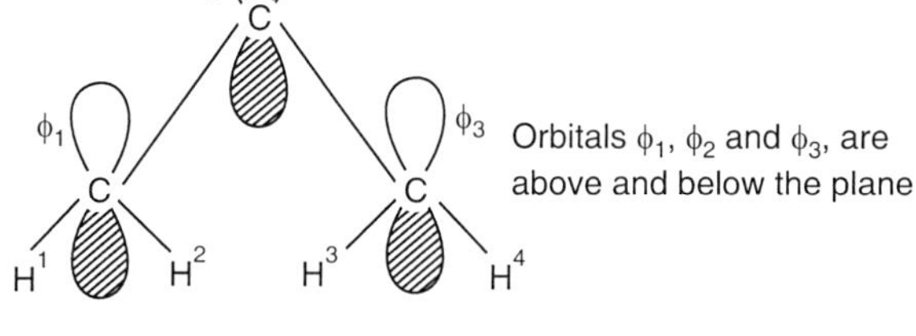


**Fig. 9.1** Allyl ion in *XZ* plane.

**Set 1:** ($\phi_1$, $\phi_3$) These orbitals ($\phi_1\phi_3$) are interchanged among each other by the operation of the group [($C_2(z)$, and $\sigma_v(xz)$].

**Set 2:** ($\phi_2$) The orbital ($\phi_2$) of this set remains invariant by all the operations of the group and is not interchanged to any orbital of the set 1 by any operation of the group.

We thus have two reducible representations. [$\Gamma_{(\phi_1, \phi_3)}$ and $\Gamma_{(\phi_2)}$].

Let us now generate $\Gamma_{red}$ (1, 3) and $\Gamma_{red}$ (2) using the rule of simple method described in chapters seven and eight. This rule is that the character for an operation of a group equals the number of the basis vectors (or bonds) unshifted by the operations of the group. If the basis vector shifts, the character is zero. We thus get the following reducible representations, [(*viz.*) $\Gamma_{red}$ (2) and $\Gamma_{red}$ (1, 3)]. These are reduced into their constituent I.Rs. These are given in the accompanying table.

**Table 9.1:** $\Gamma_{red}$ and $\Gamma_{I.R}$ of the allyl group.

| $C_{2v}$ | $E$ | $C_2$ | $\sigma_{xz}$ | $\sigma_{yz}$ | *I.Rs* |
|---|---|---|---|---|---|
| $\Gamma_{red}(2)$ | 1 | −1 | −1 | 1 | $B_2$ |
| $\Gamma_{red}(1, 3)$ | 2 | 0 | −2 | 0 | $A_2 + B_2$ |

The basis vectors ($\phi_1$, $\phi_2$, $\phi_3$) are then subjected to projection operations of I.Rs of $A_2$ and $B_2$. The result is the formation of three linear combinations. One combination belongs to $A_2$ irreducible representation, and two LCAOs belong to $B_2$ I.R. These are given below:

$$\left\{\begin{aligned} \psi_1'(A_2) &= \frac{1}{\sqrt{2}}(\phi_1 - \phi_3) \\ \psi_2'(B_2) &= \frac{1}{\sqrt{2}}(\phi_1 + \phi_3) \\ \psi_3'(B_2) &= \phi_2 \end{aligned}\right\} \quad \text{...(9.1)}$$

$\psi_2'$ and $\psi_3'$ will further interact (being of the same symmetry) and will form two different combinations, $\psi_2$ (bonding) and $\psi_2$ (antibonding). These are,

$$\psi_2(\text{bonding})_{B_2} = \frac{1}{\sqrt{2}}\left[\frac{1}{\sqrt{2}}(\phi_1 + \phi_3) + \phi_2\right] = \frac{1}{2}\left[\phi_1 + \phi_3 + \sqrt{2}\phi_2\right] \quad \text{...(9.2)}$$

$$\psi_3(\text{antibonding})_{B_2} = \frac{1}{\sqrt{2}}\left[\frac{1}{\sqrt{2}}(\phi_1 + \phi_3) - \phi_2\right] = \frac{1}{2}\left[\phi_1 + \phi_3 - \sqrt{2}\phi_2\right]$$

...(9.3)

These linearly combined basis vectors [$\psi_1(A_2)$, $\psi_2(\text{Bonding})_{B_2}$, $\psi_3(\text{antibonding})_{B_3}$] are then substituted in the secular determinant which on solving gives three roots of the energy, corresponding to the three energy levels of the allyl ion. These are given in Fig. 9.2.

## 9.2 HÜCKEL MOLECULAR ORBITAL-APPROXIMATION (HMO)

At this point, we introduce Huckel Molecular Orbital (HMO) approximation. We have indicated energy levels of allyl radical in terms of $\alpha$ and $\beta$ parameters (Fig. 9.2). We have defined SALCs of three MOs by Eqs. (9.1, 9.2 and 9.3). Each $\pi$ MO is a linear combination of basis vectors $\phi_1$, $\phi_2$ and $\phi_3$. In Eqs. (9.1–9.3), although we have already indicated the coefficients on each $\phi_i$. $\left[e.g., \frac{1}{\sqrt{2}}, \frac{1}{\sqrt{2}} \text{ and } 1\right]$ but if these coefficients, are not given, and we express the MOs in terms of unknown coefficients $C_{ij}$. The MOs can be written as follows:

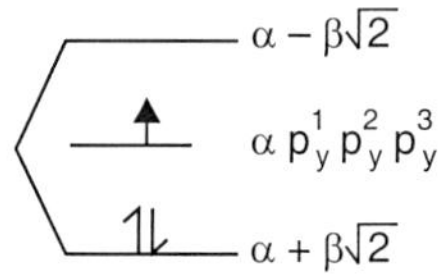


**Fig. 9.2** Energy Levels of $(C_3 \cdot H_5^-)$.

$$\psi_1 = C_{11}\phi_1 + C_{12}\phi_2 + C_{13}\phi_3 \quad \ldots(9.4)$$
$$\psi_2 = C_{21}\phi_1 + C_{22}\phi_2 + C_{23}\phi_3 \quad \ldots(9.5)$$
$$\psi_3 = C_{31}\phi_1 + C_{32}\phi_2 + C_{33}\phi_3 \quad \ldots(9.6)$$

Using any one of the above $\psi_i s$ one can get the energies of all the three MOs in the following way:

We substitute the chosen $\psi_i$ into the following energy equation.

$$E = \frac{\int \psi_i^* H \psi_i^j \partial\tau}{\int \psi_i^* \psi_j \partial\tau} \quad \ldots(9.7)$$

We minimize the energy by differentiating $E$ with respect to $C_{11}$, $C_{12}$ and $C_{13}$ and use the minimizing condition. $\left[\text{That is, } \frac{\partial E}{\partial C_{11}} = 0, \frac{\partial E}{\partial C_{12}} = 0 \text{ and } \frac{\partial E}{\partial C_{13}} = 0.\right]$ In this manner, we get three equations in terms of $Cij$. (This procedure is given in Chapter seven). As an example, suppose we choose $\psi_1$ (Eq. 9.4) as our function and substitute it in Eq. (9.7), and minimize the energy by differentiation with respect to $C_{11}$, $C_{12}$ and $C_{13}$, we get the following equations,

$$C_{11}[H_{11} - ES_{11}] + C_{12}[H_{12} - ES_{12}] + C_{13}[H_{13} - ES_{13}] = 0 \quad \ldots(9.8)$$

$$C_{21}[H_{21} - ES_{21}] + C_{22}[H_{22} - ES_{22}] + C_{23}[H_{23} - ES_{23}] = 0 \quad \ldots(9.9)$$

$$C_{31}[H_{31} - ES_{31}] + C_{32}[H_{32} - ES_{32}] + C_{33}[H_{33} - ES_{33}] = 0 \quad \ldots(9.10)$$

We have got three homogeneous linear equations known as secular equations. **For any non-trivial solutions of $\mathbf{C_{ij}}$** (*i.e.*, $C_{ij} \neq 0$), the only condition is that the following secular determinant has to be zero. That is,

$$\begin{vmatrix} H_{11}-ES_{11} & H_{12}-ES_{12} & H_{13}-ES_{13} \\ H_{21}-ES_{21} & H_{22}-ES_{22} & H_{23}-ES_{23} \\ H_{31}-ES_{31} & H_{32}-ES_{32} & H_{33}-ES_{33} \end{vmatrix} = 0 \qquad \ldots(9.11)$$

The procedure to solve (Eq. 9.11) is very difficult and cumbersome. It is at this point, Huckel has made the following approximations.

(*i*) All the Coulomb integrals $\left[\int \psi_1 H \psi_i \partial\tau\right]$ are made equal to α. In our example, $H_{11} = H_{22} = H_{33} = \alpha$ (Ionization energy of $p_y$-orbitals of carbon)

(*ii*) The only interactions of the π-orbitals are allowed between **ADJACENT-Bonded** atoms.

(*iii*) As a result of point (*ii*), the resonance integral $H_{ij} = \int \psi_i H \psi_j \partial\tau$ is written as β for adjacent (bonded) atoms. All $H_{ij}s$ for more remote atoms are made equal to zero. (*e.g.*, in allyl radical on $C_1$ and $C_3$, the resonance integral is zero).

(*iv*) The values of all $S_{ij} = \delta_{is}$ have been neglected for overlap between any two different atoms (*i.e.*, $S_{ij} = 0$ for $i \neq j$, and $S_{ij} = 1$ for $i = j$).

The above four approximations made the solution of secular equation (9.11) relatively easy to solve. As a result of HMO approximation, the secular determinant (9.11) becomes,

$$\begin{vmatrix} \alpha-E & \beta & 0 \\ \beta & \alpha-E & \beta \\ 0 & \beta & \alpha-E \end{vmatrix} = 0 \qquad \ldots(9.12)$$

(*v*) The determinant (9.12) is further simplified, if we write

$$\frac{(\alpha-E)}{\beta} = x, \qquad \ldots(9.13)$$

Finally we get, the following simplified determinant,

$$\begin{vmatrix} x & 1 & 0 \\ 1 & x & 0 \\ 0 & 1 & x \end{vmatrix} = 0 \qquad \ldots(9.14)$$

or

$$x\begin{vmatrix} x & 1 \\ 1 & x \end{vmatrix} - \begin{vmatrix} 1 & 0 \\ 0 & x \end{vmatrix} + 0\begin{vmatrix} 1 & x \\ 0 & 1 \end{vmatrix} = 0 \qquad \ldots(9.15)$$

or $x^3 - 2x = 0$, the roots of this equation are,

$$x = 0, \text{ and } x = \pm\sqrt{2} \quad \ldots(9.16)$$

Thus the energies of three orbitals are

$$\frac{\alpha - E}{\beta} = 0,$$

or $E_1 = \alpha$ and $E_2 = (\alpha - \sqrt{2}\beta)$ and $E_3 = (\alpha + \sqrt{2}\beta)$ ...(9.17)

In (Eq. 9.17) β, being interaction energy, its value should be negative, (see chapter seven). Thus the energies under the minimum conditions are $(\alpha + \sqrt{2}\beta)$, followed by α and $(\alpha - \sqrt{2}\beta)$. We have shown these energies in Fig. 9.2.

The above-calculated values of energies are then successively substituted in Eqs. (9.8–9.10). On substitution these energy values and solving them, one will obtain one set of ($C_{11}$, $C_{12}$, $C_{13}$) for one energy value. Thus for three energy values, we will obtain three sets of ($C_{11}$, $C_{12}$, $C_{13}$). These three sets will give three $\psi_{MO}$'s. That is, one for bonding $\psi_1^{(B)}$, second for non-bonding $\psi_2^{(N.B)}$ and third for antibonding $\psi_3^{\text{antibonding}}$.

In this connection, we would also like to point out that we have used $\psi_1$ (Eq. 9.4) for energy calculations. As a matter of fact, one can use any one of the three $\psi_i$'s Eqs. (9.4–9.7), the final outcome will be the same, since the secular determinant in all the three cases will be identical. (Carry out the above process, sharing with $\psi_2$ and see if the given statement is true.)

Although the above general HMO method can be used for any $C_nH_{n+2}$ molecules, but even in this method, for some $C_nH_n$ molecules, we come across energy polynomial equations of order higher than three. The solutions of the equations in those cases become very difficult. In such cases, we can calculate the energies using the SALCs in which the coefficients were obtained before the energies of the SALCs. For example, if we wish to obtain the energies of the orthonormal π M.Os of $C_5H_5$ (see Chapter on SALCs), we obtain three SALCs, *viz.*

$$\psi_1^{A_1} = \frac{1}{\sqrt{5}} = (\theta_1 + \theta_2 + \theta_3 + \theta_4 + \theta_5) \quad \ldots(9.18)$$

Only one of each pair (i = 2, or 3)

$$\psi_2^{E_1} = \frac{1}{\sqrt{7.25}}(1.62\theta_2 + \theta_3(-1)\theta_4 - 1.62\theta_5) \quad \ldots(9.19)$$

$$\psi_3^{E_1} = \frac{1}{\sqrt{7.25}}(\theta_2 - 1.62\theta_3 + 1.62\theta_4 - \theta_5) \quad \ldots(9.20)$$

Here $\phi_i^1 s$ are the π-orbitals of $C_5H_5$ molecule which are combined to form the SALC-MOs.

Now, we wish to find the energies of these MOs. The energies of these orbitals are given by,

$$E = \int \psi_i^* H \psi_i d\tau \qquad \ldots(9.21)$$

We substitute the expression for $\psi_L$ in Eq. (9.21).

For $\psi_1(A_1)$ orbital, we get,

$$E^{A_1} = \left[\frac{1}{5}\int (\phi_1 + \phi_2 + \phi_3 + \phi_4 + \phi_5)^* H (\phi_1 + \phi_2 + \phi_3 + \phi_4 + \phi_5)\partial\tau\right] \qquad \ldots(9.22)$$

$$E^{A_1} = \frac{1}{5}\underbrace{\left[\int \phi_1^* H \phi_1 \partial T + \int \phi_2^* H \phi_2 \partial T + \ldots. \int \phi_5^* H \phi_5 \partial T\right]}_{5 \text{ terms}}$$

$$+ \underbrace{\left[\int \phi_1 H \phi_2 \partial T + \int \phi_1 H \phi_3 \partial T \ldots \int \phi_i H \phi_j \partial T\right]}_{20 \text{ terms } (i \neq g)} \qquad \ldots(9.23)$$

Now, each $\int \psi_i^* H \psi_i \partial\tau$ will be equal to α and each $\int \phi_i^* H \phi_j \partial\tau$ will be equal to β. Therefore, the energy of $\psi_1^{A_1}$ from Eq. (9.23) will be equal to

$$E \text{ of } \psi^{A_1} = \frac{1}{5}(5\alpha + 10\beta)$$

$$= \alpha + 2\beta \qquad \ldots(9.24)$$

Similarly, energy of $\psi_2^{E_1}$ can also be calculated as follows:

$$\psi^{E_1} = \frac{1}{7.25}\left[\begin{matrix}\int (1.62\phi_2 + \phi_3 - \phi_4 - 1.65\phi_5) \\ H (1.62\phi_2 + \phi_3 - \phi_4 - 1.65\phi_5)\partial\tau\end{matrix}\right] \qquad (9.25)$$

$$= \frac{1}{7.25}\left[\underbrace{\int 1.62\phi_2 H 1.62\phi_2 \partial T + \int \phi_3 H \phi_3 \partial T + \ldots)}_{4 \text{ terms}} + \underbrace{\int 1.62\phi_2 H \phi_3 \partial T + \ldots}_{12 \text{ terms}}\right]$$

$$= \frac{1}{7.25}\left[(1.62)^2\alpha + (1.62)^2\alpha + \alpha + \alpha\right] + \frac{1}{7.25}[4 \times (1.62)\beta - 2\beta] \qquad \ldots(9.26)$$

$$= [\alpha + 0.6\beta] \qquad \ldots(9.27)$$

Similarly the energy for $E_2$ from Eq. (9.20) can be calculated. The value of $E_2$ will be:

$$(\alpha - 1.62\beta) \qquad \ldots(9.28)$$

The value of $\beta$ being negative, the ground state ($\psi_1^{A_1}$) will be of lowest in energy, ($\alpha + 2\beta$), followed by that of $E_1(\alpha + 0.618\beta)$ and antibonding highest orbital $E_2$ with energy ($\alpha - 1.62\beta$). (Fig. 9.3).

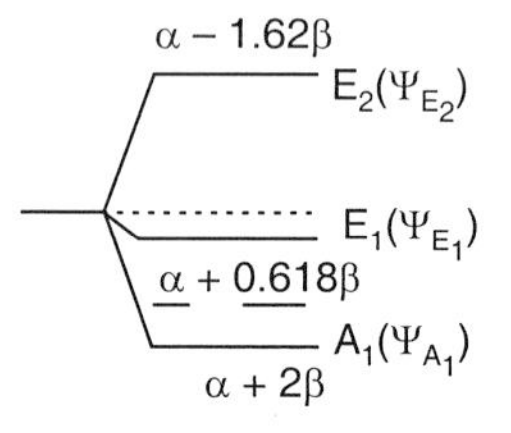


**Fig. 9.3** M.O. diagram of $C_5H_5$.

For cyclic hydrocarbons, one can calculate the energies of the MOs without adopting the above procedure. One can calculate the energy of any cyclic or acyclic system. It has, however, been found that for cyclic systems there is a pattern in the energy levels. It has therefore been possible to know the energy levels of any cyclic system without undergoing any calculations. The procedure is given below.

## Alternative Procedure

Draw a circle of radius $2\beta$. Inscribe a polygon of $n$-sides in this circle so that one vertex of the polygon is at point $A$ (lowermost point.) It will be done by making an angle $\left(\frac{\pi}{2n}\right)$ on each side of $A$. Draw the chords on each side of $A$ which make the angle $\left(\frac{\pi}{2n}\right)$, intersecting the circle at $C$ and $D$. Again draw the chords $CE$ and $DF$, which make an angle $\frac{\pi}{2n}$ with $AD$ and $AC$ intersecting the circle at $E$ and $F$. Carry on making such chords until you get $n$-sided polygon (Join $C$, and $D$, $E$ and $F$). Five membered polygon (Fig. 9.4) is shown as an example.

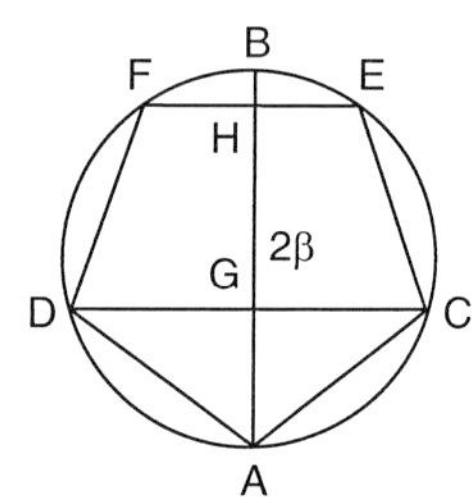


**Fig. 9.4** Five sided polygon, inscribed in the circle of radius $2\beta$.

The energies of $\pi$-MOs of $n$-sided polycyclic molecule $C_nH_n$ compounds can be obtained as given below. Join the opposite points (*e.g.*, $C$ and $D$, $E$ and $F$, *etc.*) that intersect on the circle. The lines joining these points intersect the diameter $AB$ (of $4\beta$ length) at $G$, $H$, *etc.* Then the energies of the MO's are given by the distances from $A$ to the points [*e.g.*, $G$, $H$, *etc.*] where the joining lines [*i.e.*, $CD$ and $EF$) meet $AB$ [*i.e.*, $AG$ and $AH$]. Thus in $C_5H_5$, we have the ground state orbital at $A$ (= $\alpha + 2\beta$), energy of second orbital at $G$ (= $\alpha + AG$) which is equal to $\left[\alpha + 2\cos\frac{2\pi\beta}{n}\right]$, energy of the third orbital $(\alpha + AH)$ $\left[= \alpha + \cos 2\left(\frac{2\pi}{n}\right)\beta\right]$

and the energy of last orbital $\left[\alpha + \cos n\left(\frac{2\pi}{n}\right)\beta\right] - \left[\alpha - \cos 2\left(\frac{2\pi}{n}\right)\beta\right]$.

In carbon polycyclic compounds containing delocalised π-systems, the π-orbitals are of great interest. The filled σ-orbitals are very low in energy and the unfilled σ*-orbitals are very high in energy. These π-orbitals are known as "Frontier orbitals with highest filled orbitals as HOMO and the lowest empty orbitals as LUMO". These HOMOs and LUMOs are employed in understanding the chemical reactivities (subject matter of chapter ten).

For more complicated cases, we solve the secular equation in order to get the energy values of the orbitals by using molecular symmetry factorisation. For inorganic molecules, generally "Extended Huckel MO method is used (not discussed here)."

The energies of π-orbitals in the delocalized π-electron heteroatom molecules (*e.g.*, pyridines, pyrroles, furans *etc.*) can also be obtained by using HMO approximation. However, the value of [α + β] for the heteroatom is expressed as $\alpha + h\beta$ and the $C$–$X$ ($X$ = halogen atom) bond interaction integral is expressed as $k\beta$. The values of $h$ and $k$ parameters are given in a number of books of organic chemistry. If the heteroatom contributes two electrons in π-system, the value of the parameter $h$ will be different from the case, when it contributes one electron, [*e.g.*, pyridine (2*e*), pyrrole one electron].

We give below some examples where HMO approximation is used to determine the energies of the delocalised π-MOs.

**Examples 1 and 2:** Let us take the cases of butadiene and cyclobutadiene.

**(1) Butadiene:** The molecule exists in two forms. (*cis*-and *trans*-) (Fig. 9.5)

***Trans*-butadiene:** Point group $C_{2h}$ $(E, C_2\ i, \sigma_h)$; *cis*-butadiene; point group $C_{2v}$ $(E, C_2, \sigma_{xz}, \sigma_{yz})$.

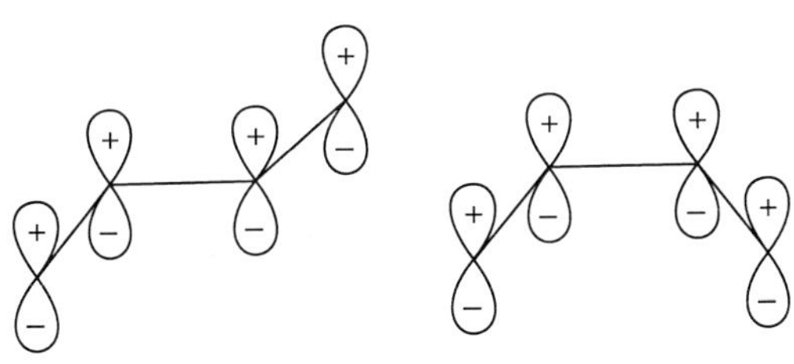


**Fig. 9.5** *Cis* and *Trans* forms of Molecular plane *YZ*.

***Trans*-Butadiene:** Basis vectors ($\phi_1$, $\phi_2$, $\phi_3$, $\phi_4$), these are $p_z(\pi)$-orbitals on $C_1$, $C_2$, $C_3$, $C_4$ atoms respectively. $\Gamma_{red}$ for MOs using $\phi_i$ as bases are given below.

**[A]**

| *Form* | *Point group* | *Represent atoms* | *E* | $C_2$ | $\sigma_{xz}$ | $\sigma_{yz}$ | *I. Rs.* |
|---|---|---|---|---|---|---|---|
| | | $\Gamma_{red}$ | 4 | 0 | 0 | –4 | $2A_u + 2B_g$ |
| *trans* | $C_{2h}$ | $\Gamma_{red}(1, 4)$ | 2 | 0 | 0 | –2 | $A_u + B_g$ |
| | | $\Gamma_{red}(2, 3)$ | 2 | 0 | 0 | –2 | $A_u + B_g$ |

For *trans*-case, $\Gamma_{red}(1, 4) = \Gamma_{red}(2, 3) = A_u + B_g$; Thus, two MOs will be of $A_u$ symmetry and two will be of $B_g$.

***Cis*-Butadiene:** Basis vectors ($\phi_1$, $\phi_2$, $\phi_3$, $\phi_4$,). These are $p_z(\pi)$-orbitals on carbon $C_1$, $C_2$, $C_3$ and $C_4$ respectively. $\Gamma_{red}$ using $\phi_i s$ as bases are given below.

**[B]**

| *Form* | *Represent atoms* | *E* | $C_2$ | $\sigma_{xz}$ | $\sigma_{yz}$ | *I.Rs.* |
|---|---|---|---|---|---|---|
| | $\Gamma_{red}$ | 4 | 0 | 0 | –4 | $2A_2 + 2B_1$ |
| *cis* ($C_{2v}$) | $\Gamma_{red}(1, 4)$ | 2 | 0 | 0 | –2 | $A_2 + B_1$ |
| | $\Gamma_{red}(2, 3)$ | 2 | 0 | 0 | –2 | $A_2 + B_1$ |

For *cis*-case, $\Gamma_{red}(1, 4) = \Gamma_{red}(2, 3) = A_2 + B_1$. Thus, two MOs will be of $A_2$ symmetry and two will be of $B_1$ symmetry.

The above division of four MOs in two sets has been carried out on the following principle. In both *cis*-and *trans*-structures, the set of two carbon atoms, (*e.g.*, ($C_1$, $C_4$) and ($C_2$, $C_3$) individually possesses $C_{2v}$ and $C_{2h}$ symmetry respectively by themselves. This important situation is applied in our calculations. Consequently carbon atoms of one set are not transformed into those of second set. So carbon atoms of one set ($C_1$, $C_4$) are not equivalent to those of the other set ($C_2$ and $C_3$). Thus each of these sets in both *cis*-and *trans*-cases can be used separately as basis for constructions of $\Gamma_{red}$ representation of the molecular group (*viz.* $C_{2V}$ and $C_{2h}$). We find $\Gamma_{red}(1, 4)$ and $\Gamma_{red}(2, 3)$ are reduced as follows.

(*Cis*-case) $\quad \Gamma_{red}(1, 4) = A_2 + B_1$ ...(9.29)

(*Trans*-case) $\quad \Gamma_{red}(2, 3) = A_u + B_g$ ...(9.30)

Thus, the expressions for MOs in the *cis*- and *trans*- butadiene can be written as

**For *cis*-case:**

$$\psi_1^{A_2}(MO) = N_1(a_1\phi_1 + a_2\phi_2 + a_3\phi_3 + a_4\phi_4). \qquad ...(9.31)$$

and
$$\psi_2^{B_2}(MO) = N_2(a_1'\phi_1 + a_2'\phi_2 + a_3'\phi_3 + a_4'\phi_4). \qquad ...(9.32)$$

**For *trans*-case:**

$$\psi_3^{A_u}(MO) = N_3(b_1\phi_1 + b_2\phi_2 + b_3\phi_3 + b_4\phi_4). \qquad ...(9.33)$$

$$\psi_4^{B_g}(MO) = N_4(b_1'\phi_1 + b_2'\phi_2 + b_3'\phi_3 + b_4'\phi_4). \qquad ...(9.34)$$

In the light of above discussion it is possible to bifurcate these four equations into two independent non-interacting components which are given below.

**For *cis*-case:**

$$\psi_1^{A_2}(MO) = N_1(a_1\phi_1 + a_4\phi_4) + N_1(a_2\phi_2 + a_3\phi_3). \qquad ...(9.35)$$

$$\psi_2^{B_1}(MO) = N_2(a_1'\phi_1 + a_4'\phi_4) + N_2(a_2'\phi_2 + a_3'\phi_3). \qquad ...(9.36)$$

**For *trans*-case:**

$$\psi_3^{A_4}(MO) = N_3(b_1\phi_1 + b_4\phi_4) + N_3(b_2\phi_2 + b_3\phi_3). \qquad ...(9.37)$$

and
$$\psi_4^{B_g}(MO) = N_4(b_1'\phi_1 + b_4'\phi_4) + N_4(b_2'\phi_2 + b_3'\phi_3). \qquad ...(9.38)$$

In the latter Eqs. (9.31–9.38), each subset belongs to the same I.Rs to which the whole set of the $\phi_i$ functions belong. For example, in (Eq. 9.35) both $(a_1\phi_1 + a_4\phi_4)$ subset, and $(a_2\phi_2 + a_3\phi_3)$ subset belong to $A_2$ I.R. of $C_{2v}$ point group. By carrying out bifurcation in this manner it is possible to construct partial molecular orbitals first from the atomic orbitals in the each of the subsets [*e.g.*, subsets $(a_1\phi_1 + a_4\phi_4)$ and $(a_2\phi_2 + a_3\phi_3)$] and then combine these into complete MO (*e.g.*, $\psi_1^{A_2}$ M.O) of Eq. (9.31). In this manner, we will get the MOs of the entire molecule for a particular I.R.

In *trans*-butadiene, we will thus get two MOs of $B_g$ symmetry and two of $A_u$ symmetry. Similarly, we will have two MOs of $A_2$ and two MOs of $B_1$ symmetry in the *cis*-butadiene. These will be our starting wave functions.

Starting with $\Gamma_{red}$, formed by the set (1, 4) of *trans*-butadiene *i.e.*, $\Gamma_{red}(1, 4)$ we write it as follows,

$\Gamma_{red}(1, 4) =$

| $C_{2h}$ | $E$ | $C_2$ | $i_i$ | $\sigma_h$ | *I.R* |
|---|---|---|---|---|---|
| $\Gamma_{red}(1, 4)$ | 2 | 0 | 0 | –2 | $A_u + B_g$ |

Its basis vectors are $\phi_1$ and $\phi_4$. Using projection operators of $A_u$ and $B_g$ I.Rs, we make SALCs of symmetry $A_u$ and $B_g$ as given below,

$$\psi^{B_g}(1,4) = \frac{1}{\sqrt{2}}(\phi_1 - \phi_4); \psi^{B_g}(2,3) = \frac{1}{\sqrt{2}}(\phi_2 - \phi_3) \quad \ldots(9.39)$$

$$\psi^{A_u}(2,3) = \frac{1}{\sqrt{2}}(\phi_1 + \phi_4); \psi^{A_4}(2,3) = \frac{1}{\sqrt{2}}(\phi_2 + \phi_3) \quad \ldots(9.40)$$

Using these SALCs of symmetry $B_g$ in the two sets and inserting these in the secular equation, we obtain MO energies. The secular determinant for $B_g$ will be (in *trans*-case),

$$\begin{vmatrix} H_{(1,4)(1,4)} - E & H_{(1,4)(2,3)} \\ H_{(2,3)(1,4)} & H_{2,3(2,3)} - E \end{vmatrix} = 0 \quad \ldots(9.41)$$

We now apply Huckel's molecular approximations (HMO) and we get,

$$H_{1,4} = \alpha; H_{(1,4)(2,3)} = H_{(2,3)(1,4)} = \beta; H_{2,3} = (\alpha - \beta), \quad \ldots(9.42)$$

Eq. 9.41 can be written as,

$$\begin{vmatrix} \alpha - E & \beta \\ \beta & (\alpha - \beta) - E \end{vmatrix} = 0 \quad \ldots(9.43)$$

or

$$\begin{vmatrix} \dfrac{\alpha - E}{\beta} & 1 \\ 1 & \dfrac{\alpha - E}{\beta} - 1 \end{vmatrix} = 0$$

or

$$\begin{vmatrix} x & 1 \\ 1 & x - 1 \end{vmatrix} = 0$$

This gives, $x = \dfrac{1 \pm \sqrt{5}}{2}$

or

$$\frac{\alpha - E}{\beta} = \frac{1 \pm \sqrt{5}}{2}$$

or

$$E_1 = |\alpha - 1.62\beta| \text{ and } E_2 = 0.62\beta \quad \ldots(9.44)$$

Similarly for $A_u$ orbital, using Eq. (9.40), we get the secular determinant for $A_u$ I.R., the same as that given in Eq. (9.41). But we have in this case $H_{23,23} = \alpha$; $H_{23,14} = \beta$; and $H_{23,23} = (\alpha + \beta)$. So our secular determinant is,

$$\begin{vmatrix} \alpha - E & \beta \\ \beta & (\alpha - \beta) - E \end{vmatrix} = 0$$

or $$\begin{vmatrix} x & 1 \\ 1 & x+1 \end{vmatrix} = 0 \qquad \ldots (9.45)$$

or $$x^2 + x - 1 = 0$$

$$x = \frac{-1 \pm \sqrt{5}}{2}$$

or $$\frac{\alpha - E}{\beta} = \frac{-1 \pm \sqrt{5}}{2}$$

or $$E = +\ 1.62\ \beta \text{ and } -\ 0.62\ \beta$$

Therefore, the energy level for *trans*-butadiene is given in Fig. (9.6).

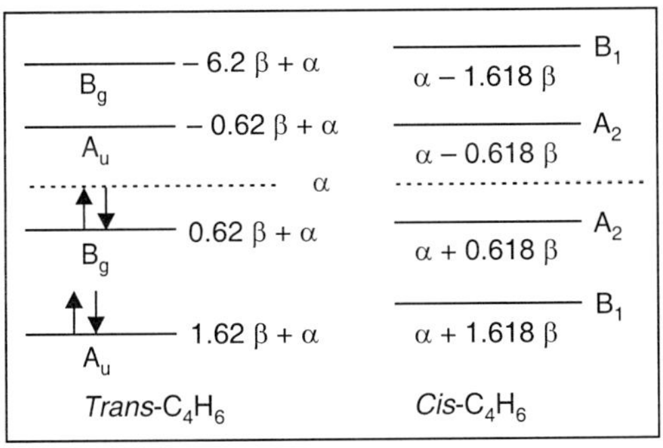


**Fig. 9.6** Energy level of *trans*- and *cis*-butadiene molecules.

For *cis*-butadiene, one can proceed in the same fashion and obtain the energy level diagram as shown in [Fig. 9.6].

**Example 3:** Cyclobutene ($C_4H_6$) (Planar) (Fig. 9.7) (point group $D_{4h}$). It contains four $p(\pi)$-orbitals one on each carbon atom. These are represented as $\phi_1$, $\phi_2$, $\phi_3$ and $\phi_4$. We first form $\Gamma_{red}$ using these $p(\pi)$-orbitals as basis. $\Gamma_{red}$ is calculated and is given below.

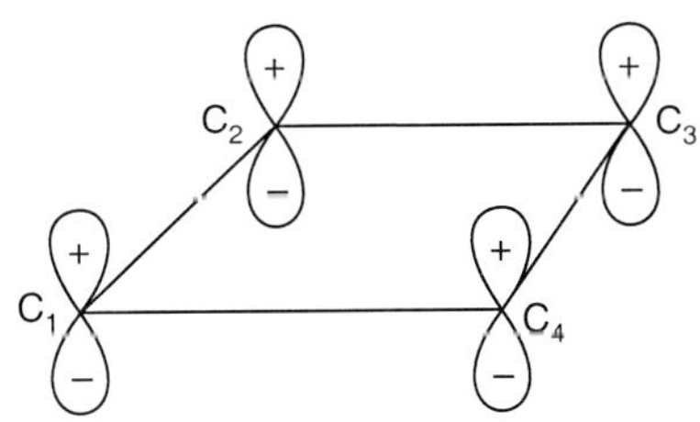


**Fig. 9.7** Energy level of *cis*-butadiene.

It contains $A_{2g} + B_{2g} + E_g$ as *I.Rs*.

| $D_{4h}$ | $E$ | $2C_4$ | $C_2$ | $2C_2'$ | $2C_2''$ | $i$ | $2S_4$ | $\sigma_h$ | $2\sigma_v$ | $\sigma_d$ | *I.Rs* |
|---|---|---|---|---|---|---|---|---|---|---|---|
| $\Gamma_{red}$ | 4 | 0 | 0 | –2 | 0 | 0 | 0 | –4 | 2 | 0 | $A_{2g} + B_{2g} + E_g$ |

By the use of projection operators for $A_{2g}$, $B_{2g}$ and $E_g$, one gets the following SALCs (*see* chapter 5).

1. SALC of for symmetry $A_{2g}$ is given by $\frac{1}{2}(\phi_1+\phi_2+\phi_3+\phi_4)$ ...(9.46)

2. SALC for $B_{2g}$ symmetry is given by $\frac{1}{2}(\phi_1-\phi_2-\phi_3-\phi_4)$ ...(9.47)

3. SALC for $E_g$ symmetry are given by $\frac{1}{\sqrt{2}}(\phi_1-\phi_3)$ and $\frac{1}{\sqrt{2}}(\phi_2-\phi_4)$

...(9.48)

**Energy Calculations:** We start with the following MO wave functions.

$\psi_1$(MO) (lowest energy state) = $C_1\phi_1 + C_2\phi_2 + C_3\phi_3 + C_4\phi_4$

Using this wave function, the secular equation is given as follows:

$$C_{11}(H_{11}-ES_{11})+C_{12}(H_{12}-ES_{12})+C_{13}(H_{13}-ES_{13})+C_{14}(H_{14}-ES_{14})=0$$

$$C_{11}(H_{21}-ES_{21})+C_{12}(H_{22}-ES_{22})+C_{13}(H_{23}-ES_{23})+C_{14}(H_{24}-ES_{24})=0$$

$$C_{11}(H_{31}-ES_{31})+C_{12}(H_{32}-ES_{32})+C_{13}(H_{33}-ES_{33})+C_{14}(H_{34}-ES_{34})=0$$

$$C_{11}(H_{41}-ES_{41})+C_{12}(H_{42}-ES_{42})+C_{13}(H_{43}-ES_{43})+C_{14}(H_{44}-ES_{44})=0$$

For non-trivial solution $C_{ij} \neq 0$, the following secular determinant is zero,

$$\begin{vmatrix} H_{11}-ES_{11} & H_{12}-ES_{12} & H_{13}-ES_{13} & H_{14}-ES_{14} \\ H_{21}-ES_{21} & H_{22}-ES_{22} & H_{23}-ES_{23} & H_{24}-ES_{24} \\ H_{31}-ES_{31} & H_{32}-ES_{32} & H_{33}-ES_{33} & H_{34}-ES_{34} \\ H_{41}-ES_{41} & H_{42}-ES_{42} & H_{43}-ES_{43} & H_{44}-ES_{44} \end{vmatrix}=0 \quad ...(9.49)$$

Using Huckel approximation and $S_{ij} = \delta_{ij}$, we get the following equation

$$\begin{vmatrix} (\alpha-E) & \beta & 0 & \beta \\ \beta & (\alpha-E) & \beta & 0 \\ 0 & \beta & (\alpha-E) & \beta \\ \beta & 0 & \beta & (\alpha-E) \end{vmatrix}=0 \quad ...(9.50)$$

Here $\alpha = H_{ii}$, $\beta = H_{\pm 1}$.

Substitute $\frac{\alpha-E}{\beta} = x$ in Eq. (9.50), we obtain

$$\begin{vmatrix} x & 1 & 0 & 1 \\ 1 & x & 1 & 0 \\ 0 & 1 & x & 1 \\ 1 & 0 & 1 & x \end{vmatrix}=0 \quad ...(9.51)$$

Solving Eq. (9.51), we get,

$$x = 0,\ x = \pm 2, \text{ and } x = 0, \textit{ i.e.}, \text{ four roots of } x,$$

For $x = 0$, we get

1. $\dfrac{\alpha - E}{\beta} = 0$ or $E = \alpha$ (for two roots *i.e.*, for energy level of double degeneracy $E$.)

2. $\dfrac{\alpha - E}{\beta} = +2, \quad \text{or} \quad E = \alpha - 2\beta$

3. $\dfrac{\alpha - E}{\beta} = -2, \quad \text{or} \quad E = \alpha + 2\beta$

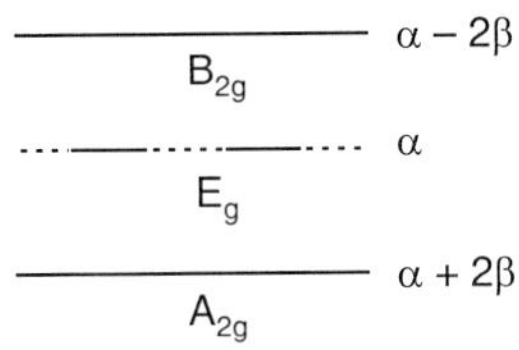


**Fig. 9.8** Energy levels of $C_4H_4$.

Hence, the energy level of $C_4H_4$ molecule is shown in [Fig. (9.8.)]

**Second Method:** In some cases of $C_nH_n$ molecules, it is not possible to solve the secular determinant. Therefore, the energy values of MOs of $C_nH_n$ cannot be obtained. In such cases we start using MO wave functions obtained as SALCs. (for $C_4H_4$ eqs. (9.44), (9.47), and (9.48) and calculate the energy values of these SALCs. The final values of $E$(energy) for such cases will be the same as those obtained by taking $(C_1\phi_1 - C_2\phi_2 - C_3\phi_3 - \phi_4\phi_4)$ as MO function. We have already exemplified it by taking the case of $C_4H_4$ in the preceding paragraph.

**Example 4(*i*):** As fourth example, we take the bonding in hexatriene molecule $(C_6H_8)$ having the point group $D_{6h}$. Analogous to other open chain polygon molecules, one can obtain, the secular equation and thereafter secular determinant in this case also. The base vectors, $\phi_1$, $\phi_2$, $\phi_3$, $\phi_4$, $\phi_5$ and $\phi_6$ are chosen. These base vectors are $2p_z$-orbitals on carbon atoms numbered 1 to 6 in Fig. (9.9).

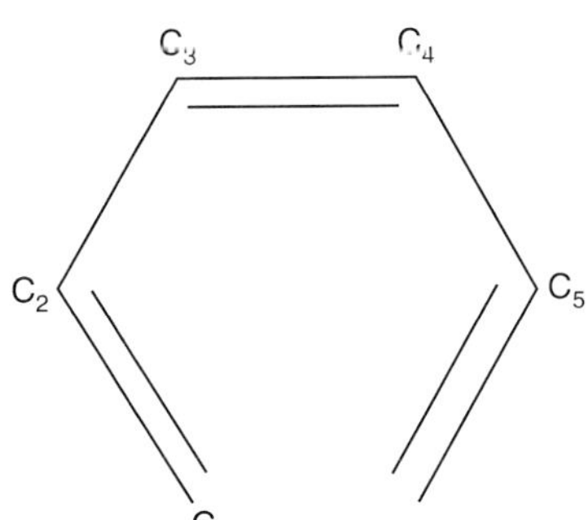


**Fig. 9.9** Shape of $C_6H_4$.

**(*ii*) Point group:** $D_{6h}$, base vectors = $(\phi_1, \phi_2, \phi_3, \phi_4, \phi_5, \phi_6)$.
Secular determinant.

$$\begin{vmatrix} y & 1 & 0 & 0 & 0 & 0 \\ 1 & y & 1 & 0 & 0 & 0 \\ 0 & 1 & y & 1 & 0 & 0 \\ 0 & 0 & 1 & y & 1 & 0 \\ 0 & 0 & 0 & 1 & y & 1 \\ 0 & 0 & 0 & 0 & 1 & y \end{vmatrix} \qquad ...(9.52)$$

$$\left[ y = \frac{\alpha - E}{\beta} \right]$$

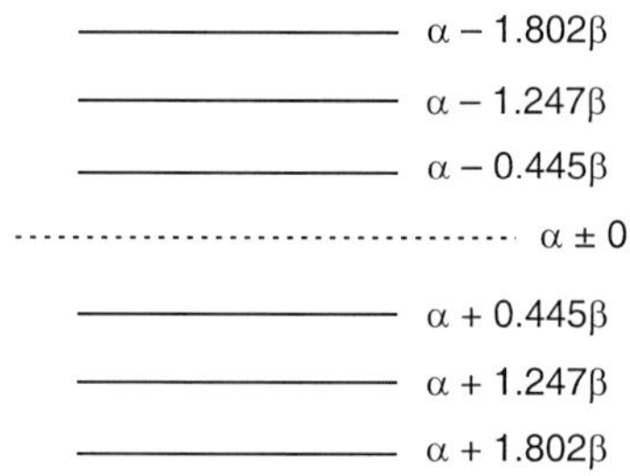


**Fig. 9.10** Energy Level diagram of $C_6H_8$.

The solution of Eq. (9.52) will give the six energy roots which are the energies of six MO's of $C_6H_8$. These are given in Fig. (9.10).

**Example 5:** We now take the case of $C_6H_6$ which is closed polyene ($C_6H_6$) [*i.e.*, benzene molecule]. We take the base vectors $\phi_1$, $\phi_2$, $\phi_3$, $\phi_4$, $\phi_5$ and $\phi_6$. Its point group is $D_{6h}$. The reducible representation $\Gamma_{red}$ generated by the operations of this group is given as.

| $D_{6h}$ | $E$ | $2C_6$ | $2C_3$ | $C_2$ | $3C'_2$ | $*3C''_3$ | $I$ | $3S_3$ | $2S_6$ | $\sigma_h$ | $3\sigma_d$ | $3\sigma_v$ | *I.R.'s* |
|---|---|---|---|---|---|---|---|---|---|---|---|---|---|
| $\Gamma_{red}$ | 6 | 0 | 0 | 0 | 2 | 0 | 0 | 0 | 0 | –6 | 0 | 2 | $A_{2u} + B_{1g} + E_{1g} + E_{2u}$ |

On reducing it into its I.R constituents, one obtains $A_{2u}$, $B_{1g}$, $E_{1g}$ and $E_{2u}$. On applying the projection operators of $A_{2u}$, $B_{1g}$, $E_{1g}$ and $E_{2u}$ (*I.Rs*) the following MOs of benzene are obtained.

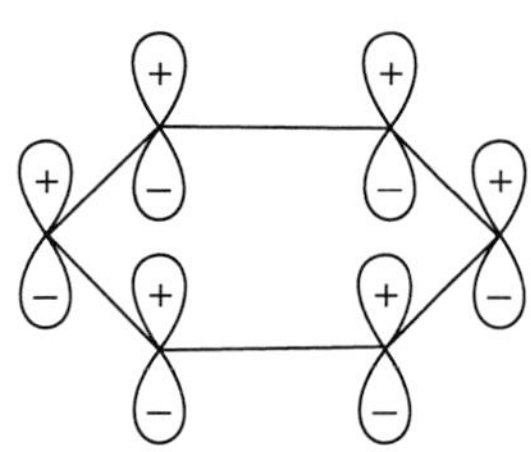


**Fig. 9.11** $2p_z$-orbitals of Benzene.

$$\left.\begin{aligned}
\psi_1^{A_{2u}} &= \frac{1}{\sqrt{6}}[\phi_1 + \phi_2 + \phi_3 + \phi_4 + \phi_5 + \phi_6] \\
\psi_2^{B_{1g}} &= \frac{1}{\sqrt{6}}[\phi_1 - \phi_2 + \phi_3 - \phi_4 + \phi_5 - \phi_6] \\
\psi_3^{E_{1g}}(a) &= \frac{1}{\sqrt{12}}[2\phi_1 + \phi_2 - \phi_3 - 2\phi_4 - \phi_5 + \phi_6] \\
\psi_4^{E_{1g}}(b) &= \frac{1}{2}[\phi_2 + \phi_3 - \phi_5 - \phi_6] \\
\psi_5^{E_{2u}}(a) &= \frac{1}{\sqrt{2}}[2\phi_1 - \phi_2 - \phi_3 + 2\phi_4 - \phi_5 - \phi_6]
\end{aligned}\right\} \quad \ldots(9.53)$$

and $$\psi_6^{E_{2u}}(b) = \frac{1}{2}[\phi_2 - \phi_3 + \phi_5 - \phi_6]$$

The energies of the MOs can now be calculated using Huckel approximation (HMO). We obtain,

For, $\psi^{A_{2u}} = \alpha + 2\beta;$

For, $\psi_2^{B_{1g}}$ as $E = \alpha - 2\beta$

For, $\psi_3^{E_{1g}}$ and $\psi_4^{E_g}$ as $E = \alpha + \beta$

And for $\psi_5^{E_{2u}}$ and $\psi_6^{E_{24}}$ as $E = \alpha - \beta$

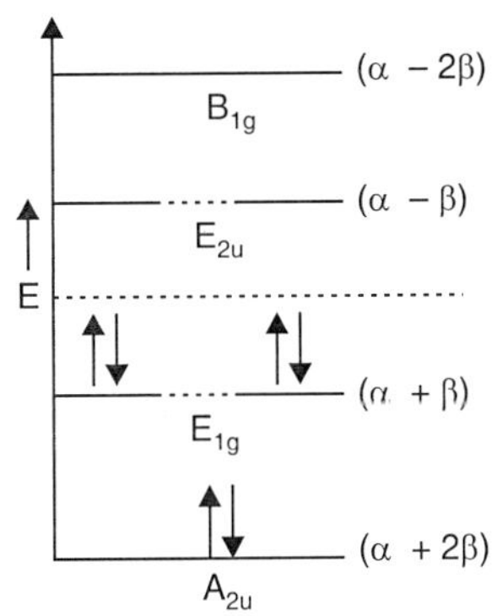


**Fig. 9.12** Energy levels of Benzene.

**Example 6:** Let us now calculate the energy level diagram for planar $C_5H_5$ molecule ($D_{5h}$ point group). Base vectors ($\phi_1$, $\phi_2$, $\phi_3$, $\phi_4$ and $\phi_5$) represent $2p_z$ orbitals of carbon atoms (No 1–5) Fig. (9.13). After obtaining $\Gamma_{red}$ for $D_{5h}$ by using these base vectors, it is reduced into its. I.R. components ($A$, $E_1$ and $E_2$). Their energies are calculated using HMO approximation. The result is as followed and is shown in [Fig. 9.13].

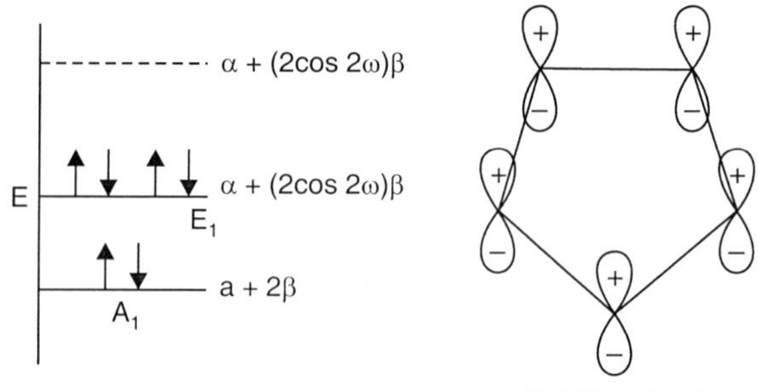


**Fig. 9.13** MO Energy of $C_5H_5$.

| *MOs* | *MO (LCAO) expression* | *Energies of MOs* |
|---|---|---|
| $\psi_1^A$ | $\frac{1}{\sqrt{5}}[\phi_1+\phi_2+\phi_3+\phi_4+\phi_5]$ | $\alpha + 2\beta$ |
| $\psi_2^{E_1}(a)$ | $\sqrt{\frac{2}{5}}(\phi_1+\phi_2\cos\omega+\phi_3\cos 2\omega)$ | $\alpha + (2\cos\omega)\beta$ |
| $\psi_3^{E_1}(b)$ | $\sqrt{\frac{2}{5}}(\phi_1\sin\omega+\phi_3\sin 2\omega-\phi_4\sin 2\omega-\phi_3\sin\omega)$ | |
| $\psi_4^{E_2}(a)$ | $\sqrt{\frac{2}{5}}(\phi_1+\phi_2\cos 2\omega+\phi_3\cos 2\omega+\phi_4\cos\omega+\phi_5\cos 2\omega)$ | $\alpha + (\cos 2\omega)\beta$ |
| $\psi_5^{E_2}(b)$ | $\sqrt{\frac{2}{5}}(\phi_2\sin 2\omega-\phi_3\sin\omega+\phi_4\sin\omega-\phi_5\sin 2\omega)$ | |

## Method of Calculating Coefficients of Basis Orbitals

In any of the above-calculated examples, we have not shown the actual calculations of the coefficients of the basis orbitals forming the $\psi_i$. One can calculate these coefficients [*viz.* $C_1$, $C_2$ ... $C_n$] in every case. This is done by substituting the energy roots [$E_1$, $E_2$ ... $E_n$] in the secular equations and solve them for the coefficients ($C_1C_2$ ... $C_n$). One obtains one set of ($C_1C_2$ ... $C_n$) for each root (value) of energy. For example, for a molecule ($C_nH_n$ or $C_nH_{n+2}$), we will be starting with '*n*' number of base vectors. These when combined linearly will give '*n*' number of SALCs which will also be the MO's of the molecule ($C_nH_n$ or $C_nH_{n+2}$), transforming according to a particular I.R. of the molecular point group. These MO's when substituted in the secular equations will give its energy roots. Thereby the '*n*' number of SALCs will give '*n*' number of energy roots. (One root for one MO of the molecule.)

The other method of calculating energy values, is by starting with linearly combined basis vectors with unknown coefficients $\left(viz., \sum_{i=1}^{n} C_i\phi_i\right)$ and substituting them in the energy equation of Schrödinger's equation, *viz.*

$$E = \frac{\int \left[\sum_{i=1}^{n} C_i\phi_i\right]^* H \left[\sum_{i=1}^{n} C_i\phi_i\right]_i \partial\tau}{\int \left[\sum_{i=1}^{n} C_i\phi_i\right]^* \left[\sum_{i=1}^{n} C_i\phi_i\right] \partial\tau} \quad \ldots(9.54)$$

This energy equation is then minimized by differentiating it with respect to $C_i$'s and equating then to zero. Thereby we get '$n$' number of homogenous linear equations in $C_{ij}$'s which are termed as secular equations. To solve these equations for the non-trivial values of $C_{ij}$'s ($C_y \neq 0$), one gets a secular determinant which has to be zero. In the process of solving the determinant, a polynomial equation of the nth order in $E$ is obtained. It is solved by using HMO approximation giving '$n$' number of roots of $E$. One by one each root of $E$ is substituted in the secular equation and a set of $C_{ij}$'s are obtained corresponding to the substituted root of $E$ in the secular equation. In this way one gets $n$-such sets of $C_{ij}$ for $n$-number of MOs. For example, for a particular value of energy $E$, we get a MO as,

$$\psi_i^{MO}(E_1) = C_{11}\phi_1 + C_{12}\phi_2 + C_{13}\phi_3 + C_{14}\phi_4 + \ldots\ldots C_{in}\phi_n \quad \ldots(9.55)$$

The set of [$C_{ij}$'s] for $E^{th}$ root of energy (obtained from the secular determinant) will give an expression (or the LCAO) for $\psi_i(E_1)$. In this equation, one finds the relative contributions of orbitals in forming the $\psi_i^{MO}(E_L)$. By the values of $(C_{ij})^2$ one can also find the sign of the orbital given by sign of $C_{ij}$. That is, if the sign is negative, the contributing $p_z$-orbital (*i.e.*, $\phi_i$) will be represented as  in place of . In this fashion one can find the nodal point in the MOs $\left[\psi_i^{MO}(E_1)\right]$. This is illustrated in Fig. (9.14) as an example.

These nodal points (or surfaces) play very important role in the relative energies of the MOs of any molecule. The rule is,

$\psi_i^{MO}(E_1) = C_{11}\phi_1 + C_{12}\phi_2 - C_{13}\phi_3 - C_{14}\phi_i$

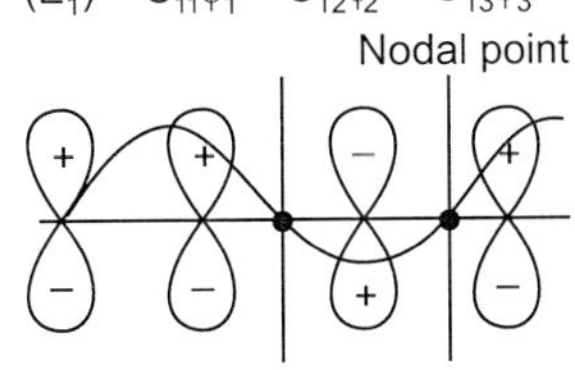


**Fig. 9.14** Nodal Point in a $\psi_i^{MO}(E_1)$.

*"For any molecule, the number of nodes in a π-type MO's increases with the increase in energy of the MO"*. Thus, if we take the case of butadiene molecule as shown in Fig. 9.6, there will be no node in the ground state wave function *i.e.*, in $\psi_{M.O.}$ $(A_u)$, one node in the second $\psi_{M.O.}(B_g)$, two nodes in the third $\psi_{M.O.}(A_u^*)$ orbital and three nodes in the fourth $\psi_{M.O.}(B_g^*)$ as shown in Fig. 9.15. When there is a MO which is double degenerate, (*E*-type I.R.), the number of nodes in both *E*-type orbitals is the same. Thus, in the case of benzene, $A_{2u}$ orbital has no node, $B_{1g}$-one node, $E_u$-two nodes, in both partner MO's and $E_2$ three nodes in both partner orbitals.

These numbers of nodes play very important role in the discussion of the role of symmetry in the chemical reactions (Woodward Hoffmann rules, *see* chapter ten).

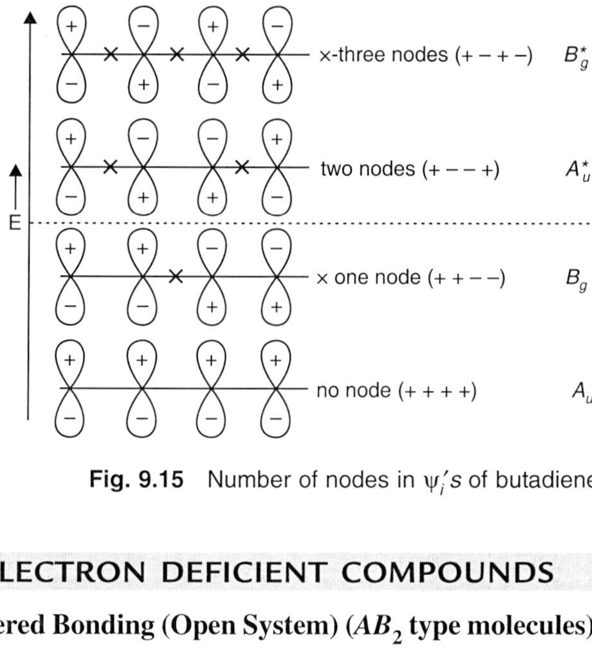


**Fig. 9.15** Number of nodes in $\psi_i$'s of butadiene.

## 9.3 ELECTRON DEFICIENT COMPOUNDS

**Three-Centered Bonding (Open System) (*AB*$_2$ type molecules):** This type of σ-bonding finds application in boron hydrides (electron deficient), zero group halides, or interhalogen molecules, (electron rich compounds)]. In these compounds two distinct types of bonding are present depending on the kind of orbitals, which participate in the bond formation. *In case I*, the central atom uses an orbital which is symmetric with respect to reflection $\sigma_v(xz)$ operation in a mirror plane perpendicular to the *B-B* line. Let this orbital be $\phi_2$. *In case II*, the central atom uses an orbital which is antisymmetric with respect the

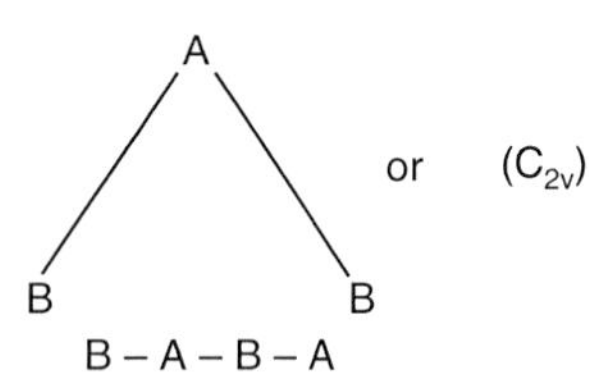


**Fig. 9.16** Three-centred bonding.

$\sigma_v(xz)$ plane. Let this orbital let $\phi_4$. In these both cases, it will be found that the final result is the same, except that in one case the system is symmetric and in the second case, it is antisymmetric with respect to reflection in a plane perpendicular to *B*-*B* line. In both cases, we will assume the symmetry of the system as $C_S$ which is a subgroup of both $C_{2v}$ and $D_{\infty h}$. In $C_S$ group, the orbital on atom *A* (to be used in case I) belongs to A type of I.R. while that in case II, the orbital on atom A belongs to *B*-type of I.R. of the group ($C_S$).

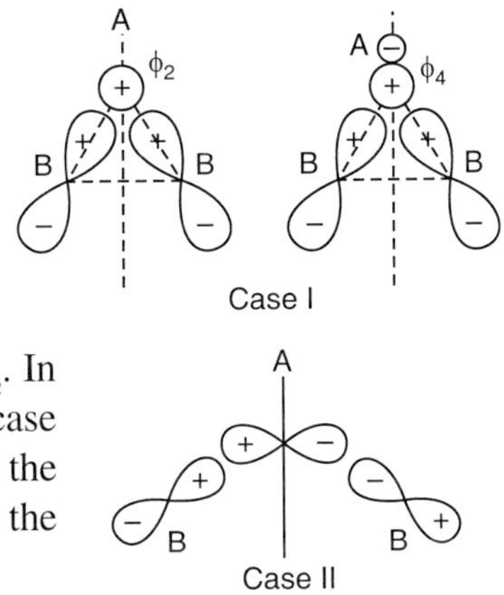


**Fig. 9.17** Symmetric and antisymmetric orbitals.

We make SALCs of the orbitals $\phi_1$ and $\phi_3$ on atoms *B*. Let us assume that the σ orbitals on *B* atoms of $AB_2$ molecule are $\phi_1$ and $\phi_3$. As usual, the two SALCs will be $\frac{1}{\sqrt{2}}(\phi_1+\phi_3)$ and $\frac{1}{\sqrt{2}}(\phi_1-\phi_3)$. $\frac{1}{\sqrt{2}}(\phi_1+\phi_3)$ belongs to *A* type of I.R. while the SALC $\frac{1}{\sqrt{2}}(\phi_1-\phi_3)$ belongs to *B* type of I.R. of $C_s$ point group. Thus the linearly combined $\frac{1}{\sqrt{2}}(\phi_1+\phi_3)$ orbital will combine with $\phi_2$ of atom *A* [both orbitals belong to the same symmetry *viz.* (A type of I.R.)] while in the second case, $\frac{1}{\sqrt{2}}(\phi_1-\phi_3)$ will combine with $\phi_4$ of atom *A* (both of *B* symmetry). Thus, we get the following secular equations.

| Case I | Case II |
|---|---|
| $C_{11}(H_{11}-E)+C_{12}H_{12}=0$ | $C'_{11}(H'_{11}-E)+C'_{12}H'_{14}=0$ |
| $C_{21}H_{21}+C_{22}[H_{22}-E]=0$ | $C'_{21}H'_{41}+C'_{22}[H'_{44}-E]=0$ |

| Case I | Case II |
|---|---|
| A-type $\begin{vmatrix}\alpha-E & \sqrt{2}\beta_{12}\\ \sqrt{2}\beta_{12} & H_{13}-E\end{vmatrix}=0$ | $\begin{vmatrix}\alpha-E & \sqrt{2}\beta_{14}\\ \sqrt{2}\beta_{14} & H_{13}-E\end{vmatrix}=0$ |

Here $\alpha=\frac{1}{4}\int\phi_1^* H\phi_L\partial\tau^*$ and

$$H'_{13}=H_{13}=\int(\phi_1+\phi_3)H(\phi_1+\phi_3)\partial\tau \quad\text{and}\quad \int(\phi_1-\phi_3)H(\phi_1-\phi_3)\partial\tau$$

$$=\int\phi_1 H\phi_1\partial\tau+\int\phi_3 H\,\phi_3\partial\tau\pm 2\int\phi_1 H\,\phi_3\partial\tau$$

$$=\alpha'+\alpha'\pm 2\beta_{13}=2\alpha'\pm 2\beta_{13}=2\alpha'+0 \qquad (\text{Here }\beta_{13}=0)$$

If we put $\alpha = 0$, and $\alpha' = 0$, we get the final secular determinant as under

Case I

$$A : \begin{vmatrix} -E & \sqrt{2}\beta_2 \\ \sqrt{2}\beta_3 & -E \end{vmatrix}$$

which gives $E = \pm\sqrt{2}\beta_2$

$E = \pm\sqrt{2}\beta_{12}(\beta_{13} = 0)$ and

$B : E = 0$ because $\beta_{13} = 0$

Case II

$A : E = 0$ because $\beta_{13} = 0$ and

$$\begin{vmatrix} -E & \sqrt{2}\beta_{12} \\ \sqrt{2}\beta_{12} & 0 - E \end{vmatrix}$$

$A : E = 0$ because $\beta_{13} = 0$

and $B = E = \pm\sqrt{2}\beta_{12}$

To get the coefficients $C_{11}$, $C_{12}$ and $C_{22}$, we substitute the values of $E(\pm\sqrt{2}\beta_{12})$ in secular equation and solve for the coefficients. We get the following expressions for MO's.

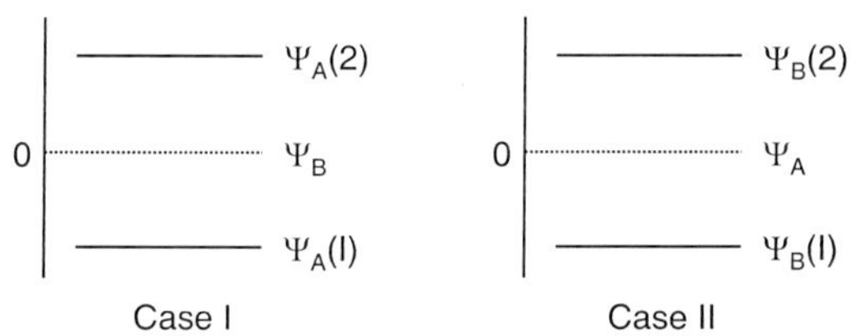


**Fig. 9.18** MOs of $AB_2$ type Molecules.

Case I

$$\psi_A(1) = \frac{1}{2}(\phi_1 + \sqrt{2}\phi_2 + \phi_3)$$

$$\psi_B(2) = \frac{1}{2}(\phi_1 - \sqrt{2}\phi_2 + \phi_3)$$

$$\psi_B(3) = \frac{1}{\sqrt{2}}(\phi_1 - \phi_3)$$

Case II

$$\psi_A(1) = \frac{1}{\sqrt{2}}(\phi_1 + \phi_3)$$

$$\psi_B(2) = \frac{1}{2}(\phi_1 + \sqrt{2}\phi_4 - \phi_3)$$

$$\psi_B(3) = \frac{1}{2}(\phi_1 - \sqrt{2}\phi_4 - \phi_3)$$

The results of the above-discussed open three-centered bonding indicate that in both cases, the energy level of MOs is the same except the type of symmetry of these levels. The energy levels are (1)-bonding, (2)-non or weakly bonding and (3) antibonding level.

Now, we have two types of molecules in which the three-centered bonding scheme can be applied. There are those molecules like $B_2H_6$ in which only one pair of electrons is available for the three-centered bond formation as in *B-H-B* bonds in $B_2H_6$. The electrons (one pair) will occupy the bonding orbital and the expression for the $\psi_{Bond}$ will be $\frac{1}{2}(\phi_1 + \sqrt{2}\phi_2 + \phi_3)$. It implies that the bonding pair of electrons is nearly equally delocalised over the three centres of the molecule. These are known as electron-deficient molecules (*i.e.*, the number of electron pairs are less than the number of bonds to be formed).

The second type of molecules having three-centered bonding are those in which there are four electrons (two pairs of electrons) which occupy the MOs. One pair will occupy the bonding orbital which is evenly distributed over three centres giving a strong bond of three centres. The second pair of electrons enters a non-bonding orbital where the electrons are centered on the two atoms *A* and $B_2$. This contributes towards the polarity of the bond because of the concentration of the electrons of the second pair on the *B* atoms of $AB_2$ (*i.e.*, ionic character of bond increases). These types of bonds are present in the zero and halogen group compounds.

**(II) The Closed Three-Centered Bonds:** In the open three-centered bonding, we have taken no interaction between two *B* atoms of $AB_2$ molecule. Because of this we have taken $\beta_{13}$ to be zero. But there are molecules where there is definite interaction between $B_1$ and $B_2$ of the $AB_2$ molecule. When the effect of $\beta_{13}$ on the result becomes too great to neglect $\beta_{13}$, we have a closed three-centered bonding. We have to include $\beta_{13}$ in the secular determinant. Now, if we go to the extreme case of including it in the secular determinant and assume its value to be equal to $\beta_{12}$ (*i.e.*, $\beta_{13} = \beta_{12}$), we have the following secular determinant

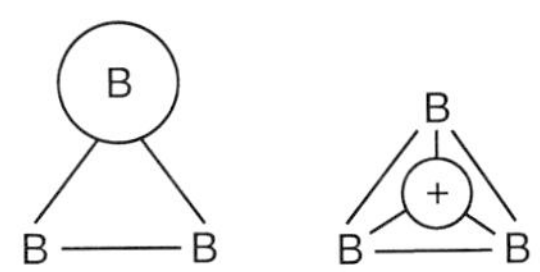


**Fig. 9.19** Closed three-Centered Bonding.

$$\begin{vmatrix} -E & \sqrt{2}\beta_{12} \\ \sqrt{2}\beta_{13} & \beta_{12} - E \end{vmatrix} = 0$$

The solution of this equation gives the roots of *E* to be equal to $(-\beta_{12})$ and $2\beta_{12}$. This difference makes the energy level of *B*-type of orbital to $(-\beta)$ in place of zero (assuming $\alpha = 0$). Thus, as the interaction between the two *B*-atoms of $AB_2$ is changing [*i.e.*, the value of $\beta_{13}$ is changing from zero to $\beta_{12}$], the energy of *B*-type of MO is also changing. This effect is illustrated in Fig. (9.20). It is seen that when $\beta_{13} = \beta_{12}$ and all atoms

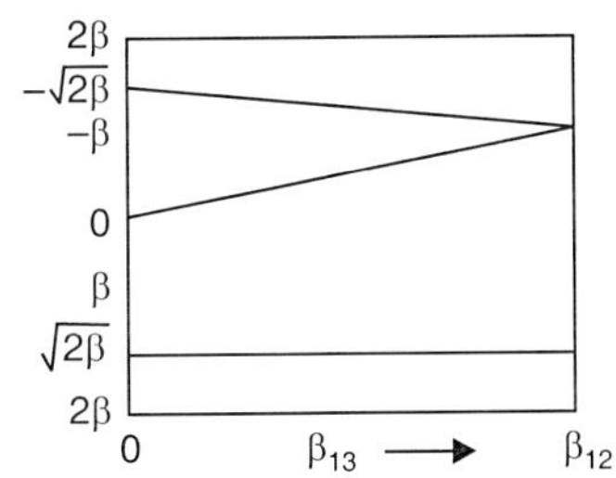


**Fig. 9.20** Effect of increasing $\beta_{13}$ on energy levels of $AB_2$.

are assigned the same α-value (*i.e.*, $B_3$ type molecules) then the symmetry of $B_3$ molecules is $C_3$. In that case, we will have the two orbitals of the same energy. That is, they will be doubly degenerate. This is analogous to the π-orbital system of cyclopropyl radical.

## 9.4 SYMMETRY OF THE ATOM

In the above-said discussion, we have talked about the symmetry of the molecules whose point groups consist of a finite number of symmetry operations. Such point groups are easy to deal with and their applications in determining the properties of molecules have been relatively easy. These molecules are composed of atoms whose symmetries have not been studied so far. We generally assume them to be spherical. The potential energy of the electrons in the field of nuclear charge is spherically symmetrical which has an infinite number of rotational and reflection symmetric operations leading to the point group of infinite order. Mathematically, such groups are difficult to deal. These groups are known as *Lie groups*. With the proposed length of the present book, we cannot discuss the properties and applications of these groups at this point. We therefore, propose that the students should refer any book of group theory in which the continuous groups are a part of the subject matter. These groups are of utmost importance in the nuclear and particle physics studies. One should, however, realise that the quantum numbers. *s*, *p*, *d*, *etc*., terms are the I.Rs. of the infinite continuous point groups. It is possible to derive the complete structure of hydrogen like atoms by the study of the continuous groups.

# 10

# Symmetry and Reactivity

Atomic and Molecular orbitals are a purely mathematical concept without any physical interpretation. It, however, provides a basis for understanding the structure of bonding and chemical changes in molecules. It also provides a concept of energy and probability of particle (electron) being present in an infinitesimal element of space. A qualitative relationship exists between the electron density in an orbital and the chemical reactivity (*e.g.*, in electrophilic substitution reactions *etc.*) There are reactions whose mechanism can be interpreted by involving "Frontier Orbitals" (*i.e.*, Highest Occupied Molecular Orbital (HOMO) and Lowest Unoccupied Molecular Orbital (LUMO). For example, K. Fukui of Japan and R.D. Woodward, R. Hoffmann of U.S. have emphasized the concept of HOMO and LUMO in certain specific reactions. In the past couple of decades there have been attempts to correlate molecular symmetry with mechanisms of chemical reactions. Bond-formation and bond-breaking in the transition states of the chemical reactions are controlled by the molecular orbital symmetries. Consequently chemical reaction mechanisms involving interactions of orbitals are expected to be related to the orbital symmetries (or molecular symmetry). One may therefore, frame certain symmetry rules in terms of orbital symmetry or bond symmetries that ought to be followed along the paths of reactions. Such attempts have been made on both theoretical and experimental basis. Three of such attempts were made by (1) Woodward and Hoffmann, who proposed some rules about conservation of orbital symmetries (2) K. Fukui and Woodward proposed Frontier orbital theory (3) Zimmermann and M.J.S. Dewar, who proposed P.M.O. theory to provide a qualitative interpretation of stereospecificity of pericyclic reactions. Later on, Pearson has framed a set of rules which must be followed during the course of reactions in general. In this chapter, we will qualitatively and briefly introduce these approaches. We will discuss some reactions based on symmetry arguments.

## 10.1 DISCUSSION OF A FEW SIMPLE REACTIONS

(*a*) ($H_2 + H^+$) Reaction: Consider the reaction.

$$H_2 + H^+ \rightarrow H_3^+ \qquad ...(10.1)$$

In the Frontier orbital concept, $1\sigma_b^2$ -orbital of $H_2$ will act as HOMO and as a base and $1\sigma_a^*$ will behave as LUMO of $H_2$. The reactant $H^+$ has empty $1s$-orbital and acts as Lewis acid in the above acid-base reaction. Electron density from HOMO ($1\sigma_b^2$) is transferred to LUMO ($1s$) of $H^+$. Since these orbitals belong to the same symmetry so they will interact and a shift of electron density from HOMO of $H_2$ to the LUMO of $H^+$ ($1s$) will take place forming $H_3^+$ molecular ion [Fig. 10.1].

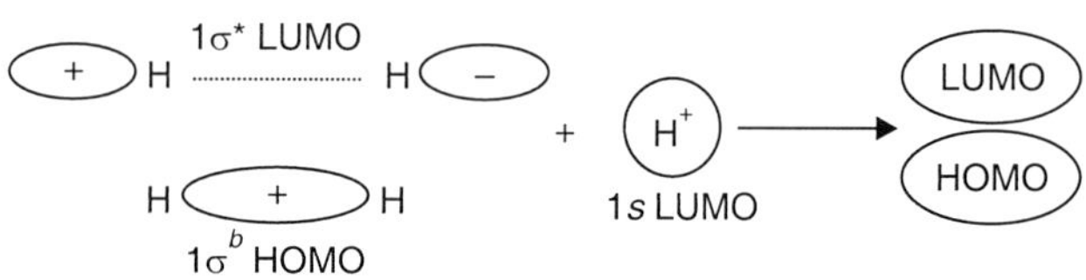


**Fig. 10.1** HOMO-LUMO Interaction of ($H_2 + H^+$).

(*b*) $[B_2H_4]^{2-} + 2H^+ \rightarrow [B_2H_6]$ ...(10.2)

Reaction: It is a hypothetical reaction.

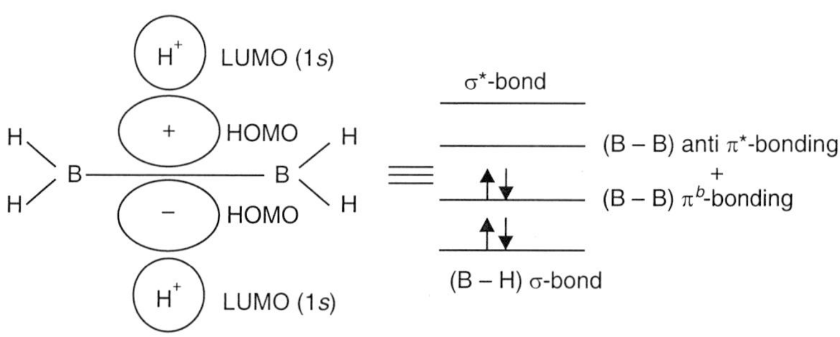


**Fig. 10.2** HOMO-LUMO Interaction in $B_2H_4 + H^+$.

$[B_2H_4]^{2-}$ is isoelectronic to $C_2H_4$, therefore, we assume that its structure is the same as $C_2H_4$. Its HOMO is $\pi^b$. Since in this reaction $H^+$ acts as an acid, so its LUMO is 1*s*. $H^+$ being Lewis acid will withdraw electron density from $\pi^b$ (HOMO) into the LUMO of $H^+$ (1*s*) in order to form $[B_2H_6]$.

One may conclude qualitatively from the above two reactions that in any acid-base reaction, the electron density shifts from HOMO of base to LUMO of acid. One can extrapolate it further that in a large number of important reactions between the orbital of reactants, (HOMO-LUMO) pairs are involved. These orbitals being least bound energetically, are most readily available for interactions with those of other molecules.

## 10.2 PEARSON'S RULES

A very large number of acid base and other different bimolecular reactions have been analysed on the HOMO-LUMO concept. The results of all these reactions have been summarised in a set of rules by R.G. Pearson. These rules have to be obeyed if the two reactants are to undergo a facile reaction. These rules are:

1. As the two reactants approach towards each other, electron density from the HOMO of one reactant (base) flow to LUMO of the other reactant (acid).
2. The HOMO of the donor and the LUMO of the acceptor must approach towards each other so as to make a net positive overlap between HOMO and LUMO.
3. The HOMO of the donor and LUMO of the acceptor must be close in energy (~5 eV).
4. The net effect of HOMO → LUMO electron density transfer must result in bond formation or/and bond-rupture during the course of reaction.
5. Besides these rules, R.G. Pearson used another approach based on symmetry termed as "*The Bond Symmetry Rule*". It says that *a reaction is allowed, if the symmetry species of the bonds that are formed is the same as the symmetry species of bonds that are broken*".
6. If HOMO is antibonding, then removing of electrons from it leads to strengthening of the bond. If LUMO is antibonding, then it leads to weakening of the involved bond.
7. While considering the effective symmetry (point 5) one must also consider the lone pairs of electrons if these become bonding or when bonding electrons become lone pairs.
8. The symmetry of the reacting system that has to be taken is the "Effective Symmetry" (Point group). This symmetry should remain invariant during the course of reaction.
9. The above rules are also applicable to unimolecular reactions.

10. For unimolecular reactions, both HOMO and LUMO belong to the same reacting species. These orbitals may not be of the same symmetry species. Consequently the reaction should not take place. In that case the symmetry rule is slightly modified. For unimolecular reaction, the electron density transfer leads to electron density change within the molecule. Consequently the nuclei in the molecule move into the region of high electron density. In that case we take the *Direct* product of the symmetry-species of HOMO and LUMO. The direct product must belong to the same symmetry species as that describing the nuclear motion that corresponds to the normal modes of vibration of the point group. (*see* point 12).
11. If the point group of the combined product (*i.e.*, product of reaction) has higher symmetry than that of the combining molecules (*i.e.*, reactants along the reaction path), the point group of the combining reactants must be a subgroup of the larger group. For example, for the dimerisation of $C_2H_4$ to $C_4H_8$ the product $C_4H_8$ belongs to $D_{4h}$. The combining reactants along the reaction path have $D_{2h}$ or $C_{2V}$ symmetry which is the subgroup of $D_{4h}$.
12. For a bond to be broken, one must promote electrons from HOMO to LUMO to remove electron density from bonding orbital or add electron to the corresponding antibonding orbital. The direct product of the symmetry species involved in the promotion of electrons must belong to the same symmetry species as the reaction coordinates.
13. The symmetry of the reaction coordinates describe the symmetry of nuclear motion involved (conrotatory or disrotatory).

To follow these rules one should know the point group to which the reacting system belongs. This is required to determine the symmetry species of the bonds to be broken and formed. We determine the effective symmetries (or the point group) of the reacting system along with the lone pairs of electrons if these are taking part in the chemical reaction or if the pairs of electrons are formed along the reaction coordinates or if the pairs of electrons of a bond becomes a lone-pair after the reaction is complete. We will illustrate it by the following examples:

**Example 1:** Consider the following reaction:

$$H_2 + C_2H_4 \rightarrow C_2H_6 \quad ...(10.3)$$

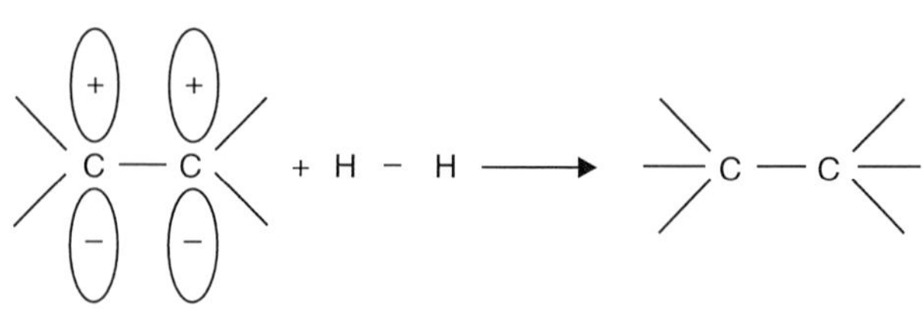

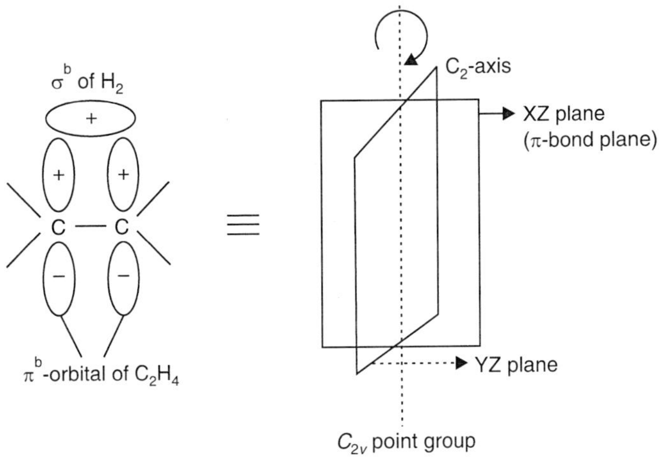


**Fig. 10.3** Symmetry of bonds.

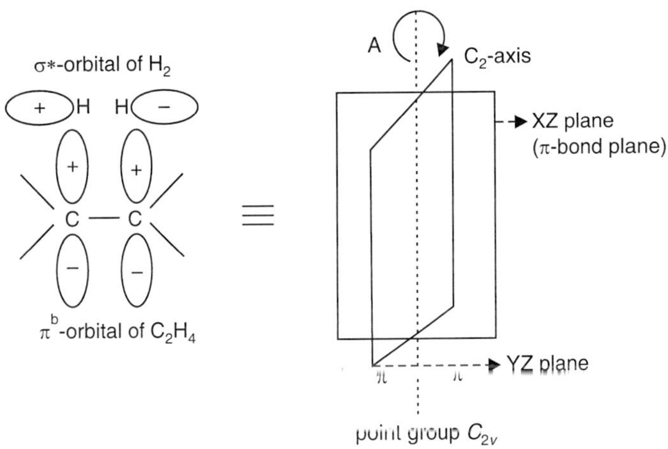


**Fig. 10.4** HOMO + LUMO of $C_2H_4$ + $H_2$.

In the reaction, one $\pi$-bond of $C_2H_4$ and one $\sigma$-bond of $H_2$ are broken and two C-H $\sigma$-bonds are formed (Fig. 10.3). If we assume the transition state as shown in (Fig. 10.3) then it is obvious, that we can take the point group to which the system (bonds-broken) belongs as $C_{2V}$ point group. Similarly, if the bonds formed are due to shifting electron density from HOMO of $C_2H_4$ ($\pi$-orbital) to the LUMO of $H_2$, then the point group to which the system belongs is also ($C_{2V}$).

Now, if we assume HOMO of $C_2H_4$ and $1s^2$-orbital of $H_2$ as LUMO, then the symmetry species of $C_{2V}$ to which these orbital belong are both $A_1$ type (*i.e.*, both belong to $A_1$ irreducible representation of $C_{2V}$). But we cannot assume any one of these orbital as LUMO, because both orbital are filled and the electron density cannot be shifted from one to another.

Now we take HOMO as π-orbital of $C_2H_4$. It belongs to $A_1$ irreducible representation of $C_{2V}$ (Fig. 10.4). As LUMO we take $1\sigma_{1s}^*$-orbital of $H_2$ which is empty. This orbital ($1\sigma_{1s}^*$) belongs to $B_2$ symmetry species of $C_{2V}$. Since π- orbital is filled while $1\sigma_{1s}^*$ is empty, electron density can flow from π to $1\sigma_{1s}^*$ empty orbital of $H_2$. In this way two bonds can be formed.

We now apply the Pearson's symmetry rule.

**Symmetry of bonds broken:** Two bonds (1) H-H bond ($1\sigma^2$) of symmetry $A_1$ point group $C_{2v.}$ (2) *C-C* π bond of $A_1$ symmetry (Fig.10.3) both belong to $A_1$ symmetry.

**Symmetry of the bonds formed:** Two (*C-H*) bonds (σ-type). To find the symmetry species of $C_{2V}$ point group, we make the linear combinations (SALC's) of these bonds, $\left(\frac{\sigma_{1s}+\sigma_{1s}}{\sqrt{2}}\right)$ and $\left(\frac{\sigma_{1s}-\sigma_{1s}}{\sqrt{2}}\right)$. These belong to $A_1$ and $B_2$ symmetry species respectively.

Thus, the symmetry species of the bonds broken are $A_1$ and these are not the same as those of bonds formed *i.e.*, $B_2$.

The reaction is forbidden because the symmetry species of the bonds (broken)/are not the same as the bonds formed.

The above example is that of bimolecular reactions. Now we take another example of *unimolecular* reaction.

**Example 2:** The following reaction is an example of *unimolecular* reaction.

$$C_4H_8 \rightarrow 2\ C_2H_4$$

In unimolecular reactions, the bonds (broken) and the bonds (formed) belong to the same molecule. Thus, HOMO and LUMO orbitals belong to the same molecule. These orbital should be of similar energies ($C_4H_8$). Thus the electron shift takes place within the same molecule (cyclobutane).

Another important point in these reactions is that during the shift of electron density, the nuclei also shift from one place of high electron density to their new place of the shifted high electron density. In these reactions if the bonds broken and bonds formed do not belong to the same symmetry, in case they belong to *different* symmetry, then the condition is that the direct product of the symmetry species of HOMO and of LUMO must belong to the *same* symmetry species, as that (symmetry species) which describe the nuclear motion (vibrational) which corresponds to one or more normal modes of vibrations of the molecule. In reaction of cyclobutane, the cyclobutane goes to the two molecules of ethylene. In both forward (unimolecular) and the backward reaction (bimolecular) the symmetry rules must be followed.

In example (under discussion), we have the energy level diagram as shown in Fig (10.5). We have assumed $C_{2v}$ point group.

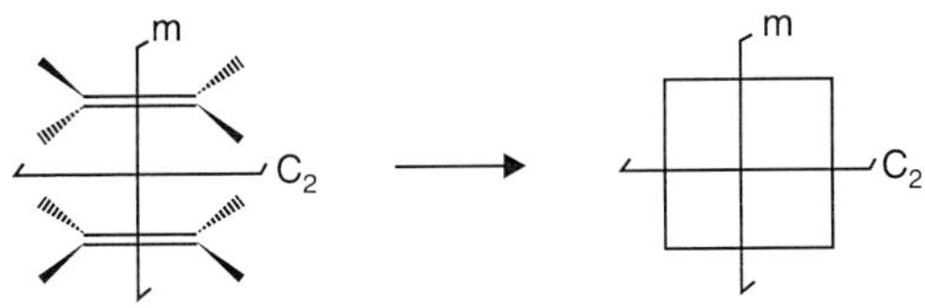


**Fig. 10.5(a)** Symmetry conversion during cycloaddition.

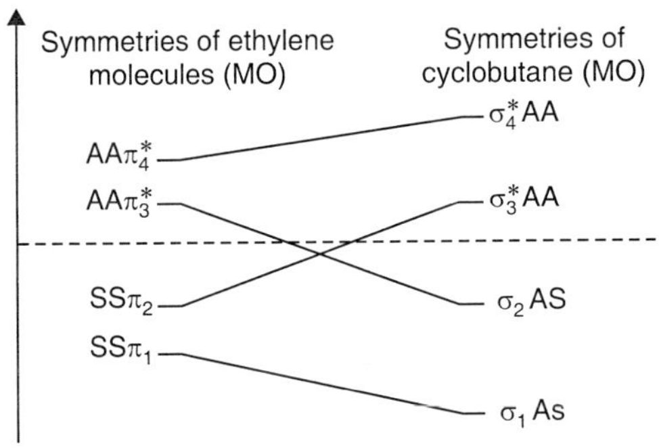


**Fig. 10.5(b)** Correlation diagram for ethylene-cyclobutane interconversion.

Let us first concentrate on cyclobutane ($C_{2V}$ symmetry), especially on the combinations of orbital between two halves (Fig. 10.5). In cyclobutane, both lower most orbitals are occupied. We have HOMO of $B_2$ symmetry species of $C_{2V}$ and LUMO of $B_1$ symmetry species. This reaction being unimolecular is the case where HOMO and LUMO belong to different symmetry species. Now according to the preceding paragraph, we take the direct product of the species of HOMO and LUMO ($B_2 \times B_1$). This is equal to $A_1$ (*i.e.*, $B_1 \times B_2 = A_1$). Therefore, if we want to have concerted reaction (*i.e.*, splitting of cyclobutane is a concerted process) we require symmetrical separation of parallel ethylene molecules with vibrational mode of $A_1$ symmetry. But the correlation diagram (Fig. 10.5) shows that the HOMO ($B_2$ species) correlate with LUMO of $B_1$ symmetry species and not with $A_1$ (*i.e.*, $\pi^*$-orbital of ethylene) which is of higher symmetry, hence the reaction is symmetry forbidden.

## 10.3 OXIDATIVE-ADDITION REACTIONS

Let us consider the following reactions which are termed as oxidative addition reactions.

$$TlCl + Cl_2 \rightarrow TlCl_3 \qquad ...(10.4)$$

$$PbCl_2 + Cl_2 \rightarrow PbCl_4 \qquad ...(10.5)$$

These reactions are shown in Figs.10.6 and 10.7. Both of these reactions are concerted reactions. We now analyse them on Pearson's symmetry rule and HUMO-LUMO rules given in the preceding paragraphs. We assume the point group of the transition state as $C_{2v}$.

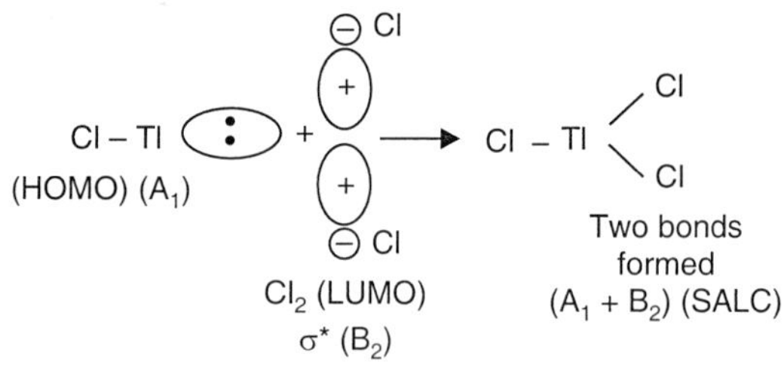


**Fig. 10.6** Bond broken and formed.

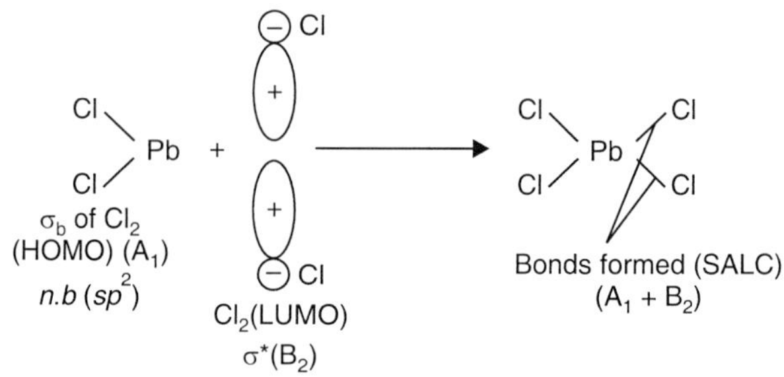


**Fig. 10.7** HOMO and LUMO of $PbCl_2$ + $Cl_2$.

Bond Broken (1) Lone pair on *Tl* or $PbCl_2$ is $C_{2v}$ point group, lone pair belongs to $A_1$ symmetry species. (2) M-Cl $\sigma_b^2$-bond having the $A_1$ symmetry under $C_{2V}$ point group–these are HOMO's *Bond formed.* Two σ-bonds (*i.e.*, $\sigma_1 + \sigma_2$) these are linearly combined and their SALCs are formed under $C_{2v}$ point group. The SALCs are $\left[\frac{\sigma_1 + \sigma_2}{\sqrt{2}}\right]$ of $A_1$ symmetry and $\left[\frac{\sigma_1 - \sigma_2}{\sqrt{2}}\right]$ of $B_2$ symmetry species.

Thus, the symmetry of bond broken $(A_1 + A_1)$ is not the same as that of the bond formed $A_1 + B_2$. Since symmetry is not identical the reaction is therefore forbidden.

One can therefore analyse any number of such reactions on the Pearson's symmetry rule.

After discussion, of these very simple reactions to illustrate the importance of (HOMO-LUMO) and symmetry rule of Pearson, we will take up the analysis of pericyclic reaction. This forms a group of organic reactions which are very important. These were not understood prior to 1960. It is only in the decade of 1960, that a group of scientists independently came out with three different models based on Huckel's MO theory to explain "Concerted Reactions". The examples of these reactions and their analysis on the basis of different models are given as follows.

## 10.4 PERICYCLIC REACTIONS

Amongst the various organic concerted reactions one extremely important set of reactions is "Pericyclic Reactions". The mechanisms of these reactions have been investigated in a great detail by three groups. (1) R.B.Woodward and R.Hoffmann, (2) K. Fukui, (3) M.J.S. Dewar during 1960 decade. They have attacked the mechanism problem by different approaches. We will discuss them separately at a later place in the section.

### 10.4.1 Definition

Pericyclic reactions are those reactions, where electrons reorganise around cycle of nuclei, with simultaneous bond-breaking and bond-forming in a concerted way. The group of such reactions is known as concerted reactions. These reactions involve transition states wherein all first order bond breaking and forming changes over a closed curve (Fig. 10.8).

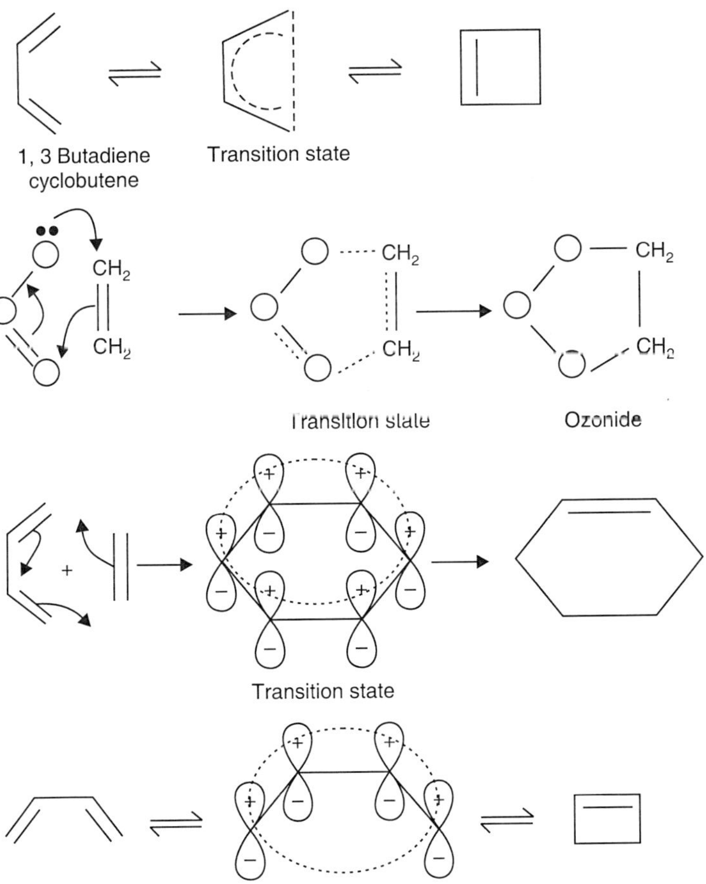


**Fig. 10.8** Transition states for some reactions showing closed curve orbital interactions.

### 10.4.2 Characteristics of Pericyclic Reactions

We have the following characteristics of the pericyclic reactions:

(*a*) Transition state is cyclic or closed curve in nature.

(*b*) In the transition state, bond-breaking and bond-formation are simultaneous; but not necessarily synchronous.

(*c*) The reaction does not proceed by a stepwise mechanism.

(*d*) No intermediates are formed.

### 10.4.3 Classification of Pericyclic Reactions

Broadly pericyclic reactions are mainly classified into the following categories:

(*a*) Electrocyclic Reactions

(*b*) Cycloaddition and Cycloreversion Reactions

(*c*) Sigmatropic Rearrangements

(*d*) Cheletropic Reactions

(*e*) Group transfer Reactions

Out of the above said pericyclic reactions, we will discuss electrocyclic and cycloaddition reactions as examples. It is impossible to discuss all the reactions in detail on the basis of various models used to analyse them because of the limited size of the book.

**(*a*) Electrocyclic Reactions:** In this category of reaction σ-and π-bonds are interconverted. For example, in the following reaction one π-bond of 1, 3-butadiene becomes one σ-bond of cyclobutene and *vice versa*.

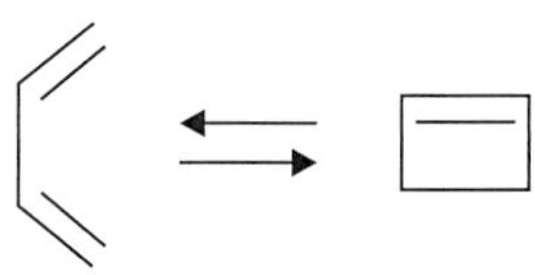

...(10.6)

The characteristic features of such reactions are:

1. These reactions require heat or light energy for initiation.
2. Their mechanisms do not involve radical or ionic intermediates. These are monomolecular concerted reactions involving a cyclic transition state.
3. These reactions are highly stereospecific (in the concerted reactions, molecular orbital of the reactants are continuously converted into those of products).

**(*b*) Cycloaddition and Cycloreversion Reactions:** In these reactions, two molecules of ethene or/and polyene react to form a cyclic product in a concerted manner. For example,

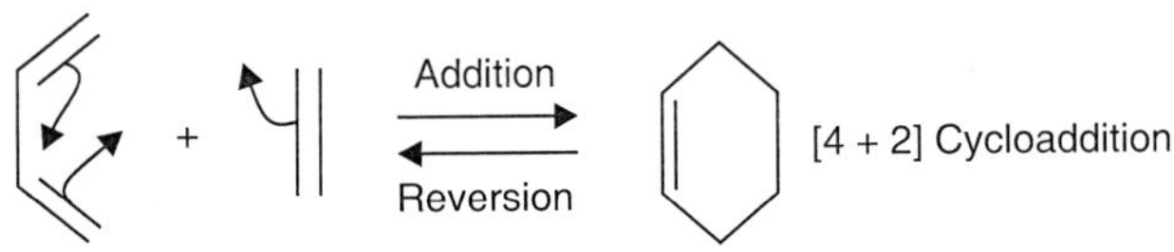

...(10.7)

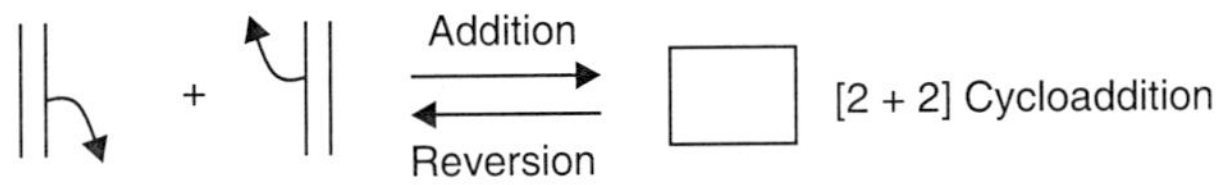

...(10.8)

These are bimolecular reactions and are classified on the basis of the number of π-electrons involved in each reactant. Thus, in example (*i*) 4π electrons are involved in diene while two π-electrons are in ethylene, so it is termed as [4 + 2] cycloaddition reaction. The back-reaction is known as cycloreversion reaction. The example (*ii*) is [2 + 2] cycloaddition reaction.

Cycloaddition or/and cycloreversion reaction are also stereospecific, initiated by heat or/and light energy. In these reactions σ-and π-orbitals are interconverted and no ionic or radical intermediates are involved. Bonds are formed and broken in a concerted single step.

## 10.5 MODELS OF ANALYSING THE REACTION

Prior to 1960, these reactions were poorly understood. In the following years, several chemists attempted to analyse them. But R.B. Woodward and Hoffmann, K. Fukui of Japan, H.C. Longuet Huggins of England, M.J.S Dewar and Zimmermann of USA provided three models independently for understanding how these reactions occur and why are they stereospecific. All the models are based on molecular orbital theory. These three models are:

1. Orbital symmetry correlation method (Woodward-Hoffmann).
2. Frontier orbital model of Fukui.
3. Perturbation Molecular orbital (PMO) model.

We describe them one by one in the following paragraphs.

### 10.5.1 Orbital symmetry Correlation Method to Control the Pericyclic Reactions

The method was suggested by R.Woodward and R. Hoffmann in 1965. The essential feature of this method is that during the concerted reactions, the participating molecular orbitals of the reactants pass smoothly into MOs of the products. For smooth passing of the reactants to products, the *symmetry characteristic* of the MOs of the reactants and those of products must correlate. This correlation is known as the *conservation of orbital symmetry*. It means that there must be one to one match between the symmetry species of the MOs of the reactants and products (not energy-wise but symmetry-species wise).One therefore, must know the symmetry species of the reactants and the products which are involved in the reactions. In a very simplified way, we may approximate the symmetry species of MOs of conjugated system by the mathematical sign (+ or –) of the various atomic orbitals involved in the conjugation process of the molecule. In Fig 10.9, we have drawn MOs for two atoms to six atoms chain. These can be calculated by HMO as discussed in chapter nine of the book. Since in the transition state of the reactant, it is essential that the symmetry of the bond breaking MOs must be identical to the symmetry of the bond forming orbitals, we have also indicated *S* and *A* as symmetric and antisymmetric respectively as the symmetry designations of MOs relative to a plane or $C_2$ axis. Symmetry plane and $C_2$ axis lie perpendicular to and in the plane of the molecule respectively. Both these symmetry elements pass through the centre of mass of the system which is also the origin of the coordinate system. We will learn later in the chapter that it is only these two symmetry transformations that are involved in the concerted reactions in the formation of products). There are *some generalizations* which play important role while discussing pericyclic reactions by orbital correlation method.

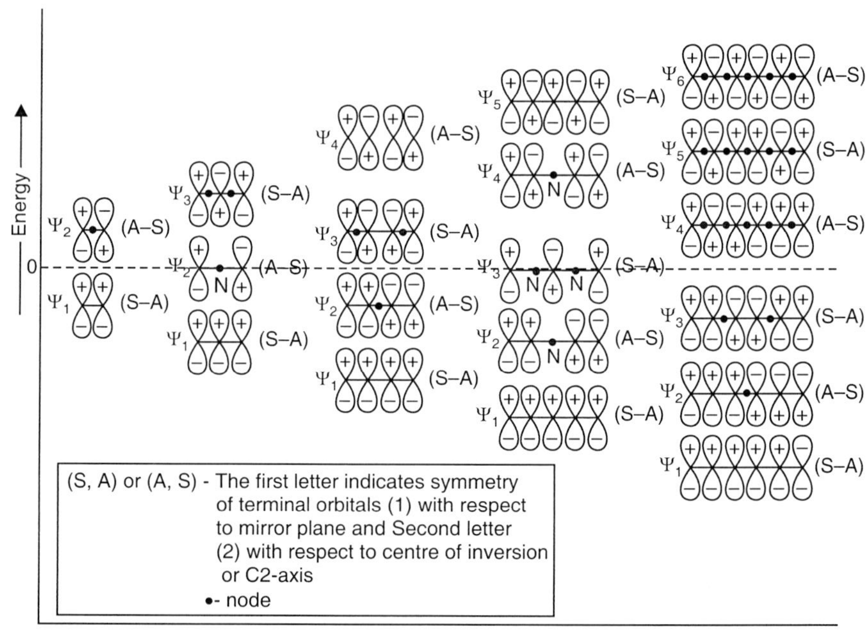


**Fig. 10.9** π-Molecular orbitals for 2 atom to 6 atom chain molecules.

These are (1) the symmetry (*S* or *A*) of the terminal orbital for a set of MOs of a molecule, alternates as the orbitals increase in energy, (2) the symmetry with respect to the mirror plane (*m*) at the centre of each molecular orbital diagram of any molecule alternates between symmetrical (*S*) and anti-symmetrical (*A*) as the MO's increase in energy (for a $C_4$-system it is *S*, *A*, *S*, *A*), (3) with respect to $C_2$ symmetry elements about the $C_2$-axis at the molecular centre and perpendicular to plane), the symmetry of the orbitals alternates between symmetrical (*S*) and antisymmetrical (*A*) as the MO's increase in energy (For a $C_5$ molecule, it is *A*, *S*, *A*, *S*, *A*).

After a very brief survey of the MOs of a few conjugated straight chain molecules, we will discuss pericyclic reactions by taking a few examples. In these reactions, Woodward and Hoffmann claimed *that orbital symmetry controls the feasibility and the stereochemical consequences of every concerted reaction.* It implies that orbital symmetry is conserved during the complete process of the reaction all the time. Essentially, we wish to devise a way by which we can determine the relative transition state energies for the particular stereochemical mode of transformation that we wish to compare during the reaction paths. For example, if we take the case of cyclisation of 1, 3, 5 hexatriene to 1, 3 cyclohexadiene, then we would like to determine if the orbital interactions in the transition state using the conrotatory rotation of the p-π orbitals on $C_1$ and $C_6$ atoms of hexatriene is energetically more favorable compared to the disrotatory motion of the same orbitals. (Two modes of the ring closure Fig. 10.10)

We would like to emphasize here that in conrotation process, rotation by 180° around the $C_2$-axis is preserved and reflection through the mirror plane is not preserved. In disrotation reflection in the vertical plane '*m*' is preserved throughout the reaction path of bond formation but $C_2$ rotation is not preserved. Since both con-and disrotations proceed via a common reactant (1, 3, 5 hexatriene, for assessment of the relative transition state, energies with these, symmetries are necessary to know of the reaction path. So we can assign the point group of the system having $C_2$ operation as $C_2$ and that having mirror plane (*m*), as $C_{s.}$ Thus, we have two approaches for determining the relative energies of transition states of the reaction. Knowledge of these energy changes indirectly gives information about the transition state which lies somewhere at a position between reactants and products of the reaction coordinate.

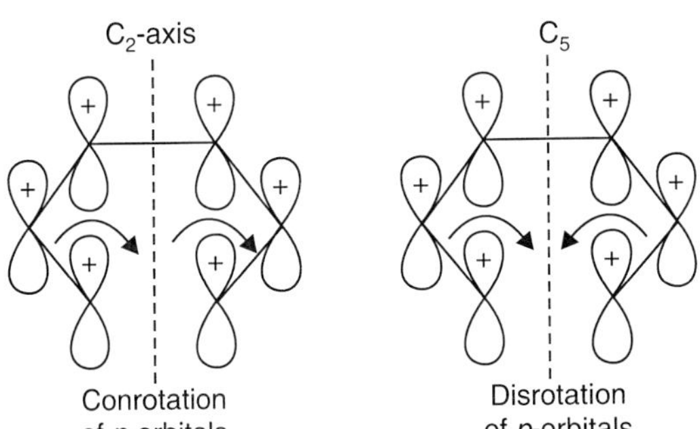


**Fig. 10.10** Conrotation and Disrotaion.

Thus, one approach concerns the preservation of mirror plane while the other preserves the $C_2$ rotations during the whole path of the reaction. In the orbital symmetry conservation model, we consider the *energy changes* of all the MOs of the reactants and the products, while in Frontier orbital approach we determine the changes in the frontier orbitals only (*i.e.*, HOMO and LUMO).

Now we describe the construction of the correlation diagrams. In this approach, we require the MOs of reactants and products. In the formation of these diagrams we follow the following steps:

1. In a given reaction, we first ascertain precisely which bonds are broken and which are formed in the transition state of the reaction. In case these bonds undergo secondary interactions, like rehybridisation, we neglect these processes.
2. We ascertain the specific geometry involved in the process of bond-breaking and bond formation. In some cases the geometry of the system is fixed because of the rigidity of bi-or polycyclic system while in some cases, there is a considerable freedom to choose the geometry. In the latter case, we assume the geometry as if the system is rigid like in the former case, because the overlap of orbitals is generally thought to be most favourable in that case.
3. While examining the system, one reduces it to a system of its maximum symmetry. That is, it possess the maximum number of symmetry elements. In this process, one replaces the symmetry-reducing group like $CH_3$ *etc*., by H atoms or even replace hetero atoms by carbon atom with which it is isoelectronic. For example, neutral nitrogen atom is replaced by $C^-$ ion or boron atom by $C^+$, *etc*.
4. We then determine or select the symmetry elements that exist throughout the reaction coordinate like $C_2$, mirror plane *etc*. We must be assured that trivial symmetry elements like *E etc*., are not chosen. All those symmetry elements are trivial which do not bisect any of the bonds formed or broken during the reaction path or which has all MOs either symmetric or antisymmetric with respect to them.
5. We then establish energy levels with the corresponding MOs for the system and classify them according to their behaviour with respect to the symmetry operation which has been chosen.
6. We carry out the above five steps for both reactants and products.
7. Finally, we connect molecular energy levels of both reactants and products which have the same symmetry (*S* with *S*, *A* with *A*). But one must keep in mind while joining the levels of like symmetry that the lines joining the same symmetry must not be *cross* (Non-crossing rule).
8. Finally we determine the path that out of the approaches which we have taken in determining the symmetry of the MOs gives lowest energy of the transition state during the reaction. Such diagrams are known as correlation diagrams.

In order to conceptual clarification of the above steps, we give three examples.

**(A) Electrocyclic Ring Opening of Cyclobutene:** The reaction is shown below.

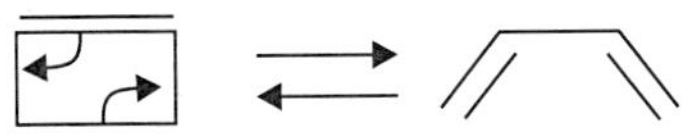

...(10.9)

Cyclobutene Butadiene

Let us now focus on the orbitals of interest that is those orbitals which are broken and formed. On cyclobutene, we have one orbital each of σ symmetry, σ* symmetry and π and π* type. That is four orbitals σ, σ*, π, π * are of interest on cyclobutene.

On butadiene we have four MO's of π-type. These orbitals have already been shown in Fig 10.9. We show them again in Fig 10.12. We now have two modes of rotation of the end orbitals viz. (*a*) conrotation (*b*) disrotation.

**Table 10.1:** Symmetries of MO's of cyclobutene and 1,4 butadiene.

| *Symmetry* | $\phi_1$ | $\phi_2$ | $\phi_3$ | $\phi_4$ | σ | π | π* | σ* |
|---|---|---|---|---|---|---|---|---|
| ***$C_2$-axis*** | *A*(*b*) | *S*(*a*) | *A*(*b*) | *S*(*a*) | *S*(*a*) | *S*(*b*) | *A*(*a*) | *A*(*b*) |
| ***Mirror plane m*** | *S*(*a*′) | *A*(*a*″) | *S*(*a*′) | *A*(*a*″) | *S*(*a*′) | *A*(*a*″) | — | *A*(*a*″) |

All these MOs will now be classified with respect to their behaviour under reflection and $C_2$ rotation around the $C_2$-axis. These two symmetry operations are appropriate ones to consider because their symmetry elements, *viz*. $C_2$-axis and mirror plane pass through the bonds to be broken or formed and represent the symmetry that is preserved during the electrocyclic transformation from reactants to products via these two modes.

We can analyze the symmetry of these orbital by two equivalent methods.

1. Symmetric orbitals are designated as (*S*) and antisymmetric as (*A*), under rotation by 180° about $C_2$-axis. It is very important to know that disrotatory mode is symmetric under reflection operation while conrotatory mode is symmetric under $C_2$-operation.

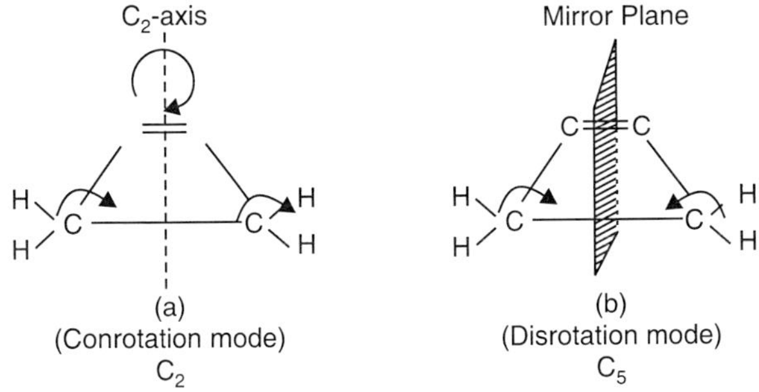


**Fig. 10.11** Two modes of converting cyclobutene to butadiene.

2. A molecule having only a plane of reflection plus identity operation belongs to $C_s$ point group. We can consider the *MOs undergoing the disrotatory mode* (having $\sigma_v$ symmetry) in terms of their behaviour with respect to reflection ($\sigma_v$) in this plane under the point group ($C_s$). The symmetric behaviour of an orbital belong to '***a'***' type of I.R. and antisymmetric by ***a''*** type of I.R. In point group $C_2$ {$E$, $C_2$} where we have $C_2$ as the only symmetry element besides $E$, symmetric behaviour with respect to $C_2$ symmetry operation belongs to I.R. **a** type and antisymmetric to ***b*** type of I.R. Thus, we can classify the MO's of butadiene and cyclobutene as ***a*** or ***b*** under conrotatory mode and as ***a'*** and ***a''*** under disrotatory mode. These classifications are given in Table 10.1.

In order to make correlations, we must know their energies. For butadiene, the relative energies are given in Fig. 10.9. In cyclobutene, we expect σ-orbitals to be of lowest and σ* to be highest energy and π and π* to be of intermediate energy, with π to be more stable than π* which, in turn, more stable than σ*. Thus, the order of stability is σ > π > π* > σ*.

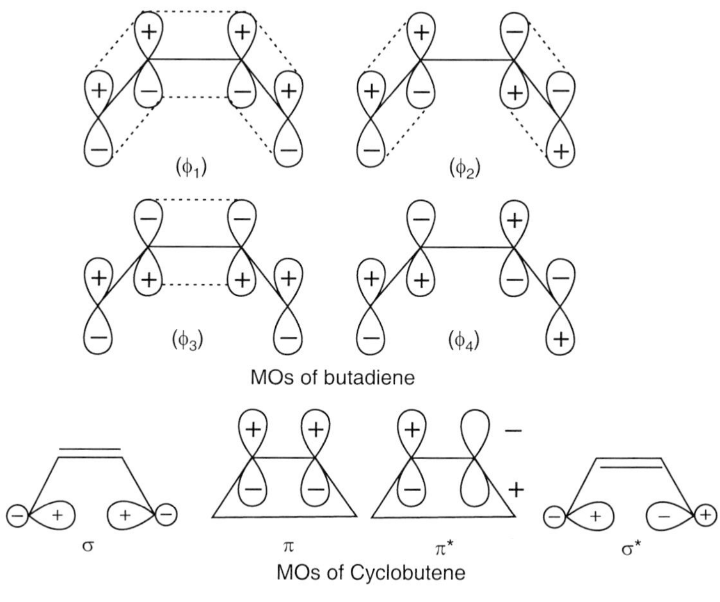


**Fig. 10.12** MOs of Butadiene and Cyclobutene.

Let us now carry out the conrotatory ring opening of cyclobutene to butadiene. We now correlate the $S$ and $A$ of reactants with $S$ and $A$ of the product keeping in mind the non-crossing role (both with respect to symmetry and energy). This is shown in Fig 10.13 (Solid lines)

Similarly we also correlate under the disrotatory mode. It is shown in Fig. 10.13 (dotted lines), these correlations conrotation (solid lines) and disrotation modes are also listed in Table 10.2.

**Table 10.2:** Orbital correlations for conrotatory and disrotatory modes.

| *Disrotation Mode* | | *Conrotation Mode* | |
|---|---|---|---|
| $\phi_1 \leftrightarrow \pi$, $\phi_2 \leftrightarrow \sigma$ | Bonding | $\phi_1 \leftrightarrow \sigma$, $\phi_3 \leftrightarrow \pi$ | Bonding |
| $\phi_3 \leftrightarrow \sigma^*$, $\phi_4 \leftrightarrow \pi^*$ | Anti-bonding | $\phi_2 \leftrightarrow \pi^*$, $\phi_4 \leftrightarrow \sigma^*$ | Anti-bonding |

Four electrons are involved in filling of orbitals and these will occupy in pairs. The ground state of butadiene is therefore is $\phi_1^2\ \phi_2^2$. If the reaction pathway is disrotatory (dotted lines) these lines will intersect the vertical line at *A*, *B*, *C* and *D* and if the pathway is conrotatory, (solid lines) these will intersect at *A′*, *B′*, *C′* and *D′* [*A* vertical line *P* is drawn (at any arbitrary position) to locate the transition state of the reaction. (Its precise location between the reactants and the products is not crucial]. Thus the energy of the transition state will depend upon the *energy changes* experienced by the individual orbital of the reactants as they transform on the reaction pathway) towards the product MO's].

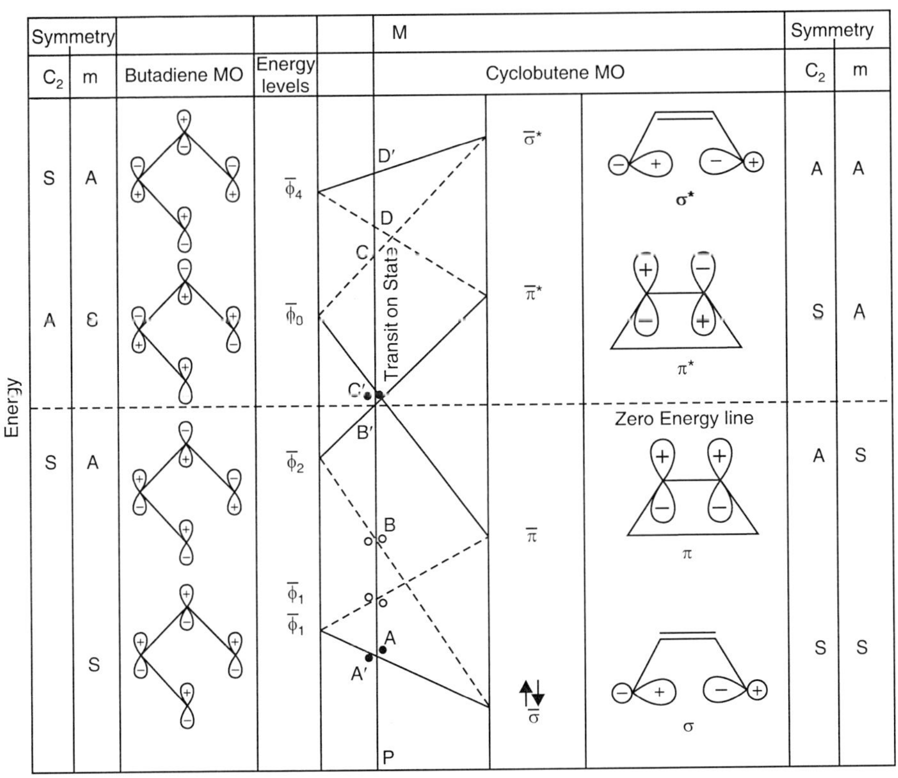


**Fig. 10.13**

Now examine the correlation indicated by dashed lines and solid lines. It is important to note that the transition state energies in the two cases are different. It will be seen that for the dotted line (disrotatory process) the energy at the transition state will be lower than that for the conrotation (solid lines). Since the energy of transition state is lower in disrotatory mode, then the disrotatory process is symmetry-allowed and the conrotatory process is symmetry-forbidden.

Another alternative way to arrive at the same conclusion is by considering the electronic state of reactants and products (Table 10.2). In the solid line pathway, we have $\phi_1^2\ \phi_2^2 \leftrightarrow \sigma^2\ (\pi^*)^2$. It suggests that ground state of butadiene correlates with the high energy excited state of cyclobutene $[(\pi^*)]$. It will produce a high energy barrier (high activation energy) and therefore, it will forbid the progress of the reaction. The alternative path will be easily feasible because the ground state of butadiene $\phi_1^2\ \phi_2^2$ correlates with the ground state of cyclobutene $(\sigma^2\pi^2)$.

In the light of the above discussion, it is concluded that, the reaction involving 4 electrons disrotatory mode is thermally allowed and the conrotatory mode is thermally forbidden.

Now, if we promote one of the electrons to the excited state photochemically, the ground state configuration of butadiene becomes ( $\phi_1^2\phi_2^1\phi_3^1$ ). This correlates with $(\pi^2\sigma^1\sigma^{*1})$ in the disrotatory mode and to $(\sigma^2\pi^{*1}\pi^1)$ configuration in the conrotatory mode. Obviously in the conrotatory mode, the change in energy to the $\pi^*$ level is counterbalanced by the stabilisation energy which the electron in $\phi_3$ orbital gets in going to $\pi$-orbital of cyclobutene. Therefore, there will be net stabilisation of the electrons of the $\phi_1^2$-orbital by going to $\sigma^2$-orbital of cyclobutene. However, in the disrotatory process, $\phi_1^2$ electrons get destabilised, $\phi_2^1$ electron gets stabilised, and $\phi_3^1$ electron is highly destabilised in going from $\phi_3^1$ to $\sigma^{*1}$. By careful examination, one will realise that there is net destabilisation energy in the disrotatory process. Thus, any attempt to correlate the ground state $(\phi_1^2\phi_2^1\phi_3^1)$ with a higher energy state $(\pi^2\sigma^1\sigma^{*1})$ will be threated by imposing a barrier in the transition state. The reaction by disrotatory mode will be forbidden. Thus, we have four electron reactions.

Thermally allowed–Conrotatory mode
Thermally forbidden–Disrotatory mode

Photochemically allowed–Disrotatory mode
Photochemically forbidden–Conrotatory mode

**(B) Electrocyclic Ring Closure of 1, 3, 5, Hexatriene to 1, 3 Cyclohexadiene:**

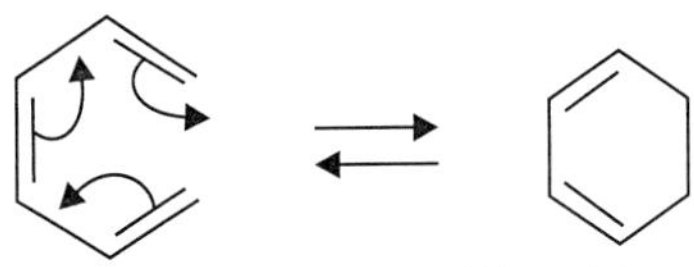

1,3,5 hexatriene 1,3 cyclohexadiene ...(10.10)

As second example, we take the closure of 1, 3, 5-hexatriene to 1, 3-cyclohexadiene. (Eq. 10.16). Unlike the previous 1, 3-butadiene case, it is a 6-electronsystem. It means that one electron is present in each of six $p\pi$-orbitals of the reactant. We analyze this reaction by the conservation of orbital symmetry model of Woodward and Hoffmann during the transformation by considering all the MOs of the reactants and the products.

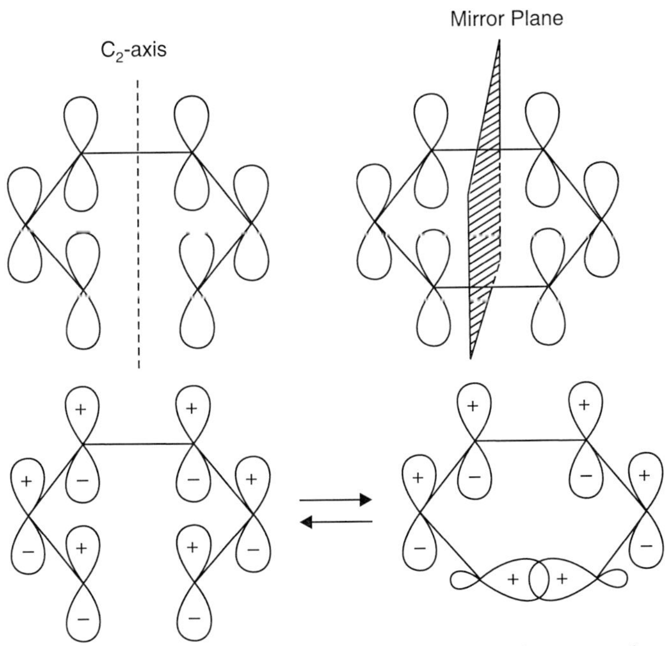


Six $\pi$-electrons
1,3,5 hexatriene

Four $\pi$-electrons and
two sigma electron
1,3 cyclohexadiene

We first recognise the bonds that are involved in the transformation. It involves transformation of six pπ-electrons of 1,3,5-hexatriene to a cyclic 1,3 cyclohexadiene containing four pπ-electron and two σ-bonds [Eq. 10.11]. We now determine the MOs of both hexatriene and cyclohexadiene. These are shown in [Fig 10.14] with proper algebraic signs (+ or –) on their pπ lobes. We also know that higher, the number of nodes present in the MO, the higher the energy of that orbital. Nodes in the MOs are shown by the dark dots in [Fig 10.14]. Thus the order of energy levels of 1, 3, 5 hexatriene is $\phi_1 < \phi_2 < \phi_3 < \phi_4 < \phi_5 < \phi_6$. The transition state of the reaction is somewhere on the reaction coordinate between hexatriene and cyclohexadiene. We have indicated this by a vertical line (*P*) in the diagram. Its precise position between the reaction and the product is immaterial for our discussion since the energy of the transition state depends critically upon the energy changes experienced by the individual molecular orbital. Clearly our both systems (*viz.* hexatriene and cyclohexadiene) have both $C_2$ as well as mirror plane as symmetry elements. Now in the transition state, the reacting pπ-orbitals of each MO has two options to react in order to form a σ-bond, either these will rotate in the same direction (conrotation) or in opposite directions (disrotation). Thus, if in the transition state, the formation of σ-bond takes place by conrotation mode, we have a certain change in energy of the system. Further in the σ-bond formation (or breaking) $C_2$ symmetry element is maintained and the mirror plane reflection (*m*) is destroyed. If σ-bond formation (or breaking) takes place by disrotation mode, $C_2$-symmetry elements is destroyed and the reflection plane is maintained and the changes in the transition state will be different from that of conrotation mode. In this way, the interaction mode (disrotation, or conrotation) will depend upon the type of energy change in transition state. One that stabilises the transition state more will be the reaction path. In Fig. 10.14, we have shown both these paths [disrotation by dotted lines (…) and conrotation by solid lines (–)]. In the same diagram, we have also shown if the MOs of the reactant and the product are symmetric (*S*) or antisymmetric (*A*) with respect to $C_2$ and (*m*). Employing the symmetries with respect to $C_2$ and (*m*) of the reactant and the products, we have drawn the correlation diagrams for the both modes according to the same method that is discussed in the example of 1, 3-butadiene to cyclobutene cyclisation. The correlations in both modes are shown in Table 10.3.

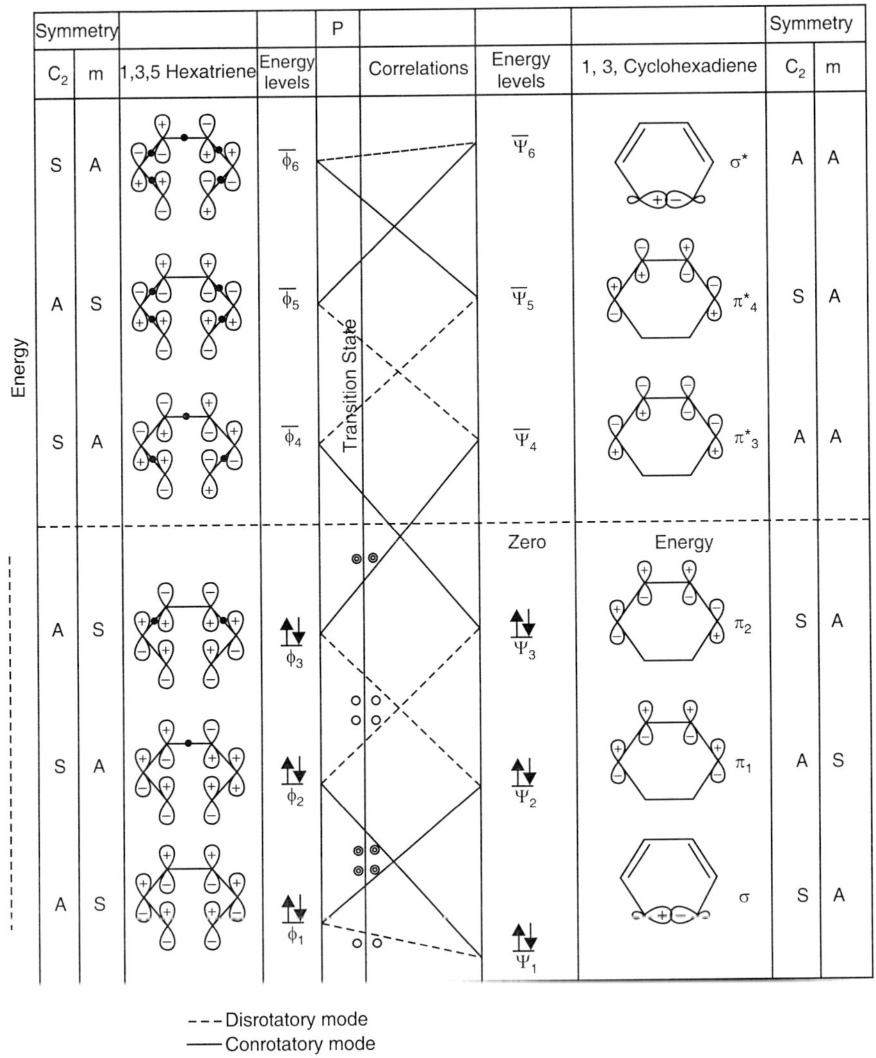


Fig. 10.14

**Table 10.3:** Correlation of 1, 3, 5 Hexatriene and Cyclohexadiene in Disrotation and Conrotation Modes

| *Disrotation* | *Conrotation* |
|---|---|
| $\phi_1^2 \leftrightarrow \sigma^2\ (\psi_1)$ | $\phi_1^2 \leftrightarrow \pi_1^2\ (\psi_2)$ |
| $\phi_2^2 \leftrightarrow \pi_2^2\ (\psi_3)$ | $\phi_2^2 \leftrightarrow \sigma^2\ (\psi_1)$ |
| $\phi_3^2 \leftrightarrow \pi_1^2\ (\psi_2)$ | $\phi_3^2 \leftrightarrow \pi_3^{2*}\ (\psi_4)$ |
| $\phi_4 \leftrightarrow \pi_4^*\ (\psi_5)$ | $\phi_4 \leftrightarrow \pi_2\ (\psi_3)$ |
| $\phi_5 \leftrightarrow \pi_3^*\ (\psi_4)$ | $\phi_5 \leftrightarrow \sigma^*\ (\psi_6)$ |
| $\phi_6 \leftrightarrow \sigma^*\ (\psi_6)$ | $\phi_6 \leftrightarrow \pi_4^*\ (\psi_5)$ |

Let us now examine the changes on the transition state in disrotatory and conrotatory modes. Ground state of 1, 3, 5 hexatriene is $(\phi_1^2\phi_2^2\phi_3^2)$. In the transition state, energy change by the disrotation will be determined by the intersection of the vertical line (*P*) with the dotted lines starting from $\phi_1$, $\phi_2$ and $\phi_1$ Similarly in the conrotatory mode, the energy of transition state will be determined by the intersection of the vertical line [*P*] with the solid lines (–) staring from $\phi_1$, $\phi_2$ and $\phi_3$. These points of intersections are indicated by dark circles (•) in the case of disrotatory mode and by crosses (*x*) in conrotatory mode. It is very clear from the relative positions of dark and hollow circles, that the disrotatory mode is more favourable than the conrotatory mode (These circles also indicate number of the electron in the transition state). It is indicated by the positions of these circles, that the net change in energy in the two modes are different and that in the disrotatory mode the energy of transition state will be lower than that in the conrotatory mode.

It is therefore concluded from the foregoing discussion that for six electron case, the disrotatory mode of interaction is thermally allowed and the conrotatory mode is forbidden.

We also examine the reaction when it is carried out photochemically, In this case, by proper radiation; we excite the electron in the HOMO of the reactant ($\phi_3$), to the LUMO of the reactant (*viz.* $\phi_4$). The ground state of the reactant, thus becomes $(\phi_1^2\phi_2^2\phi_3^1\phi_4^1)$ and our orbital correlation becomes as follows:

1. for disrotatory mode, we have $(\phi_1^2\phi_2^2\phi_3^1\phi_4^1) \leftrightarrow \sigma^2\pi_3{}^2\pi_1{}^1(\pi_5{}^*)^1$ and
2. for conrotatory mode, we have $(\phi_1^2\phi_2^2\phi_3^1\phi_4^1) \leftrightarrow \pi_1{}^2\sigma^2(\pi_3{}^*)^1(\pi_2)^1$

If we now examine the energy changes in these correlations using the Fig. 10.14, we will find that in the case of conrotatory mode, the energy at the transition state is lower than that in the disrotatory mode. Therefore, the conrotatory mode will be allowed and the disrotatory mode is forbidden.

In the above examples, we have discussed two rotatory modes of combination of terminal pπ-orbital of two conjugated systems containing 4 and 6π-electrons. Woodward and Hoffmann have studied a large number of such electrocyclic ring closure reactions, They have classified these reactions into two groups (1) Those containing $4n$ $p\pi$ electrons and to which 1,3 butadiene belongs ($n$ = 1). (2) Those systems which contain $(4n + 2)$ $p\pi$-electrons to which 1, 3, 5 hexatriene belongs [$6 = 4n + 2$ ($n = 1$)]. The preferred stereochemical mode of cyclisation within each group depends upon whether the reaction is carried out thermally or photochemically. They have thus, generalised their results is the form of "selection rules of electrocyclic reactions" which are summarized in Table 10.4.

**Table 10.4:** Woodward-Hoffmann Rules for Electrocyclic Reactions

| *No. of pπ-electrons* | *Thermally allowed Photochemically Forbidden* | *Thermally forbidden Photochemically allowed* |
|---|---|---|
| $4n$ | Conrotatory | Disrotatory |
| $4n + 2$ | Disrotatory | Conrotatory |

**(*C*) Cycloaddition Reactions:** After analysing the electrocyclic reactions on the orbital correlation model, we will now analyse cycloaddition reactions on the same model. These involve the addition of $m$ number of π-electrons of a system to another system containing $n$ number of π-electrons. Thus, addition of two π-systems will result in a new cyclic system after their addition containing $(m + n)$ number of electrons. Such reactions are known as $(m + n)$ cycloaddition reactions.

**Example 3:** An example of such a cycloaddition reaction $[m + n = 6]$ is given below.

...(10.11)

We now apply orbital correlation method to analyse the above reaction. Prior to discussion, let us first distinguish between the various different ways, by which the interaction is possible between $C_2H_4$ and $C_4H_6$ whereby various different products are obtained. Some of these different types of reactions between $C_2H_4$ and $C_4H_6$ are shown in [Fig 10.16]. Thus, when two alkenes, each of which is appropriately labelled, various different stereochemical products are possible. We will however, not give their details at this point. We will start with $(\pi^{2s} + \pi^{4s})$ as an example for analysing on the basis of orbital correlation method. The mode of this reaction is shown in (Fig 10.15).

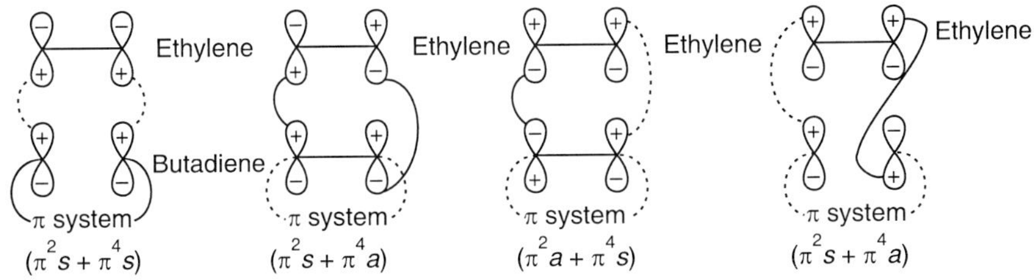


**Fig. 10.15** Orbital correlation method.

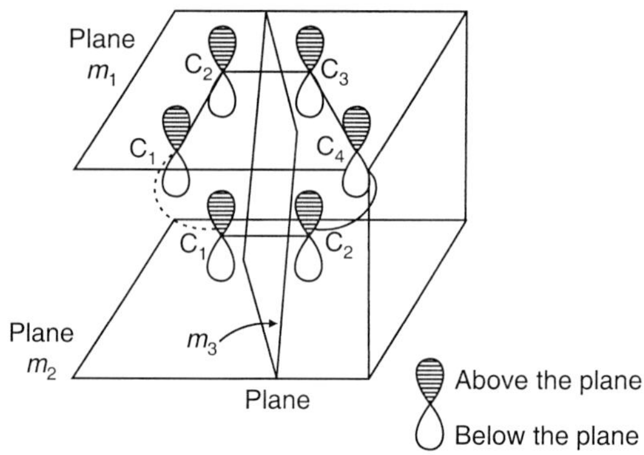


**Fig. 10.16** Reaction between $C_2H_4$ and $C_4H_6$.

We assume that the reaction is proceeding via approach of two separate parallel planes $m_1$ and $m_2$. We assume that 1, 3 butadiene is present on the plane $m_1$ and the dienophile ethylene is present on plane $m_2$ [Fig. 10.16]. There is a third plane $m_3$ that is perpendicular to the planes $m_1$ and $m_2$ and this plane is bisecting C–C bond of $C_2H_4$ $C_1 \ldots C_2$ distance and $C_2$–$C_3$ bond of 1, 3 butadiene. Further the reacting $p\pi$-lobes of $C_1$ and $C_4$ atoms of $C_4H_6$ are lying below the plane $m_1$ and those of carbon atoms of $C_2H_4$ are lying above the plane $m_2$. When the planes are at infinite distance apart, there is no reaction between the two reactants, but as these planes approach towards each other, $\pi$-MO's of the two reactants start interacting and forming the two $\sigma$-bonds. During this process, the position of the plane $m_3$ remains the same. It means that during the formation of two $\sigma$-bonds, the reflection operation $m_3$ is preserved. This is shown by viewing along the planes $m_1$ and $m_2$ of the orbitals at the beginning and the termination of interaction (Fig. 10.17).

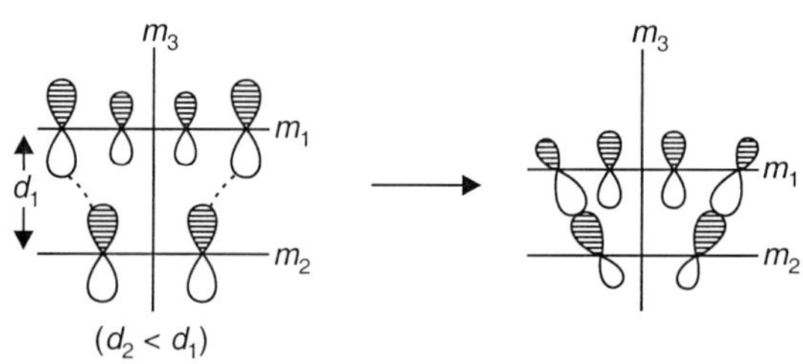


**Fig. 10.17** End view through plane $m_1$ and $m_2$ as the orbitals start interacting and termination of interaction.

We now determine those symmetry operations which remain invariant during the period of bonding and the formation of two sigma bonds between $C_2H_4$ and $C_4H_6$ form the Fig. 10.18, one should note that the only symmetry element that remains invariant throughout the reaction is the symmetry plane, $m_3$ (Reflection plane $m_3$). In accordance with the orbital conservation model, it is then necessary that this mirror symmetry is preserved throughout the path of the reaction. It implies, that the symmetry ($S$) or ($A$) of the reactant orbitals with respect to this mirror plane must correlate (or preserved) with the $S$ or $A$ symmetry of the product orbital with respect to the same plane ($m_3$). One must know the symmetries of the reactant and product orbitals and their relative energies with respect to $m_3$ plane. We have shown these in Fig 10.18. In this figure, we have shown two bonding σ-bonds as ($\sigma_1 + \sigma_2$), ($\sigma_1 - \sigma_2$) and two anti-bonding σ* bonds as, ($\sigma_1{}^* + \sigma_2{}^*$), ($\sigma_1{}^* - \sigma_2{}^*$). These bonds, in cyclohexene are generated as SALCs of the individual bonding and anti-bonding σ-bonds, formed in [Fig 10.18]. In this way, it is possible to classify each molecular orbital of cyclohexene as symmetric and antisymmetric with respect to $m_3$ plane.

There are six electrons which are involved in the transformation. These electrons are filled in the bonding orbitals of the reacting system [1, 3 butadiene + ethylene] and these are also accommodated in the bonding orbitals of cyclohexene, So there is a smooth correlation of the ground state from the reactants to the product. The reaction is therefore *thermally allowed.*

Photochemically, the situation is different. The ground state of the system in that case is $(\phi_1^2\phi_2^2\phi_3^1\phi_4^1)$ and that of cyclohexene it is $(\psi_1^2\psi_2^2\psi_3^1\psi_4^1)$. These two states do not correlate and the correlation is with the much higher energy states [*viz.* $\phi_1^2\phi_2^2\phi_3^1\phi_4^1 \;\; \psi_1^2\psi_2^2\psi_3^1\psi_4^1$]. The reaction is therefore *photochemically forbidden.*

Analogous to electrocyclic reactions, Woodward and Hoffmann also classified the cycloaddition reactions. These rules are given in Table 10.5.

**Table 10.5:** Woodward and Hoffmann Rules for ($p + q$) Cycloaddition Reactions

| **($p + q$)** | ***Thermally allowed Photochemically forbidden*** | ***Photochemically allowed Thermally forbidden*** |
|---|---|---|
| $4n$ | $(p_s + q_a)$<br>$(p_a + q_s)$ | $p_s + q_s$<br>$p_a + q_a$ |
| $4n + 2$ | $(p_s + q_s)$<br>$(p_a + q_s)$ | $p_s + q_s$<br>$p_s + q_s$ |

$S$ = supra and $a$-antara-addition. Thus, $p_s$ means that the polyene $p$ interacts in the supra-mode and $p_a$ = polyene $p$ interacts in antara-mode, and so on.

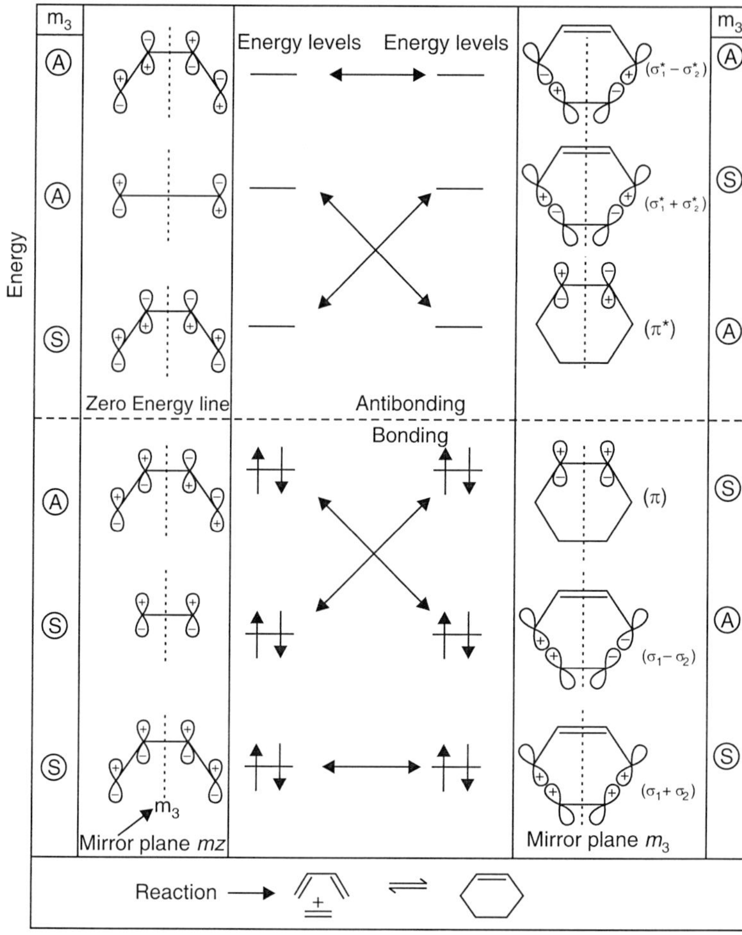


**Fig. 10.18** Correlation diagram for [$\pi^4 s + \pi^2 s$], reaction (Diels-Alder reaction).

We will not discuss and analyse any other reactions of the pericyclic category by applying orbital symmetry conservation method. We advise the students to refer to an advanced book on the topic. We will, however discuss in the following paragraphs a few pericyclic reactions on the basis of Frontier Orbital Method of Fukui, Woodward and Hoffmann. We will also discuss these reactions on the "Perturbational Molecular Orbital" method of M.J.S. Dewar and Zimmermann.

## 10.5.2 Frontier Orbital Model

In the preceding discussion, we have studied the changes in transition state energy of each MO of the reacting system as it is transformed from reactant to product. Fukui et al., have however shown that one does not have to determine these changes in all the MO's of the reactant. One can arrive at the same conclusion by studying only Frontier orbitals. These orbitals are usually HOMO

(Highest Occupied Molecular Orbital) and LUMO (Lowest Unoccupied Molecular Orbital) which is a vacant orbital. The set of these two orbitals often contribute to most of the overall changes in energy during the transformation. Let us take the example of Diels-Alder reaction, [$\pi^4s + \pi^2s$], cycloaddition of $C_2H_4$ and $C_4H_6$. The MO's of 1, 3 butadiene, ethylene and cyclohexene are shown in Fig 10.19.

**Examples**

**(I) Cycloaddition of $C_2H_4$ and $C_4H_6$ (Frontier Orbital Model):** According to this approach, we consider the interaction between HOMO of one component with LUMO of the second component. One must realise here that the consideration of only HOMO and LUMO is an approximation which gives the best result when HOMO and LUMO are of similar energy. Another approximation which is very important in this approach is that the cyclic structure (*e.g.*, 1, 3-cyclohexadiene) is considered as a cycloaddition of σ-bond to the π-system as shown below. It implies

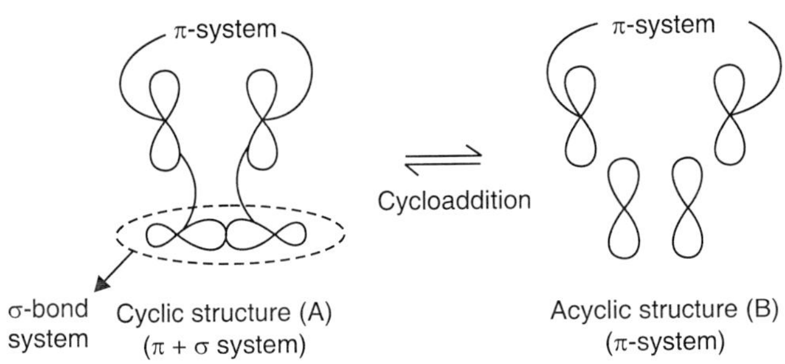


**Fig 10.19** MO's of 1, 3 butadiene, ethylene and cyclohexene.

The structure *B* which is acyclic π-bonded structure has been transformed to cyclic structure *A* having one σ bond from the π orbitals of structure *B* (That is, structure is π + one σ-bond formed from two terminal π-bonds of structure *B*). Or one σ bond is added to structure B which then get cyclised) It is similar to the addition of σ-bond to structure *B* formed by two terminal π orbitals).

We thus have the following two systems:

1. HOMO of σ-bond + LUMO of π-bond
2. HOMO of π-bond + LUMO of the σ-bond.

According to Fukui's calculations, there are two alternate important interactions (*i*) HOMO (sigma) to LUMO (π-system) and (*ii*) HOMO (π-system) to LUMO (sigma system). One of these two alternate systems is involved in the transfer of electrons and thereby forming the bond. Let us takes the case of thermal opening of cyclohexadiene to hexatriene. We have a π-and a σ-system as shown in the [Fig 10.20]. At this point it is important to realize that the *p*π-system in which we are interested is that of the reactant butadiene) and not that of the product (*viz.* hexatriene).

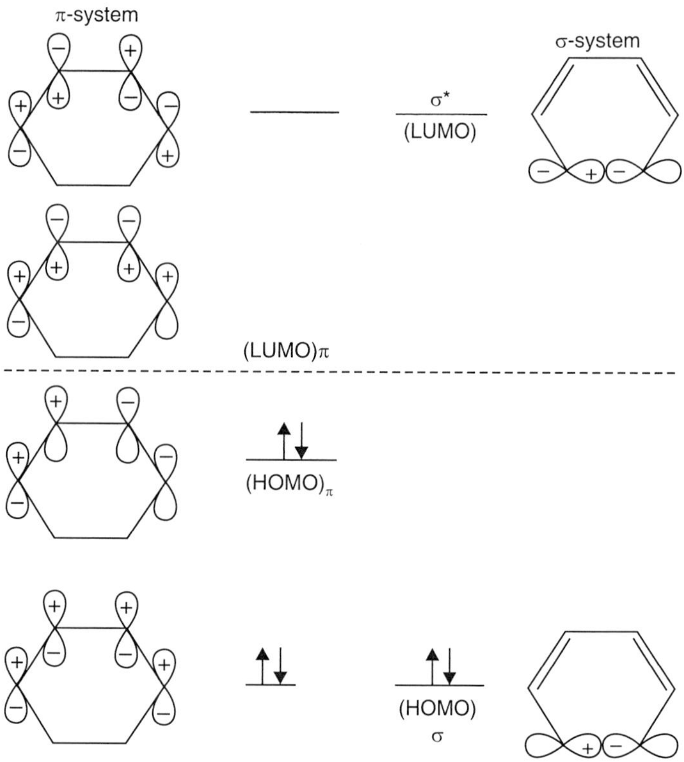


**Fig. 10.20** HOMO and LUMO of σ- and π-systems of Cyclohexadiene ⇌ Hexatriene interconversion.

We examine next (LUMO)π-(HOMO)σ-system. We know that these two systems are part of the same molecule and incorporate in the same structure as shown in Fig. 10.20. We will not examine the effect of interaction of opening of $(HOMO)_σ$ and interacting with $(LUMO)_π$ in the disrotatory and conrotatory mode. The effect is shown in Fig 10.21. In the conrotatory opening mode, we have one bonding and one antibonding interaction. It suggests no net stabilisation. However, the disrotatory opening of $(HOMO)_σ$ results in the two bonding interactions which suggested net stabilisation of the system. Therefore, the disrotatory mode is preferred over conrotatory mode.

One may also consider the interaction of $(HOMO)_π$ with $(LUMO)_σ$ one will find that in this case also the same conclusion will be reached regarding the ring opening of 1, 3-cyclohexadiene.

**(II) Electrocyclic reactions of 4π electron system-analysis using Fukui HOMO and LUMO method:** In this example, we take the ring opening of *cis*- 3, 4-dimethyl cyclohexene to produce *cis*, *trans*-2, 4-hexadiene as shown in Fig. 10.21. This is an example of thermal interconversion.

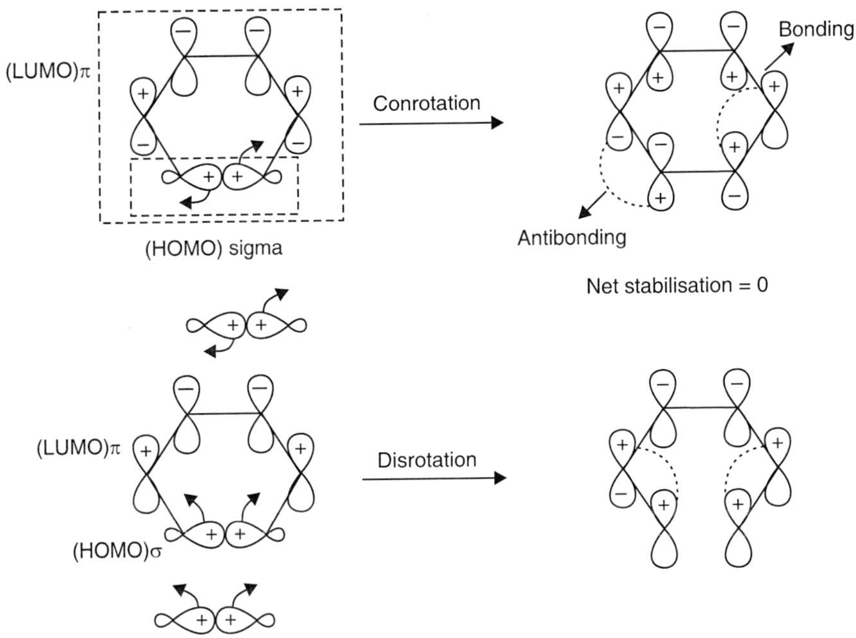


**Fig. 10.21** HOMO-LUMO interaction for conrotatory and disrotatory opening of cyclohexadiene.

This is a reversible reaction. Being a concerted reaction, the path of the forward ring opening reaction must be the same as that of backward ring closing reaction which requires heat only. We, therefore, conclude that the excited states of the hexadiene are not involved because in that case we require light or U.V. radiation, to excite the electron in the LUMO of the hexadiene (Photochemically). We, therefore, concentrate HOMO($\psi_2$) of the π-system of hexadiene). That is, how the *p*π-orbitals of the $C_2$ and $C_5$-get transformed into a σ-bond in cyclohexene.

In [Fig 10.22], we give the MO's of a 2, 4-hexadiene. The ground and the excited state which are obtained by the absorption of light you will notice that the orbitals of 2, 4-hexadiene are the same as those of 1,3-butadiene.

We now form a σ-bond between $C_2$ and $C_5$. We show the orbital of $(HOMO)_\pi$ of the hexadiene. Let us see how it forms [Fig 10.22] we start with HOMO of hexadiene molecule in the ground state. A bonding σ-type MO between $C_2$ and $C_5$ is formed when *p*-π orbitals on $C_2$ and *p*π on $C_5$ atoms rotate in the same direction (Conrotation) this type of rotation also rotate $CH_3$ groups attached to $C_2$ and $C_5$ carbon atoms in the same direction by which the two $CH_3$ groups come in *cis*-configuration.

When pπ-orbitals on $C_2$ and $C_5$ carbon atom are rotated in the disrotatory mode, the interaction between them is antibonding and we do not expect to form a σ-bond between them.

It is concluded that the thermal reaction results by conrotation in the formation of σ-bond as predicted by orbital correlation method also.

In case of photochemical reaction, the HOMO of the molecule is shown in [Fig. 10.22]. One can show easily that the disrotatory mode of transformation of *p*π-orbitals on $C_2$ and $C_5$ atoms is allowed while thermally it is forbidden.

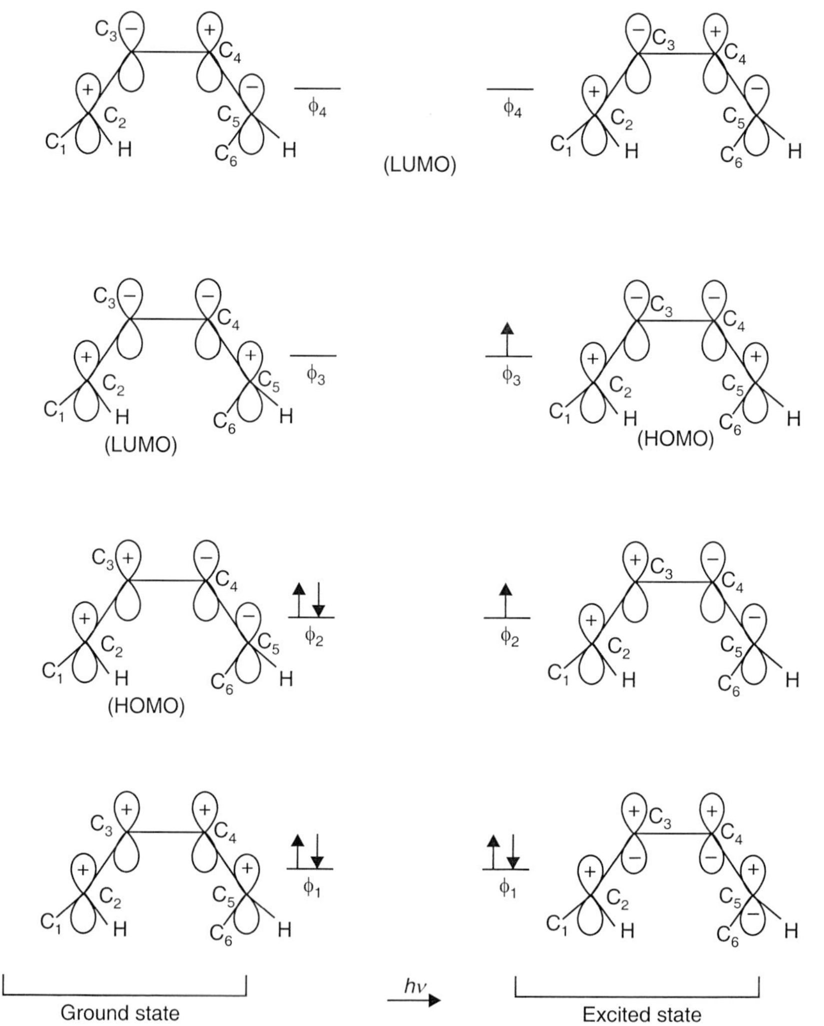


**Fig. 10.22** MO's of Hexadiene in the ground and excited state.

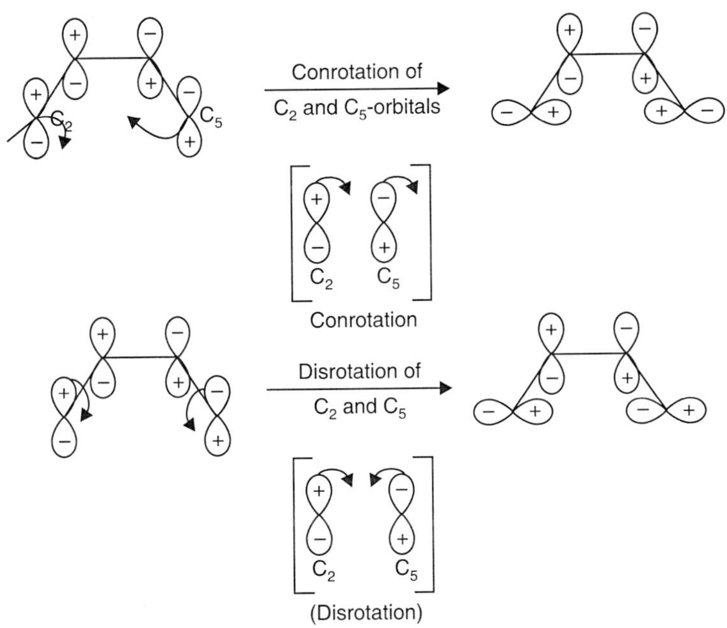


**Fig. 10.23** Conrotation and disrotation.

**(III) Electrocyclic Reaction of (4*n* + 2) π-Electrons System:** In Fig. 10.24, we show the HOMO of 1, 3, 5-hexatriene in both ground and excited states.

It is clear that in a (4*n* + 2) π electron system (6π-electrons) it is the conrotatory mode that is thermally allowed and photochemically forbidden. In the same system the disrotatory mode allowed photochemically and forbidden thermally.

The above results are obtained by using HOMO-LUMO method of Fukui, Woodward and Hoffmann. These results are identically the same as obtained by the orbital symmetry conservation method.

In ground state (Thermal) $\phi_3$

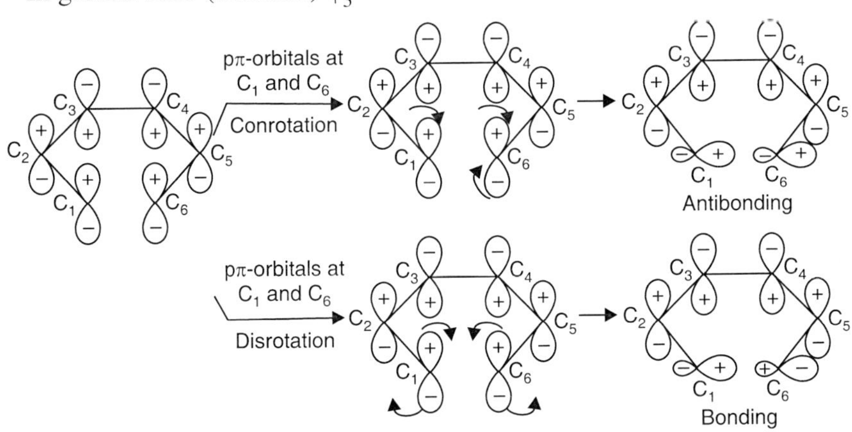

In excited state (Photochemically) $\phi_4$

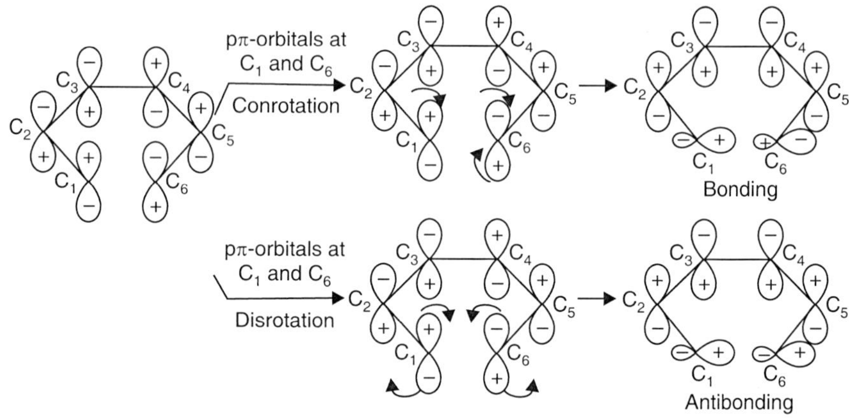


**Fig. 10.24** HOMO of 1, 3, 5 hexatriene.

### 10.5.3 The Perturbation Molecular Orbital Model [PMO Method]

We know that $(4n + 2)$ $p\pi$-electron aromatic system is particularly favourable electronic configuration while those of $(4n)$ $p\pi$-electron system are not stable and termed as anti-aromatic. This concept has been applied to explain the concerted pathway through which pericyclic reactions work. This approach has been used in the *PMO method which is also known as Mobius-Huckel method.* The fundamental concept behind this method is that a pericyclic reaction is favourable (or allowed) by a concerted mechanism if the transition state of the reaction has aromatic characteristics. This concept is largely used by M.J.S. Dewar around 1969 in his book "The MO Theory of Organic chemistry" McGraw Hill N.Y. (1969). He derived the following rules for pericyclic processes.

1. Pericyclic reactions that take place through *aromatic transition* states $(4n + 2)$ $\pi$-electron systems are *thermally allowed.*
2. Pericyclic reactions that take place through excited form of *antiaromatic transition states* $(4n\ )$ $p\pi$-electrons are *photochemically allowed.*

Although the above rules are derived using Huckel MO theory but it can be used for analysing pericyclic reactions without actual calculations. Accordingly one assigns +ve and –ve signs to the lobes of $p\pi$-orbitals of each atom of the molecule. These $p\pi$-orbitals belong only to the interacting system. We then draw reactants with the lobes of +ve high on one side and –ve/sign on the other side. Lobes representing the points of bond making are connected to each other in a way that represent. (1) suprafacial-suprafacial (disrotatory) or suprafacial antarafacial (conrotatory) motion of *p*-orbitals. Let us explain this by the example of cycloaddition of two molecules of 1, 3, 5-hexatriene is given as ahead.

**Example (*i*):** In this example reaction can take place (supra-supra) (Huckel-mode) or supra-antara (Mobius mode) without any geometrical hindrance, the bond forming pathways for two possibilities is shown in the given diagram.

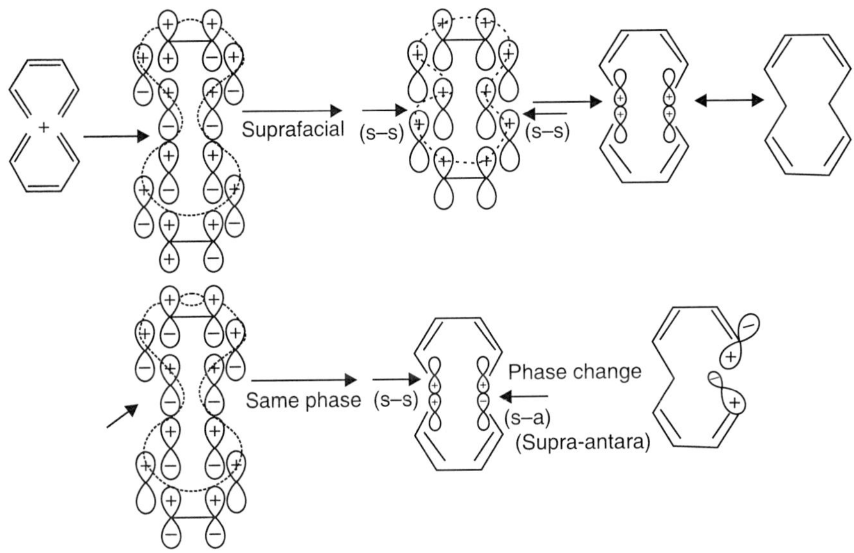


**Fig. 10.25** Bond forming pathways.

In the supra-supra mode all the +ve lobes are pointing upward. As a path is traced around the cycle array of the lobes of the same sign, no change in the phase takes place. A reaction in which there is a zero or even number of change of phases in the relation between the adjacent orbitals is known as Huckel system. A reaction in which there is an odd number of phase changes (1,3,5 *etc.*) with respect to two adjacent orbital is known as Mobius system. In the above 1,3, 5-hexatriene addition reaction supra-supra interaction between the adjacent orbitals there is no phase change (disrotatory motion) and so it belongs to Huckel system. In Huckel system, the system becomes aromatic when the reaction involves $(4n + 2)$ $p\pi$-electrons (aromatic). In case it has $4n$-electrons then it becomes antiaromatic $4n$-electrons, and the reactions will not be thermally allowed.

In the above system, we have 12-electrons $[\pi^6 s + \pi^6 s]$ addition, and so it belongs to $4n$ system, and Huckel system is not aromatic. It is antiaromatic. So the reaction is not thermally allowed through Huckel mode (supra-supra).

In the supra facial-antara facial pathway there is a twist in the cyclic array of orbital lobes (one phase change). It belongs to anti Huckel or Mobius system. Thermal reactions proceeding through Mobius pathways are allowed when the system has $4n$-electrons. Such systems are thermally allowed through Mobius pathways (*i.e.*, conrotatory pathways). So our system of (6 + 6) addition will be thermally allowed only when at proceeds through supra-antara pathways.

The general rules arrived at by the PMO method are given in Table 10.6.

**Table 10.6:** Rules for PMO Method for Thermally allowed reactions

| *No. of Phase changes* | *Type of system* | *Number of electrons* | |
|---|---|---|---|
| | | *Allowed* | *Forbidden* |
| 0, 2, 4...... | Huckel | $4n + 2$ | $4n$ |
| 1, 3, 5...... | Mobius | $4n$ | $4n + 2$ |

**Example (*ii*):** Let us take another example of electrocyclic addition *viz.* [$\pi^4s$ + $\pi^2s$]. In this example, we analyse the addition of ethylene to butadiene.

$$C_2H_4 + C_4H_8 \rightleftharpoons C_6H_{12}$$

We would like to know if the transition state of this reaction is aromatic or not (Fig 10.26). We have zero phase change and the number of $\pi$-electrons is six [*i.e.*, $(4n + 2) = 6$ or $n = 1$]. This is a Huckel mode and aromatic. It suggests that according to the above rule (1), the reaction is thermally allowed and photochemically forbidden. One must also note that changing any number of phases of atomic orbitals will lead to even number of out of phase interactions (Fig 10.26) for making it as [$\pi^4s + \pi^2s$] type of reaction.

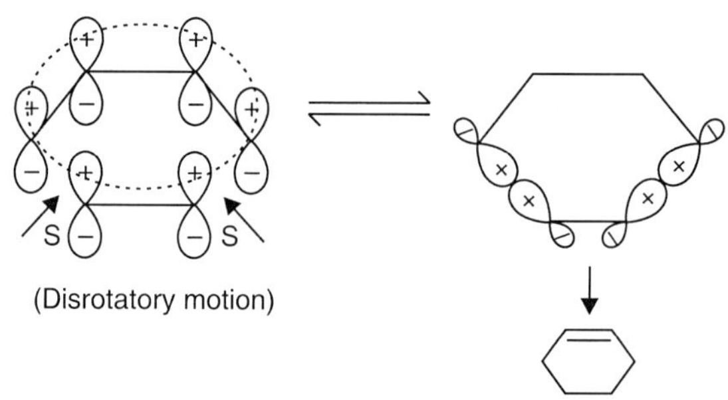


**Fig. 10.26** Transition state of [$\pi^4s + \pi^2s$].

**Example (*iii*):** Let us examine the disrotatory and conrotatory closing of butadiene to cyclobutene by using PMO method. We have drawn the transition states for these two cases in Fig 10.27.

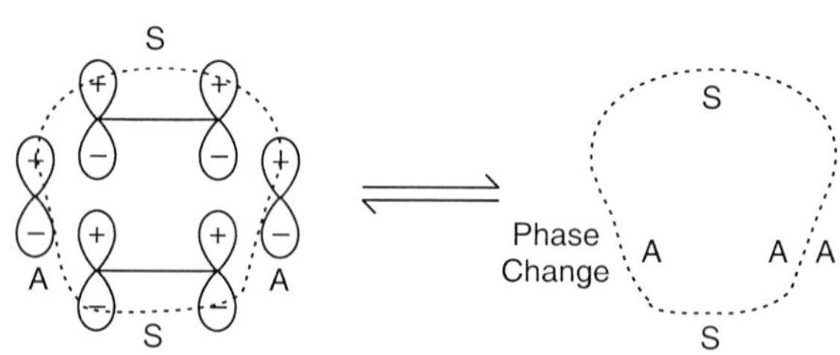


**Fig. 10.27** Transition state with phase change (2 in number [even]).

On examining the Fig. 10.27, it is obvious that the transition states in the two cases are different. The transition state for the disrotatory mode is of Huckel-type (all overlaps in phase) whereas the transition state for the conrotatory mode, it has one out of phase overlap.

One must also realize that in the conrotatory process it is impossible to avoid one out of phase interaction, so the conrotatory process belongs to anti Huckel (or Mobius) type of interaction. Even if you change the phases of individual atomic orbitals in the structure shown in the figure, the number of out of phase interaction for the Huckel (*i.e.*, disrotatory) transition state will always be even whereas the number of out of phase interactions for the anti Huckel (Mobius) transition state (conrotatory) will be odd. Since the interaction involves four electrons, it belongs to $4n$ category. So according to the above rules, cyclobutadiene is antiaromatic, if of the Huckel type of structure but aromatic if it is anti Huckel (Mobius). Thus, conrotatory process possesses an aromatic transition state and is favoured disrotatory process.

All the above three methods of analysis (orbital correlation, HOMO-LUMO, and PMO) can be used to analyze the Sigmatropic reactions. We will not be discussing them here because of (1) lack of space (2) of the fact; it does not involve any great use of symmetry properties of the molecules in general. However, if one is interested in more details about these reactions, one can consult the excellent book by Woodward (R.B.) and Hoffmann(*R*). "The conservation of orbital symmetry" Verlag Chemie, Weinheim" New York (1970).Another excellent book is by M.J.S. Dewar "The molecular orbital theory of Organic Chemistry" McGraw Hill, New York (1969). The article by K. Fukui and Fujimoto in "Mechanisms of Molecular Migrations" (Ed. Thijagrajan Vol. II, Wiley, New York (1969) is another book which will be highly beneficial to read from the standpoint of "Frontier Orbital Theory".

# 11

# Ligand Field Theory and Angular Overlap Model

The crystal field and the ligand field theories are effectively operator theories and are employed to construct multiplet states formed from atoms with $d^n$ and or $f^n$ configurations. Consequently one applies them only to systems where the ground and the lower excited energy states of atoms/ions are dominated by *d*-or *f*-orbital contributions. Consequently these theories are relevent only to Warner type transition metal complexes where the central transition metal ion is surrounded by a regular array of ligands, assumed to be a source of negative charges or dipoles with negative ends pointing towards central metal ion. It is assumed

1. that the electric charge produces electric field of symmetry identical to that of ligands,
2. that the ions are undeformable spheres,
3. that the charges are point charges located at the centre or point of each ligand and
4. that the interactions between the charges of the ligands and that of the central metal ion are solely the result of *electrostatic* attraction or repulsion. It is also assumed that the effects of the atoms which are not immediate neighbours are negligible. On these assumptions, the crystal field theory was first developed by Bethe (1929) and later by Penney and Schlapp (1932) and Van Vleck (1932).

In this chapter (Section A), we will discuss how the (*nd*), $(n+1)s$ and $(n+1)p$ orbitals of the central metal atom behave in the field of ligands and the electrons on the central metal ions interact with the field produced by ligand charges. Let us start with the field of $O_h$ symmetry.

## SECTION A

## 11.1 OCTAHEDRAL FIELDS

In the absence of any outside field *d*-, *s*- and *p*-orbitals are five-fold, one-fold and three-fold degenerate respectively. The modes of their splitting depend upon the type of symmetry of the local or external field surrounding the central metal ion. We will now illustrate as to how the energy levels of these orbitals are distributed when these are placed in the fields of different symmetries. The irreducible representations to which these orbitals belong can be seen from the character tables of the point groups, to which the symmetry of field belongs. (We have calculated the splitting of *d*-orbitals in octahedral field, when we have introduced the concept of function spaces). In some cases the I.Rs to which various orbitals span are given in Table 11.1.

**Table 11.1:** Splitting of orbitals in fields of different symmetries

| ***Symmetry of the field*** | ***nd*** | **(*n* + 1)*s*** | **(*n* + 1)*p*** | ***nf*** |
|---|---|---|---|---|
| $O_h$ | $E_g + T_{2g}$ | $A_{1g}$ | $T_{1u}$ | $A_{2u} + T_{1u} + T_{2u}$ |
| $T_d$ | $E + T_2$ | $A_1$ | $T_2$ | $A_1 + T_1 + T_2$ |
| $D_{4h}$ | $E_g + B_{2g} + b_{1g} + A_{2g}$ | $A_{1g}$ | $E_u + A_{2g}$ | $2E_u + B_{2u} + B_{1u} + A_{2u}$ |
| $D_{\infty h}$ | $E_{1g} + E_{2g} + A_{1g}$ | $A_{1g}$ | $A_1 + E_{1u}$ | – |
| $D_{3h}$ | $E'' + E' + A1'$ | $A_1'$ | $A_2'' + E'$ | $E'' + A_1'' + E' + A_2' + A_1'$ |
| $C_{3v}$ | $A_1 + 2E$ | $A_1$ | $A_1 + E$ | $2A_1 + A_2 + 2E$ |

In an octahedral field the irreducible representations to which (*nd*) orbitals belong are $E_g$ and $T_{2g}$ (Table 11.1). This is the limit of information which one can get from the symmetry of the field. It does not give any information about the relative energies of the orbitals belonging to $E_g$ and $T_{2g}$ irreducible representations. But it is obvious that the symmetry of the crystal fields (of a particular symmetry) leads to a differentiation of nd, orbitals into $T_{2g}$ and $E_g$ type of orbitals. Similar are the cases for $(n + 1)s$, $(n + 1)p$ or *nf* orbitals. We would now wish to know the sizes of these splittings of various orbitals under various symmetries of the crystal fields.

Let us take a particular example of $ML_6$ complex. We have five *d*-orbitals on the central metal ion located at the origin of an arbitrarily chosen coordinate system. First let us choose the *z*-axis of the coordinate system to coincide with the $z$-axis used to define $d$-, $s$- and $p$-orbitals. Secondly the six negative charges or dipoles of the ligands, present at the vertices of a regular octahedron, are placed on the axes of our chosen coordinate system at points ($\pm a$, 0, 0), (0, $\pm a$, 0) and (0, 0, $\pm a$). [These coordinates will form an octahedron when joined. [Fig.11.1].

We now examine qualitatively the effects of the distribution of six negative charges of the ligands around the central metal ion on metals $d$-, $s$-, and $p$-orbitals.

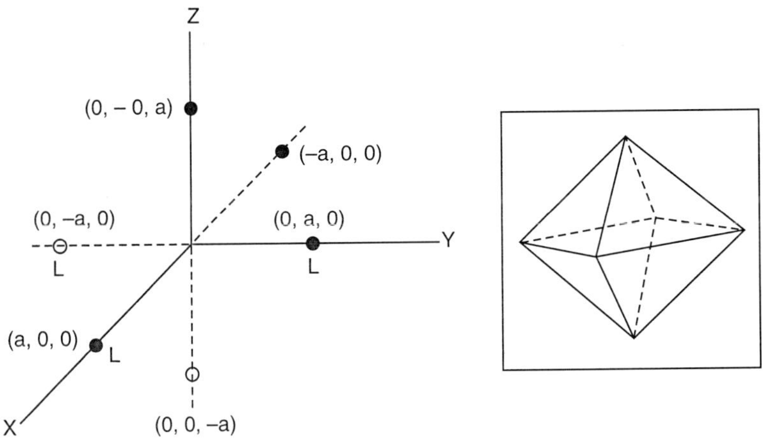


**Fig. 11.1** Coordinates of $ML_6$ octahedral.

In the first stage the discussion will be qualitative and will be based purely on electrostatic interaction between the electrons of the metal ion and those of the ligands (Assumption of crystal field model). But one must realise that the argument is not altered if the metal ligand perturbation includes some degree of covalency in the bonds). The orientation of $d$-orbitals are shown in [Fig.11.2]. If the electrons are present in the $d$-orbitals of the metal ion, we examine the

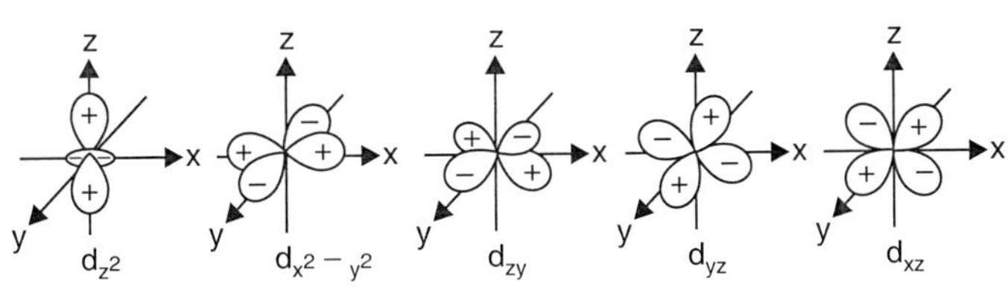


**Fig. 11.2** Shapes of $d$-orbitals.

degree to which these electrons in different d-orbitals ($d_{xy}$, $d_{yz}$, $d_{xz}$, $d_{x^2-y^2}$ and $d_{z^2}$) will like to come into the vicinity of the ligand charges. As seen in figure $d_{z^2}$ and $d_{x^2-y^2}$ orbitals have their lobes directed along the $z$- and $x$-, $y$-axes respectively (*i.e.*, directly towards the ligands). The $d_{xy}$-, $d_{yz}$- and $d_{xz}$-orbitals have their lobes (maxima) in the region in between the $(x, y)$, $(y, z)$, and $(x, z)$ axes. Consequently any electron on the central metal ion will like to avoid to be present in the $d_{z^2}$ and $d_{x^2-y^2}$; and will be more comfortable in $d_{xy}$, $d_{yz}$ or $d_{xz}$ orbitals where the inter electron repulsion between metal and ligand electrons is relatively less. This is the relative repulsion between the central metal electrons, but it must be emphasized that all the electron in whichever orbital they are present will always experience some electrostatic repulsion from the ligand charges. The only point is that their presence in $d_{z^2}$ and $d_{x^2-y^2}$ orbitals will relatively be more repulsive than that present in the $d_{xy}$-, $d_{yz}$- and $d_{xz}$-orbitals. [Fig.11.3] illustrates the effect on metal electrons by the repulsion of ligand charges. In this figure, because of the symmetry properties of $O_h$ point group, $d_{xy}$, $d_{yz}$, $d_{xz}$ form one set belonging to $T_{2g}$ I.R and all the three orbitals in this set is of equal energy. $d_{z^2}$, $d_{x^2-y^2}$ form another set belonging to $E_g$ I.R and these two orbitals are also degenerate. It is obvious from the Fig. 11.3 that all the electrons present in the $d$-orbitals

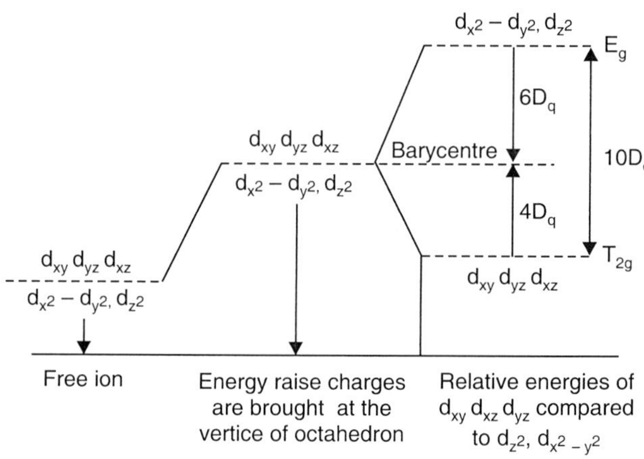


**Fig. 11.3** Relative energies of *d*-orbitals in a field of $O_h$ symmetry.

are affected by ligand electrons, but the effect on the electrons in $d_{xy}$, $d_{yz}$ and $d_{xz}$ ($T_{2g}$ set ) is lesser than those present in $d_{x^2-y^2}$ and $d_{z^2}$. Although the equal repulsive effect of the electron in $d_{xy}$, $d_{yz}$ and $d_{xz}$ (*i.e.*, their degeneracy) is understandable because they are all identical in shape. But the shapes of $d_{z^2}$, $d_{x^2-y^2}$ are not the same and even then they are degenerate. One can prove their degeneracy based on arguments on the algebraic expressions of two orbitals and their relationship to the orbitals $d_{y^2-z^2}$ and $d_{x^2-z^2}$: (Show them to be equivalent).

The second point is that the mean energy of the five $d$-orbitals is raised by the presence of six ligands. We have considered here the energy of five $d$-orbitals only, but if we consider the total energy of the system, along with six **bond energies** then there will be a net loss in energy (*i.e.*, net energy is lowered and system is thereby stabilized) due to the attraction of positive charges on metal ion and the ligand negative charges. This stabilization energy takes care of the energy by which the $d$-orbitals are a little destabilized. The result is a net stabilization energy for the coordination cluster. This is also true if one considers the covalent character in the bonds between metal and the ligands. This can be seen when we discussed the MOs of the octahedral complexes in Chapter eight where we have seen that the metal $d$-orbitals are antibonding in character and hence destabilized by complex formation, but this destabilization is more than counter-balanced by the stabilization undergone by the metal ligand bond formation using empty $s$- and $p$-orbitals of the metal ions.

## 11.2 TETRAHEDRAL COMPLEXES

Another important class of transition metal complexes is the tetrahedral class. In these complexes, the ligands occupy the alternate corners of a cube with metal ion, located at its centre which is also the origin of our chosen coordinate system (Fig. 11.4).

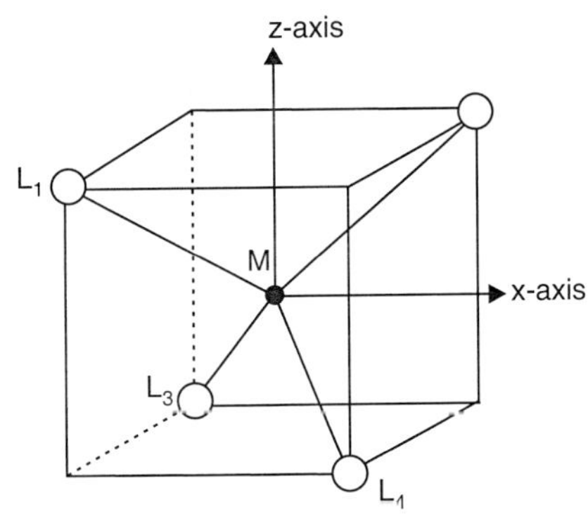


**Fig. 11.4** Tetrahedral complexes $ML_4$.

We have indicated earlier while dealing the MO theory of tetrahedral complexes that in a tetrahedral-field ($T_d$ symmetry), the five $d$ orbitals of the metal ion, are split into two groups *viz.* $T_2$ ($d_{xy}$, $d_{yz}$ and $d_{xz}$) and $E$ ($d_{x^2-z^2}$ and $d_{z^2}$). In the tetrahedral case we do not write $g$ or/and $u$ as subscripts on $T$ and $E$ because the $T_d$ point group does not have the centre of symmetry. If you draw the $d$-orbitals in the chosen coordinate system, you will notice that $T_2$-orbitals are directed more towards the negative charges of the ligands as compared to those $d$-orbitals belonging to $E$ symmetry (*i.e.*, $d_{z^2}$, $d_{x^2-y^2}$). Consequently it is the set of $T_2$-orbitals ($d_{xy}$, $d_{yz}$, $d_{xz}$) which are relatively less affected compared to those $d$-orbitals belonging to $E$ symmetry ($d_{x^2-y^2}$, $d_{z^2}$). As a result, contrary to the octahedral case, the relative energies of the electrons present in $T_2$-orbitals will be more stable than those present in $E$-orbitals. Consequently their relative energies are reversed with respect to those experienced in the field of $O_h$ symmetry. The total energies of all the electrons, like that in the case of octahedral symmetry, will also be raised in the field of ligand charges having $T_d$ symmetry (Fig. 11.5).

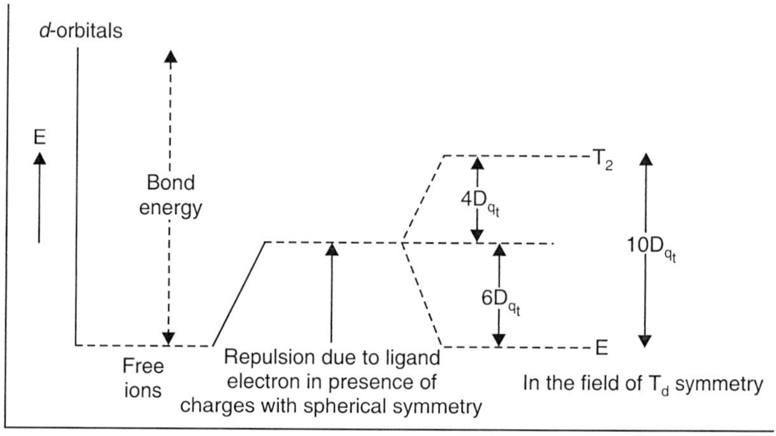


**Fig. 11.5** Splitting in the crystal field of $T_d$ symmetry.

In dealing with the stereochemistry of the other complexes, we would also like to mention cubic complexes, $ML_8$, where the ligands occupy all the eight corners of a cube. Basically a cube is composed of two interlocking tetrahedra (Fig.11.4). Therefore, we expect the *d*-orbitals to split, similar to that of tetrahedron.We, therefore, expect the splitting of *d*-orbitals of one tetrahedron the same as that of the second tetrahedron. The effect of cubic field ($ML_8$) when ligands are located at the corners of a cube is to split the *d*-orbitals as $d_{T2g}$ and $d_{Eg}$ sets with $E_g$ set, to be lower lying with respect to the orbitals of $T_{2g}$ set. (Note the restoration of "*g*" and '*u*' on *E* and *T* because cube has a centre of symmetry).

Another interesting stereochemistry is that of the icosahedron ($ML_{12}$) which we occasionally come across, especially in lanthanide and actinide complexes. The important and peculiar point in this case is that the *d*-orbitals do not split because it essentially combines both octahedral and tetrahedral fields together.

## 11.3 IMPORTANT POINTS

We consider below a few important points which should be noted while dealing with the complexes in terms of crystal or ligand field theory. These are:

1. We disregard in our usual discussions, the shift of all the levels in the average energy of all the five *d*-orbitals. We simply take into account the effect of crystal field as responsible for the splitting of various *d*-orbitals relative to each other, leaving the mean of their energies as unchanged (Fig. 11.6).

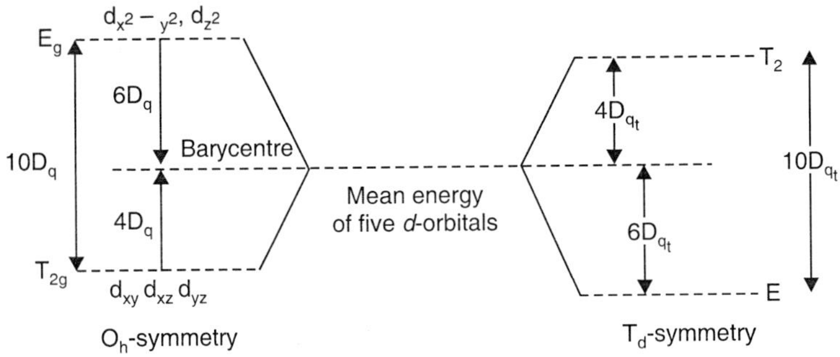


**Fig. 11.6** Splitting of *d*-orbitals in $O_h$ and $T_d$ symmetric field.

2. It is difficult to calculate the energy values of the *d*-orbitals after the interaction. In Fig. 11.6, we have indicated the splitting in terms of $10D_q$ where $D_q$ is a parameter, given by

$$D = \frac{35Z_e}{4a^5} \quad \text{and} \quad q = \frac{2e\gamma^4}{105} \qquad \text{... (11.1)}$$

and in the case of ligand acting as dipoles,

$$D_{q_{oct}} = \frac{5\mu\gamma^4}{6a^6} \qquad \text{... (11.2)}$$

Here $D_q$ is defined in a way so that the separation between $d_{Eg}$ and $d_{T2g}$ orbitals is $10D_q$. In the ligand field formalism, the parameter $10D_q$ is changed to $D_0$.

3. We have indicated the energies of $d_{T2g}$ and $dE_g$ as $4D_q$ and $6D_q$. These energies are calculated with reference to zero energy which is taken as the weighted mean energy of the five *d*-orbitals in the environment of 6 or 4 charges of the ligands assuming no splitting. The weighted mean energy is taken as the zero levels of the energy. Thus, the $d_{T2g}$ orbitals have an energy of $-4D_q$ while the energy of $dE_g$ orbitals as $+6D_q$ so that the sum of the energies of three $d_{T2g}$ orbitals $[3.(-4D_q)]$ balances that of the two *dEg* orbitals $(+2.6D_q)$. We say then, that the net sum of the energies of the five *d*-orbitals (*i.e.*, three-orbitals $d_{T2g}$ + two-orbitals of $dE_g$) must be equal to zero. It is often quoted as saying that the bari-centre (or the centre of gravity) remains the same (constant) regardless of the type of symmetry of the field. The neglect of the shift in the baricentre or the centre of gravity of *d*-orbitals is in line with the statement that the *absolute binding energies* of the complex ions have not been considered.

4. We have defined $D_q$ as a parameter which gives an idea of the **extent of splitting** of *d*-orbitals. We notice that it is a function of charge of the

metal ion ($Ze$) ($a$ or $r$ is effective radius of $d$-orbitals). As $Z$ increases, the value of '$a$' or '$r$' decreases. Besides charge on the metal ion, $D_q$ is also dependent on the field strength generated by the ligands. Which ligands produce strong and which weak field, is given empirically in the form of a series, called "Spectrochemical Series". This is given below:

$$I^- < Br^- < S_2^- < SCN^- < Cl^- < NO_2^- < F^- < OH^- < C_2O_4^{2-} < H_2O < NCS^- < CH_3CN < NH_3 < en < bipy < Phen < NO_2^- < CN^- < CO \quad ...(11.3)$$

If you wish to rationalize the series and also wish to find out the factors on which field strength generated by the ligands depend, these are just not known at the present time. On the basis of purely ionic model, it is difficult to see as to why and how a neutral ligand like CO can generate very strong field. It implies that to some extent, covalent bonding is also responsible for the generation of strong field. We see that a strong ($L \rightarrow M$)$s$ interaction increases the energy of the $E_g^*$-orbitals, thereby increasing the value of $10D_q$ (or $D_q$). A strong ($L \rightarrow M$) π-interaction, on the other hand, destabilizes $T_{2g}$ orbitals, thereby decreases the value of $10D_q$ (*See* MO of $ML_6$ with $p$-interactions Chapter eight). Therefore, a strong ($M \rightarrow L$) π-interaction lowers the energy of $T_{2g}$-orbitals thereby increasing the value of $10Dq$. We can therefore correlate the order in the above series with the relative π-donor or π-acceptor ability of the ligands. The π-acceptor ligands *i.e.*, those capable of strong ($M \rightarrow L$) π* bonding causes large splitting, whereas the π-donor ligands *i.e.*, those capable of strong ($L \rightarrow M$) π-donation cause small splittings. The ligands with intermediate $10D_q$ values have practically no π-bonding capabilities (Fig. 11.7).

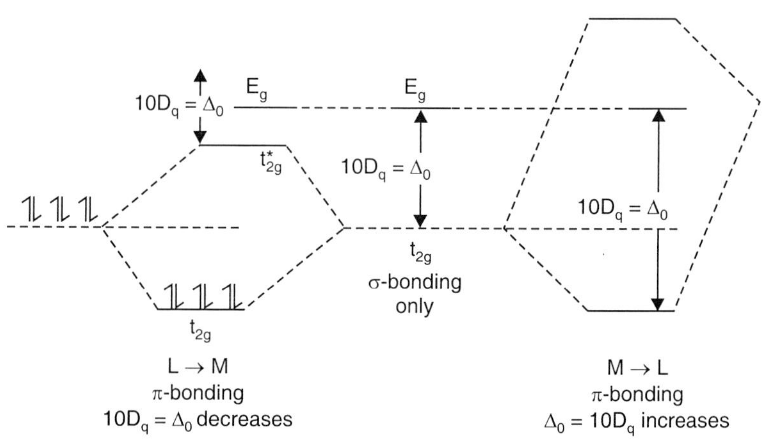


**Fig. 11.7** Effect of π-bonding on $\Delta_0 = 10Dq$ with L → M $\Delta_0$ decreases.

## 11.4 STABILISATION ENERGY

We now feed in the electrons of the metal ion starting with the $d_{T2g}$ (Oct) or with $d_E$ (Tetrahedral) orbitals. Thus each electron entering into these orbitals lowers the relative energy by $4D_q$ in octahedral ($O_h$) and $6D_q(T_2)$ for tetrahedral ($T_d$) complexes. If the electrons enter in the higher energy orbital, *e.g.*, in $dE_g$ (Oct) and $d_{T2}$ (tetrahedral), then the system is destabilised by $+4D_q$ and $+6D_q$ respectively. Thus one can calculate the ligand field stabilisation energy (L.F.S.E.) as the algebraic sum of the stabilisation and the destabilisation energies caused by the presence of electrons in $d_{T2g}$ and $d_{Eg}$ orbitals. For example, for metal ion with $d^3$ metal ion octahedral complexes, the total stabilisation energies, (L.F.S.E.) is $[3' \times (-4Dq)]$ which is equal to $-12D_q$. For the same $d^3$ system, if the complex is tetrahedral the total stabilisation energy is $[2' \times (-6D_q) + (1' \times 4D_q) = -8D_q$ (Fig. 11.8).

In this manner, the LFSE for all systems from $d^1$ to $d^{10}$ can be calculated in both $T_d$ and $O_h$ fields. These are given in Table 11.2.

You will notice two factors which control the LFSE in every case. These are: (*a*) Pairing energy and (*b*) Strength of the field

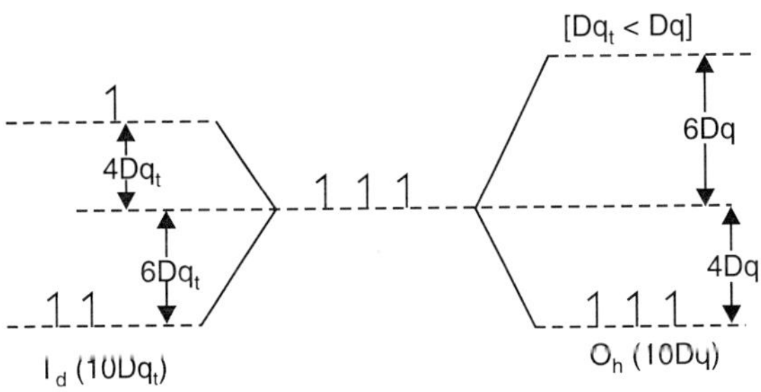


**Fig. 11.8** Arrangement of three electrons in the *d*-orbitals.

**Table 11.2:** Stabilisation Energies of Some Complexes

| | *Octahedral* | | *Tetrahedral* | |
|---|---|---|---|---|
| ***Configuration and examples*** | ***Weak field*** ($-D_q$) | ***Strong field*** ($-D_q$) | ***Weak field*** ($-D_q$) | ***Strong field*** ($-D_q$) |
| $d^0$ ($Ca^{2+}$ $Sc^{3+}$) | 0 | 0 | 0 | 0 |
| $d^1$ ($Ti^{3+}$) | 4 | 4 | 6 | 6 |
| $d^2$ ($V^{3+}$) | 8 | 8 | 12 | 12 |
| $d^3$ ($Cr^{3+}$) | 12 | 12 | 8 | No case |

*(Contd.)*

| | | | | |
|---|---|---|---|---|
| $d^5$ ($Mn^{2+}$, $Fe^{3+}$) | 0 | 20 | 0 | No case |
| $d^6$ ($Fe^{2+}$, $Co^{3+}$) | 4 | 24 | 6 | No case |
| $d^7$ ($Co^{2+}$) | 8 | 18 | 12 | 12 |
| $d^8$ ($Ni^{2+}$) | 12 | 12 | 8 | 8 |
| $d^9$ ($Cu^{2+}$) | 6 | 6 | 4 | 4 |
| $d^{10}$ ($Cu^+$) | 0 | 0 | 0 | 0 |
| $D^4$ ($Cr^{2+}$) | 6 | 16 | 4 | No case |

## 11.5 PAIRING ENERGY

In any system when two electrons are forced to pair (low spin), in an orbital, the system is destabilised by the energy used in pairing them due to electronic repulsion. This has to be subtracted from the LFSE. This repulsive energy is known as pairing energy. For example, take a $d^4$ system in an octahedral field (Fig. 11.9). Here we have two alternatives. In case (*A*), the fourth electron is paired and in the second case (*B*), the fourth electron goes to high energy *dEg* orbital. The preference for pairing or going to high energy orbital, depends upon the value of $10D_q$ which is a function of charge of the metal ion and the type of the ligand (*i.e.*, whether it generates a strong field or a weak field.) In case the field is strong and $10D_q >>$ Pairing energy the electron will prefer to pair up while if the field is week and $10D_q <$ Pairing energy, the fourth electron will prefer to go to the high energy orbitals *viz.* $E_g$-orbitals.

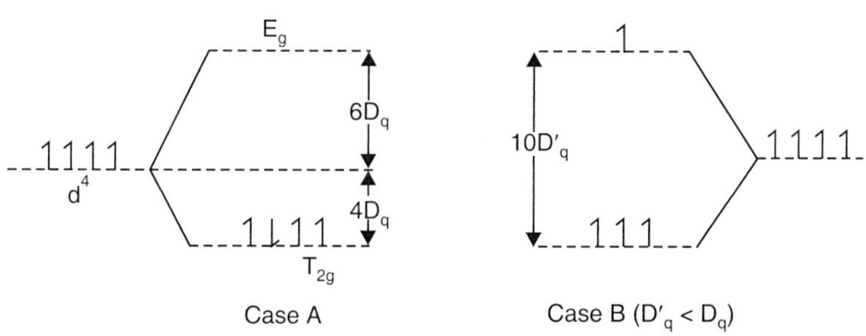


**Fig. 11.9** Strong field (case A) and week field (case B).

In tetrahedral ($T_d$) case the splitting (*viz* $10\ D_q(t)$) into $T_2$- and $E$-orbitals is generally smaller than ($10D_q(o)$). The relation is given by the following equation

$$10d_q \text{ (Tetrahedral)} \cong \frac{4}{5}\ to\ \frac{5}{6} th \text{ of } 10D_q \text{ (octahedral)} \qquad ...(11.4)$$

The consequence of the smaller splitting in the tetrahedral case is that $10D_q(tet) << 10D_q(t) <<$ Pairing energy (in most of the cases). This results in the formation of in general high spin complexes or tetrahedral complexes.

Employing all these factors into consideration, LFSE values are given in Table 11.2.

It is also important to know that $D_q$ value increases and the pairing energy decreases as we go from first transition series elements to the second and then to the third series elements. Pairing energy values for a few metal ions are given in Table 11.3.

**Table 11.3:** Pairing Energy and $10D_q$ values for some metal ions

| *ion* | *Pairing energy (KK)* | *$10D_q$ for $M(OH_2)_6$ KK* |
|---|---|---|
| $Mn^{3+}$ ($d^4$) | 26 | 21.0 |
| $Mn^{2+}$ ($d^5$) | 24 | 8.5 |
| $Fe^{3+}$ $d^5$ | 30 | 14.3 |
| $Fe^{2+}$ $d^6$ | 18 | 10.4 |
| $Co^{3+}$ $d^8$ | 23 | 20.7 |

## 11.6 DISTORTED OCTAHEDRON, SQUARE PLANAR AND LOWER SYMMETRY COMPLEXES

In a regular octahedron complex ($ML_6$), if two ligands on the *z*-axis are substituted by two different ligand then the ligand field symmetry is changed from $O_h$ to $D_{4h}$. By such substitutions, if the ligand field of *L* is strong, the *M–L* bond length shortens, and the octahedron is compressed along the *z*-axis. In the second case, if the field strength generated by *L* is weak, the bond length *M –L* becomes longer, whereby the octahedron becomes elongated along *z*-axis. In the third case, if the two ligands along the *z*-axis are totally removed, the complex becomes square planar ($ML_4$). In all the three cases, the symmetry of field changes from $O_h$ to $D_{4h}$. In the beginning of the chapter, we have given in Table 11.1, that the five *d*-orbitals in the field of $D_{4h}$ symmetry are split as $(A_{1g} + E_g + B_{2g} + B_{1g})$. Actually $T_{2g}$ ($d_{xy}$, $d_{yz}$, $d_{xz}$) orbitals are split as $E_g$ ($d_{xz}$, $d_{yz}$) and $B_{2g}$ ($d_{xy}$) and $E_g$ orbitals ($d_{z^2}$, $d_{x^2-y^2}$) $A_{1g}$ ($d_{z^2}$) and $B_{1g}$ ($d_{x^2-y^2}$). Thus, the crystal field splitting pattern as a result of tetragonal distortion of an octahedral complex deformed by either stretching or by compressing of two ligands along the *z*-axis is summarized in Fig. 11.10.

Referring to orbital splitting pattern of energy levels in the case of elongation of $M - L'$ bond, the splitting denoted by $b_1$ and $b_2$ become larger and larger. Eventually, it will lead to the removed of two axial ligands from the *z*-axis and will form a square planar complex. Practically all the $d^8$ complexes are square planar with first row transition metal ions and energy gap between $B_{1g}$ and $B_{2g}$ becomes $10D_q$.

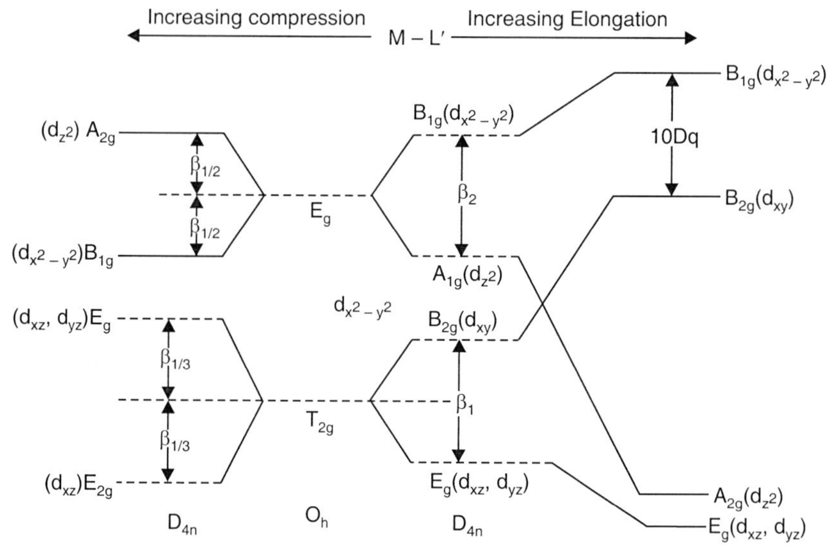


**Fig. 11.10** Splitting by tetragonal distortion as a result of elongation and compression of M – L′ along *z*-axis.

## 11.7 OCTAHEDRAL FIELD SPLITTING OF SPECTROSCOPIC TERMS

Untill now we have taken $d^1$ system and used only hydrogenic wave functions—a case when the five *d*-orbitals are split in $t_{2g}$ and $e_g$ sets. In case when we use spectroscopic terms *e.g.*, $2_D$ state arising from $d^1$ system, we will see that in this also $2_D$ state is split into $2_{T2g}$ and $2_{Eg}$ states arising out of $(t_{2g})'$ and $e_g'$ respectively. We will now learn what states arise out of *F*, *G* or *P* states which arise from $d^8$ configuration ($n > 1$). See the book of Physical Chemistry regarding formation of terms from $d^n$ electronic configurations and also chapter eight.

Instead of starting from octahedral complexes with $O_h$ symmetry, we can also start from *O* symmetry. It is just because of simplicity of calculation. It has rotation operations only ($E$, $C_4$, $C_3$, $C_2$ and $C_2'$). The characters of $G_{red}$ of various spectral terms are given in Table 11.4. It lacks the inversion centre which is present in the $O_h$ point group.

**Table 11.4:** Representations of spectral terms in $O$ point group and states these terms form

| *Term* | *L* | $\hat{E}$ | $6\hat{C}_4$ | $3\hat{C}_2$ | $8\hat{C}_3$ | $6\hat{C}_2$ | *States formed* |
|---|---|---|---|---|---|---|---|
| $S$ | 0 | 1 | 1 | 1 | 1 | 1 | $A_1$ |
| $P$ | 1 | 3 | 1 | –1 | 0 | –1 | $T_1$ |
| $D$ | 2 | 5 | –1 | 1 | –1 | 1 | $E + T_2$ |
| $F$ | 3 | 7 | – 1 | –1 | 1 | –1 | $A_2 + T_1 + T_2$ |
| $G$ | 4 | 9 | 1 | +1 | 0 | 1 | $A_1 + E + T_1 + T_2$ |
| $H$ | 5 | 11 | 1 | –1 | –1 | –1 | $E + 2T_1 + T_2$ |
| $I$ | 6 | 13 | –1 | 1 | 1 | 1 | $A_1 + A_2 + E + T_1 + T_2$ |

An octahedral crystal field reduces the (2$L$ + 1)-fold degeneracy of a spectroscopic term. For this purpose we need the appropriate reducible representation which will give the character of a rotation operation on an atomic state defined by the orbital angular momentum ($L$). It can be shown that the character $c$ for a rotation an about z-axis is given by

$$\chi_\alpha = \frac{\sin\left(L+\frac{1}{2}\right)\alpha}{\sin\frac{\alpha}{2}} \quad \text{...(11.5)}$$

Thus for identity operation, $\hat{E}$. $\alpha = 0$, for $C_4$, $\alpha = \frac{\pi}{2}$, for $C_2$, $a = p$ for $C_3$, $a = 120°$. Using these values of $a$, one can calculate the characters for $C_4$, $C_2$ and $C_3$ operations. All these values are given in Table 11.4 for various $L$ values (*viz*) when $L$ = 0, 1, 2, 3, 4, 5 ... *i.e.*, *S*, *P*, *D*, *F*, *G*, *H* ... *etc*. states respectively. (Check the values of $C_5$ of Table 11.4.)

Thus for $F$ term $L$ = 3, we get the following characters in the $O_h$ point group because:

For operation $\hat{E}$, ($\alpha = 0$), $\chi_{\hat{E}} = \dfrac{\sin\left(3+\frac{1}{2}\right).0}{\sin\frac{0}{2}} = 7$

For operation $\hat{C}_4$ ($\alpha = 90°$), $\chi_{C_4} = \dfrac{\sin\left(3+\frac{1}{2}\right)90°}{\sin\frac{0}{2}} = -1$

$$\text{For operation } \hat{C}_2 \ (\alpha = \pi), \quad \chi_{C_2} = \frac{\sin\left(3+\frac{1}{2}\right)180°}{\sin 90} = -1$$

$$\text{For operation } \hat{C}_2 \ (\alpha = \pi), \quad \chi_{C_3} = \frac{\sin\left(3+\frac{1}{2}\right)120°}{\sin 60} = 1$$

Thus, we get a reducible representation $\Gamma_{\text{red}}$ using *F* state wave functions as basis,

| ***Operations*** | $\hat{E}$ | $6\hat{C}_4$ | $8\hat{C}_3$ | $8\hat{C}_3$ | $6\hat{C}_2'$ |
|---|---|---|---|---|---|
| $\chi_R$ ***for*** $\Gamma_{red}$ | 7 | −1 | −1 | 1 | −1 |

This $\Gamma_{\text{red}}$ representation is reduced by the usual procedure (given in Chapter 10). The $\Gamma_{\text{red}}$ for point group *O* contains as components $A_2 + T_2 + T_1$ belonging to these IRs. For calculation of the energies of the orbitals, we make linear combinations of wave functions of *F*. state (seven wave functions) using projection operation of $A_1$, $T_1$ and $T_2$. These LCAO's are used as basis which will diagonalise the secular determinant into (1 × 1) (3 × 3) and (3 × 3) blocks and energy from each block components are then obtained. These will give the energy of $A_2$, $T_1$ and $T_z$ orbitals. (For details see chapter seven)

If the F-term arises from a configuration with electrons occupying *d*-orbitals, which are even functions (*g*-symmetry), then the *F*-term orbitals will split as $A_{2g}$, $T_{2g}$ and $T_{1g}$. On the other hand, if the *F*-term arises from the configuration with electrons occupying the *odd functions* (*e.g.*, *f*-orbitals), then it will split as $A_{2u} + T_{1u} + T_{2u}$. (Because the functions (*f*-orbitals) do not have a centre of symmetry and are odd functions.)

By this procedure the values and the terms (states in the $O_h$ field) for various *L* values are obtained by employing values of α in Eq. 11.5 (α will remain the same as used in the above calculations of $\Gamma_{\text{red}}$ for $L = 2$).

Employing the various terms present for an octahedral complex with a given electronic configuration, one can find the energy levels present in these spectroscopic states. Orgel has calculated the energy values and drawn energy diagrams, for all the electronic configurations and plotted the energy levels as a function of the increasing values of $D_q$. It thus implies that one can treat cases of weak, strong and also of intermediate fields in the same diagram. Since this book is on the symmetry and its applications, it is not justified to discuss these Orgel diagrams at this point.

## 11.8 SPIN ORBIT COUPLING (CONCEPT OF DOUBLE GROUPS)

In the preceeding paragraphs, we have used formula given by equation (11.5) for the calculation of $\chi_R$ for the formation of reducible representation $\Gamma_{red}$ of a particular point group using wave functions as basis for the representation without any consideration of spin-orbit interaction. (*i.e.*, *J* or *g*-values).

However, if spin-orbit interaction is significant, then L is not a good quantum number and we cannot use it in (Eq. 11.5). In that case, the total quantum number $J = (L + S)$ is a good quantum number. *J* can take both integral and half-integral values. Although when *J* is an integer (0, 1, 2, 3 ......), there is no problem of using the same (Eq. 11.5), with a change that L is replaced by *J* in (Eq. 11.5). But when *J* takes half integral values, a complication arises. In that case the values of $\chi_\alpha$ is not the same as $\chi(\alpha + 2\pi)$. [Mathematically it should be, because $\chi_E$ $(\alpha = 0)$ must be the same as $\chi_{2\pi}$ $(\alpha = 2\pi)$]. Mathematically this can be seen by the following computation.

When *J* is half-integral, we have,

$$\chi(\alpha + 2\pi) = \frac{\sin\left(J + \frac{1}{2}\right)(\alpha + 2\pi)}{\sin\left[\frac{(\alpha + 2\pi)}{2}\right]} \quad ...(11.6)$$

$$= \frac{\sin\left[\left(J + \frac{1}{2}\right)\alpha\right]}{\sin\left(-\frac{\alpha}{2}\right)} \quad ...(11.7)$$

$$= -\chi(\alpha) \quad ...(11.8)$$

[That is a rotation operation by $2\pi$ gives $\chi_a$ to be equal to $-\chi(\alpha + 2\pi)$].

In order to solve this problem, Be-the introduced a new operation *R*′, that is, rotation by $2\pi$ and which is no more equivalent to previously known $\chi_E$, (that is equivalent to the identity operation). Thus, if one adds this operation to the operations of the rotation group (*e.g.*, $C_n$, O, *etc.*), one has to expand the number of operations in any rotation group, because we will have now to add all those new operations also to the old point group. These are obtained by the direct product of $R \times R'$. These new operations are obtained by multiplying *R*′ with *R* that are already present in the rotation group. [This is similar to direct product of two subgroups (see chapter six). Here it is equivalent of [*E*, $R_1R_2$ ....] [*E*. *R*′]. Here *R*′ should commute with the operations of *R*.

This makes the number of operations of all the point groups to nearly double (not exactly). We term these point groups as "*Double groups*". For example, we have a point group with operations $\{E, 6C_4, 3C_2, 8C_3$ and $6C_2'\}$ in the rotation group O. If we add a new operation $R'$ (rotation by $2\pi$) to this group, [*i.e.*, form a group $[G_1]$ $[E, R']$ we now have the operation $4C_3^1, 4C_3^2, 3C_2, 3C_4, 3C_4^3, 6C_2^1$ $E$ (rotation by $4\pi$), $R'$ (rotation by $2\pi$), $4C_3^2R', 4C_3^1R', 3C_2R', 3C_4^3R', 3C_4^1R', 6C_2^1R'$. In the new double group $O'$, that is, $O'$ group has a total of 48 operations instead of previously existing 24 operations forming 8 classes ($E$), ($R'$), $(4C_3^1, 4C_3^2R')$, $(4C_3^2, 4C_3^1R')$, $(3C_2, 3C_2R')$, $(3C_4^1, 3C_4^3R')$ $(3C_4^3, 3C_4^1R')$, $(6C_21 \,.\, 6C_2^1R')$. This will give us eight I.R.s. Out of these eight I.R's. the five will be identical to those of the parent point group $O$, $\Gamma_1, \Gamma_2, \Gamma_3, \Gamma_4$, *etc*. or as $A_1', A_2'$ ...... *etc*. Some of the character tables of these new double groups are given along with the regular character tables in most of books on symmetry.

For half-integral $J$ values, we now have character tables of the double groups by which we determine the representations under "$O$" symmetry point group. Spanned by a state with half-integer quantum number $J$, we give in Table 11.5, the reducible representations in terms of double group. As an illustration of how the numbers in Table 11.5 are obtained, we give below one example,

We take the example for $J = \frac{1}{2}$ and calculate $\chi$ for $\alpha = 2\pi$

$$\chi \text{ for } R' = \chi(\alpha = 2\pi) = \lim_{\alpha \to 2\pi}$$

$$= \frac{\left[\sin\left(\frac{1}{2} + \frac{1}{2}\right)\alpha\right]}{\sin\frac{\alpha}{2}} \qquad ...(11.9)$$

$$\lim_{\alpha \to 2\pi} = 2\left[\frac{\cos\alpha}{\cos\alpha/2}\right] = -2$$

**Table 11.5:** Double groups for various J values

| $J$ or $S$ | $E$ | $R'$ | $4C_3$ $4C_3^2R'$ | $4C_3^2$ $4C_3^1R$ | $3C_2$ $3C_2R$ | $3C_4^1$ $3C_4^3R$ | $3C_4^3$ $3C_4^1R$ | $6C_2'$ $6C_2'R$ | Rep. in 0 or 0′ |
|---|---|---|---|---|---|---|---|---|---|
| S 0 | 1 | $G_1$ | | | | | | | $\Gamma_1$ |
| $\frac{1}{2}$ | 2 | –2 | 1 | – 1 | 0 | $(2)^{1/2}$ | $-(2)^{1/2}$ | 0 | $\Gamma_6$ |
| p 1 | 3 | | | | | | | | $\Gamma_4 = T_2$ |
| $\frac{3}{2}$ | 4 | –4 | –1 | 1 | 0 | 0 | 0 | 0 | $\Gamma_8$ |

*(Contd.)*

| | | | | | | | | | |
|---|---|---|---|---|---|---|---|---|---|
| $d$ 2 | 5 | | | | | | | | $\Gamma_3 + \Gamma_5$ $(E + T_2)$ |
| $\frac{5}{2}$ | 6 | –6 | 0 | 0 | 0 | $-(2)^{1/2}$ | $(2)^{1/2}$ | 0 | $\Gamma_7 + \Gamma_8$ |
| $f$ 3 | 7 | | | | | | | | $\Gamma_2 + \Gamma_4 + \Gamma_5$ $(A_2 + T_1 + T_2)$ |
| $\frac{7}{2}$ | 8 | –8 | 1 | –1 | 0 | 0 | 0 | 0 | $\Gamma_6 + \Gamma_7 + \Gamma_8$ |
| $g$ 4 | 9 | | | | | | | | $\Gamma_1 + \Gamma_3 +$ $\Gamma_4 + \Gamma_5$ |
| $\frac{9}{2}$ | 11 | –10 | –1 | 1 | 0 | $(2)^{1/2}$ | $-(2)^{1/2}$ | 0 | $\Gamma_6 + 2\Gamma_8$ |
| $h$ 5 | 11 | | | | | | | | $\Gamma_3 + 2\Gamma_4 + \Gamma_5$ |
| $\frac{11}{2}$ | 12 | –12 | 0 | 0 | 0 | 0 | 0 | 0 | $\Gamma_6 + \Gamma_7 + 2\Gamma_8$ |

Now we like the case of $^4D\left(2s+1=4, \text{ or } s=\frac{3}{2}\right)$. This term arises from a $d^5$ octahedral complex. In case spin-orbit interaction is ignored, this term will split $^4T_{2g}$ and $^4E_g$ which in Be-the nomenclature are $^4\Gamma_5$ and $^4\Gamma_3$ respectively in the crystal field of $O_h$ symmetry. Now we consider the same $4_D$ case when we also consider spin orbit interaction. In that case the states, $^4\Gamma_5$, $^4\Gamma_3$ further split. To know the pattern of splitting, we see in the column of $S$ of Table 11.5, for $S = \frac{3}{2}$. In the horizontal row of $S = \frac{3}{2}$, we find the symmetry species $\Gamma_8$ in the last column of the same Table 11.5. Now when the spin state $\frac{3}{2}$ interacts with the orbital part of the wave functions of symmetry species $\Gamma_3$ and $\Gamma_5$, we can obtain the states by the direct product of $\Gamma_8$ with $\Gamma_3$ and $\Gamma_5$ separately. Thus we have,

$$\Gamma_3 \times \Gamma_8 = \Gamma_6 + \Gamma_7 + \Gamma_8 \qquad ...(11.10)$$

and

$$\Gamma_5 \times \Gamma_8 = \Gamma_6 + \Gamma_7 + 2\Gamma_8 \qquad ...(11.11)$$

Accordingly, when $\Gamma_3$ and $\Gamma_5$ interacts with $\Gamma_8$, then $\Gamma_3$ and $\Gamma_5$ further split into three $(\Gamma_6 + \Gamma_7 + \Gamma_8)$ and four $(\Gamma_6 + \Gamma_7 + 2\Gamma_8)$ states respectively. (This is the case when crystal field is stronger than spin-orbit coupling.)

In case spin-orbit coupling is stronger than the crystal field interaction, we then take, $J$ as good quantum number that defines a state before interaction

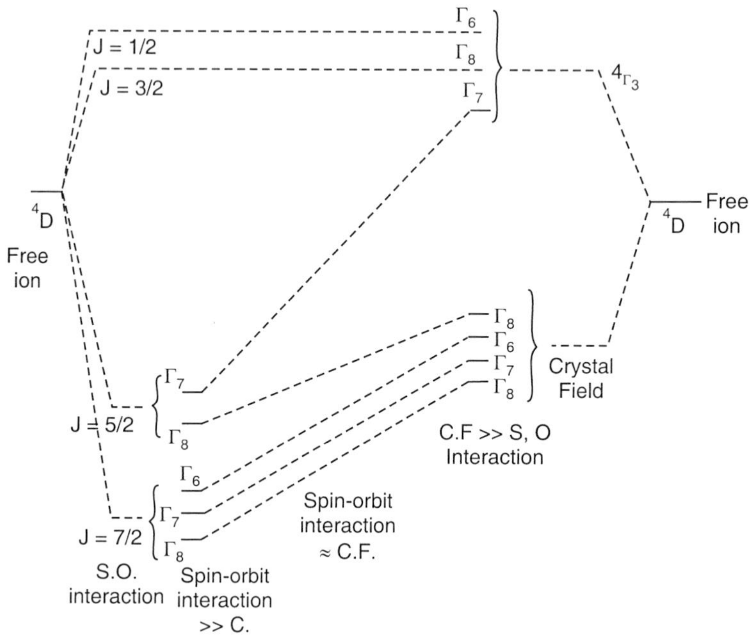


**Fig. 11.11** Relative effect of C.F. and S.O. interaction on $^4D$ state.

with crystal case, the values of $J$ are $\frac{1}{2}, \frac{3}{2}, \frac{5}{2}$ and $\frac{7}{2}$ $(J = [L + S])$. So, in the crystal field of $O_h$ symmetry, with the help of Table (11.5) we can readily see that these states split as

$$\left.\begin{aligned} J &= \frac{1}{2} \text{ as } \Gamma_6 \\ J &= \frac{3}{2} \text{ as } \Gamma_8 \\ J &= \frac{5}{2} \text{ as } \Gamma_7 + \Gamma_8 \\ \text{and } J &= \frac{7}{2} \text{ as } \Gamma_6 + \Gamma_7 + \Gamma_8 \end{aligned}\right\} \text{ from Table (11.5) (last column)} \quad ...(11.12)$$

The way these two sets of states have (weak crystal field + strong spin-orbital interaction) and (strong crystal field + weak spin-orbital interaction) are shown in Fig. 11.11.

## 11.9 CRYSTAL FIELDS OF LOWER SYMMETRIES

In previous discussions, we have studied the effect of crystal fields having $O_h$, $T_d$ and $D_{4h}$ symmetries on metal *d*-orbitals. Although most of the complexes have these symmetries, but there are a number of complexes having lower symmetries too. In these lower symmetry fields one can find the splitting of *d*-orbitals in the same manner as discussed in the first part of this chapter. But the problem is to find the energy of these split-levels. Though one can determine the energy values of energy levels, but mathematical manipulations are too complex to solve the secular equations. Another problem that one faces is that the symmetry of the fields being low, these determinants cannot be diagonalized to an extent so as to solve the energies. It is for these reasons that the various approximations have been made to solve the secular determinants and find the energies of split *d*-orbitals. One such approximation is given by Krishnamurthy and Schaap [J. Chem. Ed. T99, 46 (1969)]. They claim that it is possible to calculate the relative energies of *d*-orbitals when they are surrounded by fields of any symmetry using the relative energies for the ligand groups primarily for three symmetrices. These three symmetries correspond to (1) C.N. 1. located along the *Z*-axis (2) C.N. 2. with two ligands located in the *XY* plane at right angles to each other (along the X- and *Y*-axis) and (3) C.N. 4. with ligands located at the vertices of tetrahedron (two above and two below the *XY* plane) each at a variable angle *q* with respect to *Z*-axis. Let us suppose that we know the relative energies of these three primary groups (Table 11.6). Knowing the values of $a_i$, $b_t$ ...... $g_i$, we can find the relative energies of *d*-orbitals in the crystal field of any symmetry.

**Table 11.6:** Relative energies of *d*-orbitals

| C.N. | $d_{z^2}$ | $d_{x^2-y^2}$ | $d_{xy}$ | $d_{xz}$ | $d_{yz}$ | |
|---|---|---|---|---|---|---|
| 1 | $a_1$ | $a_2$ | $a_3$ | $a_4$ | $a_5$ | 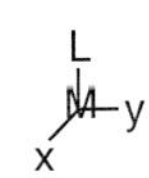 |
| 1 | $b_1$ | $b_2$ | $b_3$ | $b_4$ | $b_5$ | 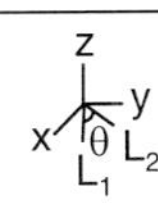 |
| 4 | $\theta = 45°$ $c_1$<br>54.74 $d_1$<br>60° $f_1$<br>62.47° $g_1$ | $c_2$<br>$d_2$<br>$f_2$<br>$g_2$ | $c_3$<br>$d_3$<br>$f_3$<br>$g_3$ | $c_4$<br>$d_4$<br>$f_4$<br>$g_4$ | $c_5$<br>$d_5$<br>$f_5$<br>$g_5$ | 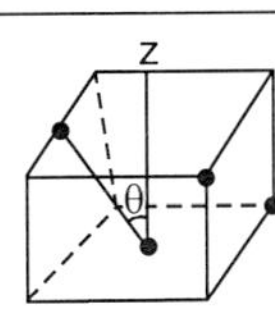 |
| The values of $a_i$, $b_i$, $c_i$, $d_i$, $t_i$ and $g_i$ electron | | | | | | |

As an illustration let us take the case of square planar complex. Ligands are on $x$- and $y$-axis. Since the effects of two ligands at $+x$ and $+y$ are the same as for two ligands at $-x$ and $-y$ axes, then the energies of $d$-orbitals for the square planar complexes are given by ($2'$ $C.N.$ = 2) case *i.e.*, the relative energies of $d_{x^2}$ $d_{x^2-y^2}$ $d_{xy}$, $d_{xz}$ $d_{yz}$ will be $2b_1$, $2b_2$, $2b_3$, $2b_4$ and $2b_5$ respectively.

For octahedral complex we add only ($2'$ $C.N.$ = 1) energies to those in the field of square planar complex. That is the relative energies of $d_{z^2}$, $d_{x^2-y^2}$, $d_{xy}$, $d_{xz}$ and $d_{yz}$ will be $(2b_1 + 2b_2)$, $2(a_2 + b_2)$, $2(a_3 + b_3)$, $2(a_4 + b_4)$, $2(a_5 + b_5)$, respectively.

We can find the energies of $d$-orbitals of any symmetry for any symmetrical arrangement of ligands (or charges) in the $XY$ plane from the group of $C.N = 2$ and also for any number of charges lying above and/or below the $XY$ plane from the group of $CN = 4$. Thus, if we have three legands or charges lying in the $XY$ plane, then the energy of the $d$-orbitals is 3/2 of these given in $CN = 2$. Thus, we have relative energies of $d_{z^2}$, $d_{x^2-y^2}$, $d_{xy}$, $d_{xz}$ and $d_{yz}$ as $\frac{3}{2}b_1$, $\frac{3}{2}b_2$, $\frac{3}{2}b_3$, $\frac{3}{2}b_4$, and $\frac{3}{2}b_5$ respectively. Sometime two or three orbitals are degenerate under certain symmetry group, *e.g.*, $D_{3h}$ or $C_{3v}$. In that case, we take the average of two or three orbitals which are degenerate. Thus, if the symmetry of three ligands is $D_{3h}$, then $d_{xy}$ and $d_{x^2-y^2}$ are degenerate and belongs to $E'$ *I.R.* In such cases, we allow for the degeneracy by taking the average energy for $d_{xy}$ and $d_{x^2-y^2}$ $n$. $\frac{3}{2}\left(\frac{b_2 + b_3}{2}\right)$. To this if we add two more ligands along the $z$-axis, we form trigonal bipyramid *i.e.* we add (2 × CN1) to $\frac{3}{2}(C.N = 3)$. In this manner, we get, the relative energies of $d_{z^2}$, $d_{x^2-y^2}$, $d_{xy}$, $d_{xz}$ and $d_{yz}$ as $\left[2(a_1) + \frac{3}{2}b_1\right]$, $\left(2a_2 + \frac{3}{2}b_2\right)$, $\left(2a_3 + \frac{3}{2}b_3\right)$, $\left(2a_4 + \frac{3}{2}b_4\right)$ and $\left(2a_5 + \frac{3}{2}b_5\right)$ respectively. In the event of two orbitals being degenerate, we take the average energy of the two degenerate orbitals.

It must be remembered that the relative energies of $d$-orbitals are in the units of $D_q$. Therefore these values change with the values of $D_q$ obtained for a particular ligand. Here $D_q$ values vary with the type of ligand *e.g.* $MX_n$ or $MY_n$ *etc*. In this manner one can obtain the relative values of $d$-orbitals for any complex with any ligand, $X$ of any symmetry.

One of the very useful applications of this approximation is that one can determine the relative energies of *d*-orbitals for complexes like *cis*- or *trans*-isomers $[MX_4Y_2]$ and their relative energy values with respect to that of $MX_6$ [or $MY_6$]. The relative energies by building up the complete using *X*- or *Y* ligands $(MX_6)$ or $(MY_6)$ and their $D_q$ values, (as it differs from ligand to ligand) [*i.e.* $D_q(X) \neq D_q(Y)$] or by subtracting the appropriate groups and adding the corresponding *Y* group. Thus, for *trans*-$[MX_4 Y_2]$ is

$$[MX_4]_{planar} + (MY_2) \text{ linear or } [MX_6]_{oct} - [MX_2]_{lin} + (MY_2) \text{ linear}$$

The above method is not an exact method because it is not taking into account the inter-electronic repulsion. However, it is a quite useful approximation because one gets qualitatively the relative energies of five *d*-orbitals in terms of $D_q$ value.

## 11.10 LIMITATIONS OF CRYSTAL FIELD AND LIGAND FIELD THEORIES

The afore-said discussion gives an outline of the qualitative description of crystal field theory which provides a useful framework for the interpretation of some of the properties of coordination compounds. But it is also true that it is a poor way to describe the effects of ligand charges on the properties of metal ions. It is entirely an electrostatic method of treating the bonding in the co-ordination compounds, while MO method given in earlier chapters is purely a covalent one. But the degree of covalency with which an electron is associated with metal ion and the ligand atom varies from zero to 1 depending on the atoms forming the *M-L* bond. The two theories are, therefore extreme ways of describing the bonding. A more powerful aspect of the crystal field theory should be that it should cover all intermediate degrees of electron sharing. Such a model is provided by the Ligand field theory. But in both these theories (*CF* and *LF*), the representation of the metal ions and the ligands is artificial, which assumes them as point-charges. There are a number of implications in the nature of interactions between the metal ions and the ligands that are ignored in the theories. Thus, it is assumed that only the immediate ligand neighbors of a metal ion are important. Sometimes to include the charges of the other metal ion and the ligands outside the first co-ordination sphere is also important. This limitation can be taken care of on the same lines as that for ion crystal field theory, but, mathematically it is quite tedious.

Besides the electrostatic nature of the crystal field there are other limitations. It is found that the parameters are changed substantially, when a metal ion is incorporated into the ligand environment. These properties of the metal ions are interelectronic repulsion energies, spin-orbit coupling and even the orbital angular momentum. It can be deduced from all these factors that the *d*-orbital splitting phenomenon is a product of not only the electrostatic interaction, but is a function of other interactions involved in the covalent bonding of the $M-L$ bond. Thus, one can imply that the qualitatives success of crystal-field theory arises because it provides effects arising due to the symmetry of ligand environment.

A theory that admits that the *d*-orbital splitting arises from chemical bonding rather than from pure electrostatic interactions only and at the same time retaining the formation of crystal fields is called "Ligand-Field Theory". Though the latter theory also uses the same potentials as $V_{oct}$, $V_{feb}$, *etc*., as does the crystal field theory; but acknowledges that the parameters and constants involved in the use of these potentials ($D_q$, $D_t$, $D_s$, *etc.*) remain but their relationship to atomic properties is sought in the nature of chemical bond rather than in classical electrostatic potential. It provides a flexibility not available in crystal-field theory.

Although in the ligand field theory, we remove some of the difficulties of the crystal-field theory, but it does not tell us how the parameters taken over from the crystal-field are related to the covalent bonding interactions between atoms. The parameters are simply taken as experimental observations related to chemical bonding in general term. In this direction, a third attempt has been made called "Angular Overlap Model" by which one can go in some way towards a chemical interpretation. It describes the *d*-orbital splittings using parameters which represent σ- and π-interactions between metal ions and each ligand separately in the complex. Even in the AOM, the model is simply a method of parameterization of experimental observations because the treatment of metal complex by proper methods of theoretical chemistry leads to results which cannot be compared with the parameters of ligand-field theory. Further the vigorous treatment of a complex by theoretically is not available or will not be available in the near future. Such parameterisation is of great value which we discuss in the next Section.

# SECTION B

## 11.11 ANGULAR OVERLAP MODEL

In Section *A* of the chapter, we have given the salient features of the crystal-field and the ligand-field theories. Although these theories explain a number of properties of complex ions, but they lack preciseness because of unsolvable difficulties in mathematical computation. This results in making a number of approximations and parameterisation in these theories. We have also described the reasons of doing so towards the end of Section *A*. It has been pointed out there in that the parameters are simply from the experimental observations related to chemical bonding only in very general terms. We have also emphasized there that a new modified ligand-field theory which goes a little deeper towards chemical interpretation is developed in the form of "The Angular Overlap Model" (AOM). It describes *d*-orbital splitting using new parameters which represent σ- and π-interactions per bond separately between metal ion and each ligand in the coordination compounds. Though it gives a separate σ-and π-interactions of each ligands parameter, but the method is again a sort of parameterizing experimental observations. Though it relates directly to the framework generally used to discuss chemical behaviour, but there are no theoretical calculations of these parameters. This section gives a very brief outline of AOM just to familiarise as to how one can parameterise in a different manner to explain the properties of complexes.

The concept of AOM is based on the concepts of MO theory for describing the metal-ligand bond, but it takes the empiricism of the ligand-field theory. It involves severe approximations. It never gives any quantitative method of calculating real physical values of energies of various energy levels of the metal or its compounds which are involved in various transitions. The resulting parameters are always derived from experimental observations.

Since it is related, though, qualitatively, with MO approach, we think it is pertinent to briefly state the concepts of MO theory that lies at the base of AOM. A few of these are given below:

(*a*) Besides $M-L$ bond being ionic, it is assumed that it contains sufficient covalent character.

(*b*) $M-L$ bond formation involves overlap between mostly the metal *s*, *p*-orbitals with the appropriate ligand orbitals. The contribution in the $M-L$ bond from metal *d*-orbitals is relatively much less.

(*c*) Because of the metal σ-and π-orbital's involvement in bonding, the charge of the metal in is reduced which causes expansion of the *d*-orbitals from that present in the naked ion.

(*d*) In the s antibonding $e_g$* and *p*-bonding orbitals, $t_{2g}$ *d*-orbital contribution is very much larger.

(*e*) The metal-ligand interaction depends solely upon the overlap between them which, in turn, depends upon the energy difference between metal ion orbitals and the ligand orbitals.

(*f*) Since *d*-orbitals in the transition metal ions are compact, the overlap between these orbitals and the ligand orbitals increases upon their expansion. Also the overlap increases with increase in principal quantum number ($n$) and decreases with the charge on the metal ions. It decreases with the increase of oxidition state of the metal ion.

(*g*) The energy gap between metal *d*-orbitals and ligand orbitals depends upon the oxidation state of the metal ion and the electronegativity of the ligands. Higher is the electronegativity of the ligand, the bigger is the energy gap. For filled ligand orbitals, a high oxidation state and low electronegativity lead to smaller energy gap while for empty ligands $\pi^*$-orbitals, the lower the oxidation state of the metal ion, the smaller is the energy gap.

## 11.12 WOLFBERG-HELMHOLTZ APPROXIMATION

The Helmholtz and Wolfberg's approximation plays a very important role in the description of AOM. We know that MOs are formed by the overlap of metal ion orbitals with the LCAOs of the ligands. These LCAOs are expressed as:

$$\psi = \sum_{g_i} C_{g_i} \phi_{g_i} \quad \text{... (11.13)}$$

where $f_{gi}$ is the $i^{th}$ wave function of the $g^{th}$ atom of the molecule. The $\psi$ wave function is an eigenfunction of the Hamiltonian operator which means that any MO wavefunction $\psi_i$ and its energy is related as

$$\hat{H}\,\psi_i = E_i\,\psi_i \quad \text{...(11.14)}$$

If the number of AO's, that one takes, is very large in order to improve the accuracy, the solution of the differential equations becomes exceedingly tedious. Therefore, at the expanse of accuracy and preciseness, we take a limited number of $f_i$'s or we use some experimental data to define them.

AOM uses these simplifications and as its basic concept, the bonding between one ligand and a $d$-orbital on the metal ion is considered (only in some special case, more than one orbital is taken into consideration). With this approximation, the MO $\psi$ beomes

$$\psi_i = C_{LM}\,\phi_M + C_{LL}\,\phi_L \quad \text{...(11.15)}$$

Employing this wave function for the formation of MO, we have the secular equation as

$$\left.\begin{aligned} C_{11}\left(H_{MM} - E\,S_{MM}\right) + \left(H_{ML} - S_{ML}E\right)C_{12} = 0 \\ C_{21}\left(H_{ML} - S_{ML}\,L\right) + \left(H_{LL} - S_{LL}E\right)C_{22} = 0 \end{aligned}\right\} \quad \text{...(11.16)}$$

The energies of MOs are obtained by solving the following equation:

$$\begin{vmatrix} H_{MM} - ES_{11} & H_{12} - ES_{12} \\ H_{LL} - ES_{21} & H_{22} - ES_{22} \end{vmatrix} = 0 \quad \text{...(11.17)}$$

For energy $E_1$ as 0, we have

$$H_{MM} = \int \phi_M \hat{H}\,\phi_M\,\partial\tau;\ H_{LL} = \int \phi_L\,\hat{H}\,\phi_L\,\partial\tau$$

$$\Psi_{ML} = \int \phi_M \hat{H} \phi \partial\tau = 0 \qquad \text{... (11.18)}$$

and $$S_{gg} = \int \phi_g \phi_g \partial\tau$$

In σ-type of MO, a single orbital on the M ion interacts with **one orbital** of the ligand giving charge density around the internuclear axis having cylinderical symmetry. In π-type MO orbitals, one orbital of metal ion interacts with one orbital of ligand orbital, giving the overlap region off-the internuclear axis with charge density having two-fold symmetry. In δ-type MO, one orbital of metal interacts with one orbital of ligand giving the overlap region of the internuclear axis with charge density of four fold symmetry. These three types of overlaps are shown in Fig. 11.12.

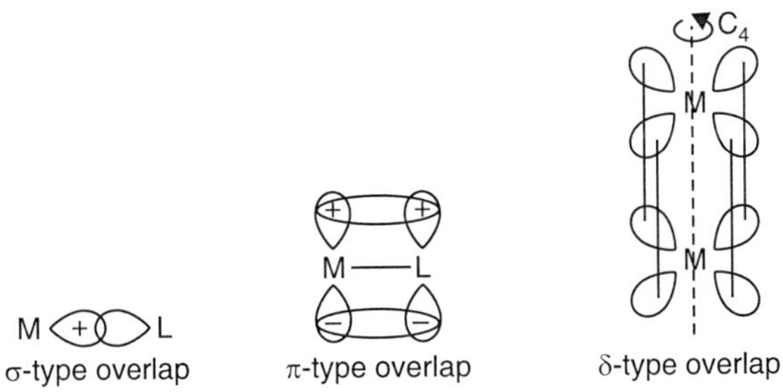


**Fig. 11.12** σ-π- and δ-type of interactions (overlap).

On solving the equation (11.17), we get,

$$(H_{MM} - E)(H_{LL} - E) - (H_{ML} - ES_{ML})(H_{LM} - ES_{ML}) = 0 \qquad \text{...(11.19)}$$

or
$$H_{MM} H_{LL} - E(H_{MM} + H_{LL}) + E^2 = \left[H_{ML}^2 - E(H_{ML}S_{ML} + H_{ML}S_{ML}) + E^2 (S_{ML})^2\right]$$

or
$$E^2(1 - S_{ML}^2) - E[H_{MM} + H_{LL} + 2H_{ML}S_{ML}] + H_{MM} H_{LL} - H_{ML}^2 = 0 \qquad \text{...(11.20)}$$

Assuming $H_{LL} \ll H_{MM}$ and $S_{ML}$, very small we obtain

$$E_a = \frac{H_{MM} + \left(H_{ML} - 2H_{ML}H_{MM}S_{ML} + H_{MM}^2 S_{ML}^2\right)}{(H_{MM} - H_{LL})}$$

$$E_b = \frac{H_{LL} - \left[\left(H_{ML}^2 - 2H_{ML}H_{LL}S_{ML} + H_{LL}^2 S_{ML}^2\right)\right]}{(H_{MM} - H_{LL})}$$

(basic equation) ...(11.21)

Here $E_b$ is bonding and $E_a$ is anti-bonding orbital energy. To obtain the energy of the bonding orbitals from the equation, it is mathematically tedius. One, therefore, has to make some approximations. Here $H_{MM}$ and $H_{LL}$ are the ionisation energies of the metal and the ligand orbital respectively in the complex. $S_{ML}$ is the overlap integral, which has also to be calculated. The problem is to calculate $H_{M-L}$ which is difficult to calculate and is not available even from experimental observations. The values of the off-diagonal element $H_{ML}$ in the above matrix or determinant is difficult to calculate. It is at this point where Wolfberg and Helmholtz made the following approximation. They approximated it as:

$$H_{ML} \cong S_{ML} \frac{(H_{MM} + H_{LL})}{2} \qquad ...(11.22)$$

Employing this approximation, one obtains

$$E_b \cong H_{LL} - \frac{H_{MM}^2 \, S_{ML}^2}{(H_{MM} - H_{LL})} \qquad ...(11.23)$$

and

$$E_a \cong H_{MM} + \frac{H_{LL}^2 \, S_{ML}^2}{(H_{MM} - H_{LL})} \qquad ...(11.24)$$

Or, we can express this energy as, $\left(\text{by substituting } \beta = \left(\frac{-H_{M\bar{M}}^2}{H_{MM} - H_{LL}}\right)\right)$

$$\left.\begin{aligned} E_b &= H_{LL} - \beta S^2 \\ \text{and } E_a &= H_{MM} + \beta S^2 \end{aligned}\right\} \qquad ...(11.25)$$

These energies are obtained as a result of interaction between two orbitals, (*viz* that of *M* and *L*). Here we have not defined the type of interaction. Now we define three types of overlap interactions. (1) σ-type (2) π-type and (3) δ-type. The difference between these lies in the difference of the symmetry of the overlap with respect to internuclear axis. In σ-type, the charge density is concentrated around the internuclear axis and has a cylindrical symmetry. In π-type of overlap, the charge is concentrated off the internuclear axis wih a charge density having two fold symmetry and there is one nodal plane containing the inter-nuclear axis. The δ-type, interaction involve an overlap region concentrated off- the inter-nuclear axis with a charge density of four fold symmetry about the inter-nuclear axis. There are two nodal planes at right angles to each other, containing the internuclear axis (Fig. 11.12).

The extent of overlap of orbital in these three types follows the order $\sigma > \pi > \delta$. Therefore when a metal orbital interacts with a ligand orbital, only these three types of interactions are possible. These interactions ($\sigma$, $\pi$ or $\delta$) are used to describe the energy changes (Fig. 11.13). Thus, in equation (11.25), $\beta S^2$ can be divided into three groups [*viz.*, $\beta S^2$ ($\sigma$-type), $\beta S^2$ ($\pi$-type) and $\beta S^2$ ($\delta$-type)]. The total bonding and antibonding energy of each $M - L$ bond can be obtained as the sum of $\beta(S^2)(\sigma) + \beta S^2(\pi) + \beta(\Sigma^2)(\delta)$, for one $M - L$ bond. That is, energy per bond is given as:

$$E_b = -E_a = \pm [\beta S^2(\sigma) + \beta S^2(\pi) + \beta S^2(\delta)] \qquad ...(11.26)$$

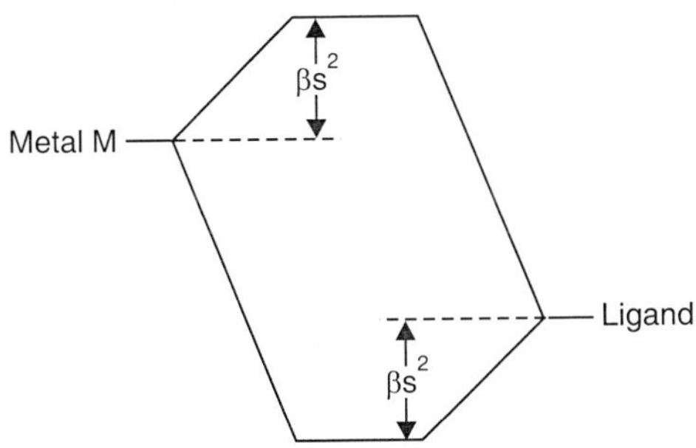


**Fig. 11.13** Total energy change ($\sigma$-, $\pi$-, $\delta$-type) interactions (*viz* b$s^2$) in single $M - L$ bond formation.

Now we have ligands, which have orbitals of only $\sigma$-symmetry which point along $M - L$ bond direction. The effect of interaction of these types of orbitals is measured by the AOM-parameter *viz* [$\beta S^2(\sigma)$]. Since the other two terms in equation 11.26 are zero.

In the other types of ligands, we have orbitals of *s*-symmetry but besides these $\sigma$-type, there are other orbitals present on the ligand atoms which can form $\pi$-bonds with the metal *p*-orbital interactions. These orbitals should be perpendicular to those of $\sigma$-type orbitals because of lateral overlap of $\pi$-symmetry orbitals of ligand and metal *p*-orbitals. Let us assume that these *p* ligand orbitals point along in the *x*- and *y*-axis of the ligand co-ordinate system as shown in Fig. (11.14). In metal ion complexes formed by the interaction with these types of ligands ($\sigma + \pi$), the effect per $M - L$ bond arising out of these *s*-and $\pi$-interactions, will be represented by two separate parameters *viz* $e_\sigma$ and $(e_{\pi x} + e_{\pi y})$, [*i.e.*, $\beta S^2(\sigma) + \beta S^2(\pi)$]. Here the third term *viz* $\beta S^2(\delta)$ is zero in equation (11.26).

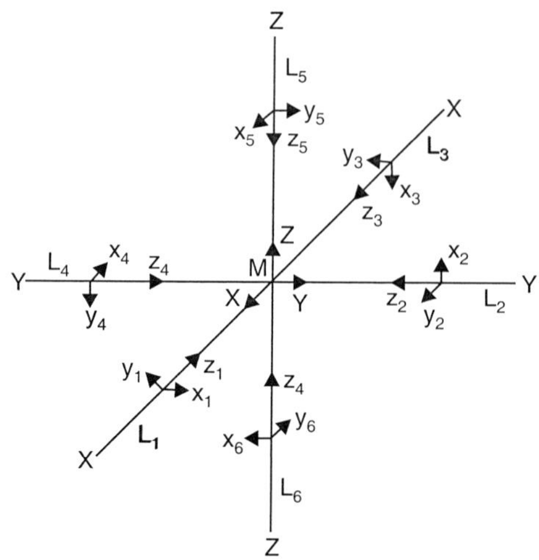


**Fig. 11.14** Global and Local co-ordinate systems of *M* and Ligands respectively.

Here $x$ and $y$ represent the orthogonal axes on the ligands which are normal to $z$-axis (*i.e.*, $M - L$ bond) and $y$-axis be perpendicular to that plane. If the molecule is planar molecule, the convention is that $x$-axis and $z$-axis are taken in the plane of molecule ($xy$-plane) and $y$-axis is taken perpendicular to ($xz$-plane). The same convention will follow here and $x$- and $z$-axes will be in the plane of ligands. As we have seen in MO treatment of metal complexes, metal $d$-orbitals ($e_g$*) are almost always expected to be antibonding with respect to $t_{2g}$-orbitals and $e_s$ is always positive, that is $\dfrac{H_{LL}^2\, S_{ML}^2}{(H_{MM} - H_{LL})}$ term will have positive value ($\beta^s$ = +ve). On the other hand, $\pi$-interaction may raise the metal $p$-orbitals ($t_{2g}$-type (*i.e.*, $d_{xy}$, $d_{yz}$, $d_{xz}$-orbitals) or it may lower these orbitals. It depends on whether the ligand $p$-orbitals are lower or higher in energy relative to metal $\delta$-orbitals (*See* Fig. 11.7). Thus, $e_\pi$ value (or $\beta S^2(\pi)$) is positive, if the ligand $\pi$-orbitals are lower and the $e_\pi$ value is negative if these $p$-ligand orbitals are higher than those of metal $\pi$-orbitals (*i.e.*, $d_{xy}$, $d_{yz}$, $d_{xz}$).

## 11.13 DERIVATION OF *d*-ORBITAL ENERGIES IN A COMPLEX USING AOM METHOD

In the above discussion, we have given the effects of one individual ligand [($M - L$) single bond] in terms of $e_\sigma$, $e_\pi$ and $e_\delta$ parameters. The total effect in a complex is thus obtained by the summation of the effects of all the $M - L$ bonds. Thus, we describe the effects of one single $M - L$ bond in terms of $e_\sigma$, $e_{\pi x}$ and $e_{\pi y}$ and then we sum them, up for all the $M - L$ bonds taking into account the geometry of the complex by means of angular overlap model.

Neglecting for the time being any complication, the δ-orbital energies in any complex are obtained by considering the geometric relationships between $x$, $y$, $z$ co-ordinate systems (centred on the ligands) used to define the $e_\sigma$, $e_{\pi x}$, $e_\pi y$ parameters of each legand and a global coordinate system $X$, $Y$, $Z$ (Fig. 11.14) centered on the metal. So the bond-direction ($z$-axis of $x$, $y$, $z$ coordinate system) is defined by two angular coordinate (1) angle $q$ between $z$-axis and the $M - L$ bond direction and (2) by the angle $f$ which is the angle between $x$-axis and the line $MN$ that is obtained by the intersection of $XY$ plane ($I$) and the plane ($II$) containing the $M - L$ bond direction and the ζ-axis Fig. 11.15. These angles will define the values of $e_\sigma$ [$\beta S^2(\sigma)$] of $M$. The values of $e_{\pi x}$ and $e_{\pi y}$ will be given by the orientation of ξ- and $y$-axes which define, the ligand $p$ coordinate systems. The orientation is defined by a final rotation $x$ about the $M - L$ bond axis which along with θ and ϕ will give the values of $e_{\pi x}$ ($\beta S^2/\pi_x$) and $e_{\pi y}$ ($\beta S^2/\pi_y$). Therefore the angular overlap factors indicate the effects of rotation through θ, ϕ and ξ angles on the metal-legand overlap integrals (*viz.* $e_\sigma$, $e_{\pi x}$ and $e_\pi y$).

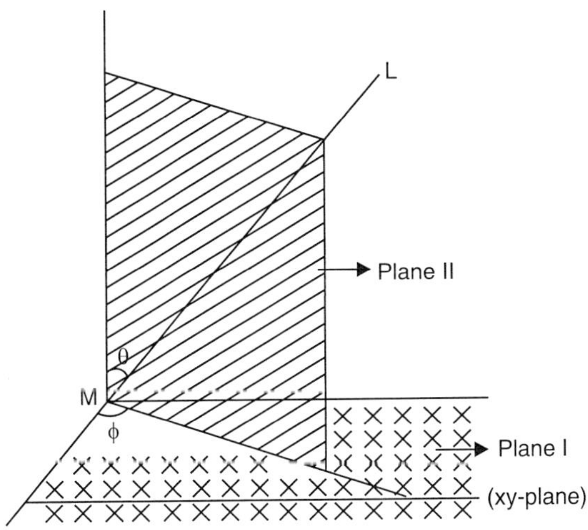


**Fig. 11.15** θ, ϕ angles formed by plane I and plan II.

These angular overlap factors which indicate the effects on the metal ligand overlap integrals $\left(\int \psi_L \psi_M \partial_N\right)$ by a rotation through the angle θ, ϕ and ξ, are given in Table (11.8). One thus calculates by using the factors given in Table 11.7 the vales of $e_\sigma$, $e_{\pi x}$ and $e_\pi y$ for each ligand in a complex and then sum them up to get the energy of δ-orbitals. This will give the energy of $d$-orbitals in a complex.

AOM is very useful for determining the energy levels of $d$-orbitals in the field of ligands of low symmetry. All we have to find is to determine the ligand θ, ϕ, and ξ coordinates and then calculate the factors from Table (11.7). Once the numerical values of these factors are known, then $H_{MM}$, $H_{ML}$, $S_{ML}$ and $H_{LL}$ can be determined in terms of $e_\sigma$, $e_{\pi x}$ and $e_{\pi y}$. These give the energy of *d*-orbitals in terms of $e_\sigma$, $e_{\pi x}$ and $e_{\pi y}$.

**Table 11.7:** A.O. factors indicating the effects on the *d*-orbitals by Angular Rotations θ, ϕ and ξ

| | $f_\sigma\,[d,(\theta, \phi, \xi)]$ | $f_{\pi x}\,[d,(\theta, \phi, \xi)]$ | $f_{\pi y}\,[d,(\theta, \phi, \xi)]$ |
|---|---|---|---|
| $d_{x^2-y^2}$ | $\frac{\sqrt{3}}{4}\cos^2\phi\,(1-2\cos 2\theta)$ | $-\sin 2\phi \sin\theta \sin\xi +$ $\frac{1}{2}\cos 2\phi \sin 2\theta \cos\xi$ | $-\sin 2\phi \sin\theta \cos\xi$ $-\frac{1}{2}\cos 2\phi \sin 2\theta \sin\xi$ |
| $d_{z^2}$ | $\frac{(1+3\cos 2\theta)}{4}$ | $\left(-\frac{\sqrt{3}}{2}\right)\sin 2\theta \cos\xi$ | $\left(-\frac{\sqrt{3}}{2}\right)\sin 2\theta \cos\xi$ |
| $d_{zy}$ | $\frac{\sqrt{3}}{4}\sin 2\phi\,(1-\cos 2\theta)$ | $\cos 2\phi \sin\theta \sin\xi +$ $\frac{1}{2}\sin 2\phi \sin 2\theta \cos\xi$ | $\cos 2\phi \sin\theta \cos\xi$ $-\frac{1}{2}\sin 2\phi \sin 2\theta \sin\xi$ |
| $d_{xz}$ | $\frac{\sqrt{3}}{2}(\cos\phi \sin 2\theta)$ | $-\sin\phi \cos\theta \sin\xi$ $+\cos\phi \cos 2\theta \cos\xi$ | $-\sin\phi \cos\theta \cos\xi$ $-\cos\phi \cos 2\theta \sin\xi$ |
| $d_{yz}$ | $\frac{\sqrt{3}}{2}\sin\phi \sin 2\theta$ | $\cos\phi \cos\theta \sin\xi +$ $\sin\phi \cos 2\theta \cos\xi$ | $\cos\phi \cos\theta \cos\xi$ $-\sin\phi \cos 2\theta \sin\xi$ |

However, in complexes of low symmetry, the *M* and *L* interaction ($H_{ML}$) involves nonzero off-diagonal matrix elements also so that the energy levels of the δ-orbitals are not given by simple equations involving $e_\sigma$, $e_{\pi x}$ and $e_{\pi y}$. Therefore, the wave function of the MO in such cases are complicated functions of various *d*-orbitals. But the favourable point is that most of the transition metal complexes belong to high symmetry and this results in the differentiation of the molecular wave functions of complexes into different sets of different symmetries *e.g.* in $O_h$ symmetry we have two sets: (1) Containing $E_g$ and the other $T_{2g}$. Therefore, their eigenvalues are simple arithmatic equations involving AOM bonding parameters. We illustrate it by taking the complexes having $O_h$ symmetry as an example.

## Octahedral Complexes

We illustrate a typical octahedral complex along with its co-ordinate system by the Fig. 11.14. The $q, f$ and $x$ angles of bond directions are in this complex.

| *Bond* | θ | φ | ξ |
|---|---|---|---|
| 1 | 0 | 0 | 0 |
| 2 | 90° | 0 | 90 |
| 3 | 90 | 90 | 90° |
| 4 | 90 | 180 | 90 |
| 5 | 90 | 270 | 90 |
| 6 | 180 | 0 | 180 |

Employing these angles, the *d*-orbital energy matrices are calculated for each $M - L$ bond in the forms of $e_\sigma$ and $e_{\pi x}$, $e_{\pi y}$. These values are then added to get the effects of the six ligands. Which are given in Table 11.7

In Table 11.7, it can easily be observed that the $d_{xy}$, $d_{xz}$ and $d_{yz}$-orbital energies are given by simple expressions in terms of $e_{\pi x}$ and $e_{\pi y}$ while those are of $d_{x^2-y^2}$ and $d_{z^2}$ are mixed through the off-diagonal element. The energy of the latter depends upon the difference in *s* interactions along the *x*- and *y*-axes (*i.e.* $e_s^1 - e_s^2$ and $e_s^3 - e_s^4$). These orbitals are therefore linear combination of these *d*-wave functions (*i.e.* $d_{z^2}$ and $d_{x^2-y^2}$) and energy is not given as simple arithmetic function. If we assume all $M - L$ bond lengths equal, the interaction can be described by just one set of bonding parameters (*viz* $e_\sigma$ and $e_\pi$) which can be given as

$$E_{(x^2-y^2)} = E_{z^2} = 3e_\sigma \text{ and } E_{xy} = E_{xz} = 2e_{\pi_x} + 2e_{\pi_y}$$

**Table 11.8**

| | $d_{x^2-y^2}$ | $d_{z^2}$ | $d_{xy}$ | $d_{xz}$ | $d_{yz}$ |
|---|---|---|---|---|---|
| $d_{x^2-y^2}$ | $0.75(e_\sigma^1 + e_\sigma^2 + e_\sigma^3 + e_\sigma^4)$ | $0.25[3^{1/2} (e_\sigma^1 + e_\sigma^3 + e_\sigma^{(2)} - e_\sigma^3)]$ | 0 | 0 | 0 |
| $d_{z^2}$ | $0.25[3/2 (+ e_\sigma^1 + e_\sigma^3 + e_\sigma^4)]$ | $e_\sigma^5 + e_\sigma^6 + 0.25 \; e_\sigma^1 + e_\sigma^2$ | 0 | 0 | 0 |
| $d_{xy}$ | 0 | 0 | $e_{\pi_x}^1 + e_{\pi_x}^4 + e_{\pi_y}^2 + e_{\pi_y}^3$ | 0<br>0 | 0<br>0 |

*(Contd.)*

| | | | | | |
|---|---|---|---|---|---|
| $d_{xz}$ | 0 | 0 | 0 | $e_{\pi_y}^1 + e_{\pi_y}^6 +$ | 0 |
| | | | | $e_{\pi_x}^5 + e_{\pi_x}^3$ | |
| $d_{yz}$ | 0 | 0 | 0 | $e_{\pi_x}^2 + e_{\pi_x}^6 +$ $e_{\pi_y}^5 + e_{\pi_y}^4$ | |

Therefore, the difference in energy of ($d_{x^2-y^2}$, $d_{z^2}$) and ($d_{xy}$, $d_{yz}$, $d_{xz}$) is equal to

$$p$$

$$E_{(d_{z^2}, d_{x^2-y^2})} - E(d_{xy}, d_{yz}, d_{xz}) = 3e_\sigma - 4e_\pi$$

$$= 10D_q$$

Thus, σ- and π-bonding parameters are independent and no δ-orbital is influenced by both types of interaction. This is in general true when all the ligands lie on the metal-centered Cartesian co-ordinate axes which are known as Ortho-Axial complexes.

If the complexes are mixed and the ligands are π-bonding which are symmetrical about $M - L$ bond axis, one obtains a simple expression for the metal complex. Thus, for the complex [$ML_5 - X$] ($C_{4v}$ symmetry), one obtains from Table 11.7.

and
$$E_{xy} = 4e_\pi(y)$$

$$E_{(x^2-y^2)} = 3e_\sigma\ (L)$$

$$E_{(z^2)} = 3e_\sigma\ (L) + e_\sigma\ (X)$$

$$E_{xz} = E_{yz} = 3e_\pi\ (L) + e_\pi\ (X)$$

Similarly for $ML_4\ X_2$ (*trans-*) ($D_{4h}$ symmetry), we get from Table 11.7

$$E_{(x^2-y^2)} = 3e_\sigma\ (L)$$

$$E_{(d_{z^2})} = e_\sigma\ (L) + 2e_\sigma\ (X)$$

$$E_{(xz)} = E_{yz} = 2e_\pi\ (L) + 2e_\pi\ (X)$$

and
$$E_{(xy)} = 4e_\pi\ (L)$$

In this manner it is possible to calculate the energies of all the metal orbitals in a complex of any symmetry. Since in the calculation and discussion of the energy values of *d*-orbitals, there is not much involvement of symmetry properties of the molecules, we will not discuss it any further. For a more detailed discussion one should refer books on chemical bonding.

It can therefore be concluded that the AOM is based on rather severe approximations similar to those in the ligand-field theory and is not truely quantitative like MO approach. No *s-p* mixing or nuclear repulsion are included in the AOM. Therefore it is by no means infallible in its predictions related to the molecular shapes. Its predictions are more reliable when the electronegativities of the ligands and the metal ions are not too different and when covalent discription is an approximate one. Finally one can summarize in totally that the AOM serves as a straight forward introduction to the MO theory.

# 12

# Application of Symmetry Properties (Molecular Vibration Analysis)

In preceding chapters, we have applied symmetry to understand the electron dynamics in atoms which aided us to reveal the basic nature of interactions of electrons in the field of nuclear charges within the atoms and molecules. Despite electronic interaction, the atoms within molecules are continuously executing vibrational motion at all temperatures, (even at absolute zero) besides molecular rotations and translations as a whole. During vibrational motion, the distances and the internal angles between bonds change periodically keeping the position of centre of mass of the vibrating molecule invariant. It does not mean that the molecules have stopped rotating or translating in space during vibrational motion, but while studying vibrational motion, we divert our attention from these motions and assume that we are sitting at the centre of mass of the molecule and at the same time we are also translating and rotating along with the molecule .We further assume that we are looking only at the vibrational motions of the atoms in the molecule (*i.e.,* internal vibratory motion of the atoms).

The vibratory motion of atoms, although rather random has a certain degree of regularity and simplicity. One can apply molecular symmetry arguments fruitfully to these motions. In this chapter, we will discuss the applications of these symmetry arguments to understand the vibrational states of the molecules, although the subject matter is quite elaborate, but the size of the book limits us to be brief in our discussion.

## 12.1 DIATOMIC MOLECULES

When a single atom (*A*) is moving, one can resolve its velocity and acceleration into three components with respect to a chosen, coordinate system. Since the atom is free to move equally in three directions, it is said to have three degrees of freedom. Its freedom of movement is however curtailed, if it is bonded to another atom *B*. (*i.e.,* in *A–B* bond because of the force acting between them). Therefore, the motion of *A–B* molecule is classified in a way so as to take this factor into account dividing the motion into three separate categories (*viz.* translation, rotation and vibration). In the vibrational motion, the average position and the orientation of the molecule in space remains fixed unlike in translational and rotational motions. It is the distance between the atoms that changes.

When we consider the translational motion of the whole molecule, we take it as if it is equivalent to a single body with its mass located at the centre of gravity and that the body is moving as a whole having three degrees of freedom. The rotation of the molecule about any axis can also be resolved into three components representing rotations about three mutually perpendicular axes passing through the centre of mass of the molecule (*e.g., x*-, *y*-, *z*-axes of the coordinate system with its origin located at the centre of mass) (this condition is however, not obligatory). In linear molecules, there is a little difference. The rotational motion of the molecule is resolved into its two components because the rotation about the molecular axis (*e.g., z*-axis) can not be perceived by any known physical process and also there is no change in the positions of atoms during rotation of molecule unlike in the usual rotation. It, therefore, is free to rotate about two mutually perpendicular axes and which are also perpendicular to the molecular axis. This limits the number of degrees of freedom of linear molecules to two instead of three for non-linear molecules.

Since every atom in an $n$-atom molecule has three degrees of freedom, we have a total of $3n$ number of degrees of freedom in any molecule. Out of these, three degrees are for translatory and three degrees for rotatory motions (two degrees for linear molecule). We are then left with only ($3n$-6) (or $3n$-5 for linear case) degrees of freedom for vibrational motions of the molecule. In the latter motion, the molecule is not moving as a whole relative to a fixed framework of the axes in space with origin of the coordinate system at the centre of mass.

According to Born-Oppenheimer approximation which allows us to consider energy changes in any system separately as vibrational, translational, rotational. These differ from each other in an order of magnitude. We will analyse the dynamics of these three motions (vibrational, translational and rotational) separately and treat them independently. Implicitly, we also ignore the changes in electronic and nuclear energies in the molecules while dealing with the vibrational motion in the molecules. (This assumption cannot, however be applied to crystalline substances).The non-linear and linear molecules thus have respectively (3*n*-6) and (3*n*-5) degrees of vibrational freedom.

Let us now analyse the vibrational motion of a diatomic molecule (one degree of vibrational freedom). We assume that one of the three coordinate axes is the molecular axis with origin at the centre of mass of the molecule (Fig. 12.1). It should be realised that our coordinate system is translating and rotating in space as a whole without any effect on the vibratory motion of the atoms in the molecules. It means that implicitly we are pretending fixed axes and the atoms in the molecule are vibrating relative to these fixed axes. It is further obvious that the movements of atoms in the *x*-and *y*-directions can be ignored because these will signify the translatory or rotatory motions. Let the masses of the atoms *A* and *B* be $m_A$ and $m_B$ respectively. Thus, the molecules *A* – *B* can be represented at any moment as shown in Fig. 12.1.

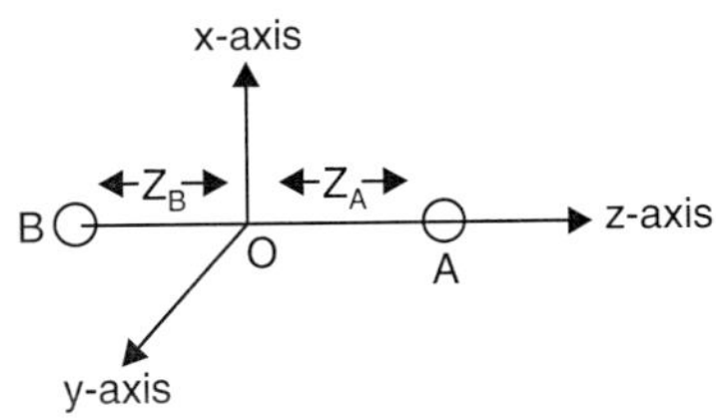


**Fig. 12.1** Cartesian Coordinate system of the diatomic molecules *A* – *B*. By convention, *z* is molecular axis $Z_B$ is negative so that the bond length is $(Z_A - Z_B)$.

Let '*O*' be the origin of coordinates system and also the centre of mass; $Z_A$ and $Z_B$ are the coordinates of the atoms *A* and *B* respectively. We represent the bond *A*–*B* by a spring and change of the distance *A*–*B* by *q*, keeping the position of '*O*' fixed. That is $q = (Z'_A - Z'_B) - (Z_A - Z_B)$. The force exerted by the spring to bring the atoms back to their equilibrium positions (*i.e.*, $q = 0$) is given by, (assuming spring to obey Hook's law)

$$\text{Force} = -fq \qquad \text{...(12.1)}$$

Let us represent the velocity and acceleration of the atoms as $\dot{q}(dq/dt)$ and $\ddot{q}(d^2q/dt^2)$ respectively and the reduced mass by (*M*). Assuming Newton's laws of classical motion to be applicable the force (*fq*) of Eq. (12.1), can be represented as,

$$[-fq] = (\text{force}) = M\ddot{q} \qquad \text{...(12.2)}$$

Here $M$ = reduced mass which equals $\left(\dfrac{m_1 m_2}{m_1 + m_2}\right)$ (It will be derived later).

Eq. (12.2) can also be written as,

$$\text{M}\ddot{q} + fq = 0 \qquad \text{...(12.3)}$$

This is second order differential equation. One of the solutions of this equation is,

$$q = A\cos(2\pi\nu t + \rho) \quad \text{...(12.4)}$$

After differentiating $q$ two times, we obtain,

$$\ddot{q} = -A4\pi^2\nu^2\cos(2\pi\nu t + \rho)$$
$$= -4\pi^2\nu^2 q \quad \text{...(12.5)}$$

Substituting Eq. (12.5) into Eq. (12.3), we get,

$$(-4\pi^2\nu^2 M + f)q = 0 \quad \text{...(12.6)}$$

Assuming $q$ to be non-zero for non-trivial solution, we get

$$(f - 4\pi^2\nu^2 M) = 0$$

or
$$\nu = \frac{1}{2\pi}\sqrt{\frac{f}{M}} \quad \text{...(12.7)}$$

This is the equation of Simple Harmonic Motion (SHM).

From Eq. (12.5), we obtain, at time $t = 0$ (the initial condition),

$$\ddot{q} = A\cos\rho \quad \text{...(12.8)}$$

Since at the initial point, we have $\ddot{q} = 0$, then

$$0 = A\cos\rho$$

or
$$\rho = \frac{\pi}{2} \quad \text{...(12.9)}$$

That is, the phase difference between the motions of two atoms is equal to $\pi/2$.

Since the two atoms $A$ and $B$ are executing SHM (Eq. 12.7), it implies that $q$ is changing regularly, periodically with a frequency $\nu$, a phase factor $\pi/2$ and with maximum amplitude $A$. Thus, we may calculate the values of $q$ at different intervals of time (*viz.* $t = \frac{1}{4\nu}, \frac{1}{2\nu}, \frac{3}{4\nu}, \frac{1}{\nu}$, etc.) and we can represent graphically the relationship between $q$ and the time $t$, (Fig. 12.2). It is obvious from Fig. 12.2 that the value $q\,[= (Z'_{A_1} - Z'_{A_2}) - (Z_{A_1} - Z_{A_2})]$ is varying periodically, with the maximum value of $A$ at $t = \left(\frac{3}{4\nu}\right), \left(\frac{6}{4\nu}\right)$ *etc.*, and minimum value at $t = \left(\frac{1}{4\nu}\right), \left(\frac{2}{4\nu}\right)$ *etc.* The motions of two atoms $A_1$ and $A_2$ in ($A_2$) diatomic molecule, are given by two cosine curves $P$ and $Q$, and at $q = 0$, by two horizontal lines $S$ and $T$.

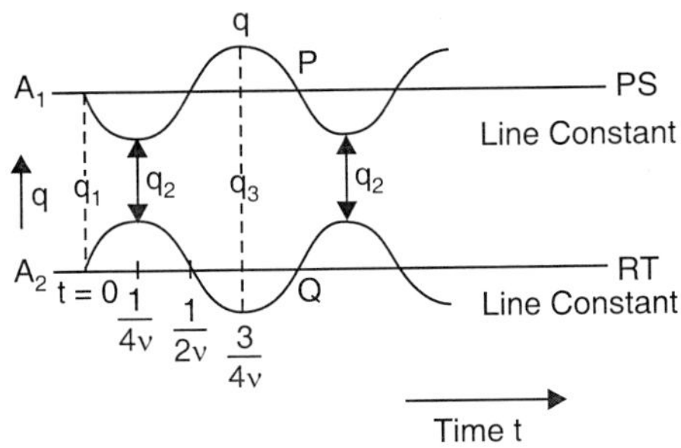


**Fig. 12.2** Periodic motions of A–A molecule.

From these curves, it is also obvious,

1. That the extensions and the compressions of both atoms ($A_1$, $A_2$) occur at the same time.
2. That we cannot determine the absolute values of amplitudes of the two atoms, at any time; only relative values can be determined ($A_1 + A_2$) and ($A_1 - A_2$) *i.e.*, sum and the difference of amplitudes of two atoms.
3. That the phase factor ρ only determines the point on the two curves at a time when we begin, and has no physical significance.
4. That ν tells the time required to execute one complete cycle of vibration (ν = cycles per second) and is proportional to '$f$' according to Eq. (12.7), where '$f$' is the force constant.

The force constant '$f$' is defined by Hook's law Eq. (12.1). It is not convenient to express it in accordance with Hook's law. We can define it alternately in terms of the vibrational potential energy as follows.

$$V = \int_{q=0}^{q} (\text{–exerted force}) \,.\, dq \qquad ...(12.10)$$

$$= -fq^2 \qquad ...(12.11)$$

Thus, by differentiating $V$ two times, the force constant $f$ becomes equal to

$$f = \frac{d^2V}{dq^2} \qquad ...(12.12)$$

The kinetic energy of the particle can be written as

$$2T = m_A \dot{z}_A^2 + m_B \dot{z}_B^2 \qquad ...(12.13)$$

We can also write

$$\dot{q} = \dot{z}_A - \dot{z}_B \qquad ...(12.14)$$

Since we have assumed that the molecule has no translational energy, we have total linear momentum of the molecule to be zero. That is,

$$m_A \dot{z}_A + m_B \dot{z}_B = 0 \qquad ...(12.15)$$

or
$$\dot{z}_B = -\frac{m_A \dot{z}_A}{m_B} \quad ...(12.15(a))$$

Substituting $\dot{z}_A$ from Eq. (12.14) in Eq. (12.15), we get,

$$\dot{z}_B = -\frac{m_A \dot{q}}{m_A + m_B} \quad ...(12.16)$$

and
$$\dot{z}_A = -\frac{m_B \dot{q}}{m_A + m_B} \quad ...(12.17)$$

Eqs. (12.16) and (12.17) are substituted in Eq. (12.13), we finally get,

$$2T = \left(\frac{m_A m_B}{m_A + m_B}\right)\dot{q}^2 \quad ...(12.18)$$

and
$$\frac{d}{dt}\left(\frac{dT}{d\dot{q}}\right) = \left(\frac{m_A m_B}{m_A + m_B}\right)\ddot{q} \quad ...(12.19)$$

Since Lagrange's equation is given as

$$\frac{d}{dt}\left(\frac{dT}{d\dot{q}}\right) + \frac{dv}{dq} = 0 \quad ...(12.20)$$

By substituting Eqs. (12.19) and (12.11) in Eq. (12.20) we finally obtain the same expression as Eq. (12.3).

or
$$\left(\frac{m_A m_B}{m_A + m_B}\right)\ddot{q} + fq = 0 \quad ...(12.21)$$

or
$$M\ddot{q} + fq = 0 \quad ...(12.22)$$

Eq. (12.7) takes the following form

$$\nu = \frac{1}{2\pi}\sqrt{\left(\frac{1}{m_A} + \frac{1}{m_B}\right)f} \quad ...(12.23)$$

We have derived an equation for ν if force constant '*f*' is given (or *f* if ν is given) for the diatomic molecules *A–B* from the basic principles. The same basic principles are applicable in the treatment of polyatomic molecules. We now state them as follows:

1. The linear and the angular momentums of the whole molecule assumed to be zero with respect to Cartesian axes respectively and with the assumption about origin of the coordinate system at the centre of mass. Although the origin is translating and rotating along with the molecule, but these restrictions allow us to study independently the energy changes with respect to changes in inter-atomic distance in the molecule.

2. The molecule is considered to be made up of particles connected among each other through spring like bonds, obeying Hook's law.
3. The vibrational dynamics of atoms in the molecule follows classical mechanics and therefore Newton's laws of motion are applicable.
4. With the above assumption, we have found that the atoms of the molecule move apart and back periodically with the same frequency (SHM). Their frequencies of moving back and forth are given by the equation (12.23).
5. Eq. (12.23) shows that the frequency of oscillation depends upon two factors, *viz.* force constant of the spring (*f*) and the reduced mass (*M*).

Although, the same principle can be extrapolated for polyatomic molecules, but the following difficulties however, arise in case of polyatomic molecules,

1. Mathematical manipulations become complicated.
2. We obtain (3*n*–6) or (3*n*–5 for linear molecules) frequencies of atoms in the molecule which will depend on the reduced mass and the force constant factors. These are hard to get for polyatomic molecules.
3. The reduced mass factor is needed because the force constant will be defined in terms of a "*MOLECULAR*" coordinate.
4. The kinetic energy (K.E.) cannot be defined in terms of molecular coordinates.
5. We have to put constraints of linear and angular momentum to be equal to zero. We can however, calculate the reduced mass factor differently by relatively an easy method.
6. The validity of the classical mechanics is assumed and thereby Hook's law in the vibrational problem of polyatomic molecules is also followed.
7. The term "*Molecular coordinate*" is defined in various forms. *Unlike* the Cartesian coordinates, the molecular coordinates do not define the positions of atoms in absolute terms. *It defines the changes in the coordinates from their respective equilibrium defined positions.* Thus, in the above description θ defines the change in the bond length (*A–B*). $[(z'_A - z'_B)-(z_A - z_B)]$ and not the actual bond length (*A–B*). Thus, in the analysis of vibrational dynamics of polyatomic molecules, the term coordinates will be defined as the changes in the bond distances from the equilibrium configuration of the molecule in the changed configuration as a result of atomic vibrations.
8. In the following analysis, we take '*n*' different "Origins" of *n* coordinate systems located at average or equilibrium positions of the each atom in the molecule. There will thus be '*n*' number of origins. There will not be any one particular coordinate system but will be *n*-different coordinate systems, with their origins at the equilibrium positions of each atom.

Directions of all the *X*-, *Y*- and *Z*- coordinate axes will respectively be parallel to each other. The changed positions of each atom is measured with respect to its own coordinate system. Thus the positions of each atom ($n$-number of them) will be given by $3n$-number of coordinates. We write them as $\{ x_1{}^1 y_1{}^1 z_1{}^1, x_1{}^2 y_1{}^2 z_1{}^2, \ldots x_1{}^b, y_1{}^b, z_1{}^b, \ldots . x_1{}^n, y_1{}^n z_1{}^n \}$ or $\{ x_1{}^1, x_2{}^1, x_3{}^1, \ldots x_2{}^1 x_2{}^2 x_2{}^3, \ldots .. x_1{}^b x_2{}^b x_3{}^b, \ldots . \text{etc.} \}$ where superscript indicates the number of atoms giving $3n$ number of coordinates for the positions of $n$-atoms. These are cartesian coordinates, which we will further explain in the following section 12.2(A). Besides these cartesian coordinates, there are other kinds of coordinates which are used in the vibrational analysis of molecules. Prior to proceeding further to analyse vibrational motions, it will be better to know about other important kinds of coordinates. We will be using them in our analysis. However, the coordinates by which one can express K.E. and potential energy by quadratic terms, without cross product terms $q_i q_j$ are most suited. So we transform them to those in which only square terms appear without cross terms. Such coordinates are Normal Coordinates (symmetry coordinates). In terms of these $3n$ coordinates, we define ($3n$-6) [or ($3n$-5) for linear molecules] coordinates. Other coordinate are internal coordinates, symmetry coordinates, and normal coordinates, besides cartesian coordinates. These are the examples of "Molecular Coordinates". A brief description of the various coordinates used in vibrational analysis of the molecules is given below.

## 12.2 COORDINATES SYSTEMS USED IN VIBRATIONAL SPECTROSCOPY

In vibrational analysis of molecules, we express K.E. and potential energies of system in various types of coordinates. Some of these are described below.

1. **Cartesian Coordinates:** These are displacement coordinates of an atom $i(\Delta x_i, \Delta y_i$ and $\Delta z_i)$ along the chosen cartesian axes $(x, y, z)$ of the molecule. There are thus $3n$ cartesian coordinates in a molecule of $n$ atoms.
2. **Internal Coordinates:** These coordinates of atomic motion in a molecule are related to the changes in (1) the relative positions of the atoms *e.g.*, changes in bond lengths (2) changes in the inter-bond angles. Usually these are broadly classified into the following different categories:

(*a*) **Bond Stretching ($\Delta r_i$):** It is described as shown in Fig. (12.3). Some times an arrow is drawn parallel to the bond indicating stretching of *A–B* bond. A change in the bond length is indicated by $r_i$(or sometimes $\Delta r_i$). In a number of books, Δ sign in not indicated and confusion is aroused in the mind of readers whether it indicates the bond length or the change in bond length. This confusion is resolved by the context of the subject matter.

(*b*) **Bending of Angles between Two Bonds (δ*BAC*):** Refer Fig. (12.3*b*). It is defined as a change in bond angle denoted by δ(angle). The displacement vector can either be perpendicular to the bonds forming the angle and in the planes in which the bonds forming the angle lie. For example, *AB* and *BC* bonds in Fig. 12.3(*b*).

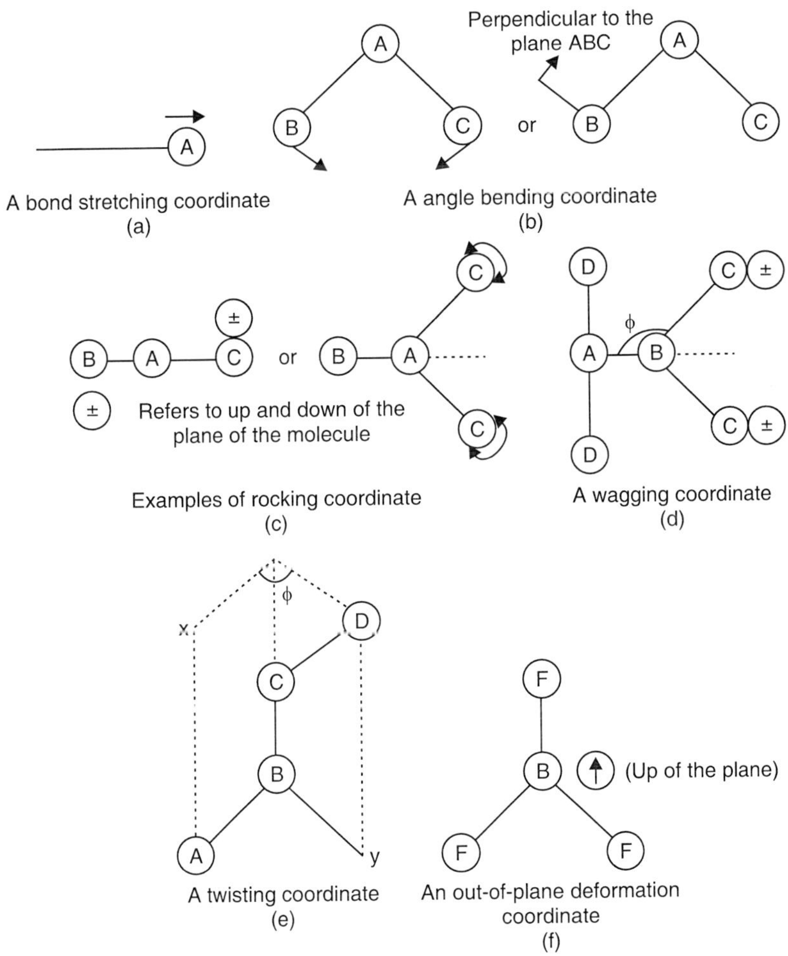


**Fig.12.3** Internal Coordinates.

(*c*) **Rocking Mode:** Refer Fig. (12.3*c*). It is a special case of bending mode. When the angle ABC is 180° in the molecule in equilibrium position and bending mode is in between two bond (*A–B*) and (*A–C*) or between ($A$–$B$) bond and a group of atoms attached to $A$–$B$ bond, then this bending mode is termed as rocking mode of vibration (Fig. 12.3*c*). In the second case, the group of atoms (*viz.* *CBC*) moves as a whole with *BC* bond and the angle *CBC* remain invariant and the angle of group to the $A$–$B$ changes. The displacements coordinate (or vector) are drawn as double headed arrows. It is important to denote it by double-headed arrow because that certain symmetry operation of the molecular group (*e.g.* $C_2$) converts this coordinate into its negative value. One should, therefore, be careful when one applies symmetry operation on this coordinate (especially at the time of forming symmetry coordinates.)The rocking mode of coordinates is represented by $\rho(BC)$.

(*d*) **Wagging Mode of Coordinate ($\rho_\omega$):** Refer Fig. (12.3*d*). A wagging coordinate is defined as a change of angle between a bond and a plane or between a bond and the folding of a plane about line through it (Fig. 12.3*d*). Here the atoms *AB*(*CC*) (*DD*) are coplanar in the equilibrium configuration of the molecule. Both the *C*-atoms move in the plane perpendicular to the molecular plane (±). Displacement above the plane is indicated by ⊕ sign and down the plane, by (–). It is represented by $\rho_\omega$, *i.e.*, $\rho_\omega(BC_2)$. Similar to rocking coordinate, the wagging coordinate changes sign by some symmetry operations of the molecular point group (*e.g.*, $\sigma_h$ operation).

(*e*) **Twisting Coordinates ($\tau$):** Refer Fig. (12.3*e*). This coordinate defines the change of angle between two planes as shown in (Fig. 12.3*e*) by change in an angle $\Delta\phi$. One should distinguish between the cases.

(1) When the two planes (*viz.* *ABCX* and *BCDY*) are coplanar (*i.e.*, $\phi$ = 180°) and,

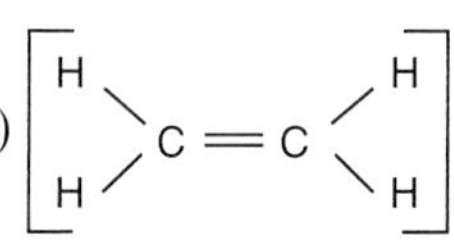


(2) When these two planes are not coplanar.

In case (1), some symmetry operations may change the sign of the coordinates (*e.g.*, in ethylene molecule, the two $CH_2$ groups relative to one another which are in the same plane).

(*f*) **Out-of Plane Deformation Coordinate ($\pi$):** Refer Fig. (12.3*f*).This coordinate is defined by the change in the angle ($\pi$), in such a fashion so that a plane, usually a symmetry plane is destroyed by that change or by the movement of atoms. This coordinate also changes sign by some operations of the point group of the molecule, (like $\sigma_h\pi = -\pi$).

Thus, we have two kinds of internal coordinates:

1. The coordinates which do not change sign of the coordinate by any operation of molecular point group. These are known as coordinates of the first kind.
2. The coordinates for which the sign is changed by some operations of the molecular point group. These are known as coordinates of the second kind.

3. **Normal Coordinates:** The normal vibrations of the atoms are so chosen that all atoms of the molecule move in phase. It means that all atoms simultaneously pass through the equilibrium configuration, at the same time, reach the extreme end of the motion at the same time and they turn around again simultaneously to pass through the equilibrium configuration and to reach the other extreme all at one time. These coordinates measure the progress of the atoms in the molecule along this path. (Fig. 12.3 for two *AB* molecules)
4. **Symmetry Coordinate:** These coordinates necessarily transform under the symmetry operations of the molecular point group according to one of the I.Rs of group. These symmetry coordinates are the SALCs of the internal coordinates. These are formed using projection operations of I.Rs. on the internal coordinates used as bases.

## 12.3 RANDOM MOTIONS

At any moment the actual motion of the atoms in a molecule are random. These types of motions cannot be studied as they are. The only way to study such non-periodic random motions is to resolve then in terms of the $(3n-6)$[(for non-linear)or$(3n-5)$for linear molecules] components of any one of the above described coordinate systems.(Example: internal, or normal symmetry coordinates).We will, however, start with the cartesian coordinates as given below.

## 12.4 KINETIC AND POTENTIAL ENERGY OF VIBRATIONS IN TERMS OF THE CARTESIAN AND OTHER COORDINATES

In cartesian coordinates, the vibrational kinetic energy of the molecule can be given as,

$$T(\text{Kinetic Energy}) = -\frac{h^2}{8\pi^2}\sum_{k=1}^{3n}\frac{\partial^2}{\partial x_k^2} \qquad \text{...(12.24)}$$

and the potential energy $V$ as,

$$V = \frac{1}{2}\sum f_{ij}x_i x_j \qquad ...(12.25)$$

Here $x_i$, $x_j$ ..*etc*., are the mass-weighted cartesian coordinates of any particle or atom of the molecule with respect to a certain orthonormal basis. Suppose, we change this basis to some other orthonormal basis, the same configuration of the molecule will be specified in terms of the new basis by $3n$ coordinates which we designate as $Q_1, Q_2 \ldots Q_{3n}$. The new coordinates are related to the original cartesian coordinates by their linear combinations as,

$$Q_k = \sum_{i=1}^{3n} a_{k_i} x_i \quad \text{and} \quad x_i = \sum_{j=1}^{3n} b_{ij} Q_j \qquad ...(12.26)$$

The matrices | $A$ | and | $B$ | in Eq. (12.27) are inverse of each other since we have assumed $x_i$ and $Q_i$ on orthonormal basis. That is,

$$| A |^{-1} = | B | \qquad ...(12.27)$$

Let us suppose that the new coordinates $Q_i'$s are such that the K.E. and the potential energy expressions (*viz. T* and *V* respectively) can be expressed as,

$$T = \frac{h^2}{8\pi^2}\sum_{k=1}^{3n}\frac{\partial^2}{\partial Q_k'^2} \qquad ...(12.28)$$

and

$$V = \frac{1}{2}\sum_{r,\,k}^{3n} V_{rk}'' Q_r' Q_k' \qquad ...(12.29)$$

where $V_{rk}'' = \sum_{ij}^{3n} f_{ij} b_{ji} b_{jr}$. These Eqs. (12.28) and (12.29) are in the same form as those in equations (12.24 and 12.25). However, we wish to take the coordinates $Q_i$'s in such a form that the cross terms in the potential energy expression (Eq. 12.29) vanishes and we get the expression in the following simplified form,

$$V = \sum_{i=1}^{3n} \lambda_i Q_i^2 \qquad ...(12.30)$$

This kind of transformation can always be found, since there is a theorem which says that it is always possible to find new coordinates $Q_1, Q_2 \ldots Q_{3n}$ which are orthonormal linear combinational of $x_1, x_2 \ldots, x_{3n}$ (mass-weighted coordinates and the potential energy $V$, can be expressed as,

$$V = \frac{1}{2}\sum_{r=1}^{3n} \lambda_i Q_r^2 \qquad ...(12.31)$$

Such coordinates are termed as Normal Coordinates.

In terms of these coordinates, we can express Hamiltonian $\hat{H}$ of the Schrödinger equations ($H\psi = E\psi$) as,

$$H = T + V = \sum_{k=1}^{3n} H_k = -\frac{h^2}{8\pi^2}\sum_{k=1}^{3n}\frac{\partial^2}{\partial Q_k^2} + \frac{1}{2}\sum_{k=1}^{3n}\lambda_k Q_k^2 \quad ...(12.31a)$$

Since in Eq. (12.31*a*), *H* is the sum of 3*k* terms in which each term involves only one normal coordinate $Q_k$ (variable), it can then, be shown that the $\psi_T$ ($T$ = total) can be taken as a product function, *viz.*

$$\psi_T = \psi_1(Q_1)\cdot\psi_2(Q_2)\ldots\ldots\ldots\psi_k(Q_k) \quad ...(12.32)$$

In such a case, we will have for each $Q_k$,

$$H_k\,\psi_k = E_k\psi_k \quad ...(12.33)$$

and the total energy of a particular state can be expressed as

$$E_T = E_1 + E_2 + E_3 + \ldots E_s$$

where ($k$ = 1, 2, 3, 4, .... $s$). In Eq. (12.31*a*) and $\lambda_k$ can be expressed as,

$$\lambda_k = (2\pi\nu k)^2 \quad ...(12.34)$$

Due to these conditions, each of the Schrödinger equation takes the form,

$$\frac{\partial^2}{\partial Q_k^2}\psi_k + (E_k - \alpha^2 Q_k^2)\psi_k = 0, \quad ...(12.35)$$

where $\alpha_k = \dfrac{4\pi^2\nu_k}{h}$ and $E_k = \left(\dfrac{8\pi^2 E_k}{h^2}\right)$

All these $k$ numbers of equations can be solved individually which are similar to those of harmonic oscillators with $E_n = \left(n + \frac{1}{2}\right)h\nu$. In this fashion, we get solutions with quantum number $n_k$ for each normal coordinate $Q_k$. The product of all $\psi_k s$ will represent the vibrational state of the whole molecule. (These $\psi_k$ states will be characterised by the set of $S$ quantum number $n_1, n_2, \ldots n_s$). The ground state will be expressed as

$$\left.\begin{aligned} E_0 &= \sum_{k=1}^{S}\frac{1}{2}h\nu_k \\ \text{and} \quad \psi_0 &= \exp\left[-\frac{1}{2}\sum_{k=1}^{S}\alpha_k Q_k^2\right]\end{aligned}\right\} \quad ...(12.36)$$

## 12.5 SYMMETRY OF NORMAL MODE OF VIBRATION

It is clear from the above discussion, that each normal coordinate is associated with one vibration and there will be $k$ $(3n - 6)$ such vibrations. Thus, any vibratory motion of atoms of a molecule will be a superposition of all these $k$-vibrations that is known as the normal modes of vibrations of the molecule. Each of these normal modes will have one frequency. When all of them are superimposed, then the superposed mode will also be periodic, but it may have a period so large, that it may not be observable.

We have drawn six normal modes of vibrations of planar $CO_3^{2-}$ ion ($D_{3h}$ point group) (Fig. 12.4). Important properties of the normal modes are:

1. Each of the vectors attached to four atoms can be resolved into three components which may be said as three cartesian components $\Delta x_1^1, \Delta x_2^1, \Delta x_3^1, \Delta x_1^2, \Delta x_2^2, \Delta x_3^2 \cdots \Delta x_1^4, \Delta x_2^4, \Delta x_3^4$.
   These are cartesian displacement vectors along $\Delta x_1, \Delta x_2, \Delta x_3$ axis. In the same way, each the three translational and three rotational vectors are also resolved into three components each. Thus, the $3n$ degrees of motions of atoms in a molecule can be represented by a suitable linear combinations of $3n$-cartesian displacement coordinates.
2. The displacement vectors of the normal modes of vibrations can be used as basis vectors related to the internal coordinates of the molecule (*viz.* changes in the bond distances and bond angles). There is no particular method of doing so. Thus one can start with the changes in bond lengths of the molecule this for the formation of as much normal modes as one can and then use the changes in bond angles for the formation of the remaining left out normal modes or one can obtain the normal mode in the reverse way *i.e.,* first by bond angles, followed by bond lengths. In $CO_3^{2-}$ ion, we first choose changes in the three C–O bond lengths and the two bond angles plus one bond distance, and two bond distances and one bond angle to make the total of six normal modes. The symmetry species (I.Rs) of these normal modes may also be determined. For this purpose, we determine the point group of molecule having a fixed number of I.Rs in its character table and carryout symmetry operation on these vectors and determine how do they change(±). Knowing these changes, one can easily determine each normal mode belongs to an I.R of the group. Sometimes two or three normal modes belong to the same I.R (Fig. 12.4).
3. The normal modes of vibrations can be expressed as functions of either in terms of cartesian coordinates or internal coordinates. In the first relationship, the cartesian components are used to determine the total number of normal modes of each symmetry types of the point group. In the second relationship internal coordinates are used mainly to gain information, how the changes in bond lengths and the bond angles

contribute to the normal modes of vibrations according to their symmetry type (I.Rs). For example in Fig. 12.4, the $A_1'$ type normal mode involves purely stretching of the three bonds and $E_1$ type involves both stretching of bond and bending of bond angles. The latter information can also be deduced from the symmetry considerations alone without even knowing the actual form of the normal mode. We will explain it in the later part of the book, how can one obtain such information.

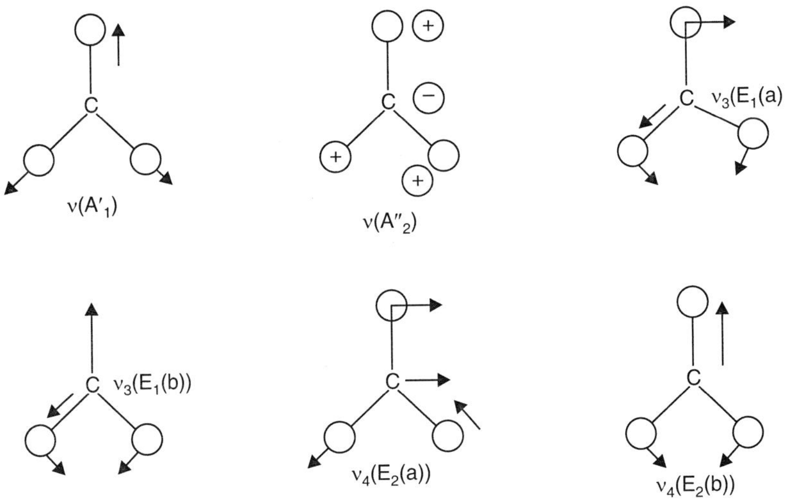


**Fig. 12.4** Normal modes of vibrations of $CO_3^{2-}$.

## 12.6 SYMMETRY COORDINATES (MORE DETAILS)

We have shown that the normal coordinates can be expressed as linear combinations of either cartesian or internal coordinates. We will see later that this relationship between the normal modes and the cartesian or internal coordinates can be utilized in determining the total number of normal modes (*viz.* $3n$-6). The second relationship (*i.e.*, between the normal modes and the internal coordinates) provides information, that is, how do the stretching and bending of bonds contribute to the normal modes, according to the symmetry types; For example, the $A_1'$ mode of vibration of $CO_3^{2-}$ (Fig. 12.5) involves only three C–O stretching coordinates, while the

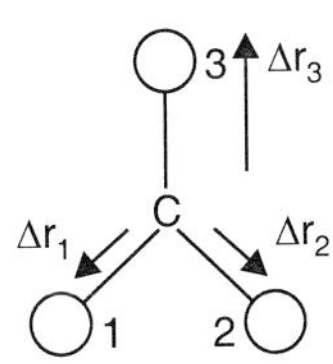


**Fig. 12.5** Stretching coordinates of $CO_3^{2-}$.

$A_2''$ vibration involves only bending coordinates, *etc*. We will learn in the following paragraph that the same information can be obtained without knowing the actual forms of the normal modes or without knowing internal coordinate vector.

As an illustrative example, let us take the $CO_3^{2-}$ ion. ($D_{3h}$ point group) and represent the three stretching coordinates as $\Delta r_1$, $\Delta r_2$ and $\Delta r_3$. Using them as basis, let us form $\Gamma_{red}^{\Delta r}$ of $D_{3h}$ point group. This is done by a shortcut method by finding the number of vectors which do not shift by the symmetry operations of $D_{3h}$ point group (*see* chapter 5). We obtain the following $\Gamma_{red}^{\Delta r}$ which is reduced to $A_1' + E'$ I.R s.

| $D_{3h}$ | $E$ | $2C_3$ | $3C_2$ | $\sigma_h$ | $2S_3$ | $3\sigma_v$ | *I.Rs* |
|---|---|---|---|---|---|---|---|
| $\chi_{(red)}$ | 3 | 0 | 1 | 3 | 0 | 1 | $A_1' + E'$ |

This implies that $A_1'$ and $E'$ types of normal modes of vibration must involve in the stretching of C–O bonds. Since there is only one $A_1'$ type of I.R. in the $\chi_{red}$, it means that it must be associated with one normal mode.

In the same way one can also start with the changes in three bond angles and form $[\Gamma_{red}^{\Delta\alpha}]$ containing the $A_1'$ and $E'$ I.Rs. But in this set, there is a lacuna because, we must realise that it is not possible to change all the three bond angles simultaneously in the plane of $CO_3^{2-}$ ion. It suggests that the three simultaneous changes in the bond angels are not independent, *i.e.*, one cannot change the three bond angles in the plane of $CO_3^{2-}$ at the same time. Thus, the normal mode belonging to $A_1'$ representation that we have found on reducing $[\Gamma_{red}^{\Delta\alpha}]$, has to be discarded as redundant coordinates. This type of redundancy is always found among the set of angle coordinates which form a closed group of atoms in a cyclic planar molecule or in the three-dimensional molecules like $AX_n$ where all $X$-atoms are bonded to one $A$ atom. We will discuss these problems in the illustrative examples later in the chapter. Thus, a set of three changes in the angles [O–C–O angles] will behave the same way by the $D_{3h}$ symmetry operations as that of C–O bonds. As a result, we obtain $E'$ representation after discarding $A_1'$(I.R.) as redundant coordinates. The in-plane bending of the O–C–O angles contribute to the normal vibrations of $E'$ symmetry. The third kind of internal coordinate must involve changes in the angles between C–O bonds and the plane of $CO_3^{2-}$ ion of $A_2''$ type. There will be other internal types of changes in the bond angles and bond-distances, but other reasons suggest that $A_2''$ type of normal mode involves the changes in angle between CO bond and the plane of $CO_3^{2-}$ ion.

In the light of above discussion, we have found that one can use the internal coordinates as basis which provides their linearly combined coordinates which transform according to each one of the I.Rs of the point group. These also span $(3n - 6)$ normal modes of genuine vibrations. To find these linearly combined internal coordinates, one may set up the SALCs employing projection operator technique, used in chapter six. These SALCs after normalisation, transform according to each of the I.Rs spanned by $(3n - 6)$ normal modes of vibrations. These are known as "symmetry coordinates" which are very useful in the mathematical computation.

## 12.7 KINETIC AND POTENTIAL ENERGY EXPRESSIONS IN VIBRATIONAL ANALYSIS

Prior to describing in detail the vibrational theory of polyatomic molecules, let us recapitulate some definitions:

(*a*) Cartesian displacement coordinates are written as $(\Delta x_1^n, \Delta y_2^n, \Delta z_3^n)$ where the values $n$ are the numbers of the atom in the polyatomic molecule. There are $3n$ cartesian coordinates.

(*b*) The internal coordinates of an $n$-atoms molecule are related to the changes in the relative positions of atom (*i.e.*, bond lengths and bond angles). There will be $(3n - 6)$ coordinates for non-linear or $(3n - 5)$ coordinates for linear molecule. These are represented by $r_i$ or $\alpha_{12}$. The rate of change of $r_i$ or $\alpha_{12}$ is written as $\dot{r}_i$ and $\Delta\dot{\alpha}$.

(*c*) The symbol $q_i$ can be written as any kind of $i$ coordinate and the symbol $Q_i$ denotes $i^{\text{th}}$ normal coordinate.

## 12.8 VIBRATIONAL K.E. AND POTENTIAL ENERGY

We assume that the vibration of any two atoms in a molecule behaves like a harmonic oscillator and thus the general quadratic potential energy expression of a general molecule is given by Eq. (12.37).

$$2V = \left[\left(\frac{\partial^2 V}{\partial q_1^2}\right)q_1^2 + \left(\frac{\partial^2 V}{\partial q_2^2}\right)q_2^2 + \left(\frac{\partial^2 V}{\partial q_3^2}\right)q_3^2\right] + \left(\frac{\partial^2 V}{\partial q_1 \partial q_2}\right)q_1 q_2$$

$$+ \left(\frac{\partial^2 V}{\partial q_1 \partial q_3}\right)q_1 q_3 + \left(\frac{\partial^2 V}{\partial q_2 \partial q_3}\right)q_2 q_1 + \ldots \qquad \ldots(12.37)$$

or $$= \left[ \sum_{i} \left( \frac{\partial^2 V}{\partial q_i^2} \right) q_i^2 + 2 \sum_{i<g} \left( \frac{\partial^2 V}{\partial q_i \partial q_j} \right) q_i q_j \right] + \ldots \qquad \text{...(12.38)}$$

Here $q_i$, $q_j$ are the coordinates and the summation $\sum_{i<j}$ include all values of $i$ and the combinations of $i$ and $j$ with $i < j$. The expression of force constant is

$$f_{ij} = \left( \frac{\partial^2 V}{\partial q_i q_j} \right) \qquad \text{...(12.39)}$$

In Eq. (12.39), if $i = j$, then $f_{ij}$ is known as the principal force constant. If $i \neq j$, then $f_{ij}$s are known as interaction constants. These are related in the way that the two coordinates $q_i$ and $q_j$ interact with each other, ($f_{ij} = f_{ji}$). Thus, the potential energy expression Eq. (12.38) can be written as,

$$2V = \sum_{1}^{i.j} f_{ij} q_i q_j \qquad \text{...(12.40)}$$

The summation $\sum_{i,\,j}$ extends to all the coordinates ($q_i$, $q_j$).

The kinetic energy ($2T$) may be written in terms of cartesian displacement coordinates ($\Delta x_1^n, \Delta x_2^n, \Delta x_3^n$) as,

$$2T = \sum_{N} m_N (\Delta \dot{x}_1^2)^n + (\Delta \dot{x}_2^2)^n + (\Delta \dot{x}_3^2)^n \qquad \text{...(12.41)}$$

$n$ is the number of the atoms whose displacement coordinates are represented by ($\Delta x_1^n, \Delta x_2^n, \Delta x_3^n$), $m_n$ is the mass of atom $n$. In terms of mass-weighted coordinates [$q_i$, $q_j$ ...], we can express kinetic energy $2T$ similar to earlier discussed case of diatomic molecule, as,

$$2T = \sum^{i,\,j} k_{ij} \dot{q}_i \dot{q}_j \qquad \text{...(12.42)}$$

where $k_{ij}$, {similar to the reduced mass factor $M$ in diatomic case [cf. Eq (12.18).] can be defined as

$$k_{ij} = \frac{\partial^2 T}{\partial \dot{q}_i \partial \dot{q}_j} \qquad \text{...(12.43)}$$

It can also be shown that, the general expression of motion in terms of Lagrange's equation, becomes

$$\left(\frac{\partial V}{\partial q_i}\right)+\frac{\partial}{\partial t}\left(\frac{\partial T}{\partial \dot{q}_i}\right)=0 \qquad ...(12.44)$$

[(force) + ($m$ × acceleration) = 0]

There will be (3$n$-6) such equations. One equation (12.44) for each coordinates $q_i$ in terms of $V$ and $T$. Since, the expression for $V$ and $T$ are given by Eqs. (12.40) and (12.42), we can differentiate $V$ and $T$ in Eqs. (12.40) and (12.42) with respect to $q_i$ and $q_j$ substituting them in Eq. (12.44). This will result in a series of simultaneous linear differential equations one for each coordinate $q_i$. The solution of these equations gives the relationship between the vibrational frequencies of atoms in the molecule, force constants and the reduced mass factor $K_{ij}$ [similar to Eq. (12.33) obtained for a diatomic molecule]. Prior to discussing a general case of obtaining a solution, we will illustrate it with a simple example of triatomic molecule which is given below in order to clarify the involved concepts in calculations.

## 12.9 EXAMPLE OF TRIATOMIC MOLECULE

We take a linear triatomic molecule $ABC$ having 4 modes [(3$n$ – 5) = 4] of vibrations. We will take two coordinates as bond stretches (*viz.* $\Delta r_1$ and $\Delta r_2$) as $A$–$B$ and $B$–$C$ bond stretches and two coordinates as bond angle deformations, which will be degenerate modes. We will discuss stretching of bonds ($\Delta r_1$ and $\Delta r_2$) separately from the bending modes. The reason for doing so will be discussed in the latter sections of the chapter. Therefore, we define two stretching internal coordinates ($\Delta r_1$ and $\Delta r_2$) in terms of cartesian coordinates as follows [Fig.12.6 ($a$, $b$, $c$)].

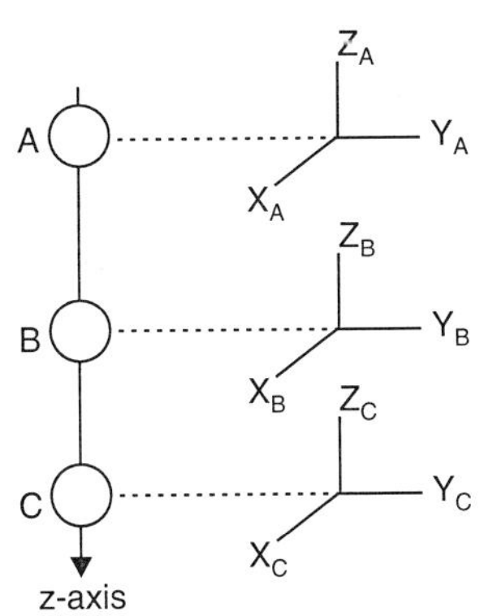


($a$) Coordinate system molecular axis is z-axis.

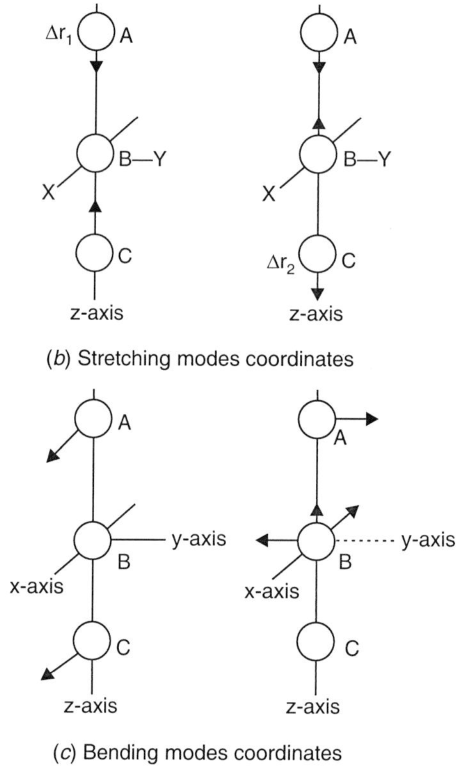


**Fig. 12.6** Coordinate system, stretching and bending modes of vibrations of linear ABC molecule.

and
$$\left.\begin{aligned}\Delta r_1 &= z_B - z_A \\ \Delta r_2 &= z_C - z_B\end{aligned}\right\} \qquad ...(12..45)$$

In terms of $\Delta r_1$ and $\Delta r_2$, the potential energy ($2V$) and K.E. ($2T$) may be written as

$$\left.\begin{aligned}2V &= \sum_{ij} f_i q_i^2 + 2\sum_{i<j} f_{ij} q_i q_j \\ &= f_{11}\Delta r_1^2 + f_{22}\Delta r_2^2 + 2f_{12}\Delta r_1 r_2\end{aligned}\right\} \qquad ...(12.46)$$

$$\left.\begin{aligned}2T &= \sum_{ij} K_{ij}\dot{q}_i\dot{q}_j \\ &= k_{11}\Delta\dot{r}_1^2 + k_{22}\Delta\dot{r}_2^2 + 2k_{12}\Delta\dot{r}_1\Delta\dot{r}_2\end{aligned}\right\} \qquad ...(12.47)$$

From Eqs. (12.46) and (12.47), we get

$$\left.\begin{aligned}\frac{dV}{d\Delta r_1} &= f_{11}\Delta r_1 + f_{12}\Delta r_2 \\ \frac{dV}{d\Delta r_2} &= f_{21}\Delta r_1 + f_{22}\Delta r_2\end{aligned}\right\} \qquad (12.48a)$$

$$\left.\begin{vmatrix}\dfrac{dV}{d\Delta r_1}\\ \dfrac{dV}{d\Delta r_2}\end{vmatrix} = \begin{vmatrix}f_{11} & f_{12}\\ f_{21} & f_{22}\end{vmatrix}\begin{vmatrix}\Delta r_1\\ \Delta r_2\end{vmatrix}\right\} \quad \text{...(12.48}b\text{)}$$

$$\left.\begin{aligned}\frac{\partial}{\partial t}\left(\frac{\partial T}{\partial \dot{r}_1}\right) &= \frac{\partial}{\partial t}(K_{11}\Delta\dot{r}_1 + K_{12}\Delta\dot{r}_2)\\ &= K_{11}\Delta\ddot{r}_1 + K_{12}\Delta\ddot{r}_2\\ \frac{\partial}{\partial t}\left(\frac{\partial T}{\partial \dot{r}_2}\right) &= \frac{\partial}{\partial t}(K_{21}\Delta\dot{r}_1 + K_{22}\Delta\dot{r}_2)\\ &= K_{21}\Delta\ddot{r}_1 + K_{22}\Delta\ddot{r}_2\end{aligned}\right\} \quad \text{...(12.49}a\text{)}$$

$$\left.\begin{vmatrix}\dfrac{\partial}{\partial t}\left(\dfrac{\partial T}{\partial \Delta\dot{r}_1}\right)\\ \dfrac{\partial}{\partial t}\left(\dfrac{\partial T}{\partial \Delta\dot{r}_2}\right)\end{vmatrix} = \begin{vmatrix}K_{11} & K_{12}\\ K_{21} & K_{22}\end{vmatrix}\begin{vmatrix}\Delta\ddot{r}_1\\ \Delta\ddot{r}_2\end{vmatrix}\right\} \quad \text{...(12.49}b\text{)}$$

Substituting Eqs. (12.48) and (12.49) in the Lagrange's equation (12.44), we get the following two equations. [There are two equations, one for each internal coordinate *viz.* ($\Delta r_1$ and $\Delta r_2$)].

$$[\Delta r_1 f_{11} + K_{11}\Delta\ddot{r}_1] + (f_{12}\Delta r_2 + K_{12}\Delta\ddot{r}_2) = 0 \quad \text{...(12.50)}$$

and
$$(f_{21}\Delta r_1 + K_{12}\Delta\ddot{r}_1 + f_{22}\Delta r_2 + K_{22}\Delta r_{22}) = 0 \quad \text{...(12.51)}$$

(In any system if we have three coordinates, we will get three equations. In general, we get *n*-such equations for *n*-coordinates.)

These equations are similar to those which we have obtained in the example of diatomic molecule, as vibrating as SHM (Eq. 12.22) and where we have taken $\Delta r_1$ to be equal to $A\cos(2\pi\nu t + \rho)$. Similarly, we will also assume that,

$$\Delta r_1 = A_1\cos(2\pi\nu t + \rho)$$

and
$$\Delta r_2 = A_2\cos(2\pi\nu t + \rho) \quad \text{...(12.52)}$$

On differentiating $\Delta r_1$ and $\Delta r_2$ twice with respect to $t$, we obtain,

$$\left.\begin{aligned}\Delta\ddot{r}_1 &= -A_1(4\pi^2\nu^2)\cos(2\pi\nu t + \rho) = -\lambda A_1\cos(2\pi\nu t + \rho)\\ \Delta\ddot{r}_2 &= -A_2(4\pi^2\nu^2)\cos(2\pi\nu t + \rho) = -\lambda A_2\cos(2\pi\nu t + \rho)\end{aligned}\right\} \quad \text{...(12..53)}$$

(where $\lambda = 4\pi^2\nu^2$), Eqs. (12.52) and (12.53) are substituted in Eqs. (12.50) and (12.51), whereby we obtain the final equations (*viz.*)

$$[f_{11}A_1\cos(2\pi\nu t + \rho) - K_{11}A_1\lambda\cos(2\pi\nu t + \rho)]$$
$$+ [f_{12}A_2\cos(2\pi\nu t + \rho) - K_{12}A_2\lambda\cos(2\pi\nu t + \rho)] = 0$$

and $$\left[f_{21}A_1\cos(2\pi\nu t+\rho)-\lambda K_{21}A_1\lambda\cos(2\pi\nu t+\rho)\right]$$
$$+\left[f_{22}A_2\cos(2\pi\nu t+\rho)-K_{22}A_2\lambda\cos(2\pi\nu t+\rho)\right]=0$$

These equations then take the final form after removing the common factor $\cos(2\pi\nu t+\rho)$. These are given below. (These are known as Secular Equations.)

$$\left.\begin{aligned}(f_{11}-\lambda K_{11})A_1+(f_{12}-\lambda K_{12})A_2=0\\(f_{21}-\lambda K_{21})A_1+(f_{22}-\lambda K_{22})A_2=0\end{aligned}\right\}\qquad ...(12.54)$$

Eq. (12.54) are homogeneous linear simultaneous equations in the unknowns $A_1$ and $A_2$. These can be solved for non-trivial solution (*i.e.*, $A_1=A_2\neq 0$) only when the following secular determinant is equal to zero, (This is known as Secular Determinant).

$$\begin{vmatrix}(f_{11}-\lambda K_{11}) & (f_{12}-\lambda K_{12})\\(f_{21}-\lambda K_{21}) & (f_{22}-\lambda K_{22})\end{vmatrix}=0\qquad ...(12.55)$$

This determinant is known as the secular determinant and Eq. (12.54) as secular equations. On solving Eq. (12.55), we obtain,

$$(f_{11}f_{22}-f_{12}^2)-[f_{11}K_{22}+f_{22}K_{11}-2f_{12}K_{12}]\lambda+[K_{11}K_{22}-K_{12}^2]\lambda^2=0\qquad ...(12.56)$$

Eq. (12.56) shows the relationship between the frequency (or $\lambda$) $[\lambda=4\pi\nu^2]$, the force constants ($f_{ij}$) and $k_{ij}$ which are specific properties of the system under study depending upon mass of the atoms and geometry of the molecule. Eq. (12.56) is a quadratic equation in $\lambda$. It will give two roots of $\lambda$. We will thus obtain two frequencies $\nu_1$ and $\nu_2$. These two roots of $\lambda$ are then substituted successively in the secular equation (Eq. 12.54), one at a time. This will result in the two sets of equations (one for $\lambda_1$ and another for $\lambda_2$). One such set is given below.

$$\left.\begin{aligned}(f_{11}-\lambda_1 K_{11})A_1+(f_{12}-\lambda_1 K_{12})A_2=0\\(f_{21}-\lambda_1 K_{21})A_1+(f_{22}-\lambda_1 K_{22})A_2=0\end{aligned}\right\}\qquad ...(12.57)$$

Although the solution of Eq. (12.57) will give a set of two values $A_1$ and $A_2$, but these will not be the unique values of $A_1$ and $A_2$. We can obtain only, a ratio of $A_1$ and $A_2$. Let us write $A_{11}$ and $A_{21}$ when we use $\lambda_1$ and $A_{22}$ and $A_{12}$ when we use $\lambda_2$ for $A_1$ and $A_2$ in the Eq. In this fashion, we obtain two sets of $A_{ik}$ (for $\lambda_1$ as $A_{11}$ and $A_{21}$ and for $\lambda_2$ as $A_{12}$ and $A_{22}$). As said above the values of these $A_{ik\text{'s}}$ will only be relative values of the amplitudes. The absolute values of $A_1$ and $A_2$ can not be obtained from these equations.

The following conclusions are thus drawn from the above description:

For $\nu=\nu_1$ coordinate $\Delta r_1=A_{11}\cos(2\pi\nu_1 t+\rho)$ and coordinates $\Delta r_2=A_{21}\cos(2\pi\nu_1 t+\rho)$

$\nu=\nu_2$, coordinate $\Delta r_1=A_{12}\cos(2\pi\nu_2 t+\rho)$ and $\Delta r_1=A_{22}\cos(2\pi\nu_2 t+\rho)$

1. For, both atoms vibrate with the same frequency and the same phase. Both atoms reach the maximum amplitude at the same time and also the minimum, but the maximum amplitudes, *viz.* $A_{11}$ and $A_{21}$ are different.

2. At any moment of time $t$, let the amplitudes $A'_{11}$ and $A'_{21}$ [= $A_i\cos(2\pi\nu_1 + \rho)$] for the frequency $\nu_1$. For the second frequency $\nu_2$, let the amplitudes be $A'_{12}$ and $A'_{22}$.

3. We now define a normal coordinate $Q_1$ for the frequency $\nu_1$ and $Q_2$ for the frequency $\nu_2$. That is,

for second frequency $\nu_2$,
$$\left.\begin{aligned} Q_1 &= (A'_{11}, A'_{21}) \\ Q_2 &= (A'_{12}, A'_{22}) \end{aligned}\right\} \qquad \text{...(12.58)}$$

Thus, at any moment of time when atoms 1 and 2 are vibrating, we have,

$$\left.\begin{aligned} \Delta r_1 &= A'_{11}Q_1 + A'_{12}Q_2 \\ \Delta r_2 &= A'_{21}Q_1 + A'_{22}Q_2 \end{aligned}\right\} \qquad \text{...(12.59)}$$

When we use these normal coordinates (*viz.* $Q_1$ and $Q_2$), it can be shown that the K.E. and the Potential energy expressions will be reduced to a particularly simple form. For Potential energy and K.E. of any vibrating system, these are given as under:

$$\left.\begin{aligned} 2V &= \sum_{i=1}^{3n-6} \lambda_i Q_i^2 \\ 2T &= \sum_{i}^{3n-6} \dot{Q}_i^2 \end{aligned}\right\} \qquad \text{...(12.60)}$$

[For linear molecules, summation is over $(3n-5)$].

We have thus shown how one type coordinates $Q_i$ can be transformed into another set ($\Delta r_i$) by a set of linear transformations. It plays a very significant role in the vibrational analysis. We are ultimately concerned with finding a transformation from one set of coordinates to another (internal coordinates to Cartesian ones, *etc.*) so that the K.E. and potential energy expressions are reduced to a sum of quadratic terms only. All cross products vanish.

Now if we let, molecule absorb one quantum of radiation of frequency $\nu_k$ then it will raise the energy of the molecule by $2V = (\lambda_K Q_K)$ and $2T = \dot{Q}_K^2$. It implies that in the absorption process only one coordinate $Q_K$ is involved which serves to give the relative displacements of all the atoms in that vibration. The displacements of the atoms can be expressed in any coordinate system (Cartesian or internal, *etc.*) because the relationship among them are known and can be inter-converted. For example, in a triatomic molecule, *A–B–C* the relative displacements of atoms are shown in Fig. 12.7 like in the thiocyanate ion. Thus the coordinates $Q_1$ and $Q_2$ are expressed as stretching or compression of both bonds whose relative values are shown by the length of arrows attached to the atoms.

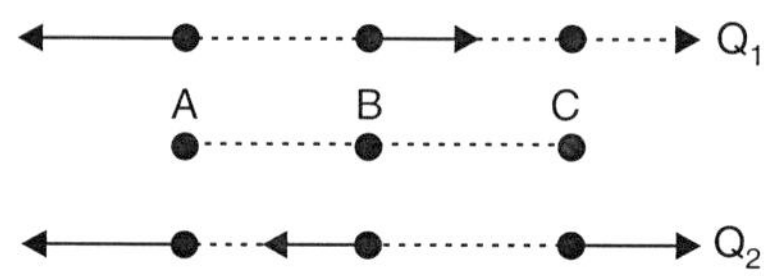


**Fig 12.7** Normal displacement of a linear triatomic molecule.

In any spectroscopic experiments, the molecule is exposed to the white light. Thereby, all the normal modes are excited simultaneously and arrow displacements of atoms signify the normal displacements. This follows from the equations (12.58) and (12.59). In these equations, there are no terms which involve more than one coordinate. Therefore, wherever all the coordinates are involved in the vibrations, it results in the state of chaotic vibrations of all atoms. This can be resolved into different normal coordinates which show the relative displacements of the atoms in particular normal coordinates (not the absolute displacements).

We wish to re-emphasize that, there is one normal mode of vibration, corresponding to each observed frequency that is described by its normal coordinate. These normal modes are completely independent to each other, but each normal mode is a mixture of various internal coordinates (bond stretching, bending *etc*). It is easy to measure the frequencies, but we are interested in understanding the nature of bonds of the molecule. We usually relate the changes in bond lengths and bond angles to the frequencies of radiation absorbed or emitted by the molecule. We do it by the use of normal coordinate analysis.

## 12.10 POLYATOMIC MOLECULES

The above procedure is based on classical mechanics. We could have used quantum mechanical principles. Prior to this, we wish to extend it to polyatomic case. We will extrapolate it very briefly in the following paragraphs:

For an $n$-atom molecule, we write K.E. as follows:

$$2T = \sum_{i=1}^{n} m_i \left[ \left( \frac{d\,\Delta x_1^i}{dt} \right)^2 + \left( \frac{d\,\Delta x_2^i}{dt} \right)^2 + \left( \frac{d\,\Delta x_3^i}{dt} \right)^2 \right] \quad \ldots(12.61)$$

and in terms of mass-weighted coordinates, $(q_1, q_2 \ldots q_{3n})$ (where $q_i = \sqrt{m_i \Delta x_i}$ ).

$$2T = \sum_{i=1}^{3n} \dot{q}_i^2 \qquad ...(12.62)$$

The potential energy of the molecule can be written as,

$$2V = \sum_{i,j=1}^{3n} f_{ij} q_i q_j \qquad ...(12.63)$$

$$f_{ij} = \left( \frac{d^2V}{dq_i q_i} \right) \qquad ...(12.64)$$

Lagrange's equation can be written as,

$$\frac{d}{dT}\left( \frac{dT}{d\dot{q}_j} \right) + \frac{dV}{dq_j} = 0 \;\; (j = 1, 2...3N) \qquad ...(12.65)$$

We will get $3n$ equations (one for each of coordinates.)

If we substitute expressions for $2T$ and $2V$ (given above). We will obtain the following $3n$ equations.

$$\ddot{q}_j = \sum_{i=1}^{3N} f_{ij} \; q_i = 0 \;\; (j = 1, 2...3n) \qquad ...(12.66)$$

This is the set of $3n$ simultaneous second order linear differential equations. For one $j$ coordinate, the solution is

$$q_j = A_j \cos(\lambda^{1/2} t + \rho) \qquad ...(12.67)$$

Where $A_j$, $\lambda$ and $\rho$ are the properly chosen constants. If these expressions are substituted in the differential equations, a following set of equations are obtained.

$$\ddot{q}_j = \sum_{j=1}^{3N} (f_{ij} - \delta_{ij}\lambda) \; A_j = 0 (j = 1, 2...3N) \qquad ...(12.68)$$

These are simultaneous secular equations ($j$ = 1, 2, $3n$). Thus, for non-trivial solutions ($A_j \neq 0$), the following secular determinant has to be zero. That is,

$$\begin{vmatrix} f_{11} - \lambda & f_{12} & f_{13} & \cdots & f_{1,3N} \\ f_{21} & f_{22} - \lambda & f_{23} & \cdots & f_{2,3N} \\ f_{31} & f_{32} & f_{33} - \lambda & \cdots & f_{3,3N} \\ \vdots & \vdots & \vdots & \ddots & \vdots \\ f_{3n,1} & f_{3n,2} & f_{3n,3} & \cdots & f_{3n,3N} - \lambda \end{vmatrix} \qquad ...(12.69)$$

The elements of this determinant are the coefficients of the unknown amplitudes, $A_j$ in the secular equation. The solution of this determinant will give an equation in λ which is of order $3n$, and so there will be $3n$ roots of λ. For each root of λ(say $\lambda_i$), the secular equation can be solved which will give a set of $3N$ $A'_{ij}s$ associated with that λ(*viz.* $\lambda_i$). Such a system of equations will not give a unique set of values of $A_{ik}$ but gives only their ratios. A convenient and unique solution may be designated by the quantities $(l)_{ik}$ which is defined as follows:

$$(l)_{ik} = \frac{A'_{ik}}{[\sum_i (A'_{ik})^2]^{\frac{1}{2}}} \qquad \ldots(12.70)$$

The solutions of actual physical problem can be obtained by putting,

$$A_{ik} = K_k \, l_{ik} \qquad \ldots(12.71)$$

where $K_k$ is constant which depends upon the initial values of $q_i$ coordinate and velocity $\dot{q}_i$.

From the above discussions, it is clear that each atom is oscillating about its equilibrium position in SHM of amplitude [ $A_{ik} = K_k \, l_{ik}$ ], frequency $(\lambda_i h\pi^2)^{1/2}$ and phase ρ. Furthermore, corresponding to each $\lambda_i$, the frequency and ρ of motion of each coordinate is the same. In this context, we define a Normal coordinate

$$Q_k = \sum_{i=1}^{3N} l_{ki} \cdot q_i \quad k = 1, 2, 3 \ldots 3N \qquad \ldots(12.72)$$

The same procedure can be carried out quantum mechanically as described in sections 12.3, and 12.4 where we have defined normal coordinates ($Q_i$'s), whatever coordinates we use in solving the vibrational problems, there remains the same problem. "It is difficult to solve the secular determinant for a molecule having more than four atoms". We, therefore have to seek some method to bring the secular equation in a block-diagonal form by some similarity transformation. Let us use matrices to express K.E. and potential energy.

### 12.10.1 Normal Coordinate Analysis (Polyatomic Molecules)

In the light of the above background, we are now in a position to discuss the vibrational analysis of a general polyatomic molecule having '$n$' atoms.

This approach is due to Wilson who described the procedure in his book "Molecular Vibrations".

The molecule contains '$n$' atoms. There will be $(3n-6)$, vibrational, three rotational and three translational modes. The total number of cartesian coordinates which describe the motion are $3n$. We write them as ($x_1^1, x_2^1, x_3^1, x_1^2, x_2^2, x_3^2, \ldots. x_1^n, x_2^n, x_3^n$). These are displacement coordinates. These are in reality $\Delta x_1$, $\Delta x_2$, $\Delta x_3$ *etc*. We can also express the potential and kinetic energies in terms of internal coordinates (*viz*. $r_1, r_2 \ldots r_{3n-6}$). These are displacement internal coordinates. (Be careful, here $r_i'\,s$ or $\Delta r_i'\,s$ *i.e.*, changes in bond lengths and bond angles).

**Potential Energy:** It is expressed as,

$$2V = \sum_{i=1}^{3n-6} f_{ij} r_i r_j \qquad \ldots(12.73)$$

where

$$f_{ij} = \frac{\partial^2 V}{\partial r_i \partial r_j}$$

In matrix notation, $2V$ is expressed as

$$2V = \left| r_1, r_2, \cdots, r_{3n-6} \right|$$

$$\begin{vmatrix} f_{11} & f_{12} & \cdots & f_{1,(3n-6)} \\ f_{21} & f_{22} & \cdots & f_{2,(3n-6)} \\ \vdots & \vdots & \ddots & \vdots \\ f_{(3n-6),1} & f_{(3n-6),2} & \cdots & f_{(3n-6)(3n-6)} \end{vmatrix} \begin{vmatrix} r_1 \\ r_2 \\ \vdots \\ r_{(3n-6)} \end{vmatrix} \qquad \ldots(12.74)$$

$$= r'Fr \qquad \ldots(12.75)$$

Here $r'$ is transpose of column matrix $r$.

**Kinetic Energy:** K.E. of the molecular vibration is expressed in terms of the cartesian coordinates. We, therefore, write it as

$$2T = \sum_{i=1}^{3n} m_i \dot{x}_i^2 \qquad \ldots(12.76)$$

In vibrational problems, we express cartesian coordinates as mass-weighted coordinates and write them as $\dot{q}_i$ which is related to $x_i$ as

$$\dot{q}_i = (m_i)^{1/2} \dot{x}_i \qquad \ldots(12.77)$$

In terms of $\dot{q}_i$'s K.E. can be written as,

$$2T = \sum_{i=1}^{3n} \dot{q}_i^2 \qquad \ldots(12.78)$$

As we saw earlier in this chapter, the time derivatives of cartesian coordinates or the $\dot{q}_i$'s are related to the time derivative of the internal coordinates by a set of linear equations. These are written as follows (Wilson's notation).

$$R_j = \sum_{i=1}^{3n} B_j \dot{q}_i \qquad \text{...(12.79)}$$

In matrix notation, it is written as

$$R = \begin{vmatrix} \dot{r}_1 \\ \dot{r}_2 \\ \vdots \\ \dot{r}_{3n-6} \end{vmatrix} = \begin{vmatrix} B_{11} & B_{12} & \cdots & B_{3n,1} \\ B_{21} & B_{22} & \cdots & B_{3n,2} \\ \vdots & \vdots & \ddots & \vdots \\ B_{1,(3n-6)} & B_{2,(3n-6)} & \cdots & B_{3n,(3n-6)} \end{vmatrix} \begin{vmatrix} \dot{q}_1 \\ \dot{q}_2 \\ \vdots \\ \dot{q}_{3n} \end{vmatrix} \qquad \text{...(12.80)}$$

$= (3n \times 3n - 6)$ order matrix

$= B\dot{q}$

(Note: In this matrix the number of rows is $(3n–6)$ and the number of column is $3n$.

Here $B$ is $(3n–6)$ row and $3n$ column transformation matrix. This equation (12.79) or (12.80) involves $\dot{q}_i$ coordinate in place of $\Delta\dot{x}$ coordinates. It is therefore expressed as

$$\dot{R}_K = \sum_{i=1}^{3n} B_{Ki} \Delta\dot{x}_k$$

$$= \sum_{i=1}^{3n} D_{Ki} \dot{q}_i \quad \text{where } D_{Ki} = B_{Ki}(m_i)^{1/2} \qquad \text{...(12.81)}$$

In matrix language,

$$\dot{R} = D\dot{q} \qquad \text{...(12.82)}$$

where, $D$ is $|3n \times (3n-6)|$ matrix and $n$ varies form 1 to $n$ atoms.

But the problem is to express $2T$ Eq. (12.78) in terms of internal coordinates $R(i.e., \dot{r}_1, \dot{r}_2 ......\dot{r}_{3n-6})$, in place of $\dot{q}$'s Eq. (12.82) which will be comparable to the expression of $2V$ Eq. (12.74). It can only be possible if we can find the inverse of matrix $D$. But the inverse of $D$ is however not possible, because $D$ is rectangular matrix.

There is however a way out to make matrix $D$ as a square matrix. This is possible by writing six additional rows as six additional conditions of vibrational motion. These six conditions are zero translational momentum and zero rotational angular momentum. Thus, by including rotational and translational momentum equal to zero, we can write $D'$ as

$$D' = \begin{vmatrix} D \\ D_0 \end{vmatrix} \qquad ...(12.83)$$

where $D_0$ is $[3n \times 6]$ order matrix for three zero translational and three zero rotational momentum. The inverse of $[D']^{-1}$ as (here $D' = 3n \times 3n$ matrix) can now be easily obtained.

Let us now suppose that $[D']^{-1} = [PP_0]$. $[D']^{-1}$ has also been divided into $P$ and $P_0$ which are respectively $| 3n \times (3n - 6) |$ and $| 3n \times 6 |$ matrices. In terms of $[D']^{-1}$ matrix, one has,

$$\dot{R} = [D']\dot{q} = \begin{pmatrix} \dot{R}' \\ \dot{r}' \end{pmatrix} = \begin{pmatrix} D \\ D_0 \end{pmatrix} \dot{q} \qquad ...(12.84)$$

where $\dot{r} = D_0 \dot{q}$ is a column matrix consisting of six zero elements. The inverse of $[D']$ (Eq. 12.83) can now be written as

$$\dot{q} = (D')^{-1} \dot{R} \qquad ...(12.85)$$

and $2T$ as,

$$2T = \dot{q}'\dot{q} = \dot{R}(D')^{-1} \times (D')^{-1}\dot{R}$$

$$= (\dot{R}'\dot{r}')\begin{pmatrix} P' \\ P_0 \end{pmatrix}(PP_0)\begin{pmatrix} \dot{R} \\ \dot{r} \end{pmatrix}$$

$$= [(\dot{R}'P') + \dot{r}'p_0'](P\dot{R} + P_0\dot{r})$$

$$= \dot{R}'P'P\dot{R} + \dot{R}'P'P_0\underline{\dot{r}} + \underline{\dot{r}}'P_0'P\dot{R} + \dot{r}'P_0'P_0 r' \qquad ...(12.86)$$

In this equation, $\dot{r}$ is a zero matrix, velocity (or linear) and angular momentum (*viz.* $m\dot{r}$ and $m\dot{r} \cdot r$) is zero for translation and angular momentum. The last three terms of Eq. (12.86) become zero. One is therefore left with,

$$T = \dot{R}'P'P\dot{R} \qquad ...(12.87)$$

This is $[(3n–6) \times (3n–6)]$ matrix. We thus have to find the matrix $P$. It can be obtained from the inverse of the matrix $\left[D' = \begin{vmatrix} D \\ D_0 \end{vmatrix}\right]$. For getting $| P |$, we write,

$$[D']^{-1}[D'] = (PP_0)\begin{pmatrix} D \\ D_0 \end{pmatrix} = [PD + P_0D_0] = | 3n \times 3n | = | A | \qquad ...(12.88)$$

and $$[D'][D']^{-1} = \begin{pmatrix} D \\ D_0 \end{pmatrix}(PP_0) = \begin{vmatrix} DP & DP_0 \\ D_0P & D_0P_0 \end{vmatrix} = | 3n \times 3n | = | A | \qquad ...(12.89)$$

In Eq. (12.89), the | $DP$ | matrix is composed of the | $D$ | matrix of order [(3$n$–6) × (3$n$)] and | $P$ | matrix, is of order [3$n$ × 3$n$–6], so |$DP$| matrix will become of [(3$n$–6) × (3$n$–6)] order which is an identity matrix. We now use the first relation, *viz.* Eq. 12.88. First transpose of [$PD + P_0D_0$] is taken which is equal to $[P'D' + P_0'D_0']$. We multiply the latter by $D$ on the LHS and by $P$ on the RHS. This result in,

$$D[D'Q'Q + DD_0'P_0']P = DP \quad ...(12.90)$$

$DP$ matrix is of (3$n$ – 6) × (3$n$ – 6) order.

It is obvious that the internal coordinates are the basis for the vibrational analysis and they are orthogonal to the overall translational and rotational coordinates. It implies the $D$ and $D_0$ are orthogonal matrices and $DD_0'$ will be zero matrices. Therefore, the term $| DD_0'P_0'P |$ in Eq. (12.90) will be zero and thus gets eliminated. We are left in Eq. (12.90), with

$$DDP'P' = \text{I |unit matrix |} \quad ...(12.91)$$

$P'P$ is obtained as

$$P'P = | DD' |^{-1} = G^{-1}$$

(we use the symbol $G$ for | $D'D$ |). We know from Eq. (12.80), that

$$\dot{R}_J = \sum_{i=1}^{3n} D_{ji}\dot{q}_i$$

where
$$D_{ji} = B_{ji}.(m)^{-1/2}$$

or
$$| \dot{R} | = | D || \dot{q} |$$

Therefore, we have

$$G_{ji} = \sum_{i=1}^{3n} \frac{B_{ji}B_{li}}{m_i} \quad (12.92)$$

By employing expression (12.92) one can calculate | $G$ | matrix in $G_{jl}$ for $j$ = 1, 2, 3, ... (3$n$–6) and $i$ = 1, 2, 3, ...3$n$–6. In case $B$, matrix is transformation matrix for cartesian and internal coordinates and can be written from the structure of the molecule.

Finally we now have,

$$2T = R'G^{-1}R \quad ...(12.93)$$

and
$$2V = R'FR \quad ...(12.94)$$

We will now use Lagrange's equations where the matrix | $G^{-1}$ | equals to $|BM^{-1}B|^{-1}$. In this expression | $M$ | is the matrix in terms of masses of the atoms in the molecule and $G$ is the symmetric square matrix of (3$n$–6) × (3$n$–6) order. Lagrange's equations is

$$\frac{d}{dt}\left[\frac{\partial T}{\partial \dot{r}_i}\right]+\frac{\partial V}{\partial r_i}=0 \qquad ...(12.95)$$

for each value of $i$. Substituting the expressions for $T$, and $V$ with $\left[F=\frac{\partial^2 V}{\partial R_i \partial R_j}\right]$, one finally obtains the expression (12.96) as given below,

$$FR = G^{-1}\ \dot{R} = 0 \qquad ...(12.96)$$

As usual and analogous to the case of di-and triatomic molecules, one substitutes for each internal coordinate $r_i$.

and
$$\left.\begin{aligned} r_i &= \left[A_{iG^{-1}}\right]\cos(2\pi\nu_i t+\rho) \\ \ddot{r}_i &= [A_{ik}][-\lambda_K]r_i \end{aligned}\right\} \qquad ...(12.97)$$

In this manner, we will obtain (3$n$–6) linear homogeneous equations in terms of $A_i(i = 1, 2, 3 ... 3n-6)$ Eq. (12.98). These equations are then solved for non-trivial solutions of $A_i(A_i \neq 0)$ by putting the secular determinant Eq. (12.99) equal to zero. That is,

$$| F-\lambda G^{-1} | A_i | = 0 \text{ (Secular equation)} \qquad ...(12.98)$$

and secular determinant, $\quad | F-\lambda G^{-1} | = 0$ (for non-trivial solutions)...(12.99)

It will give (3$n$–6) roots of $\lambda_{ki}$. Each root of $\lambda_k$[(3$n$–6) roots] will give a set of (3$n$–6) $A_{ik}$'s. This set will represent the normal coordinate associated with the value of $\lambda_K (= 4\pi^2\nu_k^2)$. In this manner, (3$n$–6) normal coordinates will be obtained. Each normal coordinate will be associated with one frequency of vibration ($\nu_k$) and also with a set of $n$ amplitudes of vibrations $A_{ki}$, (*viz.* a total of $n$ amplitudes for one $\lambda_k$).

Eq. (12.98) may also be written as

$$| FG-\lambda E| A_{ik} = 0 \qquad ...(12.100)$$

This equation is obtained by multiplying Eq.(12.98) on its LHS by the matrix | $G$ |

Eq.(12.100) is an eigenvalue equation and can also be written as,

$$GFA = A\,\Lambda \qquad ...(12.101)$$

where $\Lambda$ is a diagonal matrix of (3$n$–6) eigenvalues and $A$ is a square matrix of order (3$n$–6).

Thus, if one knows the matrices | $F$ | | $G$ |, one can calculate (3$n$–6) frequencies of vibrations and for each frequency, the corresponding eigenvector (*i.e.*, a set of $A_{nk}$) associated with a frequency $\nu_k$(or $\lambda_k$).

Although both | $F$ | and | $G$ | matrices are individually symmetrical but the matrix | $FG$ | is not symmetrical. This is taken into account while calculating $A_n$ and $\Lambda$ from the | $GF$ | matrix.

As the size of the secular determinant grows, the solutions of $A_{ik}$ and $\lambda_k$ become impossible. A number of various methods of solution has been devised. We will not discuss them here. In complicated cases, the systematic nature of Wilson's FG method is of real value (consult *Wilson's book on "Vibrations"*).

It has already been shown earlier in the chapter, that the use of symmetry coordinates which are expressed as linear combinations of the internal coordinates allow secular determinant to be factored into a number of independent determinants of lower order. We will show its use in the following paragraphs.

## 12.11 USE OF SYMMETRY COORDINATES IN CALCULATION

Relationship between the force constants and the vibrational frequencies of molecules is possible only for small molecules. For bigger molecules, one has to use the molecular symmetry for vibrational frequencies calculations. For example, the vibrational problem of $CH_4$ involves nine internal coordinates and therefore one has to solve 9 × 9 secular determinant. On the other hand, applying the symmetry of $CH_4$ ($T_d$), the problem will involve 3 × 3 secular determinant (maximum size). The basis for simplification using symmetry is as follows.

In vibrational spectroscopy, we are concerned mainly with the displacements of the atoms ($\Delta x$ or $\Delta r$), from their equilibrium positions. We assume these displacements as vectors as shown earlier in the chapter. We set up a relationship between these displacements with the symmetry operations of the molecule. We have pointed out while dealing with K.E. and potential energy of the vibrational motion (chapter 7), that these energies remain unaltered by the application of symmetry operations. It means that the potential and kinetic energy of the distorted molecule (distorted as a result of vibration), will have the same K.E. and potential energy after carrying out the symmetry operations on the distorted molecule. The symmetry operation changes only the directions of these displacements, but not its absolute magnitude. This has already been discussed in the first part of this chapter. We have shown earlier that the vibrational energy is most simply expressed in terms of the normal coordinates of the molecule as,

$$E=\left(\frac{1}{2}\dot{Q}_1^2+\frac{1}{2}\lambda_1 Q_1^2\right)+\left(\frac{1}{2}\dot{Q}_2^2+\frac{1}{2}\lambda_2 Q_2^2\right)+\ldots\left(\frac{1}{2}\dot{Q}_{3n-6}^2+\frac{1}{2}\lambda_{3n-6}\,Q_{3n-6}^2\right) \tag{12.102}$$

It is also known that $Q_i$ coordinates are expressed in terms of a set of $A_{ki}$ (amplitudes of $n$ atoms for a particular vibrational frequency $\nu_k$). This suggests that these coordinates are a function of $\Delta x^n$. In view of the Eq.(12.52), these normal coordinates are represented by displacement vectors (*i.e.*, $A_{ki}$). Thus, for the energy $E$ to remain constant in general case, where all $\lambda_i$'s ($\lambda_i \propto E$), if one performs a symmetry operation, each normal coordinate must remain unchanged (or utmost it may change sign). Thus, we can write that

$$\hat{R}Q_i \rightarrow \pm Q_i \qquad ...(12.103)$$

($\hat{R}$ = symmetry operation). Thus, the relationship between symmetry operation and the coordinate $Q_i$ is that the latter either remains the same or changes sign on performing a symmetry operation. Thus, the symmetry properties of a molecule can be used to classify the normal coordinates according to whether the coordinates transform as symmetric or antisymmetric with respect to various symmetry operations of the group. That is, we can associate I.Rs. of the group to normal coordinates.

For $n$-fold degenerate normal coordinate, we have $n$ numbers of coordinates which on the application of any symmetry operation are transformed among themselves in the form of LCAO.

Thus, total energy of a vibrating molecule is proportional to the square of either $\dot{Q}_i^2$ or $Q_i^2$ Eq. (12.102) will not have any cross terms *viz.* $\dot{Q}_i\dot{Q}_j$ or $Q_iQ_j$.We have also pointed out that these normal coordinates belong to various I.Rs of the group of the molecule. These coordinates consist of linear combinations of either cartesian or internal coordinates.

If we start our problem by expressing K.E. and potential energy in terms of cartesian or internal coordinates and make their linear combinations in such a way that each combination (*i.e.* symmetry coordinate) is either symmetric or antisymmetric with respect to the symmetry operations of the group of molecule (*i.e.* belong to one of the I.Rs. of the group), then the cross term in the expression of K.E. and potential energy will vanish. Such coordinates are known as symmetry coordinates and represented as $S_1$, $S_2$, *etc*. These are the SALCs of either cartesian or internal coordinates. These SALCs are formed using projection operators of the I.Rs. of the group. Employing these symmetry coordinates to set up K.E. and potential energy of the vibrating system one will have the secular determinant in the diagonal-blocked form, and the dimension of each block will be lower than the original determinant and will be equal to the number of the coordinates of that symmetry type. For example, if we have two coordinates of $A_2$ type of I.Rs., the block will be of (2 × 2) order. Each block can be put equal to zero which can be solved for the vibrational frequencies of that symmetry type to which the coordinate belongs.

## 12.12 USE OF SYMMETRY COORDINATES IN WILSON'S METHOD

The above discussion, indicates how the symmetry of the molecule helps in the calculation of vibrational frequencies using Lagrange's method,

As shown in section 12.18, (In Wilson's method), we express K.E. and P.E. expressions in terms of the internal coordinates $R_k$ (*e.g.*, $\Delta r_i$, $\Delta\alpha$ *etc.*) Although it allows P.E. to express quite easily, *i.e.*, (12.73 in sec. 12.8) but the K.E. expression create a little problem. Thus,

$$PE = 12V = \sum_{ij} f_{ij} R_i R_j \qquad ...(12.104)$$

But to set up K.E., we start with cartesian coordinates which are $3n$ in number. In practice we use mass-weighted cartesian coordinates ($q_i$) and start with the expression of K.E. as

$$KE = 2T = \sum_{i=1}^{3n} \dot{q}_i^2 \qquad ...(12.105)$$

We then transform the cartesian coordinates ($\Delta\dot{x}_i$) in terms of internal coordinates. That is,

$$\dot{R}_k = \sum_{i=1}^{3n} B_{ik}\dot{x}_i$$

In the matrix form we write, it as $R = |B||x|$ ...(12.106)

## 12.13 USE OF SYMMETRY COORDINATES IN THE CALCULATION OF $\lambda_K$ AND $A_{ik}$'S (QUALITATIVE AND QUANTITATIVE USE OF SYMMETRY OF THE MOLECULE)

Despite the various procedures (discussed here or elsewhere) for calculating $\lambda_K$ $A_{ik}$, one still faces the same trouble in solving the secular determinant (or eigenvalue (Eq. 100) of order higher than three). As discussed in earlier chapters, if one employs the symmetry coordinates to obtain the secular determinant in place of the usual cartesian or internal coordinates, one gets determinant in the block diagonal form with small ordered determinants on the diagonal of the secular determinant which can be easily solved for the values of $\lambda_{k.}$ These symmetry coordinates are the linear combinations of cartesian or internal coordinates of proper symmetry. (In case of bonding, the coordinates that we have used were wave functions). These symmetry coordinates are termed as SALCs (LCAO). Each coordinate is associated with one of the I.Rs. of the point group. Accordingly, these symmetry coordinates are constructed by employing the projection operators of I.Rs. and are either the same as the normal coordinates of the molecule or are closely linked with them. Thus, we have to use symmetry coordinates to solve any vibrational problem so that one may form one coordinate for one I.Rs. of the group.

Since these normal or symmetry coordinates are associated with the I.Rs of the group, we have first to find the point group and determine the I.Rs. of the group through the formation of $\Gamma_{red}$ by the chosen coordinates. If we use these coordinates in the vibrational analysis we get the secular determinant in the diagonal form. These can be easily solved. For doing so, we use the following steps:

Choose any coordinate system (either cartesian or internal).

1. Formation of $\Gamma_{red}$ using cartesian or internal coordinates of the atoms as basis.
2. Determination of I.Rs. that are contained in the $\Gamma_{red}$ formed in step one.
3. Construction of symmetry coordinates using projection operators of the I.Rs. of step three.
4. Setting up the secular equation using Lagrange's equation and determine *e.g.*,

   $| FG - \lambda E |$ secular determinant.
5. Solving the above secular determinant for $\lambda_K$, followed by the solution of normal coordinates.

In the first three steps, the symmetry of the molecule has been used qualitatively to analyse the vibrational problem, while in steps 4 and 5, the symmetry has been used quantitatively to determine the vibrational parameters of the molecule.

We give below a few examples where we have obtained the I.Rs. of the point group associated with the symmetry coordinates. (Up to step 3). We determine I.Rs. of the molecular vibrations by the methods which do not use explicitly any matrices (although they are used in deriving the results). It will also be shown how the symmetry of the normal modes are derived quantitatively and since the quantitative aspect of deriving the normal coordinates, $\lambda_K$, $A_{ik}'$ involves only matrix manipulations, we will deal with that aspect only superficially.

One is able to transform $3n$ cartesian coordinates into a set of $(3n–6)$ internal coordinates in terms of $G$ matrix which is a function of $\frac{B_{ki}B_{ki'}}{m_i}$. That is,

$$G_{kl} = \sum_{i=1}^{3n} \frac{B_{ki}B_{li}}{m_i} \qquad ...(12.107)$$

where $k$, $l$, varies from 1, 2, 3, ...$(3n–6)$

These $G$ matrices can be calculated from the tables given by Wilson in his book (Molecular Vibration by Wilson). In terms of these $G_i$'s, one can set up $G$-matrix and represent the kinetic and potential energies.

as and
$$\left.\begin{aligned} 2T &= \dot{R}'G^{-1}\dot{R} \\ 2V &= R'FR \end{aligned}\right\} \quad ...(12.108)$$

$\dot{R}'$ and $R'$ are the transposes of $\dot{R}$ and $R$ matrices. (how can one use quantitatively this information to derive the $\lambda_K$'s normal modes of vibrations and the selection rules).

Employing these expressions for $T$ and $V$ Eq. (12.108), in the Lagrange's equation to each of the internal coordinates $R_k$, one then uses vibrational type equation, for

$$R_K = A_K \cos(2\pi \nu t + \rho) \quad ...(12.109)$$

Each coordinate Eq. (12.109) seeks solutions. In this manner, the secular determinant (12.110) is formed by the coefficients of the $A_i'$s. The secular determinant is,

$$\begin{vmatrix} [F_{11} - (G^{-1})_{11}\lambda][F_{12} - (G^{-1})_{12}\lambda\ldots\ldots\ldots] \\ [F_{21} - (G^{-1})_{11}\lambda][F_{22} - (G^{-1})_{22}\lambda\ldots\ldots\ldots] \\ \vdots \\ [F_{(3n-6)1} - (G^{-1}_{(3n-6)1})\lambda]\ldots\ldots\ldots \end{vmatrix} = 0 \quad ...(12.110)$$

*i.e.*,
$$|F - (G^{-1})\lambda E| = 0 \quad ...(12.111)$$

Multiplying by $G$ on both sides of Eq. (12.111) we obtain,

$$|GF - \lambda E| = 0 \quad ...(12.112)$$

With the Eq. (12.112) one attempts to set up $F$ and $G$ matrices.

Although the calculation of $G$ matrices is an involved process; but there are a number of simplifying features of great significance. These are:

1. The $G$-matrix is a function of atomic masses and the geometry of the molecule.
2. The elements of $G$-matrix that are related to the bond-stretching are easy to calculate, but those related to angle deformation coordinates ($D\alpha$) are rather complicated.
3. G-matrix is always symmetric.
4. G-matrix is always positive and definite *i.e.*, the eigenvalues are all real and positive.
5. Each element of the $G$-matrix relates to two internal coordinates. It geometrically implies that the elements may be tabulated in terms of pair of coordinates with a given geometrical relationship.

6. Any pair of coordinates having no atom is common, will have zero $G$-matrix elements.

The tables for the $G$-matrix were first tabulated by Decius. These are given in Wilson, Decius and Cross's book on molecular vibration. Using these tables one can calculate $G$-matrix. Since $G$-matrix is symmetrical, one has to calculate nearly half of the elements.

Despite all the above simplifying rules, the calculations of elements of $G$-matrix for large molecules involve a lot of work. One can easily make the errors in the calculation and therefore, a computer program for this purpose should be used.

If one wishes to know the details, for obtaining the expressions for calculating $G$-matrix elements given in the book, by Wilson, he has to go through chapter four of the book.

By employing the calculated $F$, and $G$ matrices, one sets up the secular determinant (or the eigenvalue equation 12.112) and solves for ($3n$–6) values of $\lambda$. For each value of $\lambda$, one then determines a set of ($3n$–6) number of amplitudes for each value of $\lambda_\kappa$ ($3n$–6) normal modes of vibrations.

We give below a few examples to illustrate the above procedure.

## Example I ($H_2O$)

Point group $C_{2v}[E, C_2, \sigma_{xz}, \sigma_{yz}]$, molecular plane $yz$.

Analysis using cartesian coordinates in the following steps:

1. We have attached the coordinate system on all the three atoms (*i.e.*, on $H^1$, $H^2$, and O).
2. The three coordinate systems on the atoms are parallel to each other, thus $x^1$, $x^2 x^3$ axes are parallel, $y^1$ $y^2$ $y^3$ axes are parallel and $z^1$, $z^2$, $z^3$ axes are parallel to each other.
3. The three displacement vectors of atoms $H^1$, $H^2$, O during the vibrational motion at any instant can be resolved into ($\Delta x^1$, $\Delta y^1$, $\Delta z^1$) $\Delta x^2$, $\Delta y^2$, $\Delta z^2$) and ($\Delta x^3$, $\Delta y^3$, $\Delta z^3$). These will be cartesian coordinates.
4. In the same figure, we have also shown internal coordinates, (*viz.* $\Delta r_1$, $\Delta r_2$ and $\Delta\theta$).
5. Using any coordinate system, we can calculate the components of displacement vectors of the atoms in terms of the chosen system of coordinates.
6. Using these displacement coordinates, one can form the reducible representation of the group ($C_{2V}$ in our case) which is subsequently

reduced into its I.Rs. The symmetry coordinates corresponding to these obtained I.Rs. are then formed. These are the linear combination of cartesian or internal coordinates (SALCS). These are the normal coordinates belonging to a particular I.Rs. of the molecular point group (Fig. 12.8).

7. We use the displacement coordinates of atoms ($x_1$, $y_1$, $z_1$, $x_2$, $y_2$, $z_2$, $x_3$, $y_3$, $z_3$ instead of $\Delta x_i$, $\Delta y_i$, $\Delta z_i$, *etc.*) and form transformation matrices under the point group of $H_2O(C_{2v})$. These are given in Table (12.1).
8. We form $\Gamma_{red}$ which is given below.

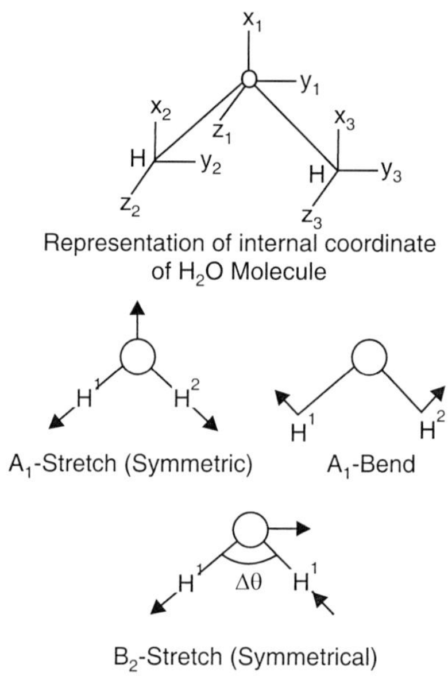


**Fig. 12.8** Coordinates and vibrational modes $H_2O$.

$$\Gamma_{red}(H_2O) = \begin{bmatrix} E & C_2 & \sigma_{xz} & \sigma_{yz} \\ 9 & -1 & +1 & 3 \end{bmatrix}$$

9. $\Gamma_{red}$ is reduced to $3A_1 + A_2 + 2B_1 + 3B_2$ IR components. ...(12.113)

Out of nine cartesian coordinates ($x_1y_1z_1$, $x_2y_2z_2$, $x_3y_3z_3$), three coordinates are involved in translational motion and three in rotational motion and the remaining three are in vibrational motion. From the character table of $C_{2v}$, we observe that the three translational coordinates belong to $A_1 + B_1 + B_2$. I.Rs. and three rotational belong to $A_1$+ $B_1$+ $B_2$. We subtract the sum of these two (*viz.* translational + rotational) from Eq. (12.113) to get vibrational I.Rs. we obtain,

$$\Gamma_{vib} = 2A_1 + B_2 \qquad ...(12.114)$$

**Table 12.1:** Transformation matrices and $\chi$'s for the stretching of bonds in $H_2O$.

| $E$ | $x_3y_3z_3$ | $x_2y_2z_2$ | $x_1y_1z_1$ | $\chi E$ | $(C_2^z)$ | $x_3y_3z_3$ | $x_2y_2z_2$ | $x_1y_1z_1$ | $\chi(C_2^z)$ | $\sigma_{yz}$ | $x_3y_3z_3$ | $x_2y_2z_2$ | $x_1y_1z_1$ | $\chi\sigma_{yz}$ | $\chi_{yz}$ |
|---|---|---|---|---|---|---|---|---|---|---|---|---|---|---|---|
| $x_3$ | 1 0 0 | 0 0 0 | 0 0 0 | | $x_3$ | -1 0 0 | 0 0 0 | 0 0 0 | | $x_3$ | -1 0 0 | | | | |
| $y_3$ | 0 1 0 | 0 0 0 | 0 0 0 | | $y_3$ | 0 -1 0 | 0 0 0 | 0 0 0 | | $y_3$ | 0 1 0 | | | | |
| $z_3$ | 0 0 1 | 0 0 0 | 0 0 0 | | $z_3$ | 0 0 1 | 0 0 0 | 0 0 0 | | $z_3$ | 0 0 1 | 0 | 0 | | |
| $x_2$ | 0 0 0 | 1 0 0 | 0 0 0 | | $x_2$ | 0 0 0 | 0 0 0 | -1 0 0 | | $x_2$ | 0 0 0 | -1 0 0 | | | |
| $y_2$ | 0 0 0 | 0 1 0 | 0 0 0 | 9 | $y_2$ | 0 0 0 | 0 0 0 | 0 -1 0 | -1 | $y_2$ | 0 0 0 | 0 -1 0 | | 3 | -1 |
| $z_2$ | 0 0 0 | 0 0 1 | 0 0 0 | | $z_2$ | 0 0 0 | 0 0 0 | 0 0 1 | | $z_2$ | 0 0 0 | 0 0 1 | | | |
| $x_1$ | 0 0 0 | 0 0 0 | 1 0 0 | | $x_1$ | 0 0 0 | -1 0 0 | 0 0 0 | | $x_1$ | | | -1 0 0 | | |
| $y_1$ | 0 0 0 | 0 0 0 | 0 1 0 | | $y_1$ | 0 0 0 | 0 -1 0 | 0 0 0 | | $y_1$ | | | 0 -1 0 | | |
| $z_1$ | 0 0 0 | 0 0 0 | 0 0 1 | | $z_1$ | 0 0 0 | 0 0 1 | 0 0 0 | | $z_1$ | | | 0 0 1 | | |

In this manner, we obtain two combination of ($x_i$, $y_i$, $z_i$) coordinates of $A$ representation, one combination of $B_2$ I.R. These combinations of the cartesian coordinate when used to solve the vibrational problem, one will obtain (from the secular determinant), one equation of quadratic type and another equation which will be simple equation of order one. The quadratic equation ($2A_1$) will give two normal modes of vibration corresponding to frequencies $\nu_1$ and $\nu_2$ and belonging to the two normal modes of $A_1$ type, while the third normal mode will be of one dimension belonging to $B_2$. It will be obtained from second equation.

## Example II ($BF_3$)

Point group $D_{3h}$ ($E$, $2C_3$, $3C_2$, $\sigma_h$, $2S_3$, $3\sigma_v$) Molecular plane, $XY$

1. **Analysis using Cartesian Coordinates:** In this case, we have twelve cartesian coordinates ($x^1y^1z^1, x^2y^2z^2, x^3y^3z^3, x^4y^4z^4$). These form the basis for a reducible representation $\Gamma_{\text{red}}$ of $D_{3h}$. The characters of this representation $\Gamma_{\text{red}}$ are the characters of the representations of the symmetry operations of the molecule on the basis of the cartesian displacement coordinates. Since the characters of all the elements of a class is the same, we only have to find the character of one element of a class. $XY$ is the plane of the molecule and $Z$-axis is parallel to $C_3$ axis and $y$-axis is taken as the bond $BF^{(1)}$. All atoms of $BF_3$ are numbered as shown in Fig. 12.8. Let us consider the symmetry operation in $\sigma_{yz}$ plane passing through atoms (1)-(4). It will have the effect on bonds $BF^2$ and $BF^3$. That is $BF^2$ will be shifted to $BF^3$ and *vice versa*. $BF^1$ will remain unshifted. Now we have a rule for calculating the $\chi_{\text{op}}$. For a given symmetry operation $R$, the contributions to $\chi_{\text{op}}$ comes only from the nuclei which remain unshifted by the operation. It contributes (+1) to $\chi_{\text{op}}$ In case the coordinates changes the sign, then the contribution to $\chi_{\text{op}}$ (–1). If it get shifted, it contributes zero. Thus, in the case of $\sigma_{yz}$ the only contribution to $\chi_{\sigma_{yz}}$ comes from the nuclei 1 and 4. It reverses the direction of the $x$-axis but leaves $x_1z_1$, $x_4z_4$ unchanged (Table 10.2) In the table, one observes that out of $(4)^2$ (= 16) sixteen blocks, only four blocks are nonzero, and out of these four blocks in accordance to the above rules only two contribute to the character of $\sigma_{yz}$. If we carry out the similar transformations for other operations, one can generalize the result and arrive at the following conclusion.

**Table 12.2**

| | $x_1\ y_1\ z_1$ | $x_2\ y_2\ z_2$ | $x_3\ y_3\ z_3$ | $x_4\ y_4\ z_4$ |
|---|---|---|---|---|
| $x_1$ $y_1$ $z_1$ | $\begin{matrix}-1 & 0 & 0\\ 0 & -1 & 0\\ 0 & 0 & 1\end{matrix}$ | 0 | 0 | 0 |
| $x_2$ $y_2$ $z_2$ | 0 | 0 | $\begin{matrix}-1 & 0 & 0\\ 0 & -1 & 0\\ 0 & 0 & 1\end{matrix}$ | 0 |
| $x_3$ $y_3$ $z_3$ | 0 | $\begin{matrix}-1 & 0 & 0\\ 0 & -1 & 0\\ 0 & 0 & 1\end{matrix}$ | 0 | 0 |
| $x_4$ $y_4$ $z_4$ | 0 | 0 | 0 | $\begin{matrix}-1 & 0 & 0\\ 0 & 1 & 0\\ 0 & 0 & 1\end{matrix}$ |

$$\chi(\sigma_{yz}) = 2$$

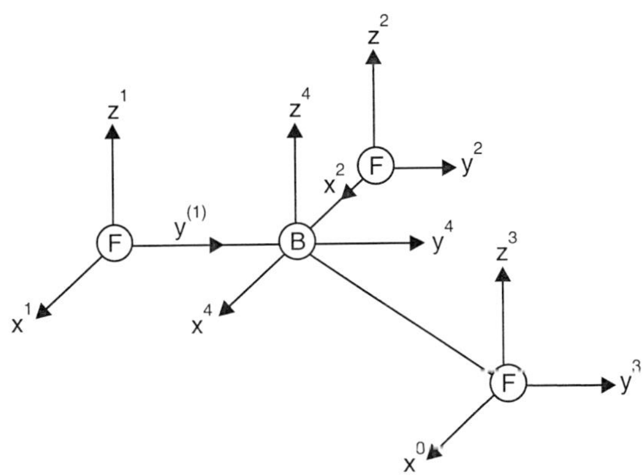


**Fig. 12.9** Cartesian displacement coordinates of $BF_3$.

"Out of the $n^2$ blocks, only $n$ will be non-zero blocks in the representation of the group". It can further be concluded that "the character of the representation of a symmetry operation with respect to any basis consisting of the cartesian coordinate is given by the characters of the representation of the operation acting on a basis [$xyz$] multiplied by the number of atoms unshifted by the operation" Mathematically it may be stated as,

$$\chi(R) = n_u \chi_{(xyz)}(R) \qquad ...(12.115)$$

$n_u$ is the number unshifted atom, and $\chi(xyz)$ is the character of the basis $x$, $y$, $z$ coordinate.

Using this formula $\Gamma_{red}$ for $BF_3$ molecule with respect to cartesian coordinates is as follows: $D_{3h}$ ($E$, $2C_3$, $3C_2$, $\sigma_h$, $2S_3$, $3\sigma_v$)

$$\Gamma_{\text{red}}\left\{\begin{array}{lccccccll} & E & 2C_3 & 3C_2 & \sigma_h & 2S_3 & 3\sigma_v & \cdots & \text{operations} \\ = & 4\times 3 & 1\times 0 & 2\times(-1) & 4\times 1 & 1\times(-2) & 2\times 1 & \cdots & (n\times\chi_{xyz}) \\ = & 12 & 0 & -2 & 4 & -2 & 2 & \cdots & \chi_{xyz} \\ = & 12 & 0 & -6 & 4 & -4 & 6 & \cdots & C_k\chi_{xyz} \end{array}\right.$$

...(12.116)

Here $C_k$ = Number of operations in the class $k$ of the point group ($D_{3h}$). The $\Gamma_{\text{red}}$ on reduction contains the following I.Rs. of the group. That is,

$$\Gamma_{\text{red}} = A_1' + A_2' + 3E' + A_2'' + E'' \quad \text{...(12.117)}$$

$\Gamma_{\text{red}}$ contains $\Gamma_{\text{tran}}$ and $\Gamma_{\text{rot}}$ besides $\Gamma_{\text{vib}}$ which have to be removed from $\Gamma_{\text{red}}$ to get the $\Gamma_{\text{vib}}$. These translation and rotation representations are given in the character table. Thus, $\Gamma_{\text{tran}}$ and $\Gamma_{\text{rot}}$ transform as $(E' + A_2'')$ and $(A_2'' + E'')$, respectively. These have to be subtracted from Eq. (107) to get $\Gamma_{\text{vib}}$. In this manner, we finally have,

$$\Gamma_{\text{vib}} = A_1' + 2E' + A'' \quad \text{...(12.118)}$$

In summary, we have the following method of finding $\Gamma_{\text{vib}}$.

1. Assign a point group to the molecule along with its defining operation.
2. Note the number of atoms unshifted by the operation of each class.
3. Multiply the number of unshifted atoms by $\chi_{xyz}$ given in the Table 12.2
4. Reduce $\Gamma_{\text{red}}$ into its I.Rs.
5. Subtract I.Rs corresponding $\Gamma_{\text{rot}}$ and $\Gamma_{\text{trans}}$ to get $\Gamma_{\text{vib}}$.
6. Thus we are concerned with $\Gamma_{\text{red}}$ only.

The above-described procedure can be written in the tabular form for $BF_3$.

| $D_{3h}$ | $E$ | $C_3$ | $C_2$ | $\sigma_h$ | $S_3$ | $\sigma_v$ | |
|---|---|---|---|---|---|---|---|
| $\Gamma_{xyz}$ | 3 | 0 | −1 | 1 | −2 | 1 | $\Gamma_A$ from character table |
| Unshifted atoms | 4 | 1 | 2 | 4 | 1 | 2 | Step2×step3 |
| $\Gamma_{xyz}\times$ no.of unshifted atoms | 12 | 0 | −2 | 4 | −2 | 2 | |
| $\Gamma_{\text{total}} = \Gamma_{\text{vib}} + \Gamma_{\text{trans}} + \Gamma_{\text{rot}}$ | 12 | 0 | −2 | 4 | −2 | 2 | |
| $\Gamma_{\text{trans}} = \Gamma_{xyz}$ instep1 | 3 | 0 | −1 | −1 | 2 | −1 | |
| $\Gamma_{\text{rot}}$ | 3 | 0 | −1 | −1 | 2 | – | Nos. in step one |
| | | | | | | | $A_1' + A_2'' + E'$ |
| $\Gamma_{\text{total}} = \Gamma_{\text{trans}} - \Gamma_{\text{rot}}$ | 6 | 0 | 0 | 4 | −2 | 2 | |
| Unmoved atoms | 4 | 1 | 2 | 4 | 1 | 2 | |
| Subtract | 1 | 1 | 1 | 1 | 1 | 1 | |
| $\Gamma_{\text{stretch}}$ | 3 | 0 | 1 | 3 | 0 | 1 | |
| $\Gamma_{\text{bend}}$ | 3 | 0 | −1 | 1 | −2 | 1 | |

$$\Gamma_{\text{stretch}} = A_1' + E' \ \Gamma_{\text{bend}} = A_2'' + E'$$

IR active $A_2''$ and $E'$ Raman active $A_1'$ and $E'$.

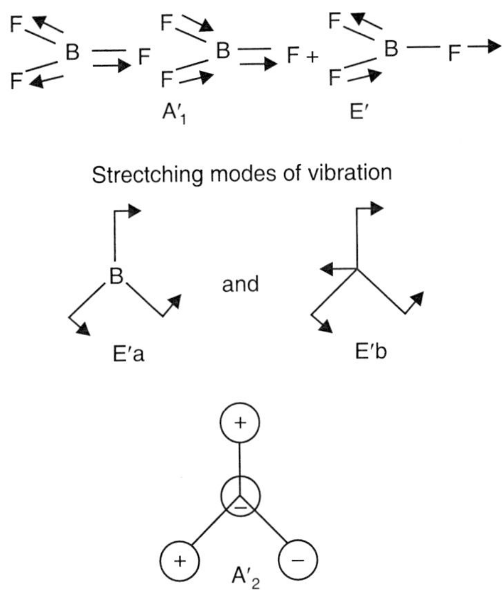


**Fig. 12.10** Various modes of $BF_3$.

## Example III ($XeF_4$)

| $D_{4h}$ | $E$ | $C_4$ | $C_2$ | $C'_2$ | $C''_2$ | $L$ | $S_4$ | $\sigma_h$ | $\sigma_v$ | $\sigma_d$ | |
|---|---|---|---|---|---|---|---|---|---|---|---|
| Unmoved atoms | 5 | 1 | 1 | 3 | 1 | 1 | 1 | 5 | 3 | 1 | |
| $\Gamma_{x, y, z}$ | 3 | 1 | −1 | −1 | −1 | −1 | −1 | 1 | 1 | 1 | |
| $\Gamma_{\text{total}}$ | 15 | 1 | −1 | −3 | −1 | −3 | −1 | 5 | 3 | 1 | (product) |
| $\Gamma_{\text{trans}}$ | 3 | 1 | −1 | −1 | −1 | −3 | −1 | 1 | 1 | 1 | (subtract) |
| $\Gamma_{\text{rol}}$ | 3 | 1 | −1 | −1 | −1 | 3 | 1 | −1 | −1 | −1 | (subtract) |
| $\Gamma_{\text{vib}}$ | 9 | −1 | 1 | −1 | 1 | −3 | −1 | 5 | 3 | 1 | |
| Unmoved atoms | 5 | 1 | 1 | 3 | 1 | 1 | 1 | 5 | 3 | 1 | |
| Subtract 1 | 1 | 1 | 1 | 1 | 1 | 1 | 1 | 1 | 1 | 1 | (subtract) |
| $\Gamma_{\text{stretch}}$ | 4 | 0 | 0 | 2 | 0 | 0 | 0 | 4 | 2 | 0 | $A_{1g} + B_{1g} + E_u$ |
| $\Gamma_{\text{bend}}$ | 5 | −1 | 1 | −3 | 1 | −3 | −1 | 1 | 1 | 1 | $A_{2u} + B_{2u} + B_{2g} + E_u$ |

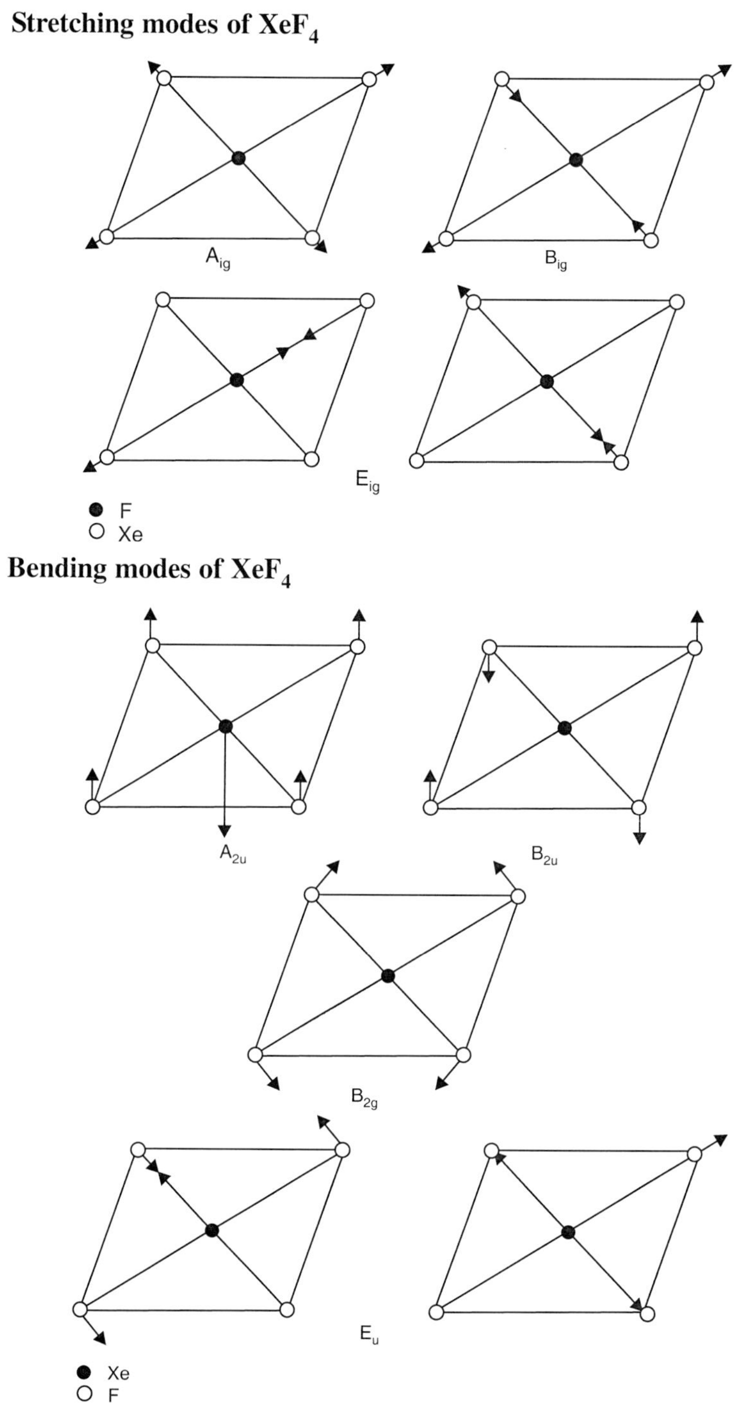


**Fig. 12.11** Vibrational modes of $XeF_4$.

## Example IV ($CO_3^{2-}$)

Vibrational modes of $CO_3^{2-}$ (square planar, $D_{3h}$)

| $D_{3h}$ | $E$ | $C_3$ | $C_2$ | $\sigma_h$ | $S_3$ | $\sigma_v$ | |
|---|---|---|---|---|---|---|---|
| $\Gamma_{x,y,z}$ | 3 | 0 | –1 | 1 | –2 | 1 | |
| Unmoved atoms | 4 | 1 | 2 | 4 | 1 | 2 | |
| $\Gamma_{total}$ | 12 | 0 | –2 | 4 | –2 | 2 | (product) |
| $\Gamma_{trans}$ | 3 | 0 | –1 | 1 | –2 | 1 | (subtract) |
| $\Gamma_{rot}$ | 3 | 0 | –1 | –1 | 2 | –1 | (subtract) |
| $\Gamma_{vib}$ | 6 | 0 | 0 | 4 | –2 | 2 | |
| Unmoved atoms | 4 | 1 | 2 | 4 | 1 | 2 | |
| | 1 | 1 | 1 | 1 | 1 | 1 | (subtract) |
| $\Gamma_{stretch}$ | 3 | 0 | 1 | 3 | 0 | 1 | (subtract from $\Gamma_{vib}$) |
| $\Gamma_{bend}$ | 3 | 0 | –1 | 1 | –2 | 1 | |

$\Gamma_{stretch} = A_1' + E'$ $\Gamma_{bend} = A_2'' + E'$

IR active $A_2''$ and $E'$ Raman active $A_1'$ and $E'$.

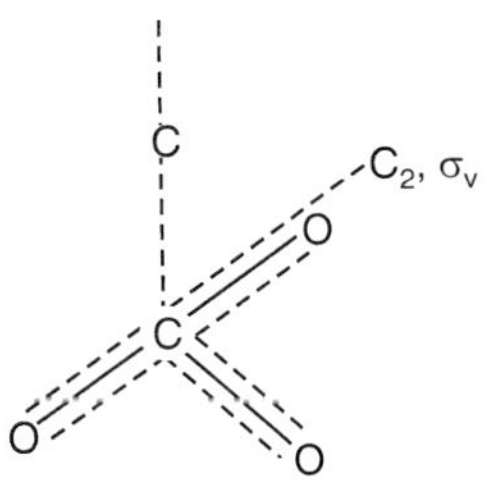


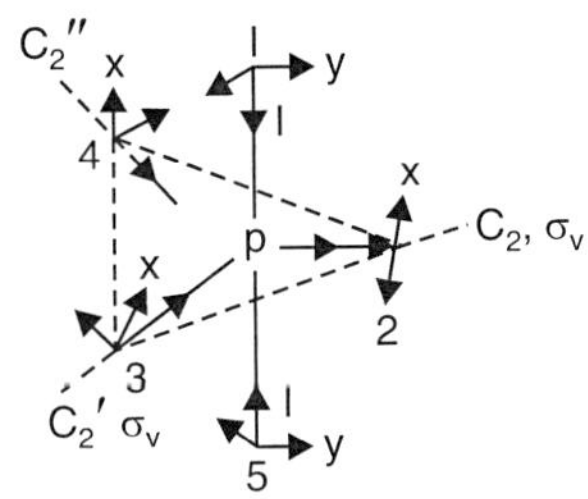


**Fig. 12.12** Coordinates of P and Cl.

## Example V ($PCl_5$) Symmetry Species for Vibrations of $PCl_5$

| $D_{3h}$ | $E$ | $2C_3$ | $3C_2$ | $\sigma_h$ | $S_3$ | $3\sigma_v$ | |
|---|---|---|---|---|---|---|---|
| $\Gamma_{x,y,z}$ | 3 | 0 | –1 | 1 | –2 | 1 | |
| Unmoved atoms | 6 | 3 | 2 | 4 | 1 | 4 | |
| $\Gamma_{total}$ | 18 | 0 | –2 | 4 | –2 | 4 | (product) |
| $\Gamma_{trans}$ | 3 | 0 | –1 | 1 | –2 | 1 | (subtract) |
| $\Gamma_{rol}$ | 3 | 0 | –1 | –1 | 2 | –1 | (subtract) |
| $\Gamma_{vib}$ | 12 | 0 | 0 | 4 | –2 | 4 | |

*(Contd.)*

| | | | | | | | |
|---|---|---|---|---|---|---|---|
| Unmoved atoms | 6 | 3 | 2 | 4 | 1 | 4 | |
| | 1 | 1 | 1 | 1 | 1 | 1 | (subtract) |
| $\Gamma_{stretch}$ | 5 | 2 | 1 | 3 | 0 | 3 | (subtract from $\Gamma_{vib}$) |
| $\Gamma_{bend}$ | 7 | –2 | –1 | 1 | –2 | 1 | |

$\Gamma_{stretch} = 2A_1' + A_2'' + E' \quad \Gamma_{bend} = A_2'' + 2E' + E''$

IR active $A_2''$ and $E'$ Raman active $A_1'$ and $E''$

Stretching modes of $PCl_5$

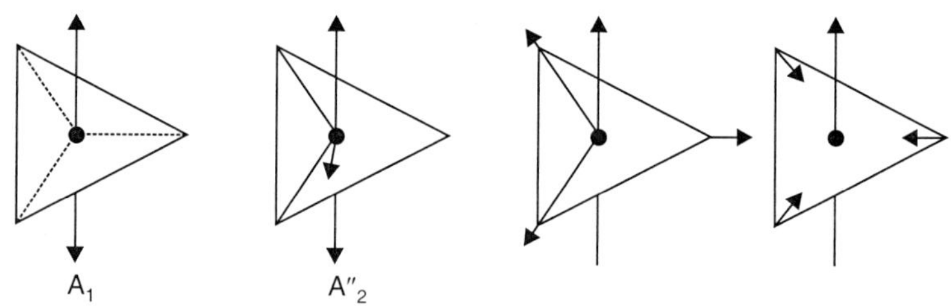


Bending modes of $PCl_5$ (In Plane)

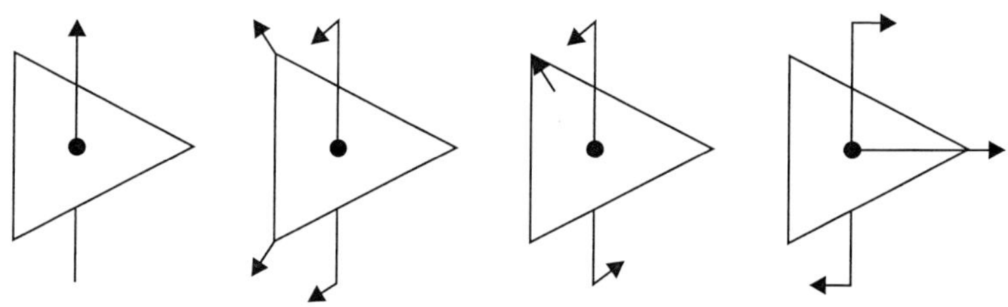

(Out of Plane)

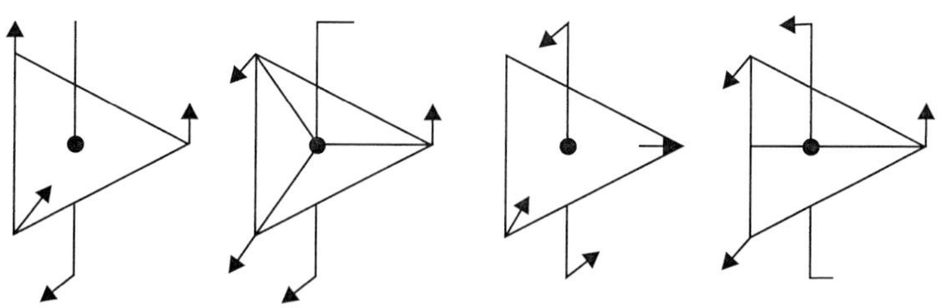

**Fig. 12.13** Vibrational modes of P and Cl in $PCl_5$.

### Example VI ($B_2H_6$)

| $D_{2h}$ | $E$ | $(C_2)_z$ | $(C_2)_y$ | $(C_2)_x$ | $i$ | $\sigma_{xy}$ | $\sigma_{xz}$ | $\sigma_{yz}$ | |
|---|---|---|---|---|---|---|---|---|---|
| $\Gamma_{x, y, z}$ | 3 | –1 | –1 | –1 | –3 | 1 | 1 | 1 | |
| Unmoved atoms | 8 | 2 | 2 | 0 | 0 | 2 | 6 | 4 | |
| $\Gamma_{total}$ | 24 | –2 | –2 | 0 | 0 | 2 | 6 | 4 | (product) |
| $\Gamma_{trans}$ | 3 | –1 | –1 | –1 | –3 | 1 | 1 | 1 | (subtract) |
| $\Gamma_{rot}$ | 3 | –1 | –1 | –1 | 3 | –1 | –1 | –1 | (subtract) |
| $\Gamma_{vib}$ | 18 | 0 | 0 | 2 | 0 | 2 | 6 | 4 | |
| $\Gamma_{vib} = 4A_g + B_{1g} + 2B_{3g} + A_u + 2B_{1u} + 2B_{2u} + 3B_{3u}$ | | | | | | | | | |

### Example VII ($C_2H_4$)

| $D_{2h}$ | $E$ | $(C_2)_z$ | $(C_2)_y$ | $(C_2)_g$ | $i$ | $\sigma_{xy}$ | $\sigma_{xz}$ | $\sigma_{yz}$ | |
|---|---|---|---|---|---|---|---|---|---|
| $\Gamma_{x, y, z}$ | 3 | –1 | –1 | –1 | –3 | 1 | 1 | 1 | |
| Unmoved atoms | 6 | 2 | 0 | 0 | 0 | 0 | 6 | 2 | |
| $\Gamma_{total}$ | 18 | 2 | 0 | 0 | 0 | 0 | 6 | 2 | (product) |
| $\Gamma_{trans}$ | 3 | –1 | –1 | –1 | –3 | 1 | 1 | 1 | (subtract) |
| $\Gamma_{rol}$ | 3 | –1 | –1 | –1 | 3 | –1 | –1 | –1 | (subtract) |
| $\Gamma_{vib}$ | 12 | 0 | 2 | 2 | 0 | 0 | 6 | 2 | |
| $\Gamma_{vib} = 3A_g + 2B_{2g} + 2B_{3g} + A_u + 2B_{1u} + 2B_{2u} + 3B_{3u}$ | | | | | | | | | |

## 12.14 REPRESENTATIONS SPANNED BY INTERNAL COORDINATES

Vectors of internal coordinates, defined earlier may also be used as a basis for the formation of $\Gamma_{red}$.

Suppose we have $r_1, r_2 ... r_{3n-6}$ internal coordinates. When we apply a symmetry operation on an internal coordinate, it has three effects on the coordinates. These are (1) it may leave it unshifted (2) it may leave it unshifted but change its sign (3) it may shift the coordinate *i.e.*, exchange it with other coordinate. We have shown earlier that only when the coordinate remains unshifted it gives a non-zero diagonal element in the representation and thereby it contributes to the character. Thus, if the coordinate remains unshifted while all other coordinates shift, then the character to the representation for that operation, will be one. The general rule is,

"The characters of representation of an operation acting on basis vectors of internal coordinates, is equal to the algebraic sum of the number of the coordinates unshifted during the operation (or multiplied by –1 if the operation changes sign)".

The above rule is further simplified if we divide the coordinates into equivalent sets. Equivalent sets are defined as follows:

"All those coordinates belong to an equivalent set if these are either permuted among themselves or left unchanged by all the symmetry operations of the group". For example, all the three B–F bonds belong to a symmetrically equivalent set. Similarly all the three $\delta(F_1BF_2)$ bending coordinates, belong to another equivalent set. Thus we have three equivalent sets. The important point about these equivalent sets are that by no operation of the group, the coordinate belonging to one equivalent sets will not exchange with any coordinate of the other set. Therefore, if we partition all the coordinates of the molecule into equivalent sets, then the $\Gamma_{red}$ also gets partitioned into the same sets

$$\Gamma_{red} = \Gamma^1_{red} + \Gamma^2_{red} + \ldots$$

where $\Gamma^1_{red}$ and $\Gamma^2_{red}$ are $\Gamma_{red}$ per equivalent sets.

Therefore, characters of any operation of the group $\Gamma_{red}$ will be the sum of the character of nonzero blocks, each of which corresponds to one symmetrically equivalent set. Thus in $BF_3$, there are three sets $(r_1r_2r_3)$, ( $\alpha_{12}a_{23}\alpha_{31}$) and for $C_3^1$operation, we will be having as shown below. It can be proved that the representation of all the operations will behave the same way. Thus, it suggests that the separation into equivalent sets of all the coordinates, is effectively a kind of reduction of a reducible representation into further smaller blocks(each block belongs to an equivalent set).Thus, if there are '$n$' number of equivalent sets then $\Gamma_{red}$ will be divided into '$n$' number of equivalent reducible representations. Thus one has to consider only one out of $n$-number of sets independently rather than the whole $\Gamma_{red}$.We may now derive the $\Gamma_{red}$ of $BF_3$ with respect to internal coordinates.

**Table 12.3** Permutation of $BF_3$ coordinates by $C_3^1$ operation

| | $r_1$ | $r_2$ | $r_3$ | $\alpha_{12}$ | $\alpha_{23}$ | $\alpha_{31}$ | $\pi$ | *eq. set* |
|---|---|---|---|---|---|---|---|---|
| $r_1$ | – | $r_2$ | – | – | – | – | – | |
| $r_2$ | – | – | $r_3$ | – | – | – | – | I |
| $r_3$ | $r_1$ | – | – | – | – | – | – | |
| $\alpha_{12}$ | – | – | – | – | $\alpha_{23}$ | – | – | |
| $\alpha_{23}$ | – | – | – | – | – | $\alpha_{31}$ | – | II |
| $\alpha_{31}$ | – | – | – | $\alpha_{12}$ | – | – | – | |
| $\pi$ | – | – | – | – | – | – | $\pi$ | III |

From the above rules, one may derive the $\Gamma_{red}$ spanned by the internal coordinates of $BF_3$.

| $D_{3h}$ | | | $E$ | $C_3^1$ | $C_{2z}$ | $\sigma_h$ | $S_3^1$ | $\sigma_{xyz}$ | |
|---|---|---|---|---|---|---|---|---|---|
| $r_1$ | $r_2$ | $r_3$ | 3 | 0 | 1 | 3 | 0 | 1 | $A_1' + E'$ |
| $\alpha_{12}$ | $\alpha_{23}$ | $\alpha_{31}$ | 3 | 0 | 1 | 3 | 0 | 1 | $A_1' + E'$ |
| | $\pi$ | | 1 | 1 | –1 | –1 | –1 | 1 | $A_2''$ |

Thus we have two equivalent sets $r_1r_2r_3$ and $\alpha_{12}\alpha_{23}\alpha_{31}$. Each set spans the same I.R *viz.* $A_1' + E'$ and the third set containing $\pi$-coordinate is not equivalent. It spans as $A_2''$.

One must notice here that $BF_3$ molecule(4atoms)should have only six coordinates $[(3n\text{-}6) = 6]$, but we have taken seven coordinates .So out of these seven coordinates there is one coordinate which is redundant. The method of finding the redundant coordinates out the assumed number of coordinates can be found. For example, in $BF_3$ it is the $\alpha_{12} + \alpha_{23} + \alpha_{31}$ symmetry coordinates which have to be redundant, because all the three angles around boron atom can not distort at the same time and the molecules also remains in the same plane. This symmetry coordinate $(\alpha_{12} + \alpha_{23} + \alpha_{31})$ spans $A_1$ type of I.R. and so one takes this coordinates as redundant.

We will not consider the redundancy and its removal in this book. We advise the readers to refer Wilson's book "Molecular Vibrations" for this topic.

After citing two examples ($H_2O$ and $BF_3$) for the formation of $\Gamma_{red}$ using cartesian and internal coordinates, it is possible to generalise the procedure of forming $\Gamma_{red}$ for any molecule. We summarise the process stepwise in the following manner.

1. Find the number of atoms in the molecule. This will give the number of vibrational modes ($3n$-6).
2. Find the point group of the molecule along with its symmetry operations.
3. Apply symmetry operations on the molecule by assigning coordinate system to all atoms.
4. Find the number of atoms which remains unshifted as a result of application of the symmetry operation with respect to coordinate system assigned to the centre of gravity of the molecule *e.g.*, on $A$ of $AX_n$ type molecules.
5. Write down the sum of the characters of the I.Rs of corresponding $\Gamma_{xyz}$ which are the same as those of $\Gamma_{Trans}$ (*Trans*-Translation). The values

of these characters are obtained from the character table of the point group. For Example for $H_2O$ case, $\Gamma_{xyz}$ equals to the sum of characters of $B_1$(for $x$-coord) + $B_2$(for $y$-coord) and $A_1$(for $z$-coordinate in $C_{2v}$ character table for each operation of $C_{2v}$ point groups).

6. Write the product of the number of atoms unshifted and the $\Gamma_{xyz}$ for all the operations of the point group separately. This set of product, for all the symmetry operations forms $\Gamma_{red}$ of the point group of the molecule.
7. Subtract the character of I.Rs for $\Gamma_{trans}$ and $\Gamma_{rot}$ separately from the characters of $\Gamma_{red}$ of point (6) for each of the symmetry operation of the point group. The characters of the operation of the point group are obtained from its character table. This will give reducible representation $\Gamma_{vib}$ of the point group.
8. Reduce $\Gamma_{red}$ into its I.Rs by the usual formula.
9. If you wish to know the reducible representation for the stretching modes only, $\Gamma_{vib}$ for the $AX_n$ type molecules, then subtract (1 no.) from each number obtained corresponding to those of the unshifted atoms in each symmetry operation (point no.4). The numbers thus obtained after subtracting one, will form the reducible representation of the stretching modes only ($\Gamma_{red}$(stretch) of the point group of the molecule).
10. Subtract the numbers obtained in step (9) from the corresponding numbers obtained in steps 6 for each operation. This will give the reducible representation for $\Gamma_{vib}$ (bending) for the bending modes of vibrations.
11. Reduce $\Gamma_{red}$ (stretch) and $\Gamma_{vib}$ (bend) separately into their respective I.Rs. of the point group.

## 12.15 FRAMEWORK AND CAGE MOLECULES

In all the above examples, the molecules are of $AX_n$ type in which an atom $A$ is present at the centre of mass of the molecule. In all such molecules, we subtract one from the number of the unshifted atoms in each symmetry operation. It is done to find the number in $\Gamma_{vib}$ (stretch) reducible representation. But, if there is no atom present at the centre of mass, then we do not subtract one from the number of unshifted atoms because there is no clear symmetry distinction between stretching and bending modes of vibration. We will take the case of $B_2H_6$ molecule for such a system. In this example, we have taken $XZ$ plane as molecular plane ($D_{2h}$ point group). We have not shown $\Gamma_{vib}$ (stretch) and $\Gamma_{vib}$(bend) separately but $\Gamma_{vib}$ (total) has been reduced into its I.Rs of the point group. We have obtained

$$\Gamma_{vib} = 4A_{1g} + B_{1g} + 2B_{2g} + 2B_{3g} + A_u + 3B_{1u} + 2B_{1u} + 3B_{3u} \quad ...(12.119)$$

which are 18 in number [*i.e.*, $18 = (3n\text{-}6) = (3 \times 8) - (6) = 18$].

We have also taken the example of ethylene [6 atom, $(3n-6)=12$ vibration] molecule with $XZ$ as molecular plane. In this example, the reducible representation $\Gamma_{vib}$ is reduced into $3A_g + 2B_{2g} + 2B_{3g} + A_u + 2B_{1u} + 2B_{2u} + 3B_{3u}$

## 12.16 SELECTION RULES

Although there are a large number of properties of the molecules where their symmetry properties are very useful qualitatively for their explanations *e.g.*, John Teller effect, stereochemistry. Selection rules for emission and absorption of photons by the molecules, one can get this information from different references. We will discuss only the absorption of light by the molecule.

It is well known that light can be absorbed or emitted by a molecule only if "the changes in vibrational state are accompanied by change in dipole moment of the molecule". It suggests that the basic nature of any nonzero interaction between a photon and a vibrating molecule demands that the work needed for the change in dipole moment is carried out by the light quantum which is absorbed or emitted by the molecule. The quantum mechanical rule is that if a molecule changes from state $A$ to state $B$, light photon can only be absorbed or emitted if the integral is not equal to zero.

$$\int \psi_A^* m \psi_B \partial\tau = M \qquad ...(12.120)$$

In this integral, the quantity '$m$' is identified as the change in dipole moment during the transition from state $A$ to state $B$, and the integral is taken over all the variables (coordinates) which are termed as $\partial\tau$ collectively.

We have shown in chapter six on direct products, that if the direct product representation of $\psi_A\, m\, \psi_B$ contains a totally symmetric representation (*viz.* $A_1$-type of the point group of the molecule) only then this integral (12.120) has a nonzero value.

In the vibrational analysis, $\psi_A$ and $\psi_B$ are functions of a number of variable internal coordinates of the molecule. The value of the integral (12.120) will be nonzero, only when the function $\psi_A\, m\, \psi_B$ is symmetric with respect to inversion of all variables (internal coordinates). That is,

$$S = \psi_A^* m \psi_B (r_1, r_2 \ldots r_{3n-6}) = \psi_A^* m \psi_B [(-r_1)(-r_2)\cdots(-r_{3n-6})] \qquad ...(12.121)$$

This condition must be true for all values of $k$. It implies that the product, $S$ must be a basis for the I.R. whose characters are +1, (*viz.* $A$-type).

For any molecule, a single normal coordinate is associated with one $\nu_k$ and since normal coordinates do not mix among each other, the energy of the molecule due to this coordinate is given as

$$E_k = (n + \frac{1}{2}) h\nu_k \qquad ...(12.122)$$

Similarly, the energy of the molecule due to other (3*n*-6) coordinates in the harmonic oscillator approximation, will be

$$E_i = (n + \frac{1}{2})h\nu_i \quad ...(12.123)$$

where *i* varies as 1, 2, 3, ... 3*n*–6. There will thus be (3*n*-6) energy values. Now, if the molecule goes from state *A* to state *B* when *n* goes from *n* to $(n + \Delta n)$ in any one normal coordinate then the changes in energy ($\Delta E_K$) of the molecule is the change of state from *A* to *B*,

$$\Delta E = \Delta n h\nu_k \quad ...(12.124)$$

In the harmonic approximation, the first selection rule for nonzero value of integral $\int \psi_A \, m \, \psi_B \, \partial T$ requires that the change of quantum number of $\psi_A$ and $\psi_B$ should be ±1 or,

$$\Delta n = \pm 1 \quad ...(12.125)$$

Quantum mechanically Eq. (12.127) may be proved but we will not do it here. One may also prove that the wave function of the normal coordinates $Q_K$ can be written as

$$\psi_{k,n} = N_{k,\,n}e^{-p_k^2/2\;H_n(p_k)} \quad ...(12.126)$$

where $$p_k = Q_k\left(\frac{4\pi\nu_k}{h}\right)^{1/2} \quad ...(12.127)$$

Now for the ground state wave function $n = 0$, $H_0\ (p_0)$ is equal to 1 so,

$$\psi_{1,\,0} = N_{1,\,0}e^{-p_k^2/2} \quad ...(12.128)$$

which is totally symmetric because $N_{1,0}$ is constant and $p_k$ is directly proportional to $Q_k$. Hence, inversion of $Q_k$ leaves $p_k$ unchanged and so $\psi_{1,0}$ is unchanged by inversion of $Q_k$. We can therefore, say that

"The wave function of the vibration at the ground state of a harmonic oscillator belongs to *A*-type of I.R. of the group (Totally symmetric),

We now reconsider the integral,

$$M = \int \psi_A m \psi_B \partial\tau.$$

For the fundamental transition, $\psi_A$ is the ground state which is totally symmetric for the normal coordinates $Q_k$. ($k$ = 1, 2, 3 ….3*n*-6). Now, if $\psi_A$ is totally symmetric, the only factor that we have to consider in the integral is ($m\psi_B$) for *M* to be totally symmetric. It suggests that the term $m\psi_B$ must belong the totally symmetric representation. One of the ways for $m\psi_B$ to be totally symmetric is that *m* and $\psi_B$ must belong to the same I.R. For non-degenerate case, it is always true, since the square of ± 1 is always (+1). It can also be proved to be true for $E$ or $T$ type of IR. Thus we can generalise the rule as follows.

"In the harmonic approximation, the fundamental transition ($\Delta n = \pm 1$, ground state $\rightarrow$ I$^{st}$ excited state)" is allowed only when the wave function of the first excited state ($\psi_B$), belongs to the same representation as the quantity ($m$) in the transition moment integral($M$).

For light absorption, the quantity '$m$' is identified with the changes in the dipole moment of the molecule. It implies that it is a vector quantity and can be resolved into three components ($x$, $y$, $z$). Now, the $x$, $y$ and $z$ coordinate spans the same representation as change of dipole moment of the molecule. The basis ($x$, $y$ and $z$) spans the same ways as $\Gamma_{trans}$, which can be looked into the character table of the group. Thus, to see if a normal vibration is infrared active or not, one sees the character table of the point group. If the normal coordinate belongs to the same representation as $x$, $y$ and $z$. the vibration associated with one of three I.Rs ( belonging to $x$, $y$ and $z$) then the normal vibration will be Infrared Active, otherwise not. In this manner group, theory gives us more specific information than merely requiring a dipole change in a particular transition to be allowed.

Suppose that the molecule does not have any dipole moment (zero dipole moment molecules). In this case, the molecule can still interact by second order process. It means that the electric field of light wave induces a dipole moment in the molecule. If the induced dipole moments in the ground and excited state are different, then there is a possibility of light absorption.

The induced dipole moment in a body by an electric field is given by Eq. (12.129) where $a$ (the polarisability tensor) replace '$m$' in the expression.

The selection rule for induced dipole transition is similar to that of dipole moment transition with a little difference. Now we have nine $a'_{ij}s$ . Therefore, we will consider nine transition moment integrals, (*viz.* $\int \psi_A \alpha_{ij} \psi_B \partial\tau$ *type*). Since these nine transition moment integrals have

$$\begin{vmatrix} \Delta_{\mu_x} \\ \Delta_{\mu_y} \\ \Delta_{\mu_z} \end{vmatrix} = \begin{vmatrix} \alpha_{xx}\alpha_{xy}\alpha_{xz} \\ \alpha_{yx}\alpha_{yy}\alpha_{yz} \\ \alpha_{zx}\alpha_{zy}\alpha_{zz} \end{vmatrix} \begin{vmatrix} \Delta E_x \\ \Delta E_y \\ \Delta E_z \end{vmatrix} \qquad \text{...(12.129)}$$

$$\alpha_{ij} = \left( \frac{\partial E_x}{\partial E_j} \right) \bar{\alpha}(\Delta E)$$

nonzero value only when these span according to the totally symmetric representations of the group. Since ground state $\psi_A$ is totally symmetric, an induced dipole transition is allowed only when the normal vibration belongs to the same representation as one or more of the polarisability components.

Now $\alpha_{ij}$ belongs to the some IR representation of the point group. These can be found in the character table of the point group. Thus one can easily find to which I.R. these polarisability components belong.

Accordingly, we have two methods for measuring the vibrational frequencies of a molecule (1) in cases where the vibrational frequencies are infrared active. (2) in other cases, where light induces dipole moments in the molecules. In the latter cases there is one condition, for the absorption of light. "The condition is that the frequency of light must be much greater than the frequency of vibration (in general, visible light is used for scattering from the molecule) Visible light is absorbed and is emitted without any change after scattering from the molecule. This is know as **Rayleigh Scattering.** If, on the other hand a quantum of vibrational energy is absorbed by the molecule, the frequency of the emitted light will be reduced by the frequency of vibration. The process is known as Raman Scattering. The scattered light with frequencies lower than the incident radiation frequencies by the energy equivalent to vibrational energy also appear along with the Rayleigh lines. These lines are known as Stokes lines. In addition the Stoke lines, there also appear lines symmetrically placed about the Rayleigh lines, on the higher energy side (higher frequencies).These are known as Anti-stokes lines. Anti-stokes lines are made up of frequency of the incident light plus the frequency of vibrations. It suggests that the molecule is vibrationally de-excited and the evolved energy is added to the Rayleigh lines. Both stokes and anti-stokes lines are of very weak intensity compared to those of Rayleigh lines.

## 12.17 THE QUANTITATIVE APPLICATIONS OF SYMMETRY

This is an extremely wide and mathematically involved topic. It involves calculations of vibrational frequencies, by solving the secular determinant, followed by the calculation of normal coordinates corresponding to each frequency.

Consider equation used to calculate the frequencies of normal vibrations.

$$|FG - \lambda E| = 0 \qquad ...(12.130)$$

The problem is to solve this determinant. If the order of the determinant is larger than three, there in no alternative except to use molecular symmetry as an aid for its mathematical computation. Although we start with these coordinates, but subsequently we have to transform them to symmetry coordinates $S_i$'s using projection operators. That is,

$$S_i = \frac{1}{h}\sum_R \chi_i(R)(Rr_j) \qquad ...(12.131)$$

where $S_i$'s are the symmetry coordinates, $h$ is order of the point group, $R$ the symmetry operation, $\xi_i(R)$ is character of the I.R to which the symmetry coordinate $S_i$ belongs and $(Rr_i)$ is a coordinate of symmetry equivalent set, obtained after the application of symmetry operation R on the internal coordinates belonging to the same equivalent set. The procedure to convert internal coordinates (or any kind of coordinates) to symmetry coordinates are given at the appropriate places in the book. In this manner we generate (3$n$-6) symmetry coordinates $S_i$'s out of (3$n$-6) internal coordinates. Each set of coordinates in one equivalent set have identical linear combination of the internal coordinates. Thus in the example of $BF_3$ we have three equivalent sets,

1. $r_1 r_2 r_3$
2. $\alpha_{12}\alpha_{23}\alpha_{31}$
3. $\pi$-coordinate

The three sets of symmetry coordinates obtained from these internal coordinate are:

Set I
$$\left.\begin{array}{l} \frac{1}{\sqrt{3}}(r_1+r_2+r_3)=S_1(A_1) \\ \left.\begin{array}{l} \frac{1}{\sqrt{6}}(2r_1-r_2-r_3)=S_2 \\ \frac{1}{\sqrt{2}}(r_2-r_3) \quad = S_3 \end{array}\right] E \end{array}\right\} \qquad ...(12.132)$$

Set II
$$\left.\begin{array}{l} \frac{1}{\sqrt{3}}(\alpha_{12}+\alpha_{31}+\alpha_{23})=S_3(A) \\ \left.\begin{array}{l} \frac{1}{\sqrt{6}}(2\alpha_{23}-\alpha_{12}-\alpha_{31})=S_4 \\ \frac{1}{\sqrt{2}}(\alpha_{12}-\alpha_{31})=S_6 \end{array}\right] E \end{array}\right\} \qquad ...(12.133)$$

Set III
$$\left.\begin{array}{l} \pi\text{-coordinates} \\ \text{on its own as} \\ \pi_{(r,\,\alpha)}=\pi_{S_i}-(A_2)\text{ type} \end{array}\right\} \qquad ...(12.134)$$

Therefore, the matrix $U$ which transforms internal coordinates to symmetry coordinates can be written as,

$$\left.\begin{array}{l} S=U(r) \\ r=U'S \end{array}\right\}\text{(Because } U \text{ is an orthogonal matrix)} \qquad ...(12.135)$$

In the case of $BF_3$,

$$\underbrace{\begin{vmatrix} S_1 \\ S_2 \\ S_3 \\ S_4 \\ S_5 \\ S_6 \\ S_7 \end{vmatrix}}_{|S_i \text{ matrix}|} = \underbrace{\begin{vmatrix} 1/\sqrt{3} & 1/\sqrt{3} & 1/\sqrt{3} & 0 & 0 & 0 & 0 \\ 2/\sqrt{3} & 1/\sqrt{6} & -1/\sqrt{6} & 0 & 0 & 0 & 0 \\ 0 & 1/\sqrt{2} & -1/\sqrt{2} & 0 & 0 & 0 & 0 \\ 0 & 0 & 0 & 1/\sqrt{3} & 1/\sqrt{3} & 1/\sqrt{3} & 0 \\ 0 & 0 & 0 & 2/\sqrt{3} & -1/\sqrt{6} & -1/\sqrt{6} & 0 \\ 0 & 0 & 0 & 0 & 1/\sqrt{2} & -1/\sqrt{2} & 0 \\ 0 & 0 & 0 & 0 & 0 & 0 & 1 \end{vmatrix}}_{U\text{-matrix}} \underbrace{\begin{vmatrix} r_1 \\ r_2 \\ r_3 \\ \alpha_{23} \\ \alpha_{31} \\ \alpha_{12} \\ \pi \end{vmatrix}}_{|r_i \alpha_{ig} \pi| \text{ matrix}} \quad ...(12.136)$$

In the above example, some of the internal coordinates belong to I.Rs. which are non-degenerate ($2A_1 + E' + A_2'$). However there are coordinates which belong to doubly degenerate vibrations, *viz.* ($E'$-Type). It is also easy to transform degenerate coordinates ($E$-type) to symmetry coordinates belong to $E$-type I.Rs. The process of transforming $E'$-type coordinates is the same as discussed in chapter nine.

However, transforming a triply degenerate coordinates ($T$-type) into symmetry coordinate which also belong to $T_2$-type of I.R is more complex. It can be shown that there are methods of such transformations from one coordinate system to another coordinates system. We thus transform the internal coordinate to sets of symmetry coordinates $S_i$. Now each symmetry coordinate belongs to one I.R of the group. It is very important to realise that these symmetry coordinates belong to only those I.R. of the point group of the molecule to which vibrational modes span. For example, in $BF_3$ molecule with $D_{3h}$ point group [($3n$-6) = 6], the vibrational modes span the representation $(A_1' + 2E' + A_2'')$. We will therefore obtain the symmetry coordinates belonging to $A_1'$, $A_2''$ and $E'$ I.Rs. We have $S_{A_2''}$, $\left[S_E^1(a), S_E^2(b)\right]$, and $\left[S_E^3(a), S_E^4(b)\right]$, types of symmetry coordinates belonging to $A_1'$, $A_2''$ and $E'$ I.Rs. We have thus $S_{A_2''}$, $\left[S_E^1(a), S_E^2(b)\right]$ and $\left[S_E^3(a), S_E^4(b)\right]$ types of symmetry coordinate.

Further Eq. (12.136) gives the matrix $U$ which transforms internal coordinates into symmetry coordinates.

## 12.18 SYMMETRY FACTORING OF G AND F MATRICES

We have the secular determinant ($GF - E\lambda$) = 0 to be solved. In this equation we have defined,

$$2V = r'F\,r$$

and from Eq. (12.136)

$$2V = S'UFU'S$$
$$= S'F_1S \qquad ...(12.137)$$

where $F_1 = U\,F\,U'$

Treating the K.E. expression in a similar way, we obtain

$$2T = \dot{r}'G^{-1}\dot{r} = \dot{S}'UG^{-1}U'\dot{S} = \dot{S}'G^{-1}\dot{S} \qquad ...(12.138)$$

where $G_1 = UG^{-1}U'$ ...(12.139)

Therefore,

$$G_1F_1 = UGU'UFU' = UGFU'.$$

Thus the eigenvalues of $G_1F_1$ will be the same as those of $GF$. But the eigenvectors of $G_1F_1$ are given by $L_1$, and the eigenvectors of $GF$ will be given as,

$$L_1^{-1}G_1F_1L_1 = L_1^{-1}UGFU'L_1 = \Lambda \qquad ...(12.140)$$

Therefore, $L = U'L_1$ ...(12.140*a*)

In all the above derivation, we have assumed (1) $U$ to be orthogonal and (2) is the matrix of the internal force constant, whose elements are defined as,

$$F_{ij} = \frac{\partial^2 V}{\partial r_i \partial r_j} \qquad (12.141)$$

while we have defined,

$$(F_{ij})_1 = \frac{\partial^2 V}{\partial S_i \partial S_j} \qquad ...(12.142)$$

In the same way, $G$ and $G_1$ are defined and $G_1$ is known as symmetric matrix. The orthogonal transformation of $G$ and $F$ matrices to $G_1$ and $F_1$ matrices greatly simplifies the normal coordinates analysis. In those cases where normal and symmetry coordinates are the same, one can easily get the following relationship.

$$\lambda_K = (G_{kk})_1\,(F_{kk})_1 \qquad ...(12.143)$$

Now, if one uses these $F_1$ and $G_1$ matrices, one can show that the $|\,G_1F_1 - \lambda E_1\,|$ determinant will be block diagonalised. For each non-degenerate representations, there is one block in $F_1$ and for each degenerate representation ($E = 2$, $T = 3$), of degeneracy ($d = 2$ or 3), there will be $d$-blocks. That is for $E$-type of representation, we will get two blocks and for $T_1$ three blocks). The order of each block is the number of normal coordinates belonging to that representation. For example, the $F_1$ matrix for $BF_3$ with the vibrations $(A_1' + A_2'' + 2E')$ will be given as Eq. (12.144).

$$\left|\begin{array}{cccc} A_1' & & & \\ & A_2'' & & 0 \\ & & \begin{array}{c}2\times 2\\ E_1\end{array} & \\ & & & \begin{array}{c}2\times 2\\ 2\times 2\end{array} \end{array}\right| = \left|\begin{array}{cccccc} F_{11} & 0 & 0 & 0 & 0 & 0 \\ 0 & F_{22} & 0 & 0 & 0 & 0 \\ 0 & 0 & F_{33} & F_{34} & 0 & 0 \\ 0 & 0 & F_{43} & F_{44} & 0 & 0 \\ 0 & 0 & 0 & 0 & F_{33} & F_{34} \\ & & & & F_{43} & F_{44} \end{array}\right| \qquad ...(12.144)$$

Thus there will be one vibration of $A_1'$ type [(1 × 1) block], one vibration $A_2$-type [(1 × 1) block, two vibration for $E$-type (2 × 2) block two in number]. Since $E$, (being of degeneracy 2) will give two blocks ($d$ = number of separate blocks), and the coefficient before $E$ is 2, will give two identical blocks of (2 × 2) order. Suppose we have I.Rs $[KA + n_1E + mA_1 + n_1E + mT]$ for the vibrations of any molecule, we will be having.

1. One block of $(K \times K)$ dimension.
2. Two blocks of $(n_1 \times n_1)$ dimensions both $(n_1 \times n_1)$ blocks will be identical.
3. Three blocks of $(m \times m)$ dimensions (All the three blocks will be identical.

Thus in general, the vibrational potential energy in terms of symmetry coordinates in given by

$$2V = S'F_1S$$

If we carry any symmetry operation of the point group on $V$. We get ($R$ = symmetry operation),

$$\begin{aligned} 2RV &= R[S'F'_1S] = 2V \\ &= [S'R'F'_1RS] \\ 2V &= [S'F'_1S] \end{aligned} \qquad ...(12.145)$$

Here $2RV = 2V$, because symmetry operation does not change the potential energy.

From Eq. (12.146) we obtain,

$$F_1R = RF_1 \qquad ...(12.146)$$

Similarly one can also prove that

$$2T = \dot{S}'[M'G^{-1}M]\dot{S} \qquad ...(12.147)$$

$$= \dot{S}'[M'G_1M]^{-1}\dot{S} \qquad ...(12.148)$$

where $M$ is given by $S = MS_1$ which shows an orthogonal change of $S_1$ to $S$, and $M$ is transformation matrix.

## 12.19 REDUNDANCY

We have not discussed removal of redundancy observed during the normal coordinate analysis of any molecule. It is a little involved process, and we will not discuss it any further. The quantitative application of symmetry is a little involved especially if there are cyclic redundancies.

In the actual vibrational analysis of any molecule, we have vibrational frequencies obtained from infrared or/and Raman spectra of the molecule. These frequencies are assigned to particular normal modes of vibration which is a rather difficult process. This quantitative use of symmetry is applied to find the symmetry coordinates *via* the use of cartesian or internal coordinates as basis for symmetry transformations. With the aid of symmetry coordinates in general, we set up *U*-matrices. The *G*-matrix is then calculated using the Wilson's tables (given at the end of his book on "Molecular Vibration"). For the formation of *F* matrix, one in general, transfer them from the other molecules or makes some appropriate guesses. This is usually the most important and critical step in the normal coordinate analysis of a molecule. One however forms *F*-matrix from the chosen force constants. Having set up *G* and *F* matrices, these matrices are then converted to $G_1$ and $F_1$ matrices by orthogonal transformation with *U* matrices which serves two purpose (1) Reduce the labor considerably in calculating force constant by diagonalysing the *GF* matrix by using *U* matrix (2) Minimising the number of force constants to be actually calculated because each symmetry coordinate is unique. Using secular determinant $|G_1F_1 - E\lambda| = 0$, one then calculates the force constants with the aid of experimental values, of $l_K$. The calculated elements of $F_1$ matrix are then refined so that the calculated frequencies, the force constant and the eigenvectors (normal coordinates) matched with the observed frequencies. This information is then used again to reanalyse the problem. One should realise that the solution of secular determinant is still a very difficult process and sometimes, some approximations are to be used.

# Annexure–I

## Discover What You Have Learnt?

### CHAPTER ONE

1. Which one of the following does not form a group?
   (*a*) Set of integers with multiplication as binary operation.
   (*b*) Set of integers with division as binary operation.
   (*c*) Set of matrices | 2 |, | 2 | ...∞ with matrix multiplication as binary operation.
   (*d*) Set $e^x$, ($x$ = positive integers) with multiplication as binary operation.

   | | |
   |---|---|
   | (*i*) *a* | (*ii*) *a*, *b* |
   | (*iii*) *a*, *c*, *d* | (*iv*) *a*, *b*, *c*, *d* |

   **Ans.** (*iv*)
2. The multiplication table of sets of elements {*E*, *a*, *b*, *c*, *d*, *f*,} is given in the text of chapter one. The set forms a group. The inverse (I) and the conjugate elements (II) of *a*, *b*, *c*, *d* and *f* are
   (*A*) (*i*) *c*, *a*, *b*, *d* and *f* respectively, (*ii*) (*a*, *b*), (*b*, *a*), (*c*, *d*, *f*) (*d*, *c*, *f*), (*c*, *d*, *f*), and *E*
   (*B*) (*i*) *c*, *a*, *b*, *d* and *f* respectively, (*ii*) (*a*, *c*), (*c*, *d*), (*a*, *b*, *c*) (*c*, *d*, *f*), (*a*, *b*, *c*), and *E*
   (*C*) (*i*) *b*, *c*, *a*, *f* and *f* respectively, (*ii*) (*a*, *d*), (*b*, *d*), ( *d*, *f*) (*a*, *b*, *c*), ( *d*, *f*, *a*), and *E*
   (*D*) (*i*) *d*, *f*, *a*, *b*, and *f* respectively, (*ii*) (*b*, *d*), (*e*, *f*), (*d*, *a*) (*a*, *b*, *d*), (*f*, *c*, *a*), and *E*

   **Ans.** *A*.

3. Given a group $G$, whose elements are positive, negative and, zero integers as elements with addition as binary operation. Prove that $H$ is a subgroup of $G$, with even integers and zero as members. Also, prove that the set of odd numbers and zero does not form a subgroup of $G$.

   **Ans.** Solve it as practice.

4. Prove that the following set of | 2 × 2 | matrices form a group. Show that this group $G^{(I)}$ is homomorphic to a matrix group $G^{(II)}$ formed by the matrices {| 1 |, | –1 |} and it is not homomorphic to a group formed by the set of matrices {|2|, | –2 |}. Prove (1) that one of the subgroups $G^{(I)}$ is an invariant subgroup and (II) that this invariant subgroup forms an identity element, that corresponds to the identity element of the group which is homomorphic to $G^{(I)}$.

   **Ans.** Solve it as practice.

5. Prove that in a group "It is impossible to have $A^2 = B^2 \neq E$", [Hint: If $A^2 \neq E$, then $A^2$ = some other element, say $C$. Similarly if $B^2 \neq E$ then $B^2$ = some other element say $D$, but $C \neq D$. So $A^2 \neq B^2$ which is not true.

   **Ans.** Solve it as practice.

6. Which one of the following statements is not true?

   (*a*) The set of all matrices is "closed" if binary operation (*o*) is matrix addition.

   (*b*) In a group multiplication table, every column and row lists each of the elements once and only once and no two rows or columns can have the same arrangements and each row or column is a rearranged list of the elements of the set. Thus, all possible rearrangements of the elements form a group multiplication table.

   (*c*) The cosets of a group are always a group.

   (*d*) If an element of a subgroup is subjected to similarity transformation by the element of a parent group $G$, the product $[x^{-1}Ax]$ is always an element of the subgroup. ($A$ is the element of a subgroup).

   Your choices are:

   (*i*) *a* (*ii*) *a*, *b*, *c*

   (*iii*) *a*, *b*, *c*, *d* (*iv*) *c*, *d*

   **Ans.** (*iii*).

7. Prove that three roots of $x$ of the equation $x^3 - 1 = 0$ forms a group with multiplication as binary operation. Find the identity elements and inverse of all the elements. [**Hints:** $x^3 - 1 = 0$, $(x - 1)(x^2 - x - 1) = 0$, roots are

$\left\{1, \left(-\frac{1}{2}-\frac{\sqrt{3}}{2}i\right) \text{and} \left(-\frac{1}{2}+\frac{\sqrt{3}}{2}i\right)\right\}$ using these roots, prove that the set forms a group. (Form a multiplication table.)

**8.** What do you mean by "*A* simple group"?

[**Hint:** *A* group having no "invariant subgroup".]

**9.** Any vector in a three-dimensional space can be written as $x(\bar{i}) + y(\bar{j}) + z(\bar{k})$ where $\bar{i}, \bar{j}, \bar{k}$ are unit vectors and $x_i$, $y_i$ and $z_i$ are the coordinates of a point in space. Prove that, the set $(x_i, y_i, z_i)$ in three-dimensional space form an Abelian group with addition as binary operation.

Can you extrapolate it to an *n*-dimensional space? What will happen when *n* goes to infinite?

**Ans.** Solve it as practice.

**10.** (*a*) In an Abelian group every element commutes with every other element of the group. Prove that every element of such a group forms a class by-itself.

(*b*) If a group has an even number of elements, then besides *E*, there has to be at least one element which is inverse of itself.

**Ans.** Solve it as practice.

## Prove the Statement

**Hint:** Without *E*, the group will have an odd number of elements. Thus, if the inverses of all members are not their own; but different elements, their number being odd, there will always remain one member which can not be paired with a different member of the group. If the members form a group, then the non-paired member has its own inverse because there has to be inverse of each element of the group.

## CHAPTER TWO

**1.** Equivalent atoms in a molecule are those which are interchanged by *A* and therefore α- and β-hydrogen atoms in naphthalene molecule are *B*. Here *A* and *B* stand respectively for:

(*i*) Symmetry operations equivalent

(*ii*) Symmetry elements, equivalent

(*iii*) Symmetry operation of the molecule, non-equivalent.

(*iv*) Symmetry elements and non-equivalent.

**Ans.** (*iii*).

**2.** Any set of symmetry *A* chosen from the set of all symmetry *A* so that any member can be transformed into each and other member of the set by the application of some symmetry *B* is said to be a set of equivalent symmetry *A*.

Here $A$ and $B$ respectively are

(*i*) Operations and operations (*ii*) Elements and elements

(*iii*) Elements and operations (*iv*) Operations and elements

**Ans.** (*iii*).

**3.** In $BF_3$ each set of three $C_2$-axes and three $\sigma_v$ planes form which type of transformation.

(*i*) Equivalent, equivalent

(*ii*) Equivalent and non-equivalent

(*iii*) Non-equivalent and equivalent

(*iv*) Non-equivalent and non-equivalent

**Ans.** (*i*).

**4.** Out of the following pairs of operations which pair/pairs do not commutes.

(1) Two rotations about the same axis.

(2) Two $C_2$-rotations about perpendicular axes.

(3) Reflections through planes which are perpendicular to each other.

(4) Rotation and reflection through a plane perpendicular to rotation axis.

(5) Inversion (*i*)
and any reflection or rotation about an axis.

Your choices are

(*i*) 1, 3, 5 (*ii*) 2, 4, 3

(*iii*) 1, 2, 3 and 4 (*iv*) None

**Ans.** (*iv*).

**5.** Prove the following relationship

$$\sigma_v \cdot C_2 = C_2 \cdot \sigma_v = i.$$

**Ans.** *See* Text.

**6.** Prove that a $S_n$ axis ($n$ = odd) generates $2n$ operations.

**Ans.** *See* Text.

**7.** Prove that a $S_n$ axis ($n$ = even) requires the presence of $C_{n/2}$-axis in the molecule.

**Ans.** *See* Text.

**8.** Prove the following relationship

(*a*) $S_{2n} = C_{nh}$ when $n$ = odd, and

(*b*) $S_{2n} = C_{n/2} \times i$ when $n$ is divisible by 4.

**Ans.** *See* Text.

**9.** Symmetry operations of the following molecules are given below. In each case one or two symmetry operations are missing find out the missing one in each case

(*i*) 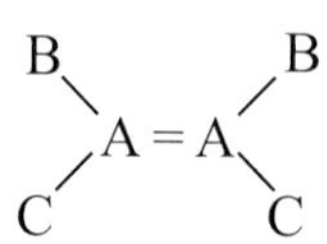

$\longrightarrow E, C_2, \sigma_h$

(*ii*) 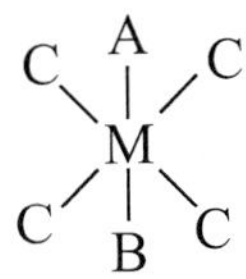 $\longrightarrow E, C_4, C_2, \sigma_v^{(I)}, \sigma_v^{(II)}$

(*iii*) 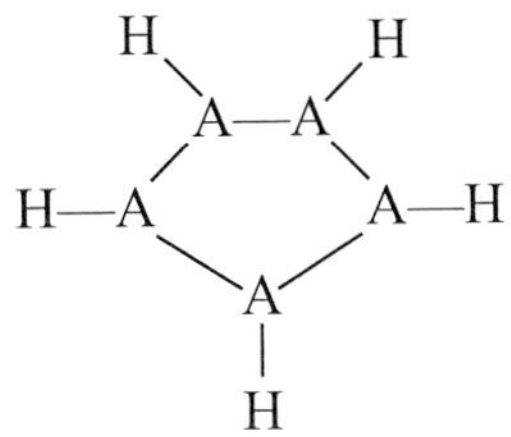 $\longrightarrow E, C_2, \sigma_h$

**Ans.** (*i*) $\sigma_V$

(*ii*) $\sigma_v^{(1)}, \sigma_v^{(2)}$

(*iii*) $5C_2 \cdot S_5$.

**10.** What are point groups of the following solids?

(*i*) Octahedron

(*ii*) $S_8$ (boat-shaped crown)

(*iii*) Cube

(*iv*) Trigonal dodecahedron (8 vertices + 12 faces)

**Ans.** (*i*) $O_h$

(*ii*) $D_{4d}$

(*iii*) $O_h$

(*iv*) $I_h$.

**11.** Which ones of the following statements are correct?

(I) The symmetry or asymmetry with respect to one operation of the group is formally the same for all the operations in the same class.

(II) If two operations of a group commute, then they belong to different classes.

(III) $E$, $\sigma_h$ and $i$ operations form the class of their own.

Your choices are

(*i*) All true

(*ii*) I and II

(*iii*) I and III

(*iv*) II and III

**Ans.** (*i*).

**12.** Define order of a class.

**Ans.** Number of operations in the class.

**13.** What is Lagrange's theorem about subgroup?

**Ans.** $\frac{g_1}{g}$ = Whole number .

**14.** Find out the even or odd symmetry of the following mathematical functions:

(*i*) $\dfrac{\sin Y}{Y}$ (*ii*) $Y^2 - Y$

(*iii*) $Y \cos Y$

**Ans.** (*i*) Even (*ii*) even (*iii*) odd.

**15.** Under what condition $S_n^n = \sigma_h$ ?

**Ans.** When $n$ = odd.

**16.** What is the point group of *cis*- and *trans*-butadiene?

**Ans.** $C_{2v}$ and $C_{2h}$.

**17.** What is the point group associated with the following set of operations $\{E, C_2, s_n, i\}$?

**Ans.** $C_{2h}$.

**18.** Match the items in List *A* with those in List *B*.

| S.No. | List *A* | List *B* |
|---|---|---|
| 1. | $CH_2Cl_2$ | (*A*) $D_{3h}$ |
| 2. | $B_2H_6$ | (*B*) $C_{3v}$ |
| 3. | Cyclohexene (chair) | (*C*) $D_{6d}$ |
| 4. | Cyclohexene (boat) | (*D*) $D_{2h}$ |
| 5. | $POCl_3$ | (*E*) $D_{3d}$ |
| 6. | $PF_5$ | (*F*) $D_3$ |
| 7. | $Cr(C_6H_6)$ (Staggered) | (*G*) $C_2$ |
| 8. | $Cr(en)_3$ | (*H*) $C_{2V}$ |
| 9. | $Cr(OX)_2(H_2O)_2$ Cis | (*I*) $C\infty_V$ |
| 10. | COS | (*J*) $D\infty_h$ |
| 11. | $CO_2$ | (*K*) $C_{3V}$ |
| 12. | $Cr(CO)_3(C_6H_6)$ | |

**Ans.** (1) *H*, (2) *D*, (3) *E*, (4) *H*, (5) *B*, (6) *A*, (7) *C*, (8) *F*, (9) *G*, (10) *I*, (11) *J*, (12) *K*.

**19.** Find the point group of the following:

(1) Cup without handle (2) Cup with a handle (3) Cup with two identical handles on the opposite sides of the rim (4) Cup with three identical handles at an angle of 120° (5) Unsharpened pencil ($S_4N_4$) (boat form) (6) $Re_2Cl_8$(Cl's are eclipsed and there is Re–Re bond) (7) $Ir_4(CO)_{12}$ with $Ir_4$ forming tetrahedron and three CO bonds on each Ir atom.

**Ans.** (1) $D_{\infty h}$, (2) $C_s$, (3) $C_s$, (4) $C_3$, (5) $D_{\infty h}$, (6) $D_{2d}$, (7) $D_{4h}$.

**20.** What is the difference between $C_4$ axis and $\bar{C}_4$-screw axis? What are the symmetry elements in both cases?

**Ans.** Rotation and Rotation plus translation.

**21.** Does the set {1, –1, *i*, –*i*} form a group? If so, to which point group does it belong.

**Ans.** (*i*) Yes (*ii*) $C_{2v}$

**22.** Fill in the blank in the following statements:

(*i*) The symmetry can be defined as an ___*A*___ in the pattern that is observed when some ___*B*___ is applied to it.

(*ii*) Although the concept of ___*C*___ has its origin in geometry, it is general enough to embrace ___*D*___ with respect to ____*D*____ of other kind. An example of ____*E*____ ______ is the _____*F*____ of electromagnetism what do the letters *A*, *B*, *C*, *D*, *E* and *F* stand for

**Ans.** (*A*) invariance (*B*) transformation

(*C*) symmetry (*D*) invariance

(*E*) non geometric (*F*) charge symmetry.

**23.** Complete the following table, regarding point group of the molecules

| S.No. | Stereochemistry | Point group |
|---|---|---|
| 1. | Linear | –A |
| 2. | I Trigonal Planar | –B |
| | II Trigonal pyramid | –C |
| | III T-shaped | –D |
| 3. | I Tetrahedron | –E |
| | II Square planar | –F |
| 4. | I Trigonal pyramid | –G |
| | II Square based pyramid | –H |
| 5. | I Octahedron | –I |
| | II Trigonal prism | –J |

**Ans.** $A - D_{\infty h}$ $B - D_{3h}$

$C - C_{3V}$ $D - C_{2V}$

$E - T_d$ $F - D_{4h}$

$G - D_{3h}$ $H - C_{4V}$

$I - O_h$ $J - D_{3h}$.

## CHAPTER THREE

**1.** Define vector space. How does the idea of vector space arise?

**Ans.** For vector space see text.

The idea of vector space arises from a consideration of the properties of ordinary three-dimensional physical space that can be put in a form which is capable of generalisation.

**2.** What do you mean by a position vector $\overline{P}$ in a physical space? How is it related to any point in the space?

**Ans.** The location of any point $P$ in space is fixed relative to any arbitrarily chosen origin $O$ and a fixed frame of reference. With respect to that fixed frame of reference, a directed line from $O$ to $P$ is drawn and its length and direction with respect to that frame of reference can be determined. The directed line $\overrightarrow{OP}$ is known as the position vector of point $P$. So one can think interchangingly a point $P$ and the position vector $OP$.

**3.** One can attach to each point ($p$, $q$, $r$...) in space forming different position vectors. What are the general geometric relations among these position vectors *e.g.*, equality, addition, subtraction, multiplication by a number scalar.

**4.** How can we generalise the ideas given in questions (1 – 3)?

**Ans.** A *general vector space* can be defined as a set of entities of any kind for which we can define different operations as given in Q. 3.

A vector space is a set of vectors, $p$, $q$, $r$...

(*i*) Which can add together like the vectors in physical space.

(*ii*) The set must have a zero vector (origin in physical space).

(*iii*) The addition of any two vectors must produce a third vector of the same space.

(*iv*) The multiplication of the vector by a scalar is defined producing vector of the same space $(a\overline{P})$.

(*v*) The multiplication by a scalar must have the following properties: $a(b\overline{P}) - ab\overline{P}\ (a+b)\,\overline{P} = a\overline{P} + b\overline{P}, O.\,\overline{P} = O, 1.\ \vec{P} = \overline{P}$ and $a(\overline{P} + \overline{Q}) = a\overline{P} + b\overline{Q}$; If all the vectors $\overline{p}, \overline{r}, \overline{q}$ ... of space have these properties, then the set constitutes a vector space.

**5.** Define a function vector space (or function space).

A function space is a vector space in which functions are the entities as defined in question 4. Thus, it is a set of functions $(f_1, f_2, f_3)$ of the same variables (*viz.* $x$, $y$, $z$...etc.).

(*i*) These functions can be added together to give a function of the same space of the same variable that is, $f_p(x, y, z\ldots) = f_1\,(x, y, z) + f_2\,(x, y, z)$.

(*ii*) It must contain a zero function. It implies that the set of zero values of $x$, $y$, $z$ must give the zero value of $f\,(x, y, z)$.

(*iii*) It we multiply any function $f(x, y, z\ldots)$ by any scalar ($a$) then the function $a\,f\,(x, y, z\ldots)$ is also a function of the same space.

(*iv*) The multiplication by scalar must also follow the rules $(a + b)f = af + bf$ ...etc.

**6.** Give an example of a function space.

**Ans.** *See* Text.

7. What do you mean by base vectors?
8. What do you mean by orthogonal base vectors?
9. How do we define base vectors in function space?
10. How do we define orthogonal base vectors in function space?

    **Ans.** Questions 6–10, See Text.

11. Dimensionality of vector space is related to the number of linearly independent vectors which one can chose in a vector space. What is the relationship?

    **Ans.** Number of linearly independent vectors equals to the dimensionality of space.

12. What is configuration space and how is it related to molecule?

    **Ans.** *See* Text.

13. The geometrical notion of perpendicularity of two vectors in general vector space is introduced by defining.

    (*i*) Scalar product of two vectors.

    (*ii*) Zero vector in vector space.

    (*iii*) Components of vectors in oblique coordinate system.

    (*iv*) Arbitrary length of any one vector in vector space.

    **Ans.** (*i*).

14. Scalar product of two vectors $\bar{P}$ is $\bar{Q}$ given by,

    (*i*) $(\bar{P}\cdot\bar{Q}) = |P||Q|\sin\theta$ (*ii*) $(\bar{P}\cdot\bar{Q}) = |P||Q|$

    (*iii*) $(\bar{P}\cdot\bar{Q}) = |P||Q|\cos\theta$ (*iv*) $(\bar{P}\cdot\bar{Q}) = \dfrac{|P|}{|Q|}$

    **Ans.** (*i*).

15. Scalar product of two vectors in function space is defined as

    (*i*) $f(x, y\ldots)\, j(x, y\ldots)$ (*ii*) $\int f(xyz)\,.\,g(xyz)\,\partial\tau_1\,\partial\iota_2\ldots$

    (*iii*) value of function at $(x, y, z\ldots)$ (*iv*) $\int_{-\infty}^{+\infty}(x, y, z\ldots)\,dx$

    **Ans.** (*ii*).

16. The conditions of two functions being orthogonal are

    (*i*) $\iiint f_i(x, y\ldots)\cdot f_j(x, y\ldots)\,\partial\tau_1\ldots = \delta_{ij}$

    (*ii*) $\iiint\ldots f_L(x, y, z\ldots)\,\partial T_1\ldots\partial\tau_2 = 1$

    (*iii*) $\iiint\ldots f_j(x, y, z\ldots)\,\partial T_1\ldots\partial\tau_2 = 0$

    (*iv*) $\iiint\ldots f_j(x, y, z\ldots)\partial T_1\ldots\partial\tau_2 = 0$ and

    $$\iiint\ldots f_j(x, y\ldots)\cdot f_i(x, y,\ldots)\,\partial\tau_1\cdot\partial\tau_2 = 1$$

    **Ans.** (*i*).

## CHAPTER FOUR

1. The basic idea behind representing symmetry operations of a molecule by a set of matrices can be enumerated as follows. Which one of the given statements are not correct.
   (*i*) Acknowledgement of symmetry operations of point group as linear operators.
   (*ii*) Acknowledgement of symmetry operations of point group as non-linear operators.
   (*iii*) Defining an *n*-dimensional space by '*n*' number of base vectors, which are not linearly independent.
   (*iv*) Defining an *n*-dimensional space by '*n*' number of linearly independent base vectors.
   (*v*) Operation of symmetry operators on these base vectors to obtain '*n*' number of matrices.
   (*vi*) Determination of the behavior of *n* matrices of point (*v*) is exactly the same fashion as symmetry operations.

   **Ans.** (*ii*) and (*iii*).

2. Suppose we have a group *G* with operation $\{E, R_1\ R_2 \ldots R_h\}$ where $(h + 1)$ is the order of a group. We also have its character table, if we take the character of the each operation for all the I.Rs of the group and put them on the diagonal of a matrix we form matrix for each operation. Will the set of matrix form a representation of the group if yes, will it be reducible or irreducible representation [Assume I.Rs are one-dimensional]. Your choices are
   (*i*) No
   (*ii*) Yes, reducible
   (*iii*) Yes, I.R
   (*iv*) Yes, reducible but in reduced form

   **Ans.** (*iv*).

3. Pick out the wrong statements among the following:
   (*i*) I.Rs of a group depend upon the basis that we choose to form I.Rs.
   (*ii*) I.Rs depend upon the dimensionality of the space.
   (*iii*) I.Rs can be obtained from reducible representation by the use of the similarity transformation.
   (*iv*) The number of reducible representation of a group is fixed.

   **Ans.** (*i*), (*ii*), (*iv*).

4. Prove that all the I.Rs of an Abelian group is one-dimensional.

   **Ans.** Because $\hat{R}_1^{-1} R_2 R_3 = R_2$ for each operation.

5. Why the identity representation of all point groups is unfaithful representation?

   **Ans.** Because the matrices of all operations are the same. For faithful representations, all matrices must be different (*i.e.,* one to one correspondence).

**6.** What $C_n$ operation will mix $p_x$ and $p_y$ base vectors; but not $d_{xy}$, and $d_{x^2-y^2}$?

**Ans.** $\left[C_4^1\right]$.

**7.** Construct the character table of $C_{4V}$.

**Ans.** *See* Text.

**8.** What is the physical significance of the character of a group?

**Ans.** The character is a number [$\chi_i$(R)] which expresses the behaviour of the symmetry on a property of a system, (*e.g.*, the vibration of a molecule) when operated by its symmetry operator ($R$) (*e.g.*, the property $V$ vibration is referred to as basis vector of the representation).

**9.** Are the basis vector (defined in Q. 8) fixed for any group representation?

**Ans.** No.

**10.** Point out examples of a few basis vectors.

**Ans.** (*i*) One can define all kinds of basis vectors. Some of them have no connection with our original object *e.g.*, molecules such as non-symmetry operations, like translations along the $z$-axis $T_z$, or rotation about any axis by an arbitrary amount referred to $R_{x,\, y \text{ or } z}$ .

(*ii*) Other basis vector include actual vectors such as dipole moment, stretching of a bond.

(*iii*) More elaborate vectors consist of the three cartesian displacement vectors of each $n$-number of atoms in a molecule or the wave functions in a molecule or even tensors like polarisability, *etc*.

**11.** What do you mean by the equivalent representations?

**Ans.** Two representations of a point group by $| n \times n |$ matrices are said to be equivalent {*viz*. $[O^g(R)]$ and $[O^f(R)]$} if a matrix $A$ with its inverse $A^{-1}$ exists for which the relation $\lfloor O^g(R) \rfloor = AO^f(R)A^{-1} \rfloor$ holds good for every operation $R$.

**12.** The characters $c_i$ of a reducible representation $G$ of the indicated point group for various classes of $G$ in the order of classes given in their character table are given as follows. Prove that $G$ (reducible) are reduced in the manner as given below.

| S.No. | Point group | $\chi_i$ for $\Gamma_{red}$ for each class | Γ(I.R.) |
|---|---|---|---|
| 1. | $C_{2v}$ | 4, –2, 0, –2 | $A_2 + 2B_1 + B_2$ |
| 2. | $C_{3h}$ | 4, 1, 1, 2, –1, 1 | $A' + A'' + E$ |
| 3. | $D_{4h}$ | 4, 2, 0, –2, –2, 2, 0, –2, 0, 0 | $A_{2g} + A_{2u} + E_g$ |
| 4. | $D_{4d}$ | 6, 0, –2, 0, –2, 0, 0 | $E_1 + E_2 + E_3$ |
| 5. | $D_{6h}$ | 30, –4, 0, 2, –2, –2, 0, 0, 0, 12. 0, 4 | $2A_{1g} + A_{2g} + 2B_{2g} + E_{1g}$ $+ 4E_{2g} + A_{2u} + 2B_{1u}$ $+ 2B_{2u} + 3E_{1u} + 2E_2$ |

*(Contd.)*

| | | | |
|---|---|---|---|
| 6. | $C_{2V}$ | 32, 3, –3, 32 | $16A_1 + 16B_2$ |
| 7. | $C_{2V}$ | 12, –2, –2, 12 | $6A_1 + 6B_2$ |
| 8. | $D_{2h}$ | 24, 0, 0, 0, 0, 0, 0, 0... | $3A_{1g} + 3B_{1g} + 3B_{2g} + 3B_{3g} + 3A_{1u} + 3B_{1u} + 3B_{2u} + + 3B_{3u}$ |
| 9. | $T_d$.. | 9, 0, 1, –1, 3 ... | $A_1 + E + 2T_2$ |
| 10. | $O_h$... | 15, 0, –1, 1, 1, –3, 0, 5, –1, 3... | $A_{1g} + E_g + T_{2g} + 2T_{1u} + T_{2u}$ |

**13.** What is the orthogonality rule applied in the formation of character table of a group?

**Ans.** The orthogonality rule states that the sum of the products of characters of two I.Rs is equal to zero.

**14.** Why does each operation of a cyclic group form a class by itself?

**Ans.** Because cyclic groups are Abelian.

**15.** We have two matrices | $A$ | and | $B$ | which are $\begin{vmatrix} 1 & 0 \\ 0 & 0 \end{vmatrix}$ and $\begin{vmatrix} 0 & 0 \\ 0 & 1 \end{vmatrix}$ and

$|A| \cdot |B| = |C| = \begin{vmatrix} 0 & 0 \\ 0 & 0 \end{vmatrix}$. What does this signify?

**Ans.** $|A| \cdot |B| = 0$ does not mean $|A| = 0$ and $|B| = 0$.

## CHAPTER FIVE

**1.** Given a reducible representation which arises from the operations of a point group in a given vector space, one faces two problems when applying symmetry of a system to solve any problem. What are two problems that one faces?

**Ans.**

(*i*) To decide which of the I.Rs of $G$ will appear in the reduction of $\Gamma_{red}$.

(*ii*) To find a new basis for the vector space in which matrices are in the completely reduced form.

**2.** How do we solve these problems?

**Ans.** The first problem is solved by using the Eq. $a_i = \frac{1}{h}\sum_R \chi^*_{red}\chi^{I.R}$ (*see* text of chapter four). The second problem is solved by using projection operators (*see* chapter) for details.

**3.** What is a projection operator?

**Ans.** Projection operator $P_{k'} = h\sum_R \chi_k(R)\hat{O}_R$ where $\chi_k(R)$ is the character of the $P^{th}$ operator; $k = k^{th}$ I.R. of the group.

**4.** Write the projection operators for $C_{3V}$ point group for the I.Rs $A_1$, $A_2$ and $E$.

**Ans.** Solve for practice.

**5.** Use the projection operators to obtain bases for the I.Rs of $C_{3V}$ point group from the function.

$$F(x, y, z) = ax + by + cz + k \text{ (constant)}$$

Here $a$, $b$, $c$ are constant coefficients.

**Ans.** Projection operators for I.Rs of $C_{3V}$ point group are

$$P_{A_1} = \frac{1}{6}[E + C_3^1 + C_3^2 + \sigma_V^{(1)} + \sigma_V^{(2)} + \sigma_V^{(3)}]$$

$$P_{A_2} = \frac{1}{6}[E + C_3^1 + C_3^2 - \sigma_V^{(1)} - \sigma_V^{(2)} - \sigma_V^{(3)}]$$

$$P_E = \frac{1}{3}[2E - C_3^1 - C_3^2].$$

We have for,

$$E = \begin{vmatrix} 1 & 0 & 0 \\ 0 & 1 & 0 \\ 0 & 0 & 1 \end{vmatrix}, C_3^1 = \begin{vmatrix} -\frac{1}{2} & -\frac{\sqrt{3}}{2} & 0 \\ \frac{\sqrt{3}}{2} & -\frac{1}{2} & 0 \\ 0 & 0 & 1 \end{vmatrix}, C_3^2 \begin{vmatrix} -\frac{1}{2} & \frac{\sqrt{3}}{2} & 0 \\ -\frac{\sqrt{3}}{2} & -\frac{1}{2} & 0 \\ 0 & 0 & 1 \end{vmatrix}$$

$$\sigma_V^{(1)} \begin{vmatrix} 1 & 0 & 0 \\ 0 & -1 & 0 \\ 0 & 0 & 1 \end{vmatrix}, \sigma_V^{(2)} \begin{vmatrix} \frac{1}{2} & -\frac{\sqrt{3}}{2} & 0 \\ -\frac{\sqrt{3}}{2} & -\frac{1}{2} & 0 \\ 0 & 0 & 1 \end{vmatrix}, \sigma_V^{(3)} \begin{vmatrix} \frac{1}{2} & \frac{\sqrt{3}}{2} & 0 \\ \frac{\sqrt{3}}{2} & -\frac{1}{2} & 0 \\ 0 & 0 & 1 \end{vmatrix}$$

Thus, one obtains,

$$\hat{E} \cdot F = ax + by + cz + k$$

$$C_3^1 F = \left(-\frac{a}{2} - \frac{\sqrt{3}}{2}b\right)x + \left(\frac{\sqrt{3}}{2}a - \frac{b}{2}\right)y + cz + k$$

$$C_3^2 F = \left(-\frac{a}{2} - \frac{\sqrt{3}}{2}b\right)x + \left(-\frac{\sqrt{3}}{2}a - \frac{b}{2}\right)y + cz + k$$

$$\sigma_V^{(1)} F = -ax + by + cz + k$$

$$\sigma_V^{(2)} F = \left(\frac{a}{2} - \frac{\sqrt{3}}{2} b\right) x + \left(-\frac{\sqrt{3}}{2} a - \frac{b}{2}\right) y + cz + k$$

$$\sigma_V^{(3)} F = \left(\frac{a}{2} + \frac{\sqrt{3}}{2} b\right) x + \left(\frac{\sqrt{3}}{2} a - \frac{b}{2}\right) y + cz + k$$

For the basis $A_1$, after summing the above expression, we get,

$$P_{A_1} F = \frac{1}{6}[6cz + 6k] = (cz + k)$$

Similarly one can calculate using projection operator of $A_2$

$$P_{A_2} F = \frac{1}{6}[0] = 0$$

and $$P_E F = \frac{1}{3}[3ax + 3by] = ax + by$$

Thus, we have got basis functions out of the given function $F(ax + by + cz + k)$ which transform as $A_1$, $A_2$ and $E$ *viz.* $(z + k)$, 0 and $(ax + by)$ which are linear combination of two functions *viz.* $ax$ and $by$].

**6.** Use projection operators of $C_{2V}$ and $D_{2h}$ point group I.Rs to obtain the bases for the I.Rs of these groups from the function f $(x, y, z) = ax + by$, $cz + k$ *by* following the process given in Q. 5.

**Ans.** Solve for practice.

**7.** Let there be $g_1, g_2, g_3$ and $g_4$ orthonormal base vectors of a four-dimensional vector space forming a $\Gamma_{red}$ representation of $D_4$ point group. The effects of symmetry operations on these vectors are given below.

| | $E$ | $C_4$ | $C_4^3$ | $C_2$ | $C_{2a}$ | $C_{2b}$ | $C''_{2a}$ | $C''_{2b}$ |
|---|---|---|---|---|---|---|---|---|
| $g_1$ | $g_1$ | $g_2$ | $g_4$ | $g_3$ | $-g_4$ | $-g_2$ | $-g_1$ | $-g_3$ |
| $g_2$ | $g_2$ | $g_3$ | $g_1$ | $g_4$ | $-g_3$ | $-g_1$ | $g_4$ | $-g_2$ |
| $g_3$ | $g_3$ | $g_4$ | $g_2$ | $g_1$ | $-g_2$ | $-g_4$ | $-g_3$ | $-g_1$ |
| $g_4$ | $g_4$ | $g_1$ | $g_2$ | $g_2$ | $-g_1$ | $-g_3$ | $-g_2$ | $-g_4$ |

Show that $\Gamma_{red} = A_2 + B_1 + E$. Using the projection operators of $A_2$, $B$, and $E$ prove that the orthonormal bases are

(*i*) $$F^{A_2} = \frac{1}{2}(g_1 + g_2 + g_3 + g_4)$$

$$F^{B_1} = \frac{1}{2}(g_1 - g_2 + g_3 - g_4)$$

$$F^E = \frac{1}{\sqrt{2}}(g_1 - g_3) \text{ and } \frac{1}{\sqrt{2}}(g_2 - g_4)$$

**8.** Take $p_z$-orbitals of carbon atoms of cyclopropane as base vectors. Using projection operators of $C_3$ point group make the linear combinations of $p_z$-orbitals which transform as the I.Rs. of the $C_3$ point group.
For other examples of using Projection operators to obtain SALCs *see* chapters that follow chapter five.

## CHAPTER SIX

**1.** Prove that
(*i*) $A \otimes B \neq B \otimes A$ ($\otimes$ means Direct Product)
(*ii*) $\text{Tr}[(A) \otimes B] = \text{Tr}[(B) \otimes A]$ (Tr = Trace)
(*iii*) $\text{Tr}(A) \otimes Tr(B) = \text{Tr}[A \otimes B]$
Here | $A$ | and | $B$ | are the following matrices

$$A| = \begin{vmatrix} a_{11} & a_{12} \\ a_{21} & a_{22} \end{vmatrix} \quad \text{and} \quad |B| = \begin{vmatrix} b_{11} & b_{12} & b_{13} \\ b_{21} & b_{22} & b_{23} \\ b_{31} & b_{32} & b_{33} \end{vmatrix}.$$

**2.** Define the Direct Product of two Representations.
**Ans.** *See* Text.

**3.** Out of a number of rules which apply to form Direct Product Representations, the following four rules are given. Which ones of these are odd ones.
(*i*) The D.P. of two I.Rs gives the representations of the same group.
(*ii*) The D.P. of two I.Rs is always reducible representation which has to be reduced to a sum of I.Rs. of the group.
(*iii*) The D.P. of any one-dimensional I.R with itself always give the same representation and is never an Identity representation that is $\Gamma_L \otimes \Gamma_i = \Gamma_i$ and is never equal to $A_1$.
(*iv*) The D.P. of two-, three- or four-dimensional representation with itself always gives a reducible representation. In this reducible representation, these are always present one I.R of $A_1$ type.
(*v*) The same rule (*iv*) is also true if the direct product is taken between two multiple dimensional representations of different types, *e.g.*, $E \otimes T$ or $T_1 \otimes T_2$ etc.
**Ans.** The odd ones are (*ii*), (*iii*) and (*v*).

**4.** Which ones in the following D.P. are not correct?
(*i*) $A \otimes A = A$ (*ii*) $A \otimes B = A$
(*iii*) $A \otimes T = A + E$ (*iv*) $B \otimes E = E \otimes B = E$
**Ans.** (*ii*), (*iii*).

**5.** The D.P. of two I.Rs with the same and different sub- and superscript are given below. Which ones among them are correct?

(*i*) $1 \otimes 1 = 1$ (*ii*) $1 \otimes 2 = 2$

(*iii*) $2 \otimes 2 = 1$ (*iv*) $2 \otimes 1 = 2$

(*v*) $(1) \otimes (11) = (11)$ and $(1) \otimes (1) = (11) \otimes (11) = (1)$ (1 = dash superscript)

(*vi*) For $D_2$ and $D_{2h}$ point groups, we have

$$1 \otimes 2 = 3$$

$$3 \otimes 1 = 2$$

**Ans.** All.

**6.** Which ones of the following D.P's are correct?

(*i*) For all groups $A \otimes E_i = E_i$ (*i* can be any number)

(*ii*) $B \otimes E_1 = E_2$ and $B \times E_1 = E_2$; $B \otimes E_3 = E_1$ except for $D_d$ where $B \otimes E_1 = E_5$ and $B \otimes E_2 = E_4$; $B \otimes E_3 = E_3$; $B \otimes E_4 = E_2$ and $B \otimes E_5 = E_1$ and For $D_{4d}$ and $S_8$, we have, $B \otimes E_1 = E_3$ and $B \times E_3 = E_1$.

**Ans.** Solve for practice.

**7.** Similar to multiplication table, one can also form D.P. table. For $C_{2h}$ such a table is given below, in which some of the entries are missing. Which are the missing ones?

| $C_{2h}$ | $A_g$ | $A_u$ | $B_g$ | $B_u$ |
|---|---|---|---|---|
| $A_g$ | $A_g$ | $A_u$ | $B_g$ | $B_u$ |
| $A_u$ | $A_u$ | $A_g$ | (1) | (2) |
| $B_g$ | $B_g$ | $B_u$ | $A_g$ | $A_u$ |
| $B_u$ | $B_u$ | (3) | (4) | (5) |

**Ans.** (1) $B_u$ (2) $B_g$ (3) $B_g$ (4) $A_u$ (5) $A_g$.

**8.** Which ones of the following statements are not correct?

(*i*) Invariably all D.P. representations among one or other types of I.Rs give one-dimensional representations.

(*ii*) All one-dimensional representation give one-dimensional I.Rs when one makes D.P. of one-dimensional group I.R. of any group.

(*iii*) The D.P. of one-dimensional I.R. by itself of one group gives I.R of other group.

(*iv*) The D.P. of I.R of *u* type with itself gives '*g*' type of I.R. of the same group.

**Ans.** (*i*) and (*iii*).

**9.** Which ones of the following (D.P.) are not correct for $C_{4V}$ point group.

Character Table of $C_{4V}$ — Direct Product of I.Rs of $C_{4V}$

| $C_{4V}$ | $E$ | $C_2$ | $2C_4$ | $2\sigma_V$ | $2\sigma_d$ | *D.P.* | $E$ | $C_2$ | $2C_4$ | $2\sigma_V$ | $2\sigma_d$ |
|---|---|---|---|---|---|---|---|---|---|---|---|
| $A_1$ | 1 | 1 | 1 | 1 | 1 | (1) $A_1 \times A_2 \times B_1$ | 1 | 1 | –1 | –1 | 1 |
| $A_2$ | 1 | 1 | 1 | –1 | –1 | (2) $E \times E \times E \times E$ | 16 | –16 | 0 | 0 | 0 |
| $B_1$ | 1 | 1 | –1 | 1 | –1 | (3) $B_1 \otimes E$ | 2 | 2 | 0 | 0 | 0 |
| $B_2$ | 1 | 1 | –1 | –1 | 1 | (4) $A_2 \times B_1 \times B_2$ | 1 | 1 | 1 | 1 | 1 |
| $E$ | 2 | –2 | 0 | 0 | 0 | (5) $B_1 \times B_2 \times E$ | 2 | –2 | 0 | 0 | 0 |

**Ans.** (*ii*) and (*iii*).

**10.** Under what condition (*S*) $\int_{-\infty}^{+\infty} f_A \cdot f_B \partial t \neq 0$.

**Ans.** It is the product of two functions like $(x_i\ x_j)$ etc. The value of this integral will be zero unless the integrant to which the molecule belongs, or, unless some term is invariant when operated by the symmetry operations of the molecule. How do we determine that the reducible representation formed by $f_A \cdot f_B$ contains $A_1$ type of I.R. of the group? This we do it by the D.P. That is, if $f_A$ forms a basis of some representation of the group (say $\Gamma_1$) and $f_B$ to $\Gamma_2$ of the same group. Then $(\Gamma_1 \otimes \Gamma_2)$ forms a basis for the reducible representation of the same group (say $\Gamma_3$). If $\Gamma_3$ on reducing it to I.Rs of the group contains $A_1$ type of I.R. Then $\int_{-\infty}^{+\infty} f_A \cdot f_B \partial t \neq 0$.

(or $\Gamma_3 = \Sigma$ I.Rs of the group)

**11.** For the indicated point groups, calculate the D.P. of the following:

(*i*) B ⊗ B (for all groups)

(*ii*) In $C_{2h}$ point group $B_g \otimes B_u$

(*iii*) In $C_{4V}$ point group $A_2 \otimes B_2$

(*iv*) In $D_{2h}$ $B \otimes B$.

**Ans.** (*i*) *A* except for $D_2$ and $D_{2h}$ (*ii*) $A_u$

(*iii*) $B_1$ (*iv*) *B*.

**12.** In tetrahedral complex of $3d^2$ system, the two electrons can give either $E^2$ or $E^1T_2^1$ or $E^0T_2^2$ configurations. When interelectronic interactions is taken into account, these levels are split in various levels. What are those levels?

$$E_1 \otimes E_1 = \Gamma_{\text{red}} = 4, 1, 4, 0, 0 = A_2 + A_1 + E$$

$$E_1 \otimes T_2 = \Gamma_{\text{red}} = 6, 0, -2, 0, 0 = T_1 + T_2$$

$$T_2 \otimes T_1 = \Gamma_{red} = 9, 0, +1, -1, -1 = A_2 + E + T_1 + T_2$$

**Ans.** $E^2 = A_2 + A_1 + E$

$E^1T^1 = T_1 + T_2$

$T^2 = A_2 + E + T_1 + T_2.$

**13.** What is the physical significance of Direct Product?

**Ans.** Direct Product is usually applicable in a number of calculations. We generally come across in the calculation of integrals of the type $\int \phi_1 \, A \, \phi_2 \, \partial\tau$ (*e.g.*, in selection rules matrix elements, *etc.*). In such calculations one tries to know if such integrals are zero or has some value. This can be easily obtained by calculating the D.P. of the symmetry species of $\phi_1$, $A$ and $\phi_2$. If the I.R arising out of D.P. contains $A_1$ spices as one of the components, then the value of integral is different from zero. In case it does not contain $A_1$ species, the value is zero. Hence without calculation, one finds if the integral is zero or not.

**14.** The D.P. principle can be used to enlarge any group by taking the D.P. of two groups. What do you mean by the D.P. of two groups $[G_1 \otimes G_2]$? What are the conditions for the application of this principle?

**Ans.** The D.P. of group $G_h^1(G_1, G_2, \ldots G_h)$ and $G_k^2(G_1, G_2, \ldots G_k)$ [$h$, $k$ are the order of groups] is defined by a group $G_{h\times k}^3$ of order $(h \times k)$. In $G_{hk}^3$ the elements are derived by taking the product of each element of $G_h^1$ with every element of $G_k^2$ [*i.e.*, $G_h^1(i) \,.\, G_k^2(j)$]. The condition for forming $G_{hk}^3$ are

($i$) $G_h$ and $G_h^2$ have no common element except $E$, (Identity) and

($ii$) Every element of $G_h^1$ commutes with all the elements of $G_k^2$.

Thus, $$\{G_h^1 \,.\, G_k^2\} = \{E \,.\, E, E, G_h^1(1)\, G_k^2(1),$$
$$G_h^1(1) \,.\, G_{k,}^2\,(2) \ldots\} = [E, G_h^1\; G_h^1(2) \ldots G_h^1(h) \text{ etc.}]$$

**15.** What is scalar Matrix?

**Ans.** Scalar Matrix is $\lambda I = \lambda \begin{vmatrix} 1 & 0 & 0 & \ldots \\ 0 & 1 & 0 & \ldots \\ 0 & 0 & 1 & \ldots \end{vmatrix}$ Here $\lambda$ = constant.

## CHAPTER SEVEN

**1.** What is a Hermitian Scalar product?

**Ans.** In our computation works, we often come across complex numbers. In such cases the ordinary scalar product is changed to Hermitian product in which $v_i^* \cdot v_j$ is used in place of $v_i \cdot v_j$. These modifications of the scalar product prevent non-zero vectors from having zero length. Sometimes $\overset{*}{v_i} \cdot v_j$ expressed as $v_i \cdot v_j$ without any astrics. For example, it $v_i = (i, 1, 0)$ its length should be $\sqrt{(i\ 10)\ (i\ 10)} = \sqrt{-1+1+0} = 0$, but if we take Hermitian scalar product, then its length = $\sqrt{(-i, 1, 0)\ (i, 1, 0)} = \sqrt{1+1+0} = \sqrt{2}$.

2. What is a Hermitian matrix?

**Ans.** If we have a matrix $|A|$, then Hermitian matrix is $|A^T|^*$, *i.e.*, transpose of $A$ ($A^T$) with its elements changed to complex conjugate of the elements of $|A|$.

3. What are the properties of Hermitian matrix?

**Ans.** Properties are (*i*) if H is a Hermitian matrix then $|H^T|^* = H$, (*ii*) Similarly transformation of a Hermitian matrix gives Hermitian matrix, (*iii*) All eigenvalues of a Hermitian matrix are real, (*iv*) Eigenvectors belonging to different eigenvalues of a Hermitian matrix are orthogonal, and (*v*) A Hermitian matrix can be diagonalised by a unitary transformation.

4. Prove that if $H$ (Hamiltonian) is Hermitian; its eigenfunctions belonging to different eigenvalues must be orthogonal.

**Ans.** We have,

$$H\psi = E_1\psi$$

So $$\int \psi_2^* H \psi_1 \partial\tau = E_1 \int \psi_1^* \psi_2 \partial\tau = E_1 S_{12} \text{ and } \int \psi_1^* H \psi_2 = E_2 S_0$$

Thus $$H_{12} = E_2\ S_{12} \text{ and } H_{12}^* = E^*\ S_{12}^*.$$

We thus have $$H_{12} - H_{21}^* = E_1\ S_{12} - E_2^*\ S_{12}^*$$

But $$S_{12} = S_{12}^*$$

So $$H_{12} - H_{21}^* = [E_1 - E_2^*]\ S_{12}$$

But $$E_1 - E_2^* \neq 0$$

So, $$S_{12}^* = S_{12} = \int \psi_2^* \psi_1 \partial\tau = 0$$

5. Prove that eigenvalues of a Hermitian operator are real.

**Ans.** *See* Text.

6. Prove that set of all non-singular matrices that commute with $H$ forms a group.

**Ans.** *See* text.

7. Use the character table to find the representations for *f*-orbitals in $C_{4V}$, $C_{4h}$, $D_{4h}$, $T_d$ and $O_h$.

**Ans.** *See* Text.

8. Prove that the potential and kinetic energy remain invariant in a system when a symmetry operation of the system is applied. Use the vibrational and electronic properties of the system.

**Ans.** *See* Chapter 12.

**9.** Why we cannot have a hybridised state of an atom in an unbound state. For example, we have electronic configuration of $C$ as $1s^2$, $2s^2$, $2p^2$ and not $1s^2$ $(SP^3)$?

**Ans.** *See* Text.

**10.** What is the basic idea behind hybridisation process? Explain.

**Ans.** *See* Text.

**11.** What are the transformation properties of AO of atom $A$ in $AX_n$ type molecules?

**Ans.** In $AX_n$ molecules, atom A holds a unique position because it lies on all the symmetry elements of the point group of $AX_n$. The atomic orbitals of $A$ have a radial part, unaffected by the symmetry operations of $AX_n$ point group. Only the angular part of $\Psi$-function depends upon the symmetry operation $[A(\theta, \phi)]$. $(B)$ $[A(\theta, \phi)]$ will be changed by the symmetry operations of $AX_n$ point group. Thus, the transformation properties of the orbitals of $A$ will be a function of the operations of the point group of $AX_n$. These transformation properties are given in the character table of the point group. For example, for $T_d$ point group the transformation properties of $s$-orbital is $A$; $(p_x, p_y, p_z)$, $T_2$; $(d_{x^2-y^2}, d_{z^2})$, $E$; $f_{xyz}$, $A$;$(f_{x^3}, f_{y^3}, f_{z^2})$, $T_2$; and rest of $f$-orbitals, $T_1$).

**12.** How are the wave functions of multi-electron atom's orbitals are transformed in $AX_n$ type molecule?

**Ans.** The basic principle for multi-electron system is the same as those for single electron system. In such cases, the wave function is equal to product of wave functions of single electron. However, the angular part of wave functions of all the electrons are the same, but the radial part differs because of mutual electron repulsion. Therefore because of the same angular part of the wave function of all electrons, the symmetry properties (transformation behaviour) can be taken as the same as that of single electron wave function of $H$-atom.

**13.** Explain the hybridisation scheme for $\sigma$-orbitals in tetrahedral $AX_4$ molecules.

**Ans.** We wish to know which atomic orbital on atom $A$ of $AX_4$ are needed to form a set of four $\sigma$-bonds, that is, orbitals on atom $A$ whose lobes are directed to the atoms $X$ *i.e.*, towards the corners of tetrahedron. This set of four orbitals will form a basis for the representation of the symmetry group of $AX_4$. These hybrids are regarded as vectors $r_1$, $r_2$, $r_3$ and $r_4$. The set of vectors will form a $\Gamma_{red}$ representation of the group. On applying the symmetry operations of the group, we get $\Gamma_{red}$ as [4 $(E)$, 1 $(8C_3)$, $O$, $(3C_2)$, $O(4S_4)$ and $2(6\sigma_d)$ which transform as $A_1$ and $T_2$ components. It means that the four orbitals on A atom which we need to combine to form hybrid orbitals must be such that one orbital must be of $A_1$ symmetry and three orbitals of $T_2$ symmetry species. These orbitals are $S$ and $p_x p_y p_z$ (or $d_{xy}$ $d_{yz}$ $d_{xz}$) orbitals respectively. Hence, the hybrid orbitals can be formed either as $sp^3$ or $sd^3$. The symmetry properties cannot differentiate between them ($sp^3$ of $sd^3$). These can be differentiated only on energy basis.

**14.** On the energy basis which orbitals (*p* or *d*-orbitals) of the underlined atoms in the following molecules/ions should be used to form tetrahedral hybrids.

$\underline{Mn}O_4^-, \underline{Cr}O_4^{2-}, \underline{C}Cl_4, [\underline{P}Cl_4]^-$

(*i*) $sp^3, sp^3, sp^3, sp^3$ (*ii*) $sd^3, sd^3, sp^3, sp^3$

(*iii*) $sd^3, sd^3, sp^3, sd^3$ (*vi*) $sd^3, d^3s, d^3s, sp^3$

**Ans.** (*ii*).

**15.** In $AB_5$ type molecule (trigonal bipyramid), we take five-bonds around atom *A*, ($D_{3h}$ symmetry). These five σ-bonds are used as basis for the formation of $\Gamma_{red}$ which contain I.Rs. $2A_1' + A_2'' + E$, *i.e.*,

$$\Gamma_{red}(5, 2, 1, 3, 0, 3) = 2A_1' + A_2'' + E$$

The atomic orbitals on atom *A* which transform as $A_1'$, $A_2''$ and *E* are (from the character table of $D_{3h}$) $A'(s, d_{z^2})$, $A_2''$ $(p_z)$ and $E(p_x, p_y)$ or $(d_{xy}, d_{x^2-y^2})$. Thus, we have the following hybrid schemes.

(*a*) $ns$ $(n + 1)s, p_z, p_x, p_y$ (*b*) $nd_{z^2}, (n + 1)d_{z^2}, p_z$

(*c*) $ns, (n + 1)s, p_z, d_{xy}, d_{x^2-y^2}$ (*d*) $nd_{z^2}, (n+1)d_{z^2}, p_z, p_x, p_y$

(*e*) $s, d_{z^2} (n+1)\ d_{z^2} p_z, d_{xy}, d_{x^2-y^2}$ (*f*) $s, d_{z^2} p_z, p_x, p_y$ $(dsp^3)$

(*g*) $s, d_{z^2}, p_z, d_{xy}, d_{x^2-y^2}$ $(d^3sp)$

Which one of the above combinations will be most appropriate on energy of the orbital basis for $PF_5$ and $MoCl_5$ molecules?

$PF_5$, $MoCl_5$

**Ans.** [(*e*) and (*f*) combined]

**16.** In $C_{4v}$ symmetry of $AB_5$ molecule, we have the following possibilities of hybridisation.

(*a*) $dsp^3$ (With $d = d_{x^2-y^2}$), $sd^4$, $spd^3$, $sd^2p^2$, $pd^4$, $p^3d^2$

In $AB_5$ ($D_{3h}$) molecule we also have $dsp^3$ ($d = d_{z^2}$).

Which one of the geometries of $AB_5$ ($C_{4v}$ or $D_{3h}$) will be obtained?

**Ans.** Which one of geometries of $AB_5$ ($C_{4v}$ or $D_{3h}$) of $AB_5$ will be obtained will depend upon whether $d_{x^2-y^2}$ or a $d_{z^2}$ is used for compound formation, possibly on energy reasons similar problem arises in other molecules also like in $AB_3$, $AB_4$ (planar), and $AB_6$ (octahedron).

**17.** What is the difference between a $\sigma$- and a $\pi$-bond?

**Ans.** The difference lies in terms of the number of nodal planes possessed by each atom $A$. Nodal planes or surfaces are the locii of all points at which the wave function has zero amplitude. As a result of this, the sign of $\Psi$-function changes on passing form one side of the surface to the other. A $\sigma$-bond or orbital is defined as one having no nodal plane or surface containing the bond axis. This concept extends further to $\delta$-orbital or bond which contain two nodal surfaces intersecting along the bond axis.

**18.** How are one $s$-and three $p$-orbitals of $B$ in $BF_3$ linearly combined to form $sp^2$ type of hybrid orbitals on $B$.

**Ans.** *See* Text.

**19.** What are the limitations of hybridisation scheme?

**Ans.** *See* Text.

**20.** What hybridisation scheme should be applicable in the following molecules?
(*i*) $AB_6$ (Anti-trigonal bipyramid) (*ii*) $AB_8$ (anti-square bipyramid) (*iii*) $AB_{12}$ (Icosahedron).

**Ans.** *See* Text.

## CHAPTER EIGHT

**1.** When we apply Schrödinger equation ($H\psi = E\psi$) to any system, we usually use in the equation Hamiltonian in the form of general or arbitrary functions and differential operators. We wish to know its eigenfunctions and eigenvalues, to obtain the Hamiltonian in the matrix form, we take as basis set $\psi_m = \sum_{i=1}^{p} a_{mi}\phi_i$. Where $m$ = 1, 2, 3...$p$. Using this function how do we calculate its eigenfunctions and eigenvalues ($E_m$).

**Ans.** Using $\phi_i'^s$ as basis set, we have,

The matrix element of the Hamiltonian $H_{ij}^{\phi} = \int \phi_i^* H \phi_j \partial\tau$.

Here $\phi_i'^s$ is not a complete set, *i.e.*, $p < n$ (which may be infinite). It only gives approximate eigenfunction and approximate eigenvalues.

Thus,
$$H\left[\sum_{j=1}^{p} a_{mj}\phi_j\right] = E_m\left[\sum_{j=1}^{p} a_{mj}\phi_j\right] \qquad ...(1)$$

(Here $E'$ is an approximate eigenvalue > $E_m$)

Multiplying Eq. (1) by $\phi_i^*$, we get after integrating it,

$$\int \phi_i H \sum_{j=1}^{p} a_{mj}\phi_j \partial\tau = E_m \int \phi_i^* \sum_{j=1}^{p} a_{mj}\phi_j \partial\tau$$

$$\sum_{j=1}^{p} a_{mj} H_{ij}^{\phi} \cong \sum_{j=1}^{p} a_{mj} E_m S_{ij} \qquad ...(2)$$

or

$$\sum_{j=1}^{p} (H_{y}^{\phi} - E_m S_{ij}) a_{ij} = 0$$

(This Eq. is obtained after multiplying by $\phi_i^*$ ).

In this fashion, we can multiply equation (1) by all $\phi_i^*$ (1, 2, 3... $p$) and integrate. We thus get $p$-equations of the type (2). Using these equations, we get, $p$-values of $E_m$ and for each $E_p$, we get one set of $a_{ij}$ [each set will give one $\psi_m = \sum_{j=1}^{p} a_{mj}\phi_i$ and so we get $p$-vectors which will be eigenfunctions.

**2.** What three conditions are to be satisfied for each eigenfunctions?

**Ans.** (1) Single valued (2) at ∞ it goes to zero and (3) $\int_{\infty} \psi_i^* \psi_i \partial\tau = 1$.

**3.** What are symmetry properties of Hamiltonian?

**Ans.** Hamiltonian operator is invariant under all the symmetry operation of the point group of the system to which Hamiltonian operator belongs.

**4.** What effects the symmetry of the Hamiltonian has on Hamiltonian matrix?

**Ans.** Hamiltonian matrix takes a diagonal form when its basis functions belong to the I.Rs. the group of molecule.

**5.** On what principle the answer of question 4 is based?

**Ans.** Suppose the basis set of functions $\phi_i$ are a "Symmetry Adapted set". It implies that these are the basis functions for the I.Rs of the symmetry group of $H$. Then the matrix element (integral)".

$H_{ij} = \int \phi_i^* H \, \phi_j \partial\tau$ is zero unless $\phi_i$ and $\phi_j$ are the same functions or are from two different set of bases which belong to the same I.Rs. and are corresponding partners in these two sets. This principle helps us in diagonalising secular determinant.

**6.** What is the basic idea in taking the wave functions to solve the Schrödinger equation by MO theory?

**Ans.** (*i*) All orbitals in a molecule extend over the entire molecule. It implies that the electrons occupying those orbitals are delocalised over the whole molecule.

(*ii*) Therefore, it is assumed that MOs can be written as linear combination of AO's on various atoms of the molecule. $\psi_{MO}$ thus can be expressed as $\psi_k = \sum_i C_{iK}\phi_i$, when $\psi_k$ are the $K^{th}$ MO and $\phi_i$, the AO of atom $i$.

**7.** Why are symmetry properties useful in MO theory?

**Ans.** Since MO is spread over the whole molecule. Its symmetry properties make it possible to determine the symmetry properties of the MO wave functions. Because of this it is possible to draw many useful conclusions about the bonding without actual computation. By the use of symmetry properties of the MO wave functions, it is possible to find which matrix element is zero without calculation and thus reduce the labour of calculation.

**8.** By using LCAO MO′s one can form a secular equation from which energy levels and the actual wave function of the MO can be calculated. What is the procedure of finding these out?

**Ans.** *See* Text.

**9.** What are Huckel approximations?

**Ans.** In order to calculate the energies of various MO′s one has to solve a secular equation of the type $\begin{vmatrix} H_{11} - ES_{11} \ldots & H_{1n} - ES_{1n} \\ H_{n1} - ES_{n1} & H_{nn} - ES_{nn} \end{vmatrix} = 0$. The process to calculate the various matrix elements $H_{ij}$ and $S_{ij}$, is quite tedious and one usually takes a very long time. For simplifying the process, one makes various approximations. The most drastic ones are those given by Huckel. These approximations are (1) All $H_{ij}$ and $S_{ij}$ are zero except the $i^{th}$ and $g^{th}$ wave functions (*viz.* $\psi_i$ and $\psi_j$) are on adjacent atoms. (2) It is applied to calculate the energies of π-orbitals where $H_{ii} = \alpha_i$ *i.e.*, energy of the atom $H_{ij} = \beta$, energy interaction between orbitals on adjacent atoms. Calculations are further simplified if $\alpha = 0$ and β is taken as unit of energy which measures the amount of stabilisation and destabilisation of MOs with respect to the ground state of atomic orbitals [*viz.* α which is taken as in zero state].

**10.** How is the secular equation is factored by the use of symmetry properties of the molecule?

**Ans.** *See* Text.

**11.** Take the example of tetramethylene ethane. Use Huckel method to calculate delocalisation energy, bond orders, and spin multiplicity.

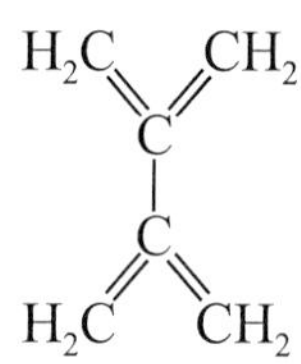


**Ans.** This question is for practice.

[**Hint:** *See* Cotton's book on symmetry and group theory.]

**12.** Calculate and discuss bonding in the π-system of allyl ion $H_2C \overset{\cdots}{=} C \overset{\cdots}{=} CH_2$ (Three centered bonding).

**Ans.** This is similar to diborane. Calculate as given in text.

**13.** *Nb*$_2$ is diamagnetic? Explain its diamagnetism on the basis of MO theory. What is its bond order?

**Ans.** $\sigma_g^2 \pi_i^\phi, \delta_g^\psi$, bond order 5.

**14.** $(Mo)_2$ is diamagnetic? What is its bond order? Explain on the basis of Mo theory. Consider two *sp* hybrids on each Mo atoms to combine to give four M.O.'s.

**Ans.** Bond order 6 with two π, 2δ-bonds + one σ-bond (strong) + one weak σ-bond from $s.d_z$ hybrid.

**15.** Bonding in $NF_5$ is:

(*i*) $C_{4V}$ symmetry square pyramid

(*ii*) Trigonal bipyramid $D_{3h}$

(*iii*) $[NF_4]^+ F^-$ $(Sp^3)$ Tetrahedral $(NF_4)^+$

(*iv*) $(NF_3)^{2+} F_2^-$ planar $(NF_3)$ $Sp^2$

**Ans.** (*iii*).

**16.** The bond orders of $C_2^-$, $Cl_2$ and $Cl_2^+$ are:

(*i*) $\frac{1}{2}, 1, 1.5$ (*ii*) 1, 1.5, 2

(*iii*) 1.5, 2, 2.5 (*iv*) 0, 0.5, 1.5

**Ans.** (*i*).

**17.** Give a MO description of π-bonding in $ClO_3^+$. Give a qualitative energy level diagram.

**Ans.** Point group $C_{3v}$, *d*-orbitals $d_{z^2}$, $[d_{(x^2-y^2)} d_{xy}]$, $(d_{xz}, d_{yz})$ [*p*-orbitals, $(A_2 + E), p_x$; $(A_i + E), p_y$; combination of oxygen orbitals], $A_2$. Group order is non-bonding; $p_y$ group orbitals combine with and $(d_{xz} \cdot d_{yz})$. The $p_x(e)$ group orbitals combine with $(d_{x^2-y^2}, d_{xy})$.

**18.** Give MO description of σ-orbitals of $SF_4$.

**Ans.** Point group $C_{2V}$, $\Gamma\sigma = 2A_1 + B_1 + B_2$. The non-bonding lone pair (bonding) is $A_1$ which is $(S + p_z - d_{z^2})$ hybrid orbital. One $A_1$-orbital $(S + p_z - d_{z^2})$ hybrid is used to bond axial *F* atoms. The other $\sigma A_1$-orbital $(S - p_z + d_{z^2})$ hybrid is used to bind the equatorial *F*-atoms. The *S* (sulphur) $p_z$-orbital ($B_1$ species I.R.) is used for axial *F*-atoms and $p_y(B_2$. I.R) is equatorial *F* atoms.

## CHAPTER NINE

**1.** $[C_4O_4]^{2-}$ is 4-member cyclic molecule (planar). The σ-hybridisation for the carbon orbital in 4-membered ring is

(*i*) *Sp* (*ii*) $Sp^2$

(*iii*) $Sp^3$ (*iv*) Pure *p*-orbitals ($P_z$ and $P_y$)

**Ans.** (*ii*).

**2.** In $\left[C_4O_4\right]^{2-}$, π-bonds between oxygen and carbon are formed by the overlap of

(*i*) $p_x$ (*ii*) $p_y$

(*iii*) $p_z$ (*iv*) $(s + d_{z^2})$

**Ans.** (*iii*).

**3.** Give MO treatment for the π-bonding involving all eight atoms of $[C_4O_4]^{2-}$.

**Ans.** Assuming a planar ring, the point group of the molecule is $D_{4h}$. Taking $p_z$-orbitals on carbon and oxygen atoms each, for the formation of π-bond, the $\Gamma_{red}$ representations are for $C$ and $O$ atom $p_z$-orbitals separately. I.R are contained in this $\Gamma_{red} = A_{2u} + B_{2u} + E_g$. Using projecting operators for $A_{2u}$, $B_{2u}$, and $E_g$, we get MOs. As,

$$A_{2u} = \phi_1 + \phi_2 + \phi_3 + \phi_4 \qquad (\phi_i = p_z\text{-orbitals})$$
$$B_{2u} = \phi_1 - \phi_2 + \phi_3 - \phi_4$$
$$E_g = \phi_1 - \phi_3 \text{ and } \phi_2 - \phi_3$$

MOs. are:

$$A_{2u}(C) \pm A_{2u}(O) = A_{2u} + A_{2u}^*$$
$$B_{2u} \pm B_{2u} = B_{2u} + B_{2u}^*$$

and

$$E_g \pm E_g = E_g + E_g^*$$

Energy levels is

$A_{2u}^2 < E_g^2 < B_{2u}^2 < B_{2u}^* < E_g^{*0} + A_{2u}^*$ (Two unpaired electrons in $E_g$ level).

**4.** Give the π-MO of benzene molecule.

**Ans.** *See* Text.

**5.** Give the π MOs. $[C_4H_4]^{2-}$.

**Ans.** Same as ques 3. except that hydrogen atoms will use 1*s*-orbitals rather then $p_z$-orbitals of oxygen in oxygen in $[C_4O_4]^{2-}$. This is six π-electron system. So, energy level of $[C_4H_4]^{2-}$ is $A_{2u}^2 < E_j^2 < B_{2u}^0 < E_{2u}^* < E_j^0 + A_{2u}^*$.

**6.** Give the MO for the *M*-ligand interaction in $[M(C_4H_4)_2^{2-}]$ using π-MO of $[C_4H_4]^{2-}$.

**Ans.** Point group $D_{4h}$. Ligand Orbitals *see* text for $[Cr(C_6H_6)_2]$

$$a_{2u} \pm a_{2u} \rightarrow a_{1g} + a_{1u}$$
$$e_g \pm e_g \rightarrow e_u + e_g$$
$$B_{2u} \pm B_{2u} \rightarrow B_{1g} + B_{2u}$$

Metal orbitals in $D_{4h}$

$$s, d_{z^2} - A_{1j} \qquad p_z - A_{2u}$$

$$d_{x^2-y^2} - B_{1j} \qquad p_z, p_y - E_u$$

$$d_{xy}, d_{yz} - E_j \qquad d_{xy} - B_{2j}$$

(From character table of $D_{4h}$)

Now no metal $B_{2u}$-orbital, so ligand $B_{2u}$ = non-bonding

No ligand $B_{2g}$-orbital, so metal $d_{xy}$ = non-bonding

Thus, we will be having bonding orbitals ($a_u$, $a_{2u}$, $e_u$, $e_g$, $b_{1g}$). Non-bonding ($B_{2u}$, $B_{2g}$) and anti-bonding $[a_{1u}^*, a_{2u}^*, e_u^*, e_g^*, B_{1g}^*]$.

**7.** Give the π-MO description of butadiene [$CH_2$=CH—CH=$CH_2$] (for both *cis*- and *trans*-isomers).

**Ans.** *See* Text.

**8.** Explain the term Coulomb, resonance and overlap integral.

**Ans.** *See* Text.

**9.** Organic compounds containing delocalised π-system and hetero atoms like *N*, *O*, *S* can be treated by the Huckel molecular orbital method; but with a slight modifications. What are the modifications and why should these modifications arise?

**Ans.** *See* Text.

**10.** Give the MO description for $B_2H_6$. How does it differ from the case where 3-centred bond is formed in a ring compound?

**Ans.** *See* Text.

## CHAPTER TEN

**1.** In the study of concerted reactions and the formation of stereoisomers of various molecules, we talk about the symmetries of the molecules, match the list in column 1 with the items on column 2 regarding these symmetries.

| List 1 | List 2 |
|---|---|
| (I) Concerted reaction | (*A*) Symmetry of the molecule as a whole |
| (II) Stereochemistry | (*B*) Symmetry of the atomic orbital |

Pick the right choice out of the following ones.

(*a*) I-A, II- B (*b*) I-A, II-A

(*c*) I-B, II-A (*d*) I-B, II-B

**Ans.** (*c*).

**2.** Pick correct statements defining concerted reactions.

(*i*) These reactions involve single transition state.

(*ii*) These reactions involve multiple transitions states.

(*iii*) Only ionic intermediates are formed.

(*iv*) Only covalent intermediates are formed.

(*v*) Both ionic and covalent intermediates can be formed.

(*vi*) Only radical intermediates can form.

**Ans.** (*i*).

**3.** If a concerted reaction is not allowed according to the selection rules, it implies that,

(*i*) These reactions can never take place.

(*ii*) These reactions can take place under different reaction conditions, because their activation energies are very high.

(*iii*) These reactions can easily take place because their activation energies are low.

(*iv*) These reactions are independent of the values of activation energies.

**Ans.** (*ii*).

**4.** Classify pericyclic reactions, why are these known as pericyclic reactions.

**Ans.** *See* Text.

**5.** Give the symmetries of orbitals which actually participate in the Diels-Alder's reactions.

**Ans.** *See* Text.

**6.** Which symmetry or symmetries are conserved when the reaction goes through con- and disrotatory process?

**Ans.** *See* Text.

**7.** In thermal and photochemical reactions, the orbital rotate as,

(*i*) Conrotation

(*ii*) Disrotation

(*iii*) Conrotation in thermal and disrotation in photochemical reaction.

(*iv*) Both rotate differently.

**Ans.** (*iv*).

**8.** What is the basic difference between the orbital correlation approach and the frontier orbital approach while discussing the mechanism of cyclo-addition reactions?

**Ans.** *See* Text.

**9.** Which of the following statements are correct?

(*a*) Will all overlaps can bring about reactions. There is no condition which has to be observed for any effective overlap.

(*b*) All pericyclic reactions can be described as "electrophilic" or "nucleophilic" reactions.

Give reasons for your answer.

(*i*) (*a*) is correct (*ii*) Only (*b*) is correct

(*iii*) Both are wrong. (*iv*) Both are correct.

**Ans.** (*iii*) *See* Text.

**10.** In what ways Pearson's rules of chemical reactions are related to the frontier orbital approach? Point out the similarities and dissimilarities between the two models?

**Ans.** *See* Text.

**11.** What do you mean by Huckel type of aromaticity and antiaromaticity of hydrocarbons? How does it differ from Mobius type of aromaticity? How are these related with the PMO. model used to explain pericyclic reactions (Dewar's model).

**Ans.** *See* Text.

**12.** What do you understand by "Odd-Alternant Hydrocarbons"?

**Ans.** Whenever in a molecule, the number of "starred atom" (of question 11) are different from the number of "unstarred atoms, the A.H. has at least as many non-bonding molecular orbitals (NBMO) as the difference between starred and unstarred atoms. One such case is where this difference is one. Such molecules are known as "Odd-Alternant Hydrocarbons (OAH) *e.g.*, allyl radical $[\dot{C}_3 H_7]$.

**13.** What is the full form of PMO method?

**Ans.** Perturbational Molecular Orbital method.

**14.** What is the basic concept of PMO method?

**Ans.** It involves direct examination of the "Transition state" of the reaction or transformation process. The rules are the following:

(*i*) "Thermal electrocyclic reactions" take place *via* "Aromatic transition" state.

(*ii*) Photochemical reactions (Pericyclic) take place through excited states or forms of "Antiaromatic transition state".

**15.** What do you mean by "Correlation Diagram"? Point out the position of the transition state in a "Correlation Diagram"? Do we show it by a fixed line?

**Ans.** *See* Text.

## CHAPTER ELEVEN

**1.** CFT (Crystal field theory), LFT (Ligand field theory) and MO theory are

(*i*) All are perimetric theory.

(*ii*) CFT and LFT are perimetric and MO is not.

(*iii*) CFT is parametric, LFT and MO are not.

(*iv*) None of the above theories are parametric theory.

**Ans.** (*ii*).

2. What is the electronic configuration of $Cr^{3+}$ in a tetrahedral complex $[CrA_4]^{3+}$ on the basis of crystal field model?

   **Ans.** [Cr]. $3d_{xy}^1\ 3d_{yz}^1\ 3d_{xz}^1\ 3d_{z^2}^0\ 3d_{x^2-y^2}^0\ 4s^0 4p^0$.

3. Are the 3*d*-orbitals of $Na^+$ split in a crystal of NaCl? Give reasons for your answer?

   **Ans.** In a NaCl crystal, each sodium ion is surrounded by six $Cl^{-1}$ion (–ve charges) having $O_h$ symmetry. Although there are no electrons in the 3*d*-orbitals of $Na^+$, but its symmetry is changed from spherical (*R*3) to $O_h$, which will lift the degeneracy of 3*d*-orbital and split into $t_{2g}$+ $e_g$.

4. If we take $A^-$ ion [*e.g.*, $Cl^-$] and surround it by six positive ions [*e.g.*, $Na^+$] will the 3*d*-orbitals of $Cl^-$ be split into $t_{2g}^* + e_g$.

   **Ans.** In principle yes; but in this case in $O_h$ field, $e_g$ orbitals will be lower in energy than $t_{2g}$-orbitals.

5. Give an expression for *Dq*. In *Dq*, what *D* and *q* stand for?

   **Ans.** *See* Text.

6. How do the atomic orbitals of an atom in $3d^1$ system differ from that of $3d^n$ ($n < 10$) atomic orbitals?

   **Ans.** *See* Text.

7. How *d*-orbitals of a central atom *A* split in a molecule $AX_5$ having $D_{3h}$ symmetry? What are their relative symmetries? (Use crystal field model)

   **Ans.** *See* Text.

8. In a linear field of *B* in $AB_2$ molecule, how are *p*-orbitals split. Using crystal field model, what are their relative symmetries?

   **Ans.** *See* Text.

9. In what way Ligand field differs from Crystal field theory?

   **Ans.** LFT is a modification of CFT in which we drop the CF's assumption that "the partially filled electron shell is one consisting of pure *d*-orbitals". In LFT we allow partial overlap of *d*-orbitals with ligand orbitals.

10. What effects the partial atomic *d*- and ligand orbitals have on metal *d*-orbitals?

    **Ans.** (*i*) For delocalisation of non-bonding *d*-orbitals over ligand orbitals interelectronic *d*-orbital repulsion decreases.

    (*ii*) We abandon all hopes of making a *priori* calculations of orbital splitting by a pure point charge electrostatic model using pure *d*-orbitals wave functions and crystal field parameters like effective charge, dipoles of the ligands.

M-L distance, radial part of wave function $R_{(r)}$ lose their physical significance.

(*iii*) *Energy level of $d^n$ configurations.* The energy level separations becomes as adjustable parameters.

**11.** Using the following formula prove that the '$l$' orbitals ($l = 6$) in a octahedral field are split as $A_{1g} + A_{2g} + E_g + T_{1g} + 2T_{2g}$.

$$\chi(\alpha) = \frac{\sin(l + 1/2)\alpha}{\sin \alpha/2} (\alpha \neq 0)$$

**12.** For $d^8$ system, the ground state of the ion without any ligand interaction is $^3F$. Prove that when ligands start interacting in an octahedral field of ligands, the atomic orbitals split as $^3T_{1g}, {}^3T_{2g}, {}^3A_{2g}$.

**13.** Complete the following table:

| **Orbital symm.** | $O_h$ | $D_{4h}$ | $D_3$ | $D_{5d}$ |
|---|---|---|---|---|
| $d$ | $E_g + T_{2g}$ | $A_{1g} + X_1 + B_{2g} + E_g$ | $A_1 + X_2$ | $A_1 + B_1 + X_4 + E$ |
| $p$ | $T_{1u}$ | $A_{1u} + X_3$ | $A_2 + E$ | $X_4 + E$ |

**Ans.** $X_1 = B_{1g}$, $X_2 = E$
$X_3 = E_u$, $X_4 = B_2$

**14.** What is the difference between Sugano diagram and Orgel diagram?

**Ans.** *See* Text.

**15.** Match the terms in List I to those in List II.

| **List I** | **List II** |
|---|---|
| 1. $10D_q$ | (*a*) Energy level difference in symmetry $O_h$ |
| 2. $^3F$ | (*b*) $^3A_2 + {}^3T_1 + {}^3T_2$ |
| 3. Mag. Moments $L$ of two electrons | (*c*) One band around 600 mμ |
| 4. Covalent character | (*d*) Spectrochemical series |

Your choices are

(*A*) (1) *a*, (2) *b*, (3) *c*, (4) *d*, (*B*) (1) *a*, (2) *b*, (3) none, (4) none,

(*C*) (1) *b*, (2) *a*, (3) *c*, (4) *d*, (*D*) (1) None, (2) *c*, (3) *d*, (4) *a*

**Ans.** (B).

**16.** How one modifies LFT by AOM?

**Ans.** See Text.

## CHAPTER TWELVE

**1.** If you are sitting on the centre of mass of a molecule *e.g.*, $AX_y$ at a temp. zero degree Kelvin, which of the following molecular motions you are watching and how many motions can you watch?

(*i*) Translational motions, 3 in number along *x*-, *y*- and *z*-axis.

(*ii*) Rotational motions, 3 in number about *x*-, *y*-, and *z*-axis.

(*iii*) Vibrational motion $(3n - 6)$ in number, all different depending on vibrational frequency.

(*iv*) All the three motions, $(3n)$ in number ($n$ = number of atoms in the molecule.

**Ans.** (*iii*).

**2.** The order of energy associated with

(*i*) Translation (*ii*) Rotational

(*iii*) Vibrational

(*iv*) Electronic motions of a molecule at a temperature T°K is.

(*a*) (*i*) > (*ii*) > (*iii*) > (*iv*) (*b*) (*i*) < (*ii*) < (*iii*) < (*iv*)

(*c*) (*i*) < (*iv*) < (*iii*) < (*ii*) (*d*) (*iii*) < (*iv*) < (*ii*) < (*i*)

**Ans.** (*b*).

**3.** We assume that the vibrational potential energy of stretching or contracting a bond between two atoms can be approximated by

(*i*) Hook's law (*ii*) Coulomb's law

(*iii*) Exchange forces (*iv*) van der Waals forces.

**Ans.** (*i*).

**4.** It has been established that harmonic oscillation approximation is not valid for the higher vibrational levels *i.e.*, $\nu = \frac{1}{2\pi}\sqrt{\frac{k}{\mu}}\left[\text{or } V = \frac{1}{2}k(\Delta r)^2\right]$. But one can still find relations like $V = \frac{1}{2}k'(\Delta r)^2$ to describe the potential energy of vibrations at higher energy levels (*i*) How is $k'$ related to $k$? (*ii*) As energy level $n$ increases, how does $k'$vary? (*iii*) What does $k'$ signify?

**Ans.** (*i*)$k' < k$, *i.e.,* force constant decreases as $n$ increases.

(*ii*) As the value of $k$ increases, then unlike $k$, $k'$ decreases with $n$.

(*iii*) As $n \to \infty$, $k' \to 0$, *i.e.,* the molecule dissociates as $n \to \infty$.

**5.** The following are statements regarding the relationship between force constants of $k$ diatomic molecule and vibrational frequencies. Point out the wrong statement.

(*i*) The stretching frequencies of isotopes (*e.g.,* $H_2$, $D_2$, $T_2$) depend upon masses while stretching frequencies do not.

(*ii*) The force constant of a bond is always proportional to bond strength.

(*iii*) Bending force constant is always higher than stretching force constants.

(*iv*) Force constant is directly proportional to ν.

**Ans.** (*ii*) and (*iii*).

**6.** In a potential energy vs. bond length curve of a diatomic molecule the force constant and dissociation energies respectively are.

(*i*) Tangent at a point and depth of the curve from zero energy value.

(*ii*) Depth of the curve in both cases.

(*iii*) Tangent at a point in both cases.

(*iv*) Depth of curve form zero energy line and tangent at a point.

**Ans.** (*i*).

**7.** The vibrations of atoms (nuclei) in a molecule are random and highly disorderly. How can one define the vibration of a nucleus?

**Ans.** Classically the displacement of a nucleus from its equilibrium position is the vectorial sum of the displacements associated with the spatial vibration in coordinate system in which the nucleus moves in straight line and in-phase.

**8.** In question 7, we have used the word in-phase. What do you understand by this word?

**Ans.** The in-phase property of the vibrational motions of the nuclei of a molecule means that all nuclei in the molecule pass through their equilibrium point and reach their maxima at the same time and simultaneously. Such vibrations are known as the Normal, or Fundamental modes. The number of such modes are $(3n - 6)$.

**9.** Explain the concept of normal modes in detail.

**Ans.** *See* Text.

**10.** Does the total energy of the molecule vary during vibrations?

**Ans.** Although the *KE* and *PE* vary separately during the vibration of atoms in a molecule; but the total, (*KE* + *PE*) remains invariant.

**11.** In a diatomic molecule, we have only two atoms that are vibrating and vibrations are defined as stretching or compression of a bond. But in polyatomic molecule, we have '*n*' number of atoms which are vibrating. How do we define the vibrations in that case?

**Ans.** We define in terms of normal modes after we define a particular coordinate system placed on every atom. For further detail see text and question 7.

**12.** Besides normal modes of vibrations, are there any other coordinate system in which we define the vibrational motions of atoms in molecules?

**Ans.** Yes. (*i*) Cartesian coordinate system (*ii*) Internal coordinate system. As a matter of fact any coordinate system can be used to define the vibrational motions *e.g.*, symmetry coordinate system.

**13.** Explain "symmetry coordinate system".

**Ans.** *See* Text.

**14.** Define potential energy of vibrations in terms of cartesian coordinate system.

**Ans.** *See* Text.

**15.** Define two important properties of normal modes of vibration.

**Ans.** The important properties are

(*a*) Each vector attached to an atom of the molecule in a normal coordinate and which represents an instantaneous 'atomic displacement may be regarded as a resultant' of three basic vectors attached to that atom (*e.g.*, $\Delta x$, $\Delta y$, $\Delta z$).

(*b*) Each of the normal mode forms a basis for or belongs to an I.R. of the molecular point group.

**16.** Name two coordinates systems in which the normal coordinate is defined.

**Ans.** Cartesian and Internal coordinate system.

**17.** How do we determine the symmetry type (I.R.) of normal coordinate?

**Ans.** This can be achieved by using the character table of the point group of the molecule.

**18.** Sketch the normal modes of vibrations of $AB_3$ molecules and find their symmetry species (I.R.).

**Ans.** *See* Text.

**19.** Give stepwise procedure of normal coordinate analysis of a molecule. Find the normal coordinates of $AB_4$ ($T_d$) molecules.

**Ans.** *See* Text (for $AB_4$ type $A_1 + E + 2T_2$).

**20.** Calculate the isotope shift by substituting $^{12}$C by $^{13}$C in CO molecule.

**Ans.** $$\frac{\nu_{^{12}CO}}{\nu_{^{13}CO}} = \sqrt{\frac{\mu_{^{12}CO}}{\mu_{^{13}CO}}} \quad \text{so,} \quad \nu_{^{13}CO} = \nu_{^{12}CO}\sqrt{\frac{\mu_{^{13}CO}}{\mu_{^{12}CO}}}$$

$$\nu_{^{12}CO} = 2143\ \text{cm}^{-1} \text{ shift } \Delta\nu = 47.4\ \text{cm}^{-1}.$$

**21.** What do you mean by "Fermi Resonance"? How does it correlate with overtone and combination bands?

**Ans.** *See* Text.

**22.** How do we find the in-plane and out-of-plane vibrational bands in a $AB_4$ planar molecule?

**Ans.** *See* Text.

**23.** How do we determine "Which I.R. bands are I.R. active and which are Raman-active"?

**Ans.** By the use of character table of molecular point group; for method *see* text.

**24.** The molecular symmetry of benzene is $D_{6h}$.

Determine:

(*i*) the symmetry types of I.R. vibrations.

(*ii*) the activities of each type of vibrations.

(*iii*) all the in-plane vibrations to whether they involve C—H stretching, C—C stretching or angle bending.

(*iv*) what vibrations involve out of plane angle bending.

**Ans.** *See* Text.

**25.** Set up the symmetry coordinates for the five in-plane vibrations of the *trans*- $N_2F_2$ molecule.

**Ans.** *See* Text.

# REFERENCES

## CHAPTER ONE TO SIX

1. M.Tinkham, *Group Theory and Quantum Mechanics*, McGraw Hill, New York, 1964.
2. G.F. Koster. J.O. Dimmock, et. al. *Properties of Thirty Two Point Groups.* MIT Press, Cambridge Massachusetts, 1963.
3. R. McWeeny, *Symmetry: An Introduction to Group Theory and its Applications,* MacMillan, New York, 1963.
4. D.C. Harris and M.D. Bertolucei, *Symmetry and Spectroscopy.* McGrawHill, NY (1960).
5. L.H. Hall, *Group Theory and Symmetry in Chemistry*, McGraw Hill, New York, 1969.
6. R.L. Flurry Jr., *Symmetry Groups*, Prentice Hall, Englewood Cliffs, New Jersey, 1980.
7. F.A. Cotton, *Chemical Application of Group Theory*. 2$^{nd}$ed. Wiley Inter Science, New York, 1971.
8. L.M. Fallicov, *Group Theory and Its Physical Applications*, The Chemistry of Chicago Press, Chicago, 1966.
9. H.Hammermesh, *Group Theory and its Applications to Physical Problems*. Addition-Wesley Publishing Company, Reading, Mass, 1962.
10. E.P.Wignor, *Group Theory and its Application to Quantum Mechanics of Atomic Spectra Academic Press,* New York, 1959.
11. David S. Schonland, *Molecular Symmetry: An Introduction to Group Theory and its Uses in Chemistry*, D. Vein Nostrand Company Ltd. London, 1965.
12. V. Hein, *Group Theory in Quantum Mechanics: An Introduction to its Present Usage*, Pergamon Press Oxford, 1969.
13. C.I. Ballhausen, *Introduction to Ligand Field Theory*, McGraw Hill, New York, 1962.
14. J.S. Griffith, *The Theory of Transition Metal Ions*. The Cambridge University Press, 1961.
15. E.L. Albasiny and J.R.A. Cooper, *Molecular Physics*, 4, 353 (1961)
16. Bodie E. Douglas and Charles A. Hollingsworth, *Symmetry in Bonding and Spectra: An Introduction,* Academic Press Inc., New York, 1985.

17. H.H. Jaffe and Milton Orchin, *Symmetry in Chemistry*, John Wiley & Sons, Inc., New York, 1967.
18. John R. Dyer, *Applications of Absorption Spectroscopy of Organic Compounds,* Prentice Hall of India Pvt. Ltd., New Delhi, 1969.

## CHAPTER SEVEN

1. P.W. Atkins, *Molecular Quantum Mechanics*, Oxford University Press, New York, 1970.
2. W. Kaugmann, *Quantum Chemistry*, Academic Press, New York, 1957.
3. L. Pauling and E.B.Wilson Jr., *Introduction to Quantum Mechanics*, McGraw Hill, New York, 1935.
4. F.L. Pilar, *Elementary Quantum Chemistry*, McGraw Hill, New York, 1965.

## CHAPTER EIGHT

1. C.L. Ballhausen and H.B. Gray, *Molecular Orbital Theory*, Benzamin-Cummings, New York, 1965.
2. R.L. Dekock and H.B. Gray, *Chemical Structure and Bonding,* Benzamin-Cummings, Menlo Park, California, 1980.
3. J.P. Lowe, *Quantum Chemistry*, Academic Press, New York, 1972.

## CHAPTER NINE

1. M.J.S. Dewar, *The Molecular Orbital Theory of Organic Chemistry*, McGraw Hill, New York, 1969.
2. B.F.G. Johnson, *Transition Metal Clusters*, Wiley InterScience, New York, 1980.
3. J.P. Lowe, *Quantum Chemistry*, Academic Press, New York, 1978.
4. A. Liberces, *Introduction to Molecular Orbital Theory*, Holt, New York, 1966. (Also for chapter seven)

## CHAPTER TEN

1. R.G. Pearson, *Symmetry Rules for Chemical Reactions*, Wiley InterScience, New York, 1976.
2. R.B. Woodward and R. Hoffmann, *The Conservation of Orbital Symmetry*, Verlag Chemie, Gmbh, Weinheim/Bergstrasse, 1970.
3. K. Fukui, *Account Chem. Res. 197*, Angew Chem. Int. Edu., 1982, 11. 801.

## CHAPTER ELEVEN

1. B.N. Figgis, *Introduction to Ligand Fields,* Wiley, New York, 1966.
2. M.Gerlock and R.C. Slade, *Ligand Field Parameters*, Cambridge University Press, New York, 1961.
3. L.E. Orgel, *An Introduction to Transition Metal Chemistry, Ligand Field Theory,* Methuen, London, 1960.
4. H.L. Schlafer and G. Glieman, *Basic Principles of Ligand Field Theory*, Wiley InterScience, NewYork, 1969.
5. A.P.B. Lever, *Inorganic Electronic Spectroscopy*, 2nd ed., Elsevier, Amsterdam, 1984.

## CHAPTER TWELVE

1. E.B. Wilson Jr., J.C. Decius, and P.C. Cross, *Molecular Vibrations*, McGraw Hill, New York, 1955.
2. K. Nakamoto, *Infrared and Raman Spectra of Inorganic and Coordination Compounds*, 3rded. Wiley InterScience, New York, 1978.
3. T.R. Conley, *Infrared Spectroscopy*, Allyn and Bacon, Boston, 1972.
4. G. Herzberg, *Infrared and Raman Spectra of Polyatomic Molecules*, Van Nostrand Reinhold New York, 1945.

# Character Tables

**A.** Non-Axial Groups

| $C_1$ (1) | $E$ | $h = 1$ |
|---|---|---|
| A | 1 | |

| $C_s = C_h$ ($m$) | $E$ | $\sigma_h$ | $h = 2$ | |
|---|---|---|---|---|
| $A'$ | 1 | 1 | $x, y, R_z$ | $x^2, y^2$ $z^2, xy$ |
| $A''$ | 1 | −1 | $z, R_x, R_y$ | $yz, xz$ |

| $C_i = S_2$ (1) | $E$ | $i$ | $h = 2$ | |
|---|---|---|---|---|
| $A_g$ | 1 | 1 | $R_x, R_y, R_Z$ | $x^2, y^2, z^2$ $xy, xz, yz$ |
| $A_u$ | 1 | −1 | $x, y, z$ | |

**B.** $C_{nv}$ Groups

| $C_{2v}$, $2mm$ | $E$ | $C_2$ | $\sigma_y$ | $\sigma'_y$ | $h = 4$ | |
|---|---|---|---|---|---|---|
| $A_1$ | 1 | 1 | 1 | 1 | $z, z^2, x^2, y^2$ | |
| $A_2$ | 1 | 1 | –1 | –1 | $xy$ | $R_z$ |
| $B_1$ | 1 | –1 | 1 | –1 | $x, xz$ | $R_y$ |
| $B_2$ | 1 | –1 | –1 | 1 | $y, yz$ | $R_x$ |

| $C_{3v}$, $3m$ | $E$ | $2C_3$ | $3\sigma_v$ | $h = 6$ | |
|---|---|---|---|---|---|
| $A_1$ | 1 | 1 | 1 | $z, z^2, x^2 + y^2$ | |
| $A_2$ | 1 | 1 | –1 | | $R_z$ |
| $E$ | 2 | –1 | 0 | $(x, y), (xy, x^2 - y^2)\ (xz, yz)$ | $(R_x, R_y)$ |

| $C_{4v}$, $4mm$ | $E$ | $C_2$ | $2C_4$ | $2\sigma_v$ | $2\sigma_d$ | $h = 8$ | |
|---|---|---|---|---|---|---|---|
| $A_1$ | 1 | 1 | 1 | 1 | 1 | $z, z^2, x^2 + y^2$ | |
| $A_2$ | 1 | 1 | 1 | –1 | 1 | | $R_z$ |
| $B_1$ | 1 | 1 | –1 | 1 | –1 | $x^2 - y^2$ | |
| $B_2$ | 1 | 1 | –1 | –1 | 1 | $xy$ | |
| $E$ | 2 | –2 | 0 | 0 | 00 | $(x, y), (xz, yz)$ | $(R_x, R_y)$ |

| $C_{5v}$ | $E$ | $2C_3$ | $2C_5^2$ | $5\sigma_v$ | $h = 10, \alpha = 72°$ | |
|---|---|---|---|---|---|---|
| $A_1$ | 1 | 1 | 1 | 1 | $z, z^2, x^2 + y^2$ | |
| $A_2$ | 1 | 1 | 1 | –1 | | $R_z$ |
| $E_1$ | 2 | $2 \cos \alpha$ | $2 \cos 2\alpha$ | 0 | $(x, y), (xz, yz)$ | $(R_x R_y)$ |
| $E_2$ | 2 | $2 \cos \alpha$ | $2 \cos 2\alpha$ | 0 | $(xy, x^2 - y^2)$ | |

| $C_{6v}$, $6mm$ | $E$ | $C_2$ | $2C_3$ | $2C_6$ | $3\sigma_d$ | $3\sigma_v$ | $h = 12$ | |
|---|---|---|---|---|---|---|---|---|
| $A_1$ | 1 | 1 | 1 | 1 | 1 | 1 | $z, z^2, x^2 + y^2$ | |
| $A_2$ | 1 | 1 | 1 | 1 | –1 | 1 | $R_z$ | |
| $B_1$ | 1 | –1 | 1 | –1 | –1 | 1 | | |
| $B_2$ | 1 | –1 | 1 | –1 | 1 | –1 | | |
| $E_1$ | 2 | –2 | –1 | 1 | 0 | 0 | $(x, y), (xz, yz)$ | $(R_x, R_y)$ |
| $E_2$ | 2 | 2 | –1 | –1 | 0 | 0 | $(xy, x^2 - y^2)$ | |

| $C_{\infty v}$ | $E$ | $2C_\phi$† | $\infty\sigma_v$ | $h = \infty$ | |
|---|---|---|---|---|---|
| $A_1(\Sigma^+)$ | 1 | 1 | 1 | $z, z^2, x^2 + y^2$ | |
| $A_2(\Sigma^-)$ | 1 | 1 | –1 | | $R_z$ |
| $E_1(\Pi)$ | 2 | $2 \cos \phi$ | 0 | $(x, y), (xz, yz)$ | $(R_x, R_y)$ |
| $E_2\ (\Delta)$ | 2 | $2 \cos 2\phi$ | 0 | $(xy, x^2 - y^2)$ | |

† There is only member of this class if $\phi = \pi$.

## C. $D_n$ Groups

| $D_2$, 222 | $E$ | $C_2^z$ | $C_2^y$ | $C_2^x$ | $h = 4$ | |
|---|---|---|---|---|---|---|
| $A_1$ | 1 | 1 | 1 | 1 | $x^2, y^2, z^2$ | |
| $B_1$ | 1 | 1 | –1 | –1 | $z, xy$ | $R_z$ |
| $B_2$ | 1 | –1 | 1 | –1 | $y, xz$ | $R_y$ |
| $B_3$ | 1 | –1 | –1 | 1 | $x, yz$ | $R_x$ |

| $D_3$, 32 | $E$ | $2C_3$ | $3C'_2$ | $h = 6$ | |
|---|---|---|---|---|---|
| $A_1$ | 1 | 1 | 1 | $z^2, x^2 + y^2$ | |
| $A_2$ | 1 | 1 | –1 | $z$ | $R_z$ |
| $E$ | 2 | –1 | 0 | $(x, y), (xz, yz), (xy, x^2 - y^2)$ | $(R_x, R_y)$ |

| $D_4$, 422 | $E$ | $C_2$ | $2C_4$ | $2C'_2$ | $2C''_2$ | $h = 8$ | |
|---|---|---|---|---|---|---|---|
| $A_1$ | 1 | 1 | 1 | 1 | 1 | $z^2, x^2 + y^2$ | |
| $A_2$ | 1 | 1 | 1 | –1 | –1 | $z$ | $R_z$ |
| $B_1$ | 1 | 1 | –1 | 1 | –1 | $x^2 - y^2$ | |
| $B_2$ | 1 | 1 | –1 | –1 | 1 | $xy$ | |
| $E$ | 2 | –2 | 0 | 0 | 0 | $(x, y), (xz, yz)$ | $(R_x, R_y)$ |

## D. $D_{nh}$ Groups

| $D_{3h}$, $\bar{6}2m$ | $E$ | $\sigma_h$ | $2C_3$ | $2S_3$ | $3C'_2$ | $3\sigma_v$ | $h = 12$ | |
|---|---|---|---|---|---|---|---|---|
| $A'_1$ | 1 | 1 | 1 | 1 | 1 | 1 | $z^2, x^2 + y^2$ | |
| $A'_2$ | 1 | 1 | 1 | 1 | –1 | –1 | | $R_z$ |
| $A''_1$ | 1 | –1 | 1 | –1 | 1 | –1 | | |
| $A''_2$ | 1 | –1 | 1 | –1 | –1 | 1 | $z$ | |
| $E'$ | 2 | 2 | –1 | –1 | 0 | 0 | $(x, y), (xy, x^2 - y^2)$ | |
| $E''$ | 2 | –2 | –1 | 1 | 0 | 0 | $(xz, yz)$ | $(R_x, R_y)$ |

| $D_{4h}$, 4/mmm | $E$ | $2C_4$ | $C_2$ | $2C'_2$ | $2C''_2$ | $i$ | $2s_4$ | $\sigma_h$ | $2\sigma_v$ | $2\sigma_d$ | $h = 16$ | |
|---|---|---|---|---|---|---|---|---|---|---|---|---|
| $A_{1g}$ | 1 | 1 | 1 | 1 | 1 | 1 | 1 | 1 | 1 | 1 | $x^2 + y^2, z^2$ | |
| $A_{2g}$ | 1 | 1 | 1 | –1 | –1 | 1 | 1 | 1 | –1 | –1 | | $R_z$ |
| $B_{1g}$ | 1 | –1 | 1 | 1 | –1 | 1 | –1 | 1 | 1 | –1 | $x^2 - y^2$ | |
| $B_{2g}$ | 1 | –1 | 1 | –1 | 1 | 1 | –1 | 1 | –1 | 1 | $xy$ | |
| $E_g$ | 2 | 0 | –2 | 0 | 0 | 2 | 0 | –2 | 0 | 0 | $(xz, yz)$ | $(R_x, R_y)$ |
| $A_{1u}$ | 1 | 1 | 1 | 1 | 1 | –1 | –1 | –1 | –1 | –1 | | |
| $A_{2u}$ | 1 | 1 | 1 | –1 | –1 | –1 | –1 | –1 | 1 | 1 | $z$ | |
| $B_{1u}$ | 1 | –1 | 1 | 1 | –1 | –1 | 1 | –1 | –1 | 1 | | |
| $B_{2u}$ | 1 | –1 | 1 | –1 | 1 | –1 | 1 | –1 | 1 | –1 | | |
| $E_u$ | 2 | 0 | –2 | 0 | 0 | –2 | 0 | 2 | 0 | 0 | $(x, y)$ | |

| $D_{5h}$ | $E$ | $2C_3$ | $2C_5^2$ | $5C_2$ | $\sigma_h$ | $2s_5$ | $2s_5^3$ | $5\sigma_v$ | $h = 20$ | $\alpha = 72°$ |
|---|---|---|---|---|---|---|---|---|---|---|
| $A'_1$ | 1 | 1 | 1 | 1 | 1 | 1 | 1 | 1 | $x^2 + y^2, z^2$ | |
| $A'_2$ | 1 | 1 | 1 | –1 | 1 | 1 | 1 | –1 | | $R_z$ |
| $E'_1$ | 2 | 2 cos α | 2 cos 2α | 0 | 2 | 2 cos α | 2 cos 2α | 0 | $(x, y)$ | |
| $E'_2$ | 2 | 2 cos 2α | 2 cos 2α | 0 | 2 | 2 cos 2α | 2 cos α | 0 | $(x^2 - y^2, xy)$ | |
| $A''_1$ | 1 | 1 | 1 | 1 | –1 | –1 | –1 | –1 | | |
| $A''_2$ | 1 | 1 | 1 | –1 | –1 | –1 | –1 | 1 | $z$ | |
| $E''_1$ | 2 | 2 cos α | 2 cos 2α | 0 | –2 | –2 cos α | –2cos 2α | 0 | $(xz, xy)$ | $(R_x, R_y)$ |
| $E''_2$ | 2 | 2 cos 2α | 2 cos α | 0 | –2 | –2 cos 2α | –2cos α | 0 | | |

| $D_{\infty h}$ | $E$ | $2C_\phi$ | $\infty C'_2$ | $i$ | $2iC_\infty$ | $iC'_2$ | $h = \infty$ | |
|---|---|---|---|---|---|---|---|---|
| $A_{1g}$ $(\Sigma^+_g)$ | 1 | 1 | 1 | 1 | 1 | 1 | $z^2, x^2 + y^2$ | |
| $A_{1u}$ $(\Sigma^+_u)$ | 1 | 1 | 1 | –1 | –1 | –1 | | |
| $A_{2g}$ $(\Sigma^-_g)$ | 1 | 1 | –1 | 1 | 1 | –1 | $z$ | |
| $A_{2u}$ $(\Sigma^-_u)$ | 1 | 1 | –1 | –1 | 1 | 1 | | $R_z$ |
| $E_{1g}$ $(\Pi_g)$ | 2 | 2cos ϕ | 0 | 2 | –2 cosϕ | 0 | $(xz, yz)$ | $(R_x, R_y)$ |
| $E_{1u}$ $(\Pi_u)$ | 2 | 2cos ϕ | 0 | –2 | 2 cosϕ | 0 | $(x, y)$ | |
| $E_{2g}$ $(\Delta_g)$ | 2 | 2cos 2ϕ | 0 | 2 | 2 cos 2ϕ | 0 | $(xy, x^2 - y^2)$ | |
| $E_{2u}$ $(\Delta_u)$ | 2 | 2cos 2ϕ | 0 | –2 | –2 cos 2ϕ | 0 | | |
| ⋮ | | | | | | | | |

**E.** Higher Order Groups

| $T_d$, 43m | $E$ | $8C_3$ | $3C_2$ | $6\sigma_d$ | $6S_4$ | $h = 24$ | |
|---|---|---|---|---|---|---|---|
| $A_1$ | 1 | 1 | 1 | 1 | 1 | $x^2 + y^2 + z^2$ | |
| $A_2$ | 1 | 1 | 1 | –1 | –1 | | |
| $E$ | 2 | –1 | 2 | 0 | 0 | $(3z^2 - r^2, x^2 - y^2)$ | |
| $T_1$ | 3 | 0 | –1 | –1 | 1 | | $(R_x, R_y, R_z)$ |
| $T_2$ | 3 | 0 | –1 | 1 | –1 | $(x, y, z), (xy, xz, yz)$ | |

| $O_h$ $(m3m)$ | $E$ | $8C_3$ | $6C_2$ | $6C_2$ | $3C_2$ $(= C_4^2)$ | $i$ | $6S_4$ | $8S_6$ | $3\sigma_h$ | $6\sigma_h$ | $h = 48$ | |
|---|---|---|---|---|---|---|---|---|---|---|---|---|
| $A_{1g}$ | 1 | 1 | 1 | 1 | 1 | 1 | 1 | 1 | 1 | 1 | | $x^2 + y^2 + z^2$ |
| $A_{2g}$ | 1 | 1 | –1 | –1 | 1 | 1 | –1 | 1 | 1 | –1 | | |
| $E_g$ | 2 | –1 | 0 | 0 | 2 | 2 | 0 | –1 | 2 | 0 | | $(2z^2 - x^2 - y^2,$ |
| $T_{1g}$ | 3 | 0 | –1 | 1 | –1 | 3 | 1 | 0 | –1 | –1 | $(R_x, R_y, R_z)$ | $x^2 - y^2)$ |
| $T_{2g}$ | 3 | 0 | 1 | –1 | –1 | 3 | –1 | 0 | –1 | 1 | | $(xy, yz, xy)$ |
| $A_{1u}$ | 1 | 1 | 1 | 1 | 1 | –1 | –1 | –1 | –1 | –1 | | |
| $A_{2u}$ | 1 | 1 | –1 | –1 | 1 | –1 | 1 | –1 | –1 | 1 | | |
| $E_u$ | 2 | –1 | 0 | 0 | 2 | –2 | 0 | 1 | –2 | 0 | | |
| $T_{1u}$ | 3 | 0 | –1 | 1 | –1 | –3 | –1 | 0 | 1 | 1 | $(x, y, z)$ | |
| $T_{2u}$ | 3 | 0 | 1 | –1 | –1 | –3 | 1 | 0 | 1 | –1 | | |

| $I$ | $E$ | $12C_5$ | $12C_5^2$ | $20C_3$ | $15C_2$ | $h = 60$ | |
|---|---|---|---|---|---|---|---|
| $A$ | 1 | 1 | 1 | 1 | 1 | | $z^2 + y^2 + z^2$ |
| $T_1$ | 3 | $\frac{1}{2}(1+\sqrt{5})$ | $\frac{1}{2}(1-\sqrt{5})$ | 0 | –1 | $(x, y, z)$ | |
| $T_2$ | 3 | $\frac{1}{2}(1-\sqrt{5})$ | $\frac{1}{2}(1+\sqrt{5})$ | 0 | –1 | $(R_x, R_y, R_z)$ | |
| $G$ | 4 | –1 | –1 | 1 | 0 | | |
| $G$ | 5 | 0 | 0 | –1 | 1 | | $(2z^2 - x^2 - y^2, x^2 - y^2, xy, yz, zx)$ |

# Annexure–II

## Character Tables for Chemically Important Symmetry Groups

**1.** The Non-axial Groups

| $C_1$ | $E$ |
|---|---|
| $A$ | 1 |

| $C_s$ | $E$ | $\sigma_h$ | | |
|---|---|---|---|---|
| $A'$ | 1 | 1 | $x, y, R_z$ | $x^2, y^2$<br>$z^2, xy$ |
| $A''$ | 1 | −1 | $z, R_x, R_y$ | $yz, xz$ |

| $C_i$ | $E$ | $i$ | | |
|---|---|---|---|---|
| $A_g$ | 1 | 1 | $R_x, R_y, R_Z$ | $x^2, y^2, z^2$<br>$xy, xz, yz$ |
| $A_u$ | 1 | −1 | $x, y, z$ | |

**2.** The $C_n$ Groups

| $C_2$ | $E$ | $C_2$ | | |
|---|---|---|---|---|
| $A$ | 1 | 1 | $z, R_z$ | $x^2, y^2, z^2, xy$ |
| $B$ | 1 | −1 | $x, y, R_x, R_y$ | $yz, xz$ |

| $C_3$ | $E$ | $C_3$ | $C_3^2$ | | $\varepsilon = \exp(2\pi i/3)$ |
|---|---|---|---|---|---|
| $A$ | 1 | 1 | 1 | $z, R_z$ | $x^2 + y^2, z^2$ |
| $E$ | $\left\{\begin{matrix}1 \\ 1\end{matrix}\right.$ | $\begin{matrix}\varepsilon \\ \varepsilon^*\end{matrix}$ | $\left.\begin{matrix}\varepsilon^* \\ \varepsilon\end{matrix}\right\}$ | $(x, y)(R_x, R_y)$ | $(x^2 - y^2, xy)\ (yz, xz)$ |

The $C_n$ Groups (*continued*)

| $C_4$ | $E$ | $C_4$ | $C_3$ | $C_4^3$ | | |
|---|---|---|---|---|---|---|
| $A$ | 1 | 1 | 1 | 1 | $z, R_z$ | $x^2 + y^2, z^2$ |
| $B$ | 1 | –1 | 1 | –1 | | $x^2 - y^2, xy$ |
| $E$ | 1 | $i$ | –1 | $-i$ | $(x, y)\ (R_x, R_y)$ | $(yz, xz)$ |
| | 1 | $-i$ | –1 | $i$ | | |

| $C_5$ | $E$ | $C_5$ | $C_5^2$ | $C_5^3$ | $C_3^4$ | | $\varepsilon = \exp(2\pi i/5)$ |
|---|---|---|---|---|---|---|---|
| $A$ | 1 | 1 | 1 | 1 | 1 | $z, R_z$ | $x^2 + y^2, z^2$ |
| $E_1$ | 1 | $\varepsilon$ | $\varepsilon^2$ | $\varepsilon^{2*}$ | $\varepsilon^*$ | $(x, y)\ (R_x, R_y)$ | $(yz, xz)$ |
| | 1 | $\varepsilon^*$ | $\varepsilon^{2*}$ | $\varepsilon^2$ | $\varepsilon$ | | |
| $E_2$ | 1 | $\varepsilon^2$ | $\varepsilon^*$ | $\varepsilon$ | $\varepsilon^{2*}$ | | $(x^2 - y^2, xy)$ |
| | 1 | $\varepsilon^{2*}$ | $\varepsilon$ | $\varepsilon^*$ | $\varepsilon^2$ | | |

| $C_6$ | $E$ | $C_6$ | $C_3$ | $C_2$ | $C_3^2$ | $C_6^5$ | | $\varepsilon = \exp(2\pi i/6)$ |
|---|---|---|---|---|---|---|---|---|
| $A$ | 1 | 1 | 1 | 1 | 1 | 1 | $z, R_z$ | $x^2 + y^2, z^2$ |
| $B$ | 1 | –1 | 1 | –1 | 1 | –1 | | |
| $E_1$ | 1 | $\varepsilon$ | $-\varepsilon^*$ | –1 | $-\varepsilon$ | $\varepsilon^*$ | $(x, y)$ | $(xz, yz)$ |
| | 1 | $\varepsilon^*$ | $-\varepsilon$ | –1 | $-\varepsilon^*$ | $\varepsilon$ | $(R_x, R_y)$ | |
| $E_2$ | 1 | $-\varepsilon^*$ | $-\varepsilon$ | 1 | $-\varepsilon^*$ | $-\varepsilon$ | | $(x^2 - y^2, xy)$ |
| | 1 | $-\varepsilon$ | $-\varepsilon^*$ | 1 | $-\varepsilon$ | $-\varepsilon^*$ | | |

| $C_7$ | $E$ | $C_7$ | $C_7^2$ | $C_7^3$ | $C_7^4$ | $C_7^5$ | $C_7^6$ | | $\varepsilon = \exp(2\pi i/7)$ |
|---|---|---|---|---|---|---|---|---|---|
| $A$ | 1 | 1 | 1 | 1 | 1 | 1 | 1 | $z, R_z$ | $x^2 + y^2, z^2$ |
| $E_1$ | 1 | $\varepsilon$ | $\varepsilon^2$ | $\varepsilon^3$ | $\varepsilon^{3*}$ | $\varepsilon^{2*}$ | $\varepsilon^*$ | $(x, y)$ | |
| | 1 | $\varepsilon^*$ | $\varepsilon^{2*}$ | $\varepsilon^{3*}$ | $\varepsilon^3$ | $\varepsilon^2$ | $\varepsilon$ | $(R_x, R_y)$ | $(xz, yz)$ |
| $E_2$ | 1 | $\varepsilon^2$ | $\varepsilon^{3*}$ | $\varepsilon^*$ | $\varepsilon$ | $\varepsilon^3$ | $\varepsilon^{2*}$ | | $(x^2 - y^2, xy)$ |
| | 1 | $\varepsilon^{2*}$ | $\varepsilon^3$ | $\varepsilon$ | $\varepsilon^*$ | $\varepsilon^{3*}$ | $\varepsilon^2$ | | |
| $E_3$ | 1 | $\varepsilon^3$ | $\varepsilon^*$ | $\varepsilon^2$ | $\varepsilon^{2*}$ | $\varepsilon$ | $\varepsilon^{3*}$ | | |
| | 1 | $\varepsilon^{3*}$ | $\varepsilon$ | $\varepsilon^{2*}$ | $\varepsilon^2$ | $\varepsilon^*$ | $\varepsilon^3$ | | |

| $C_8$ | $E$ | $C_8$ | $C_4$ | $C_8^3$ | $C_2$ | $C_8^5$ | $C_4^3$ | $C_8^7$ | | $\varepsilon = \exp(2\pi i/8)$ |
|---|---|---|---|---|---|---|---|---|---|---|
| $A$ | 1 | 1 | 1 | 1 | 1 | 1 | 1 | 1 | $z, R_z$ | $x^2 + y^2, z^2$ |
| $B$ | 1 | –1 | 1 | –1 | 1 | –1 | 1 | –1 | | |
| $E_1$ | 1 | $\varepsilon$ | $i$ | $-\varepsilon^*$ | –1 | $-\varepsilon$ | $-i$ | $\varepsilon^*$ | $(x, y)$ | $(xz, yz)$ |
| | 1 | $\varepsilon^*$ | $-i$ | $-\varepsilon$ | –1 | $-\varepsilon^*$ | $i$ | $\varepsilon$ | | |
| $E_2$ | 1 | $i$ | –1 | $-i$ | 1 | $i$ | –1 | $-i$ | $(R_x, R_y)$ | $(x^2 - y^2, xy)$ |
| | 1 | $-i$ | –1 | $i$ | 1 | $-i$ | –1 | $i$ | | |
| $E_3$ | 1 | $-\varepsilon$ | $i$ | $\varepsilon^*$ | –1 | $\varepsilon$ | $-i$ | $-\varepsilon^*$ | | |
| | 1 | $-\varepsilon^*$ | $-i$ | $\varepsilon$ | –1 | $\varepsilon^*$ | $i$ | $-\varepsilon$ | | |

**3.** The $D_n$ Groups

| $D_2$ | $E$ | $C_2(z)$ | $C_2(y)$ | $C_2(x)$ | | |
|---|---|---|---|---|---|---|
| $A$ | 1 | 1 | 1 | 1 | | $x^2, y^2, z^2$ |
| $B_1$ | 1 | 1 | –1 | –1 | $z, R_z$ | $xy$ |
| $B_2$ | 1 | –1 | 1 | –1 | $y, R_y$ | $xz$ |
| $B_3$ | 1 | –1 | –1 | 1 | $x, R_x$ | $yz$ |

| $D_3$ | $E$ | $2C_3$ | $3C_2$ | | |
|---|---|---|---|---|---|
| $A_1$ | 1 | 1 | 1 | | $x^2 + y^2, z^2$ |
| $A_2$ | 1 | 1 | –1 | $z, R_z$ | |
| $E$ | 2 | –1 | 0 | $(x, y)\ (R_x, R_y)$ | $(x^2 - y^2, xy)\ (xz, yz)$ |

| $D_4$ | $E$ | $2C_4$ | $C_2\ (=C_4^2)$ | $2C_2'$ | $2C_2''$ | | |
|---|---|---|---|---|---|---|---|
| $A_1$ | 1 | 1 | 1 | 1 | 1 | | $x^2 + y^2, z^2$ |
| $A_2$ | 1 | 1 | 1 | –1 | –1 | $z, R_z$ | |
| $B_1$ | 1 | –1 | 1 | 1 | –1 | | $x^2 - y^2$ |
| $B_2$ | 1 | –1 | 1 | –1 | 1 | | $xy$ |
| $E$ | 2 | 0 | –2 | 0 | 0 | $(x, y)\ (R_x, R_y)$ | $(xz, yz)$ |

| $D_5$ | $E$ | $2C_5$ | $2C_5^2$ | $5C_2$ | | |
|---|---|---|---|---|---|---|
| $A_1$ | 1 | 1 | 1 | 1 | | $x^2 + y^2, z^2$ |
| $A_2$ | 1 | 1 | 1 | –1 | $z, R_z$ | |
| $E_1$ | 2 | 2 cos 72° | 2 cos 144° | 0 | $(x, y)\ (R_x, R_y)$ | $(xz, yz)$ |
| $E_2$ | 2 | 2 cos 144° | 2 cos 72° | 0 | | $(x^2 - y^2, xy)$ |

| $D_6$ | $E$ | $2C_6$ | $2C_3$ | $C_2$ | $3C_2'$ | $3C_2''$ | | |
|---|---|---|---|---|---|---|---|---|
| $A_1$ | 1 | 1 | 1 | 1 | 1 | 1 | | $x^2 + y^2, z^2$ |
| $A_2$ | 1 | 1 | 1 | 1 | –1 | –1 | $z, R_z$ | |
| $B_1$ | 1 | –1 | 1 | –1 | 1 | –1 | | |
| $B_2$ | 1 | –1 | 1 | –1 | –1 | 1 | | |
| $E_1$ | 2 | 1 | –1 | –2 | 0 | 0 | $(x, y)\ (R_x, R_y)$ | $(xz, yz)$ |
| $E_2$ | 2 | –1 | –1 | 2 | 0 | 0 | | $(x^2 - y^2, xy)$ |

**4.** The $C_{nv}$ Groups

| $C_{2v}$ | $E$ | $C_2$ | $\sigma_v(xz)$ | $\sigma_v'(yz)$ | | |
|---|---|---|---|---|---|---|
| $A_1$ | 1 | 1 | 1 | 1 | $z$ | $x^2, y^2, z^2$ |
| $A_2$ | 1 | 1 | –1 | –1 | $R_z$ | $xy$ |
| $B_1$ | 1 | –1 | 1 | –1 | $x, R_y$ | $xz$ |
| $B_2$ | 1 | –1 | –1 | 1 | $y, R_x$ | $yz$ |

| $C_{3v}$ | $E$ | $2C_3$ | $3\sigma_v$ | | |
|---|---|---|---|---|---|
| $A_1$ | 1 | 1 | 1 | $z$ | $x^2+y^2, z^2$ |
| $A_2$ | 1 | 1 | –1 | $R_z$ | |
| $E$ | 2 | –1 | 0 | $(x, y)$ $(R_x, R_y)$ | $(x^2-y^2, xy)$ $(xz, yz)$ |

| $C_{4v}$ | $E$ | $2C_4$ | $C_2$ | $2\sigma_v$ | $2\sigma_d$ | | |
|---|---|---|---|---|---|---|---|
| $A_1$ | 1 | 1 | 1 | 1 | 1 | $z$ | $x^2+y^2, z^2$ |
| $A_2$ | 1 | 1 | 1 | –1 | –1 | $R_z$ | |
| $B_1$ | 1 | –1 | 1 | 1 | –1 | | $x^2-y^2$ |
| $B_2$ | 1 | –1 | 1 | –1 | 1 | | $xy$ |
| $E$ | 2 | 0 | –2 | 0 | 0 | $(x, y)$ $(R_x, R_y)$ | $(xz, yz)$ |

| $C_{5v}$ | $E$ | $2C_5$ | $C_5^2$ | $5\sigma_v$ | | |
|---|---|---|---|---|---|---|
| $A_1$ | 1 | 1 | 1 | 1 | $z$ | $x^2+y^2, z^2$ |
| $A_2$ | 1 | 1 | 1 | –1 | $R_z$ | |
| $E_1$ | 2 | 2 cos 72° | 2 cos 144° | 0 | $(x, y)$ $(R_x, R_y)$ | $(xz, yz)$ |
| $E_2$ | 2 | 2 cos 144° | 2 cos 72° | 0 | | $(x^2-y^2, xy)$ |

| $C_{6v}$ | $E$ | $2C_6$ | $2C_3$ | $C_2$ | $3\sigma_v$ | $3\sigma_d$ | | |
|---|---|---|---|---|---|---|---|---|
| $A_1$ | 1 | 1 | 1 | 1 | 1 | 1 | $z$ | $x^2+y^2, z^2$ |
| $A_2$ | 1 | 1 | 1 | 1 | –1 | –1 | $R_z$ | |
| $B_1$ | 1 | –1 | 1 | –1 | 1 | –1 | | |
| $B_2$ | 1 | –1 | 1 | –1 | –1 | 1 | | |
| $E_1$ | 2 | 1 | –1 | –2 | 0 | 0 | $(x, y)$ $(R_x, R_y)$ | $(xz, yz)$ |
| $E_2$ | 2 | –1 | –1 | 2 | 0 | 0 | | $(x^2-y^2, xy)$ |

## 5. The $C_{nh}$ Groups

| $C_{2h}$ | $E$ | $C_2$ | $i$ | $\sigma_h$ | | |
|---|---|---|---|---|---|---|
| $A_g$ | 1 | 1 | 1 | 1 | $R_z$ | $x^2, y^2, z^2, xy$ |
| $B_g$ | 1 | –1 | 1 | –1 | $R_x, R_y$ | $xz, yz$ |
| $A_u$ | 1 | 1 | –1 | –1 | $z$ | |
| $B_u$ | 1 | –1 | –1 | 1 | $x, y$ | |

| $C_{3h}$ | $E$ | $C_3$ | $C_3^2$ | $\sigma_h$ | $S_3$ | $S_3^5$ | | $\varepsilon = \exp(2\pi i/3)$ |
|---|---|---|---|---|---|---|---|---|
| $A'$ | 1 | 1 | 1 | 1 | 1 | 1 | $R_z$ | $x^2+y^2, z^2$ |
| $E'$ | 1 | $\varepsilon$ | $\varepsilon^*$ | 1 | $\varepsilon$ | $\varepsilon^*$ | $(x, y)$ | $(x^2-y^2, xy)$ |
| | 1 | $\varepsilon^*$ | $\varepsilon$ | 1 | $\varepsilon^*$ | $\varepsilon$ | | |
| $A''$ | 1 | 1 | 1 | –1 | –1 | –1 | $z$ | |
| $E''$ | 1 | $\varepsilon$ | $\varepsilon^*$ | –1 | $-\varepsilon$ | $-\varepsilon^*$ | $(R_x, R_y)$ | $(xz, yz)$ |
| | 1 | $\varepsilon^*$ | $\varepsilon$ | –1 | $-\varepsilon^*$ | $-\varepsilon$ | | |

| $C_{4h}$ | $E$ | $C_4$ | $C_2$ | $C_4^3$ | $i$ | $S_4^3$ | $\sigma_h$ | $S_4$ | | |
|---|---|---|---|---|---|---|---|---|---|---|
| $A_g$ | 1 | 1 | 1 | 1 | 1 | 1 | 1 | 1 | $R_z$ | $x^2 + y^2, z^2$ |
| $B_g$ | 1 | –1 | 1 | –1 | 1 | –1 | 1 | –1 | | $x^2 - y^2, xy$ |
| $E_g$ | 1 | $i$ | –1 | $-i$ | 1 | $i$ | –1 | $-i$ | $(R_x, R_y)$ | $(xz, yz)$ |
| | 1 | $-i$ | –1 | $i$ | 1 | $-i$ | –1 | $i$ | | |
| $A_u$ | 1 | 1 | 1 | 1 | –1 | –1 | –1 | –1 | $z$ | |
| $B_u$ | 1 | –1 | 1 | –1 | –1 | 1 | –1 | 1 | | |
| $E_u$ | 1 | $i$ | –1 | $-i$ | –1 | $-i$ | 1 | $i$ | $(x, y)$ | |
| | 1 | $-i$ | –1 | $i$ | –1 | $i$ | 1 | $-i$ | | |

| $C_{5h}$ | $E$ | $C_5$ | $C_5^2$ | $C_5^3$ | $C_5^4$ | $\sigma_h$ | $S_5$ | $S_5^7$ | $S_5^3$ | $S_5^9$ | | $\varepsilon = \exp(2\pi i/5)$ |
|---|---|---|---|---|---|---|---|---|---|---|---|---|
| $A'$ | 1 | 1 | 1 | 1 | 1 | 1 | 1 | 1 | 1 | 1 | $R_z$ | $x^2 + y^2, z^2$ |
| $E_1'$ | 1 | $\varepsilon$ | $\varepsilon^2$ | $\varepsilon^{2*}$ | $\varepsilon^*$ | 1 | $\varepsilon$ | $\varepsilon^2$ | $\varepsilon^{2*}$ | $\varepsilon^*$ | $(x, y)$ | |
| | 1 | $\varepsilon^*$ | $\varepsilon^{2*}$ | $\varepsilon^2$ | $\varepsilon$ | 1 | $\varepsilon^*$ | $\varepsilon^{2*}$ | $\varepsilon^2$ | $\varepsilon$ | | |
| $E_2'$ | 1 | $\varepsilon^2$ | $\varepsilon^*$ | $\varepsilon$ | $\varepsilon^{2*}$ | 1 | $\varepsilon^2$ | $\varepsilon^*$ | $\varepsilon$ | $\varepsilon^{2*}$ | | $x^2 - y^2, xy$ |
| | 1 | $\varepsilon^{2*}$ | $\varepsilon$ | $\varepsilon^*$ | $\varepsilon^2$ | 1 | $\varepsilon^{2*}$ | $\varepsilon$ | $\varepsilon^*$ | $\varepsilon^2$ | | |
| $A''$ | 1 | 1 | 1 | 1 | 1 | –1 | –1 | –1 | –1 | –1 | $z$ | |
| $E_1''$ | 1 | $\varepsilon$ | $\varepsilon^2$ | $\varepsilon^{2*}$ | $\varepsilon^*$ | –1 | $-\varepsilon$ | $-\varepsilon^2$ | $-\varepsilon^{2*}$ | $-\varepsilon^*$ | $(R_x, R_y)$ | $(xz, yz)$ |
| | 1 | $\varepsilon^*$ | $\varepsilon^{2*}$ | $\varepsilon^2$ | $\varepsilon$ | –1 | $-\varepsilon^*$ | $-\varepsilon^{2*}$ | $-\varepsilon^2$ | $-\varepsilon$ | | |
| $E_2''$ | 1 | $\varepsilon^2$ | $\varepsilon^*$ | $\varepsilon$ | $\varepsilon^{2*}$ | –1 | $-\varepsilon^2$ | $-\varepsilon^*$ | $-\varepsilon$ | $-\varepsilon^{2*}$ | | |
| | 1 | $\varepsilon^{2*}$ | $\varepsilon$ | $\varepsilon^*$ | $\varepsilon^2$ | –1 | $-\varepsilon^{2*}$ | $-\varepsilon$ | $-\varepsilon^*$ | $-\varepsilon^2$ | | |

| $C_{6h}$ | $E$ | $C_6$ | $C_3$ | $C_2$ | $C_3^2$ | $C_6^5$ | $i$ | $S_3^5$ | $S_6^5$ | $\sigma_h$ | $S_6$ | $S_3$ | | $\varepsilon = \exp(2\pi i/6)$ |
|---|---|---|---|---|---|---|---|---|---|---|---|---|---|---|
| $A_g$ | 1 | 1 | 1 | 1 | 1 | 1 | 1 | 1 | 1 | 1 | 1 | 1 | $R_z$ | $x^2 + y^2, z^2$ |
| $B_g$ | 1 | –1 | 1 | –1 | 1 | –1 | 1 | –1 | 1 | –1 | 1 | –1 | | |
| $E_{1g}$ | 1 | $\varepsilon$ | $-\varepsilon^*$ | –1 | $\varepsilon$ | $\varepsilon^*$ | 1 | $\varepsilon$ | $-\varepsilon^*$ | –1 | $-\varepsilon$ | $\varepsilon^*$ | $(R_x, R_y)$ | $(xz, yz)$ |
| | 1 | $\varepsilon^*$ | $-\varepsilon$ | –1 | $-\varepsilon^*$ | $\varepsilon$ | 1 | $\varepsilon^*$ | $-\varepsilon$ | –1 | $-\varepsilon^*$ | $\varepsilon$ | | |
| $E_{2g}$ | 1 | $-\varepsilon^*$ | $-\varepsilon$ | 1 | $-\varepsilon^*$ | $-\varepsilon$ | 1 | $-\varepsilon^*$ | $-\varepsilon$ | 1 | $-\varepsilon^*$ | $-\varepsilon$ | | $(x^2 - y^2, xy)$ |
| | 1 | $-\varepsilon$ | $-\varepsilon^*$ | 1 | $-\varepsilon$ | $-\varepsilon^*$ | 1 | $-\varepsilon$ | $-\varepsilon^*$ | 1 | $-\varepsilon$ | $-\varepsilon^*$ | | |
| $A_u$ | 1 | 1 | 1 | 1 | 1 | 1 | –1 | –1 | –1 | –1 | –1 | –1 | $z$ | |
| $B_u$ | 1 | –1 | 1 | –1 | 1 | –1 | –1 | 1 | –1 | 1 | –1 | 1 | | |
| $E_{1u}$ | 1 | $\varepsilon$ | $-\varepsilon^*$ | –1 | $-\varepsilon$ | $\varepsilon^*$ | –1 | $-\varepsilon$ | $\varepsilon^*$ | 1 | $\varepsilon$ | $-\varepsilon^*$ | $(x, y)$ | |
| | 1 | $\varepsilon^*$ | $-\varepsilon$ | –1 | $-\varepsilon^*$ | $\varepsilon$ | –1 | $-\varepsilon^*$ | $\varepsilon$ | 1 | $\varepsilon^*$ | $-\varepsilon$ | | |
| $E_{2u}$ | 1 | $-\varepsilon^*$ | $-\varepsilon$ | 1 | $-\varepsilon^*$ | $-\varepsilon$ | –1 | $\varepsilon^*$ | $\varepsilon$ | –1 | $\varepsilon^*$ | $\varepsilon$ | | |
| | 1 | $-\varepsilon$ | $-\varepsilon^*$ | 1 | $-\varepsilon$ | $-\varepsilon^*$ | –1 | $\varepsilon$ | $\varepsilon^*$ | –1 | $\varepsilon$ | $\varepsilon^*$ | | |

## 6. The $D_{nh}$ Groups

| $D_{2h}$ | $E$ | $C_2(z)$ | $C_2(y)$ | $C_2(x)$ | $i$ | $\sigma(xy)$ | $\sigma(xz)$ | $\sigma(yz)$ | | |
|---|---|---|---|---|---|---|---|---|---|---|
| $A_g$ | 1 | 1 | 1 | 1 | 1 | 1 | 1 | 1 | | $x^2, y^2, z^2$ |
| $B_{1g}$ | 1 | 1 | –1 | –1 | 1 | 1 | –1 | –1 | $R_z$ | $xy$ |
| $B_{2g}$ | 1 | –1 | 1 | –1 | 1 | –1 | 1 | –1 | $R_y$ | $xz$ |
| $B_{3g}$ | 1 | –1 | –1 | 1 | 1 | –1 | –1 | 1 | $R_x$ | $yz$ |
| $A_u$ | 1 | 1 | 1 | 1 | –1 | –1 | –1 | –1 | | |
| $B_{1u}$ | 1 | 1 | –1 | –1 | –1 | –1 | 1 | 1 | $z$ | |
| $B_{2u}$ | 1 | –1 | 1 | –1 | –1 | 1 | –1 | 1 | $y$ | |
| $B_{3u}$ | 1 | –1 | –1 | 1 | –1 | 1 | 1 | –1 | $x$ | |

| $D_{3h}$ | $E$ | $2C_3$ | $3C_2$ | $\sigma_h$ | $2S_3$ | $3\sigma_v$ | | |
|---|---|---|---|---|---|---|---|---|
| $A_1'$ | 1 | 1 | 1 | 1 | 1 | 1 | | $x^2 + y^2, z^2$ |
| $A_2'$ | 1 | 1 | –1 | 1 | 1 | –1 | $R_z$ | |
| $E'$ | 2 | –1 | 0 | 2 | –1 | 0 | $(x, y)$ | $(x^2 - y^2, xy)$ |
| $A_1''$ | 1 | 1 | 1 | –1 | –1 | –1 | | |
| $A_2''$ | 1 | 1 | –1 | –1 | –1 | 1 | $z$ | |
| $E''$ | 2 | –1 | 0 | –2 | 1 | 0 | $(R_x, R_y)$ | $(xz, yz)$ |

| $D_{4h}$ | $E$ | $2C_4$ | $C_2$ | $2C_2'$ | $2C_2''$ | $i$ | $2S_4$ | $\sigma_h$ | $2\sigma_v$ | $2\sigma_d$ | | |
|---|---|---|---|---|---|---|---|---|---|---|---|---|
| $A_{1g}$ | 1 | 1 | 1 | 1 | 1 | 1 | 1 | 1 | 1 | 1 | | $x^2 + y^2, z^2$ |
| $A_{2g}$ | 1 | 1 | 1 | –1 | –1 | 1 | 1 | 1 | –1 | –1 | $R_z$ | |
| $B_{1g}$ | 1 | –1 | 1 | 1 | –1 | 1 | –1 | 1 | 1 | –1 | | $(x^2 - y^2)$ |
| $B_{2g}$ | 1 | –1 | 1 | –1 | 1 | 1 | –1 | 1 | –1 | 1 | | $xy$ |
| $E_g$ | 2 | 0 | –2 | 0 | 0 | 2 | 0 | –2 | 0 | 0 | $(R_x, R_y)$ | $(xz, yz)$ |
| $A_{1u}$ | 1 | 1 | 1 | 1 | 1 | –1 | –1 | –1 | –1 | –1 | | |
| $A_{2u}$ | 1 | 1 | 1 | –1 | –1 | –1 | –1 | –1 | 1 | 1 | $z$ | |
| $B_{1u}$ | 1 | –1 | 1 | 1 | –1 | –1 | 1 | –1 | –1 | 1 | | $(x^2 - y^2, xy)$ |
| $B_{2u}$ | 1 | –1 | 1 | –1 | 1 | –1 | 1 | –1 | 1 | –1 | | |
| $E_u$ | 2 | 0 | –2 | 0 | 0 | –2 | 0 | 2 | 0 | 0 | $(x, y)$ | |

| $D_{5h}$ | $E$ | $2C_5$ | $2C_5^2$ | $5C_2$ | $\sigma_h$ | $2S_5$ | $2S_5^3$ | $5\sigma_v$ | | |
|---|---|---|---|---|---|---|---|---|---|---|
| $A_1'$ | 1 | 1 | 1 | 1 | 1 | 1 | 1 | 1 | | $x^2 + y^2, z^2$ |
| $A_2'$ | 1 | 1 | 1 | –1 | 1 | 1 | 1 | –1 | $R_z$ | |
| $E_1'$ | 2 | 2 cos 72° | 2 cos 144° | 0 | 2 | 2 cos 72° | 2 cos 144° | 0 | $(x, y)$ | |
| $E_2'$ | 2 | 2 cos 144° | 2 cos 72° | 0 | 2 | 2 cos 144° | 2 cos 72° | 0 | | $(x^2 - x^2, xy)$ |
| $A_1''$ | 1 | 1 | 1 | 1 | –1 | –1 | –1 | –1 | | |
| $A_2''$ | 1 | 1 | 1 | –1 | –1 | –1 | –1 | 1 | $z$ | |
| $E_1''$ | 2 | 2 cos 72° | 2 cos 144° | 0 | –2 | –2 cos 72° | –2 cos 144° | 0 | $(R_x, R_y)$ | $(xy, yz)$ |
| $E_1''$ | 2 | 2 cos 144° | 2 cos 72° | 0 | –2 | –2 cos 144° | –2 cos 72° | 0 | | |

| $D_{6h}$ | $E$ | $2C_6$ | $2C_3$ | $C_2$ | $3C_2'$ | $3C_2''$ | $i$ | $2S_3$ | $2S_6$ | $\sigma_h$ | $3\sigma_d$ | $3\sigma_v$ | | |
|---|---|---|---|---|---|---|---|---|---|---|---|---|---|---|
| $A_{1g}$ | 1 | 1 | 1 | 1 | 1 | 1 | 1 | 1 | 1 | 1 | 1 | 1 | | $(x^2 + y^2, z^2)$ |
| $A_{2g}$ | 1 | 1 | 1 | 1 | –1 | –1 | 1 | 1 | 1 | 1 | –1 | –1 | $R_z$ | |
| $B_{1g}$ | 1 | –1 | 1 | –1 | 1 | –1 | 1 | –1 | 1 | –1 | 1 | –1 | | |
| $B_{2g}$ | 1 | –1 | 1 | –1 | –1 | 1 | 1 | –1 | 1 | –1 | –1 | 1 | | |
| $E_{1g}$ | 2 | 1 | –1 | –2 | 0 | 0 | 2 | 1 | –1 | –2 | 0 | 0 | $(R_x, R_y)$ | $(xz, yz)$ |
| $E_{2g}$ | 2 | –1 | –1 | 2 | 0 | 0 | 2 | –1 | –1 | 2 | 0 | 0 | | $(x^2 - y^2, xy)$ |
| $A_{1u}$ | 1 | 1 | 1 | 1 | 1 | 1 | –1 | –1 | –1 | –1 | –1 | –1 | | |
| $A_{2u}$ | 1 | 1 | 1 | 1 | –1 | –1 | –1 | –1 | –1 | –1 | 1 | 1 | $z$ | |
| $B_{1u}$ | 1 | –1 | 1 | –1 | 1 | –1 | –1 | 1 | –1 | 1 | –1 | 1 | | |
| $B_{2u}$ | 1 | –1 | 1 | –1 | –1 | 1 | –1 | 1 | –1 | 1 | 1 | –1 | | |
| $E_{1u}$ | 2 | 1 | –1 | –2 | 0 | 0 | –2 | –1 | 1 | 2 | 0 | 0 | $(x, y)$ | |
| $E_{2u}$ | 2 | –1 | –1 | 2 | 0 | 0 | –2 | 1 | 1 | –2 | 0 | 0 | | |

| $D_{8h}$ | $E$ | $2S_8$ | $2C_8^3$ | $2C_4$ | $C_2$ | $4C_2'$ | $4C_2''$ | $i$ | $2S_8$ | $2S_8^3$ | $2S_4$ | $\sigma_h$ | $4\sigma_d$ | $4\sigma_v$ | | |
|---|---|---|---|---|---|---|---|---|---|---|---|---|---|---|---|---|
| $A_{1g}$ | 1 | 1 | 1 | 1 | 1 | 1 | 1 | 1 | 1 | 1 | 1 | 1 | 1 | 1 | | $x^2 + y^2, z^2$ |
| $A_{2g}$ | 1 | 1 | 1 | 1 | 1 | –1 | –1 | 1 | 1 | 1 | 1 | 1 | –1 | –1 | $R_z$ | |
| $B_{1g}$ | 1 | –1 | –1 | 1 | 1 | 1 | –1 | 1 | –1 | –1 | 1 | 1 | 1 | –1 | | |
| $B_{2g}$ | 1 | –1 | –1 | 1 | 1 | –1 | 1 | 1 | –1 | –1 | 1 | 1 | –1 | 1 | | |
| $E_{1g}$ | 2 | $\sqrt{2}$ | $-\sqrt{2}$ | 0 | –2 | 0 | 0 | 2 | $\sqrt{2}$ | $-\sqrt{2}$ | 0 | –2 | 0 | 0 | $(R_x, R_y)$ | $(xz, yz)$ |
| $E_{2g}$ | 2 | 0 | 0 | –2 | 2 | 0 | 0 | 2 | 0 | 0 | –2 | 2 | 0 | 0 | | $(x^2 - y^2, xy)$ |
| $E_{3g}$ | 2 | $-\sqrt{2}$ | $\sqrt{2}$ | 0 | –2 | 0 | 0 | 2 | $-\sqrt{2}$ | $\sqrt{2}$ | 0 | –2 | 0 | 0 | | |
| $A_{1u}$ | 1 | 1 | 1 | 1 | 1 | 1 | 1 | –1 | –1 | –1 | –1 | –1 | –1 | –1 | | |
| $A_{2u}$ | 1 | 1 | 1 | 1 | 1 | –1 | –1 | –1 | –1 | –1 | –1 | –1 | 1 | 1 | $z$ | |
| $B_{1u}$ | 1 | –1 | –1 | 1 | 1 | 1 | –1 | 1 | 1 | 1 | –1 | –1 | –1 | 1 | | |
| $B_{2u}$ | 1 | –1 | –1 | 1 | 1 | –1 | 1 | –1 | 1 | 1 | –1 | –1 | 1 | –1 | | |
| $E_{1u}$ | 2 | $\sqrt{2}$ | $-\sqrt{2}$ | 0 | –2 | 0 | 0 | –2 | $-\sqrt{2}$ | $\sqrt{2}$ | 0 | 2 | 0 | 0 | $(x, y)$ | |
| $E_{2u}$ | 2 | 0 | 0 | –2 | 2 | 0 | 0 | –2 | 0 | 0 | 2 | –2 | 0 | 0 | | |
| $E_{3u}$ | 2 | $-\sqrt{2}$ | $\sqrt{2}$ | 0 | –2 | 0 | 0 | –2 | $\sqrt{2}$ | $-\sqrt{2}$ | 0 | 2 | 0 | 0 | | |

## 7. The $D_{nd}$ Groups

| $D_{2d}$ | $E$ | $2S_4$ | $C_2$ | $2C_2'$ | $2\sigma_d$ | | |
|---|---|---|---|---|---|---|---|
| $A_1$ | 1 | 1 | 1 | 1 | 1 | | $x^2 + y^2, z^2$ |
| $A_2$ | 1 | 1 | 1 | –1 | –1 | $R_z$ | |
| $B_1$ | 1 | –1 | 1 | 1 | –1 | | $x^2 - y^2$ |
| $B_2$ | 1 | –1 | 1 | –1 | 1 | $z$ | $xy$ |
| $E$ | 2 | 0 | –2 | 0 | 0 | $(x, y); (R_x, R_y)$ | $(xz, yz)$ |

| $D_{3d}$ | $E$ | $2C_3$ | $3C_2$ | $i$ | $2S_6$ | $3\sigma_d$ | | |
|---|---|---|---|---|---|---|---|---|
| $A_{1g}$ | 1 | 1 | 1 | 1 | 1 | 1 | | $x^2+y^2, z^2$ |
| $A_{2g}$ | 1 | 1 | –1 | 1 | 1 | –1 | $R_z$ | |
| $E_g$ | 2 | –1 | 0 | 2 | –1 | 0 | $(R_x, R_y)$ | $(x^2-y^2, xy), (xz, yz)$ |
| $A_{1u}$ | 1 | 1 | 1 | –1 | –1 | –1 | | |
| $A_{2u}$ | 1 | 1 | –1 | –1 | –1 | 1 | $z$ | |
| $E_u$ | 2 | –1 | 0 | –2 | 1 | 0 | $(x, y)$ | |

| $D_{4d}$ | $E$ | $2S_8$ | $2C_4$ | $2S_8^3$ | $C_2$ | $4C_2'$ | $4\sigma_d$ | | |
|---|---|---|---|---|---|---|---|---|---|
| $A_1$ | 1 | 1 | 1 | 1 | 1 | 1 | 1 | | $x^2+y^2, z^2$ |
| $A_2$ | 1 | 1 | 1 | 1 | 1 | –1 | –1 | $R_z$ | |
| $B_1$ | 1 | –1 | 1 | –1 | 1 | 1 | –1 | | |
| $B_2$ | 1 | –1 | 1 | –1 | 1 | –1 | 1 | $z$ | |
| $E_1$ | 2 | $\sqrt{2}$ | 0 | $-\sqrt{2}$ | –2 | 0 | 0 | $(x, y)$ | |
| $E_2$ | 2 | 0 | –2 | 0 | 2 | 0 | 0 | | $(x^2-y^2, xy)$ |
| $E_3$ | 2 | $-\sqrt{2}$ | 0 | $\sqrt{2}$ | –2 | 0 | 0 | $(R_x, R_y)$ | $(xz, yz)$ |

| $D_{5d}$ | $E$ | $2C_5$ | $2C_5^2$ | $5C_2$ | $i$ | $2S_{10}^3$ | $2S_{10}$ | $5\sigma_d$ | | |
|---|---|---|---|---|---|---|---|---|---|---|
| $A_{1g}$ | 1 | 1 | 1 | 1 | 1 | 1 | 1 | 1 | | $x^2+y^2, z^2$ |
| $A_{2g}$ | 1 | 1 | 1 | –1 | 1 | 1 | 1 | –1 | $R_z$ | |
| $E_{1g}$ | 2 | 2 cos 72° | 2 cos 144° | 0 | 2 | 2 cos 72° | 2 cos 144° | 0 | $(R_x, R_y)$ | $(xz, yz)$ |
| $E_{2g}$ | 2 | 2 cos 144° | 2 cos 72° | 0 | 2 | 2 cos 144° | 2 cos 72° | 0 | | $(x^2-y^2, xy)$ |
| $A_{1u}$ | 1 | 1 | 1 | 1 | –1 | –1 | –1 | –1 | | |
| $A_{2u}$ | 1 | 1 | 1 | –1 | –1 | –1 | –1 | 1 | $z$ | |
| $E_{1u}$ | 2 | 2 cos 72° | 2 cos 144° | 0 | –2 | –2 cos 72° | –2 cos 144° | 0 | $(x, y)$ | |
| $E_{2u}$ | 2 | 2 cos 144° | 2 cos 72° | 0 | –2 | –2 cos 144° | –2 cos 72° | 0 | | |

| $D_{6d}$ | $E$ | $2S_{12}$ | $2C_6$ | $2S_4$ | $2C_3$ | $2S_{12}^5$ | $C_2$ | $6C_2'$ | $6\sigma_d$ | | |
|---|---|---|---|---|---|---|---|---|---|---|---|
| $A_1$ | 1 | 1 | 1 | 1 | 1 | 1 | 1 | 1 | 1 | | $x^2+y^2, z^2$ |
| $A_2$ | 1 | 1 | 1 | 1 | 1 | 1 | 1 | –1 | –1 | $R_z$ | |
| $B_1$ | 1 | –1 | 1 | –1 | 1 | –1 | 1 | 1 | –1 | | |
| $B_2$ | 1 | –1 | 1 | –1 | 1 | –1 | 1 | –1 | 1 | $z$ | |
| $E_1$ | 2 | $\sqrt{3}$ | 1 | 0 | –1 | $-\sqrt{3}$ | –2 | 0 | 0 | $(x, y)$ | |
| $E_2$ | 2 | 1 | –1 | –2 | –1 | 1 | 2 | 0 | 0 | | $(x^2-y^2, xy)$ |
| $E_3$ | 2 | 0 | –2 | 0 | 2 | 0 | –2 | 0 | 0 | | |
| $E_4$ | 2 | –1 | –1 | 2 | –1 | –1 | 2 | 0 | 0 | | |
| $E_5$ | 2 | $-\sqrt{3}$ | 1 | 0 | –1 | $\sqrt{3}$ | –2 | 0 | 0 | $(R_x, R_y)$ | $(xz, yz)$ |

## 8. The $S_n$ Groups

| $S_4$ | $E$ | $S_4$ | $C_2$ | $S_4^3$ | | |
|---|---|---|---|---|---|---|
| $A$ | 1 | 1 | 1 | 1 | $R_z$ | $x^2 + y^2, z^2$ |
| $B$ | 1 | –1 | 1 | –1 | $z$ | $x^2 - y^2, xy$ |
| $E$ | $\begin{cases}1\\1\end{cases}$ | $i$<br>$-i$ | –1<br>–1 | $-i$<br>$i$ | $(x, y); (R_x, R_y)$ | $(xz, yz)$ |

| $S_6$ | $E$ | $C_3$ | $C_3^2$ | $i$ | $S_6^5$ | $S_6$ | | $\varepsilon = \exp(2\pi i/3)$ |
|---|---|---|---|---|---|---|---|---|
| $A_g$ | 1 | 1 | 1 | 1 | 1 | 1 | $R_z$ | $x^2 + y^2, z^2$ |
| $E_g$ | 1<br>1 | $\varepsilon$<br>$\varepsilon^*$ | $\varepsilon^*$<br>$\varepsilon$ | 1<br>1 | $\varepsilon$<br>$\varepsilon^*$ | $\varepsilon^*$<br>$\varepsilon$ | $(R_x, R_y)$ | $(x^2 - y^2, xy); (xz, yz)$ |
| $A_u$ | 1 | 1 | 1 | –1 | –1 | –1 | $z$ | |
| $E_u$ | 1<br>1 | $\varepsilon$<br>$\varepsilon^*$ | $\varepsilon^*$<br>$\varepsilon$ | –1<br>–1 | $-\varepsilon$<br>$-\varepsilon^*$ | $-\varepsilon^*$<br>$-\varepsilon$ | $(x, y)$ | |

| $S_8$ | $E$ | $S_8$ | $C_4$ | $S_8^3$ | $C_2$ | $S_8^5$ | $C_4^3$ | $S_8^7$ | | $\varepsilon = \exp(2\pi i/8)$ |
|---|---|---|---|---|---|---|---|---|---|---|
| $A$ | 1 | 1 | 1 | 1 | 1 | 1 | 1 | 1 | $R_z$ | $x^2 + y^2, z^2$ |
| $B$ | 1 | –1 | 1 | –1 | 1 | –1 | 1 | –1 | $z$ | |
| $E_1$ | 1<br>1 | $\varepsilon$<br>$\varepsilon^*$ | $i$<br>$-i$ | $-\varepsilon^*$<br>$\varepsilon$ | –1<br>–1 | $-\varepsilon$<br>$-\varepsilon^*$ | $-i$<br>$i$ | $\varepsilon^*$<br>$\varepsilon$ | $(x, y); (R_x, R_y)$ | |
| $E_2$ | 1<br>1 | $i$<br>$-i$ | –1<br>–1 | $-i$<br>$i$ | 1<br>1 | $i$<br>$-i$ | –1<br>–1 | $-i$<br>$i$ | | $(x^2 - y^2, xy)$ |
| $E_3$ | 1<br>1 | $-\varepsilon^*$<br>$-\varepsilon$ | $-i$<br>$i$ | $\varepsilon$<br>$\varepsilon^*$ | –1<br>–1 | $\varepsilon^*$<br>$\varepsilon$ | $i$<br>$-i$ | $-\varepsilon$<br>$-\varepsilon^*$ | | $(xz, yz)$ |

## 9. The Cubic Groups

| $T$ | $E$ | $4C_3$ | $4C_3^2$ | $3C_2$ | | $\varepsilon = \exp(2\pi i/3)$ |
|---|---|---|---|---|---|---|
| $A$ | 1 | 1 | 1 | 1 | | $x^2 + y^2 + z^2$ |
| $E$ | 1<br>1 | $\varepsilon$<br>$\varepsilon^*$ | $\varepsilon^*$<br>$\varepsilon$ | 1<br>1 | | $(2z^2 - x^2 - y^2, x^2 - y^2)$ |
| $T$ | 3 | 0 | 0 | –1 | $(R_x, R_y, R_z); (x, y, z)$ | $(xy, xz, yz)$ |

| $T_h$ | $E$ | $4C_3$ | $4C_3^2$ | $3C_2$ | $i$ | $4S_6$ | $4S_6^5$ | $3\sigma_h$ | | $\varepsilon = \exp(2\pi i/3)$ |
|---|---|---|---|---|---|---|---|---|---|---|
| $A_g$ | 1 | 1 | 1 | 1 | 1 | 1 | 1 | 1 | | $x^2 + y^2 + z^2$ |
| $E_g$ | 1<br>1 | $\varepsilon$<br>$\varepsilon^*$ | $\varepsilon^*$<br>$\varepsilon$ | 1<br>1 | 1<br>1 | $\varepsilon$<br>$\varepsilon^*$ | $\varepsilon^*$<br>$\varepsilon$ | 1<br>1 | | $(2z^2 - x^2 - y^2, x^2 - y^2)$ |
| $T_g$ | 3 | 0 | 0 | –1 | 1 | 0 | 0 | –1 | $(R_x, R_y, R_z)$ | $(xz, yz, xy)$ |
| $A_u$ | 1 | 1 | 1 | 1 | –1 | –1 | –1 | –1 | | |
| $E_u$ | 1<br>1 | $\varepsilon$<br>$\varepsilon^*$ | $\varepsilon^*$<br>$\varepsilon$ | 1<br>1 | –1<br>–1 | $-\varepsilon$<br>$-\varepsilon^*$ | $-\varepsilon^*$<br>$-\varepsilon$ | –1<br>–1 | | |
| $T_u$ | 3 | 0 | 0 | –1 | –1 | 0 | 0 | 1 | $(x, y, z)$ | |

| $T_d$ | $E$ | $8C_3$ | $3C_2$ | $6S_4$ | $6\sigma_d$ | | |
|---|---|---|---|---|---|---|---|
| $A_1$ | 1 | 1 | 1 | 1 | 1 | | $x^2 + y^2 + z^2$ |
| $A_2$ | 1 | 1 | 1 | −1 | −1 | | |
| $E$ | 2 | −1 | 2 | 0 | 0 | | $(2z^2 - x^2 - y^2, x^2 - y^2)$ |
| $T_1$ | 3 | 0 | −1 | 1 | −1 | $(R_x, R_y, R_z)$ | |
| $T_2$ | 3 | 0 | −1 | −1 | 1 | $(x, y, z)$ | $(xy, xz, yz)$ |

| $O$ | $E$ | $8C_3$ | $3C_2\ (=C_4^2)$ | $6C_4$ | $6C_2$ | | |
|---|---|---|---|---|---|---|---|
| $A_1$ | 1 | 1 | 1 | 1 | 1 | | $x^2 + y^2 + z^2$ |
| $A_2$ | 1 | 1 | 1 | −1 | −1 | | |
| $E$ | 2 | −1 | 2 | 0 | 0 | | $(2z^2 - x^2 - y^2, x^2 - y^2)$ |
| $T_1$ | 3 | 0 | −1 | 1 | −1 | $(R_x, R_y, R_z)$; $(x, y, z)$ | |
| $T_2$ | 3 | 0 | −1 | −1 | 1 | | $(xy, xz, yz)$ |

| $O_h$ | $E$ | $8C_3$ | $6C_2$ | $6C_4$ | $3C_2\ (=C_4^2)$ | $i$ | $6S_4$ | $8S_6$ | $3\sigma_h$ | $6\sigma_d$ | | |
|---|---|---|---|---|---|---|---|---|---|---|---|---|
| $A_{1g}$ | 1 | 1 | 1 | 1 | 1 | 1 | 1 | 1 | 1 | 1 | | $x^2 + y^2 + z^2$ |
| $A_{2g}$ | 1 | 1 | −1 | −1 | 1 | 1 | −1 | 1 | 1 | −1 | | |
| $E_g$ | 2 | −1 | 0 | 0 | 2 | 2 | 0 | −1 | 2 | 0 | | $(2z^2 - x^2 - y^2, x^2 - y^2)$ |
| $T_{1g}$ | 3 | 0 | −1 | 1 | −1 | 3 | 1 | 0 | −1 | −1 | $(R_x, R_y, R_z)$ | |
| $T_{2g}$ | 3 | 0 | 1 | −1 | −1 | 3 | −1 | 0 | −1 | 1 | | $(xy, xz, yz)$ |
| $A_{1u}$ | 1 | 1 | 1 | 1 | 1 | −1 | −1 | −1 | −1 | −1 | | |
| $A_{2u}$ | 1 | 1 | −1 | −1 | 1 | −1 | 1 | −1 | −1 | 1 | | |
| $E_u$ | 2 | −1 | 0 | 0 | 2 | −2 | 0 | 1 | −2 | 0 | | |
| $T_{1u}$ | 3 | 0 | −1 | 1 | −1 | −3 | −1 | 0 | 1 | 1 | $(x, y, z)$ | |
| $T_{2u}$ | 3 | 0 | 1 | −1 | −1 | −3 | 1 | 0 | 1 | −1 | | |

## 10. The Groups $C_{\infty v}$ and $D_{\infty h}$ for Linear Molecules

| $C_{\infty v}$ | $E$ | $2C_\infty^*$ | . . . | $\infty\,\sigma_v$ | | |
|---|---|---|---|---|---|---|
| $A_1 \equiv \Sigma^+$ | 1 | 1 | . . . | 1 | $z$ | $x^2 + y^2, z^2$ |
| $A_2 \equiv \Sigma^-$ | 1 | 1 | . . . | −1 | $R_z$ | $(xz, yz)$ |
| $E_1 \equiv \Pi$ | 2 | $2 \cos \Phi$ | . . . | 0 | $(x, y)\ (R_x, R_y)$ | $(x^2 - y^2, yz)$ |
| $E_2 \equiv \Delta$ | 2 | $2 \cos 2\Phi$ | . . . | 0 | | |
| $E_3 \equiv \Phi$ | 2 | $2 \cos 3\Phi$ | . . . | 0 | | |
| | . . . | . . . | . . . | . . . | | |

| $D_{\infty h}$ | $E$ | $2C_\infty^*$ | . . . | $\infty\,\sigma_v$ | $i$ | $2S_\infty^*$ | . . . | $\infty C_2$ | | |
|---|---|---|---|---|---|---|---|---|---|---|
| $\Sigma_g^+$ | 1 | 1 | . . . | 1 | 1 | 1 | . . . | 1 | | $x^2 + y^2, z^2$ |
| $\Sigma_g^-$ | 1 | 1 | . . . | −1 | 1 | 1 | . . . | −1 | $R_z$ | |
| $\Pi_g$ | 2 | $2 \cos \Phi$ | . . . | 0 | 2 | $-2 \cos \Phi$ | . . . | 0 | $(R_x, R_y)$ | $(xz, yz)$ |
| $\Delta_g$ | 2 | $2 \cos 2\,\Phi$ | . . . | 0 | 2 | $2 \cos 2\,\Phi$ | . . . | 0 | | $(x^2 - y^2, yz)$ |
| . . . | . . . | . . . | . . . | . . . | . . . | . . . | . . . | . . . | | |
| $\Sigma_u^+$ | 1 | 1 | . . . | 1 | −1 | −1 | . . . | −1 | $z$ | |
| $\Sigma_u^-$ | 1 | 1 | . . . | −1 | −1 | −1 | . . . | 1 | | |
| $\Pi_u$ | 2 | $2 \cos \Phi$ | . . . | 0 | −2 | $2 \cos \Phi$ | . . . | 0 | $(x, y)$ | |
| $\Delta_u$ | 2 | $2 \cos 2\,\Phi$ | . . . | 0 | −2 | $-\,2 \cos 2\,\Phi$ | . . . | 0 | | |
| . . . | . . . | . . . | . . . | . . . | . . . | . . . | . . . | . . . | | |

**11.** The Icosahedral Groups*

| $I_h$ | $E$ | $12C_5$ | $12C_5^2$ | $20C_3$ | $15C_2$ | $i$ | $12S_{10}$ | $12S_{10}^3$ | $20S_6$ | $15\sigma$ | | |
|---|---|---|---|---|---|---|---|---|---|---|---|---|
| $A_g$ | 1 | 1 | 1 | 1 | 1 | 1 | 1 | 1 | 1 | 1 | | $x^2 + y^2 + z^2$ |
| $T_{1g}$ | 3 | $\frac{1}{2}(1+\sqrt{5})$ | $\frac{1}{2}(1-\sqrt{5})$ | 0 | –1 | 3 | $\frac{1}{2}(1-\sqrt{5})$ | $\frac{1}{2}(1+\sqrt{5})$ | 0 | –1 | $(R_x, R_y, R_z)$ | |
| $T_{2g}$ | 3 | $\frac{1}{2}(1-\sqrt{5})$ | $\frac{1}{2}(1+\sqrt{5})$ | 0 | –1 | 3 | $\frac{1}{2}(1+\sqrt{5})$ | $\frac{1}{2}(1-\sqrt{5})$ | 0 | –1 | | |
| $G_g$ | 4 | –1 | –1 | 1 | 0 | 4 | –1 | –1 | 1 | 0 | | |
| $H_g$ | 5 | 0 | 0 | –1 | 1 | 5 | 0 | 0 | –1 | 1 | | $(2z^2 - x^2 - y^2, x^2 - y^2, xy, yz, xz)$ |
| $A_u$ | 1 | 1 | 1 | 1 | 1 | –1 | –1 | –1 | –1 | –1 | | |
| $T_{1u}$ | 3 | $\frac{1}{2}(1+\sqrt{5})$ | $\frac{1}{2}(1-\sqrt{5})$ | 0 | –1 | –3 | $-\frac{1}{2}(1-\sqrt{5})$ | $-\frac{1}{2}(1+\sqrt{5})$ | 0 | 1 | $(x, y, z)$ | |
| $T_{2u}$ | 3 | $\frac{1}{2}(1-\sqrt{5})$ | $\frac{1}{2}(1+\sqrt{5})$ | 0 | –1 | –3 | $-\frac{1}{2}(1+\sqrt{5})$ | $-\frac{1}{2}(1-\sqrt{5})$ | 0 | 1 | | |
| $G_u$ | 4 | –1 | –1 | 1 | 0 | –4 | 1 | 1 | –1 | 0 | | |
| $H_u$ | 5 | 0 | 0 | –1 | 1 | –5 | 0 | 0 | 1 | –1 | | |

"For the pure rotation group $I$, the outlined section in the upper left is the character table; the $g$ subscripts should, of course, be dropped and $(x, y, z)$ assigned to the $T_1$ representation."

***Source:*** B.E. Douglas and C.A. Hollingsworth, "Symmetry in Bonding & Spectra: An Introduction on", 1985, Orlando, FL; Academic Press, pp. 391–408.

# Index

## A



## B



## C

## T



## V



## W